Praise for *The Inventor's Desktop Companion*

D1498559

THE
INVENTOR'S
DESKTOP
COMPANION

To my wife, Sheryl, with whom for the past twenty-eight years I have peered into life's mysteries; a woman of upright will, great resolve, and honesty who spends her many talents like a millionaire intent on going broke. Thanks for your love, friendship, unlimited patience, sage advice, and energy of mind.

To my daughter, Bettie, who possesses intelligence, creativity, passion, strength, courage, and confidence in uncommon measures. Continue to seize the moment, keep curiosity alive, and hold onto your ideals: they are jewels.

Thanks to you both for the time and space required to do another book. No more books (at least for a while).

THE *I*NVENTOR'S

I OP

C ON

Th ____ ting

an

R

608
L668

VISIBLE
INK
PRESS

DETROIT · WASHINGTON, D.C.

THE INVENTOR'S DESKTOP COMPANION

Published by Visible Ink Press™
a division of Gale Research Inc.
835 Penobscot Building
Detroit, MI 48226-4094

Visible Ink Press™ is a trademark of Gale Research Inc.

Most Visible Ink Press books are available at special quantity discounts when purchased in bulk by corporations, organizations, or groups. Customized printings, special imprints, messages, and excerpts can be produced to meet your needs. For more information, contact Special Markets Manager, Gale Research Inc., 835 Penobscot Bldg., Detroit, MI 48226. Or call 1-800-776-6265.

Art Director: Cynthia Baldwin

Front cover photo: Vincent J. Ricardel

Back cover photo: Jeff Slate

ISBN 0-7876-0490-9

Contents

1

How to Realize Your Full Potential 1

Information Is Power • Formal Education Is No Guarantee • Levy's 10 Commandments of Business • Basic Types of Protection

2

Inventors Beware: Invention Marketing Companies 8

Who Are These Parasites? • How to Tell the Pros from the Cons • Never Pay an Up-Front Fee • Direct Mail Tactics • How Do They Get Your Name? • How to Protect Yourself • Deceptive Trade Practices • FTC Guidelines • The FTC Takes Action • What To Do If You Are a Victim • Meanwhile on Capitol Hill • How to Find an Honest Broker • Invention Consultants • Wisconsin Innovation Service Center

3

The Patent & Trademark Office 26

History of the PTO • The Mandate of the PTO • What Does It Cost to Run the PTO? • How Many People Work at the PTO? • How Much Work Do They Handle? • General Information and Correspondence

continued

Energy-Related Invention Program • Department of Energy • The Innovative Concepts Program • National Innovation Workshops • National Inventors Conference • Invention Evaluation Request Form

Introduction

Of more than five million patents granted in the United States since 1790, a few have had enormous impact on our lives while at the same time bringing fame and fortune to their inventors. The electric lamp, the transistor radio, the internal combustion engine, and the telephone come to mind quickly. Some inventions have had little or no impact on our lives, such as the Pet Rock and Cabbage Patch Dolls, but have brought fame and fortune to their inventors. But the vast majority of inventions dreamed up during the last two hundred years have created no recognition or financial gain for their inventors. Why? They simply were never commercialized. And they've left no trace. You can bet that when the inventors applied for their patents, the inventions probably seemed like terrific ideas. But then something happened, or, more accurately, didn't happen.

What didn't happen was the sale and nothing happens until something is sold. Just as invention begins with resistance, so does the sale. When the resistance is overcome, the idea finds its successful structure, and the sale is made. *The Inventor's Desktop Companion* is all about overcoming resistance.

We are all laboratories, conducting experiments with thought and extending that thought to function and utility. And what we can accomplish is bewildering. All it usually takes is a jolt out of inertia and the rut of habit. It is not enough to sit by and wait for things to happen. Success comes to those who are proactive, not reactive; to people who possess what I call the **dare to go.** The ultimate responsibility is, after all, taking responsibility.

With this book, you'll learn how to:

- Avoid invention marketing rip-off schemes
- Protect your invention through patenting, trademarking, and/or copyrighting
- License your concepts to manufacturers
- Make the best deal
- Find public and private funding for research and development
- Dramatically reduce your legal expenses
- Find expert advice and support—via associations, publications, and databases

In addition to learning how to license and protect your inventions, it is my hope that *The Inventor's Desktop Companion* will encourage you to:

- Trust yourself more and recognize, accept, and take responsibility for the mutuality of events. Part of the adventure is recognizing and capitalizing on consequences.

- Forbid yourself to be deterred by poor odds just because your mind has calculated that the opposition is too great. If it were easy, everyone would do it.

- Take your chances, not someone else's. The rewards are greater. Ask questions. If you don't ask the question, the answer is an automatic no.

- Have the courage to make mistakes. Mistakes are the by-product of experimentation. The price of never being wrong is never being right.

- Resist the herd instinct. Be yourself and be faithful to your own muse. Never give up your individuality.

- Look for opportunities, not guarantees. Surely you've heard the one about death and taxes.

- See rejection as a rehearsal before the big event. There can be no success without failure. Learn the value of teamwork and how much people contribute to each other's success.

- And finally, learn to enjoy the hunt. For it is here, in the moment of transition, in the rushing to a goal, that the power resides.

The Better Idea and How to Market It

You don't have to be a rocket scientist to be an inventor. This book is written for the ordinary person with an extraordinary idea and the yen to market it. After all, it's a classic American landscape: real fortunes being made by plain folks with the itch, the courage, the indefatigable entrepreneurial spirit, and, of course, the better idea.

Apply Your Personal Spin

A person's judgment cannot be better than the information upon which it is based. And information by itself is worthless. Information, when combined with insight, becomes the thing that greases the wheel. I have based *The Inventor's Desktop Companion* upon my own personal experience. As such, it is based on life as it happens and not theoretical or hypothetical situations. My hope is that you'll look upon the knowledge and information imparted in this book as a chart, draw your own lines on that chart, and use it to successfully navigate the sometimes stormy waters of protecting and commercializing inventions.

Getting the Most out of *The Inventor's Desktop Companion*

I have designed this book to be a friendly and helpful companion to accompany you throughout your journey—starting with the **Contents** page, which lists not only the chapter names and page numbers but also the major points discussed in each; and ending with the **Index,** a comprehensive listing of key people, topics, and organizations created with your needs in mind. In between are nearly two dozen chapters packed with the inspiration and the nuts and bolts information that you'll need to bring your ideas to fruition. And don't forget the **Glossary** (Appendix 1): a journey into foreign territory is always easier when you know the language.

Protecting your intellectual property, whether you are seeking a patent, trademark, or copyright, necessarily involves a lot of paperwork and the difficulty of dealing with bureaucracies. To smooth your path I have provided **dozens of forms** that you can photocopy, fill out, and send in to the appropriate government agency. To further assist your travels through the maze that is the Patent and Trademark Office, consult Appendix 2: **PTO Phone List**; this directory, organized by subject, will lead you to the key players at the PTO.

Protecting your invention can also be costly. While I can do nothing to alleviate patent fees, I can help reduce your legal expenses. I have provided several **samples of legal agreements** that you can use as templates for your own transactions, saving your lawyer hours of work and you thousands of dollars.

The rest is up to you!

Acknowledgements

This edition of *The Inventor's Desktop Companion,* like the first in 1991, depended heavily upon the cooperation and assistance of a great many people. During the course of my research I contacted independent inventors, educators, federal and state government officials, patent attorneys, corporate executives, editors of professional publications, and members of inventor organizations nationwide. Their warm reception, hospitality, and good cheer made my assignment one that I'll always remember fondly.

Oscar Mastin, a former public affairs officer at the Patent and Trademark Office, has been a friend for many years and an invaluable source of assistance to me and to inventors nationwide. Before being reassigned, he provided to me much of the updated background material for the patenting sections and he filled in the blanks on many a fax I sent to his office.

Another friend of many years, George P. Lewett, Director, Office of Technology Innovation (OTI) at the National Institute of Standards and Technology (NIST), and Dr. Gilbert M. Ugiansky, Deputy Director, OTI, unselfishly spent an entire afternoon briefing me on their work, making copies of reports and speeches, suggesting leads, and checking my facts and figures for accuracy.

Fred L. Hart, Director, Inventions and Innovation Division at the Department of Energy (DOE), responded, as always, above and beyond the call of duty, to get me anything I needed for the section about DOE inventor programs; not the least of which was an eight-page report he wrote to help me understand DOE's technology development support programs.

Herbert C. Wamsley, Executive Director, and Ada E. Barnes, Associate Executive Director, Intellectual Property Owners Association, were extremely cooperative in helping me to gather certain facts and figures and to understand the intellectual property ramifications of GATT. All I had to do was place a call and within minutes my fax machine rang, delivering the material I requested.

I owe a debt of appreciation to many people at the Patent and Trademark Office, including: Ed Kazenske, Executive Assistant to the Commissioner; John D. Hassett, Director, Office of General Services; Gerald Goldberg, Director, Group 2400; Lou Massel, Editor, *The Manual of Patent Examining Procedure*; Carole A. Shores, Director, Center for Patent and Trademark Information; Martha Crocket, Director, PTDLs; Jeff Look, Esq., Trademark Examiner, Law Office 10; Lynette Samuels, Public Service Desk, PTO; Don Kelly, Group Director, Group 320; Jeff Nase, Supervisory Petition Examiner; J. Michael Thesz, Special Program Examiner; Ruth Ann Nyblod, Director, Project XL; Elizabeth Weimar, Patent Examiner, Group 180; Jeff Alderson, Office of Information Systems; Maureen Brown, Computer Systems Analyst; James R. Lynch, Comptroller; Cameron Weiffenbach, Director, Enrollment and Discipline; Gary Robinson, International Technical Specialist; and Jennifer Bahr, Primary Examiner, Art Unit 3304.

At the Copyright Office, thanks to Dorothy Schrader, General Counsel; and Mary Berghaus Levering, Associate Register for National Copyright Programs.

At the Library of Congress, my gratitude goes to Helen Dalrymple, Public Affairs Specialist, and David Levy, Esq., General Counsel's Office; and Boris Bohun-Chudyniv, Chief Systems Analyst, FedLink.

Also of assistance were, in no order of priority, Kathie Scarrah, press secretary to U.S. Senator Joe Lieberman (D-CT); Alison Jewett, NIST; Renee Edwards, Director, Public Affairs, National Technical Information Service; Howard Shapiro, Press Officer, Federal Trade Commission; Phoebe D. Morse, Regional Director, Boston Regional Office, Federal Trade Commission; Maureen Wood, Office of Technology Evaluation and Assessment; Marianne K. Clarke, Senior Policy Analyst, Center for Policy Research and Analysis, National Governor's Association; Jeff Norris and Joyce Hamaty, Public Affairs Specialists, National Science Foundation; Richard Sparks, Program Manager, Defense Technical Information Center; Program Analysts Claire Dondero and Bob Connolly, Small Business Administration, Office of Technology; Gil Young, Director, Minnesota Department of Trade and Economic Development; Mark A. Spikell, Ed.D., Professor, George Mason University; G. Thomas Cator, Executive Director, Association of Small Business Development Centers; Michael Blommer, AIPLA; Leonard Williamson, NCIPLA; John T. Farady, President, Affiliated Inventors Foundation; Joanne M. Hayes, Publisher and Editor, *Inventors' Digest*; Jan Kosko, Public Affairs Officer, NIST; Ray Watts, former editor of the *Inventor-Assistance Program News*; Robin Conger, Battelle Pacific Northwest Laboratories; and George E. Harvill, Greentree Information Services, the best and most diligent patent and trademark searcher I know.

To Ron Weingartner, Associate Vice President of R&D at Milton Bradley, my good buddy and co-author, thanks for allowing me to pull some material from our book, *Inside Santa's Workshop: How Toy Inventors Develop, Sell, and Cash in on Their Ideas*. (Henry Holt, 1990).

No book of mine would be complete without a special thanks to Michael Ross, Vice President and Editorial Director of World Book Publishing, my closest friend and the smartest person I know; one who is always there with a bit of wisdom on just about anything.

At Gale Research and Visible Ink Press, thanks again to Elizabeth A. Geiser for taking the phone call that started this enterprise many moons ago; to Marty Connors, former publisher of Visible Ink Press for his friendship, wise counsel, invaluable contributions, confidence and voicemail; and to my new Visible Ink Press editor, Judy Galens, who helped reorganize this body of work to make it even more reader-friendly, and who braved a February snowstorm to visit us "inside the Beltway." Invaluable editorial assistance was provided by Dean D. Dauphinais, Kathleen Dauphinais, Kevin Hillstrom and Laurie Collier Hillstrom of The Northern Lights Writers Group, Sean McCready, Paulette Petrimoulx, and Les Stone.

And for their friendship during this and every project, a special thanks to Cap "The Capper" Weinberger; Richard "Mad Socks" Maddocks, Hank "The Bear" Jacob; Kenny "Killer" Richards; Richie "MauMau" Malmed; Mimi "The Meemster" Malmed; and Haig der Marderosian, wherever you are.

The Beck Rules

THE INVENTOR'S DESKTOP COMPANION

CHAPTER 1

How to Realize Your Full Potential

*If you think you can do it, you're right. If you think
you can't do it, you're still right.*

—HENRY FORD

"Heavier-than-air flying machines are impossible," the well known British mathematician and physicist William Kelvin assured everyone back in 1895.

"Man can never tap the power of the atom," said Nobel Laureate and physicist Robert Andrews Millikan, credited with being the first to isolate the electron and measure its charge.

"Everything that can be invented has been invented," stated another man of vision, Charles H. Duell, Director of the U.S. Patent Office in 1899.

TRW listed these observations in a 1985 *Wall Street Journal* ad that was tagged with the line: "There's no future in believing something can't be done. The future is in making it happen." What the people quoted above did not understand, quite obviously, is what inventors from high-tech labs to basement workshops across America are proving everyday:

- There is no future in the word impossible.
- Results first. Theory second.
- Alert optimism beats conservative skepticism.
- Celebrate failure. Failure breeds success.

The April 1994 edition of *The Optimist* magazine gives some examples of people who overcame rejection and failure. In his first three years in business, Henry Ford went broke three times. Dr. Seuss's first children's book was rejected by 23 publishers. The 24th publisher sold six million copies. In 1902, a young poet had his poems rejected by *Atlantic Monthly* as being "too vigorous"; but Robert Frost persevered. Michael Jordan was cut from his high school basketball team. The University of Bern rejected a Ph.D. dissertation, saying that it was irrelevant and fanciful. Albert Einstein was disappointed, but not down for the count.

More recently, 45-year-old George Foreman, trailing badly in points, reclaimed his heavyweight champion title with a stunning knockout of 26-year-old Michael Moorer on November 5, 1994, at 2 minutes and 3 seconds into the 10th round. "Anything you desire you can make happen. . . . Look at me tonight," said Foreman.

Biro Gets to the Point

Tired of the ink stains made by his fountain pen, Hungarian journalist, painter, and inventor Laszlo Biro (1899–1985), along with his brother Georg, developed a pen with a ball point that wrote without leaving ink blotches. Although not the first ballpoint pen, it was the best design and Laszlo is credited as the pen's inventor.

During World War II the ballpoint pen became standard equipment in military aircraft— only the ballpoint would work at high altitudes—and with this came much recognition for the innovative writing instrument.

In 1945 a New York City department store introduced a ballpoint pen to the public and sold 10,000 pens in one day at $12.50 each. Today billions are sold worldwide, many for pennies apiece.

A Great Idea That Stuck

In 1948, George de Mestral, a Swiss engineer, took a hunting trip into the Alps. As he hiked around the mountains, he picked burrs off his socks and pants. De Mestral wondered why they stuck to his clothing; when he looked closely at the burrs on his clothes, he realized that tiny hooks on the burrs were catching to thread loops in the clothing.

A light bulb went off in de Mestral's head and eight years later Velcro® was born. Velcro® was composed of two pieces of material, one with hooks and one with loops. When pressed to one another, they stuck together, separating with a rip and ready to be used again and again. Today the patent on Velcro® has expired, but the trademark is alive and the technology is used in all kinds of applications from shoes to space suits.

Nowhere in the world do people have more freedom and encouragement to innovate and be different than in America, a nation born with a dream. "The moment an American hears the word 'invention,'" observed a visitor in the late 1820s, "he pricks up his ears."

It has been said that Americans are men and women with "new eras" in their brains. Our history is replete with examples of independent and courageous individuals who succeeded by doing things differently: people who believed in themselves and their ideas.

Alexis de Tocqueville, the French writer who visited America in 1831, wrote about Americans, "They have all a lively faith in the perfectibility of man, they judge that the diffusion of knowledge must necessarily be advantageous, and the consequences of ignorance fatal; they all consider society as a body in a state of improvement, humanity as a changing scene, in which nothing is, or ought to be, permanent; and they admit that what appears to them today to be good, may be superseded by something better tomorrow.

"America is a land of wonders in which everything is in constant motion and every change seems an improvement. The idea of novelty is there indissolubly connected with the idea of amelioration. No natural boundary seems to be set to the efforts of man; and in his eyes what is not yet done is only what he has not yet attempted to do."

On the occasion of the 150th anniversary of the Patent Act of 1836, then vice president George Bush, speaking at the National Museum of American History, said "It takes a special kind of independence to invent something. You put yourself and your ideas on the line. And maybe some people will say that you're crazy or that you're impractical, but for [over] two centuries, millions of Americans have ignored the ridicule.

"They've worked on ideas. From those ideas, they've started businesses. And many of those businesses have grown and are our greatest industrial companies, companies like Xerox, Ford Motor Company, American Telephone and Telegraph and Apple Computer. Think of what America would be like if the skeptics had silenced the inventors."

This was never more true than in the cases of the following inventive Americans, people who dared to be different and refused to swap incentive for security. Many of their names have become well known.

In 1923, Clarence "Bob" Birdseye, the Father of Frozen Food, got a patent on quick-freezing. Dr. William "Billy" Scholl patented his first arch support in 1904 at the age of 22. Beginning in 1896 and over a 47-year career, one-time slave George Washington Carver developed more than 300 products from peanuts alone. On a cold day in January 1839, rubber was vulcanized by Charles Goodyear. Yankee tinkerer Eli Whitney transformed the South with the cotton gin he invented in 1792. New Englander Edwin H. Land demonstrated his first instant cameras in 1947.

Many inventors are known for more than one product. William Lear, best known for the Learjet, also received a patent on the first car radio (his company eventually became Motorola) and the 8-track tape system. Peter Cooper, inventor of the famous locomotive "Tom Thumb," was awarded in 1845 the first patent for a gelatin dessert, which later became known as Jell-O. George Westinghouse not only invented air brakes for railroad cars, but he is also credited with a type of gas meter and a pipeline system that safely conducted natural gas into homes.

Some Americans are well known for something other than their inventions. Actress Hedy Lamarr patented a sophisticated anti-jamming device to foil Nazi radar during World War II. After the patent expired, Sylvania adapted the invention, and it is still used in satellite communication. Author Samuel L. Clemens (better known as Mark Twain) patented a pair of suspenders. Abraham Lincoln was awarded a patent for a device to "buoy vessels over shoals." Zeppo Marx patented a wrist-worn heart alarm. Novelist John Dos Passos is listed as co-inventor of a toy pistol that blows bubbles. Actress Julie Newmar patented a type of panty hose. Singer Edie Adams patented a ring-shaped cigarette and cigar holder. Confederate General James E.B. Stuart patented a method of attaching sabers to belts. Escape artist Harry Houdini held a patent on a diver's suit (permitting escape). And Bette Nesmith, mother of Michael Nesmith of The Monkees, invented Liquid Paper typewriter correction fluid.

Other Americans are not as well known as their inventions. In 1893 Witcomb L. Judson filed the first patent on a "slide fastener" for shoes (commonly known as the zipper). During the Civil War, Martha Coston invented a flare for which the U.S. government paid her $20,000. Northam Warren, after graduating from Detroit College of Pharmacy, originated the first liquid cuticle remover (Cutex) in 1911. Newspaper editor Carlton Magee invented the parking meter in Oklahoma City in 1932. Walter Hunt patented the safety pin in 1849. Joseph F. Glidden protected his idea for barbed wire in 1874. A bright idea was struck by Alonzo D. Phillips in 1836 when he patented the friction match. Before Elijah McCoy invented the automatic lubricator in the early 1870s, machines had to be shut down for lubrication.

Not all inventions are so conventional. Two virtually unknown American inventors, Philip Leder of Chestnut Hill, Massachusetts and Timothy A. Stewart of San Francisco, California made history on April 12, 1988 when U.S. Patent No. 4,736,866, entitled "Transgenic Non-Human Mammals" was issued. The Harvard University researchers were awarded the first patent covering an animal. Their technique introduces activated cancer genes into early-stage mice embryos. The resulting mice are born with activated cancer genes in all their cells. These mice are extremely sensitive to cancer-causing chemicals, developing tumors quickly if exposed even to small amounts of such chemicals. The resulting value to medical and scientific research should be considerable.

As diverse as the aforementioned individuals are, these inventors share many things in common, characteristics that you too will need to fulfill your aspirations and see your inventions patented, licensed, manufactured, and commercialized. Perhaps the most important characteristic shared by these people is that, while others reflected on their ideas, these inventors answered the call.

Information Is Power

This is not a theoretical book. It is based upon my empirical experience. Over the past 17 years, I've personally dreamed-up, invented, co-invented, or developed thousands of concepts, more than 90 of which have been licensed and commercialized.

Although I specialize in inventing and licensing toys and games, there are common denominators in my work that apply to inventing and licensing anything. Therefore, whether you have invented an elevator winding device, a gizmo for attaching tag pins, a tape drive mechanism, a three-dimensional digitizer, a path game, a U-joint mount, or a better mouse trap, *The Inventor's Desktop Companion* will give you an added edge, and the inspiration to go for it.

As Inventive As He Is Honest

In May of 1849, when Abraham Lincoln was a Congressman from Illinois, he received Patent No. 6,469 for "A Device for Buoying Vessels over Shoals." The invention consisted of a set of bellows attached to the hull of a ship just below the waterline. On reaching a shallow place the bellows were inflated and the ship, thus buoyed, was expected to float clear. Lincoln whittled the prototype of the invention with his own hands.

Lincoln's appreciation of inventions was later to be of great service to the nation. John Ericcson's *Monitor*, the ironclad ship which defeated the *Merrimac*, would never have been built except for Lincoln's insistence, nor would the Spencer repeating rifle have been adopted for use by the Army.

Humorist Leaves His Mark

On December 19, 1871, Mark Twain received Patent No. 121,992 for "An Improvement in Adjustable and Detachable Straps for Garments," otherwise known as suspenders. Twain, who later lost a fortune investing money on the inventions of others, actually received three U.S. Patents: the second was in 1873 on his famous "Mark Twain's Self-Pasting Scrap-book," and the third was in 1885 for a game to help people remember important historical dates.

In Twain's novel, *A Connecticut Yankee in King Arthur's Court,* his character Sir Boss remarks that "a country without a patent office and good patent laws is just like a crab and can't travel anyway but sideways and backways."

Twain would have agreed with the statement "formal education is no guarantee." In fact, he once said, "Never let formal education get in the way of your learning."

Formal Education Is No Guarantee

Thomas Alva Edison received 1,093 patents, four being issued posthumously. Perhaps the greatest inventor in history, Edison was a grammar school dropout who had just three months of formal education. Edwin H. Land, inventor of the Polaroid camera, Bill Gates, founder and CEO of MicroSoft and one of the nation's richest individuals, and R. Buckminster Fuller, the social theorist, all dropped out of Harvard. The fact that you may not have a formal education is not everything. The world is full of intelligent failures, my dad would remind me.

Some experts think many schools squelch creativity in the way they perpetuate entrenched thinking. According to *Business Week,* tests show that a child's creativity plummets 90% between ages five and seven. By the age of 40 most adults are about two percent as creative as they were at age five.

Creative people usually don't have dull, predictable childhoods. Instead, childhood is marked by "exposure to diversity," says Dean K. Simonton, a psychologist at the University of California at Davis. And an exceptionally high IQ may mean something in some fields, but studies show that the threshold for creativity is an IQ of about 130. After that, *Business Week* reports, "IQ doesn't make much difference—such non-intellectual traits as values and personality become more important."

My university degree is in television and cinematography with a minor in English. I have an advanced degree in "High Shtik" learned on the job at Paramount Pictures and Avco Embassy Pictures.

My philosophies and social training came primarily from my mother and grandmother, two of the strongest, most savvy women I have ever met (along with my wife and daughter) who taught me to push the envelope, even at the risk of failure. I was brought up to believe I could accomplish anything I put my mind to. In college, my wrestling coach, Jim Peckham, drilled me on the importance of "earning the right to win." Bottom line: You can't score on defense. Get the ball!

Levy's 10 Commandments of Business

To be successful in the exciting enterprise of product development and commercialization it takes a lot more than a good idea, a strong patent, and some luck. An understanding of the following ten points is critical for the inventor who wants to beat what can sometimes seem like insurmountable odds.

1) **Don't take yourself too seriously.** Don't take your idea too seriously either. The world will probably survive without your idea. Industry will probably survive without your idea. You may need it to survive, but no one else does.

2) **The race is not always to the swift . . . but to those who keep running.** It is a mistake to think anything is made overnight other than baked goods and newspapers. Levy's first corollary is: Nothing is as easy as it looks. Levy's second corollary is: Everything takes longer than you think.

    When the Convair Co. could not find a way to stop San Diego's night fog from rusting Atlas missile parts they were manufacturing, it put out a public plea for assistance.

    Norm Larsen, a local chemist, responded with 39 formulas. But it was his 40th that held the answer, producing a petroleum-based chemical that gets under water and displaces it through the pores of metal. Larsen's invention became WD–40; last year, sales for this product topped $100 million. By the way, in case you have ever wondered what WD–40 stands for, it is "Water Displacement—40th formula."

It is not speed that separates winners from losers. It is perseverance. The salesperson driving 30 miles at 4:15 p.m. to make one last sales call before 5:00 p.m.; the actor auditioning for the hundredth time; the writer facing a keyboard every day, creating twenty-five pages to get four or five that are keepers; and the athlete never quitting the team: all exhibit perseverance. That is the quality required to hit the heights of personal achievement.

3) **You can't do it all yourself.** Never forget the sage words of John Donne, "No man is an island, entire of itself; every man is a piece of the continent, a part of the main."

The success I have had is the result of unselfish, highly talented, and creative partners and associates willing to face the frustrations, rejections, and seemingly open-ended time frames that are inherent to any product development and licensing exercise. I have also been lucky to meet and work with very creative, understanding, and courageous corporate executives willing to believe in me and gamble on our concepts.

It is the cross-pollination and subsequent synergism that results in success, success in which all parties share. For if any link in this complex and often serpentine chain breaks, an entire project could flag.

4) **Keep your ego under control.** Creative and inventive people, according to profile, hate to be rejected or criticized for any reason. They are usually critical of others. And they are extremely defensive where their creations are concerned.

An idea can turn to dust or magic, depending on the talent that rubs against it. I have always found that my concepts are enhanced by the right touch. Working together or in competition, other people contribute time and time again to making an idea more useful or marketable. Share an idea, get back a better one.

Unchecked egocentricity can be the source for major failure in the development and licensing of new concepts. Arrogance has no place in the process.

5) **You'll always miss 100% of the shots you don't take.** Don't be afraid to make mistakes. If you don't put forth the effort, you won't fail, but you won't succeed either. Inaction will keep opportunities from coming your way. If Edison had stopped at 30 filaments, we'd still be in the dark.

Trust yourself. Never permit yourself to be deterred by poor odds because your mind has calculated that the opposition is too great. In a conversation I once had with Dr. Erno Rubik, inventor of Rubik's Cube, I asked him why children are so good at solving his puzzles and adults often don't even try. "Because no one has told the children that they cannot do it," he explained.

Taped to the top of my computer monitor is this quote by Theodore Roosevelt: "Far better it is to dare mighty things, to win glorious triumphs, even though checkered by failure, than to take rank with those poor spirits who neither enjoy much nor suffer much, because they live in the gray twilight that knows not victory nor defeat."

The Real McCoy

The oil lubricator was invented by Elijah McCoy, son of runaway slaves. Born in Canada, he studied engineering in Scotland, and then worked for a railroad in Michigan as an oilman.

Back in those days, trains had to stop at regular intervals to be lubricated. Crew members would run around with oil cans squirting oil between the train's moving parts. Without periodic lubrication, parts would rub together, overheat, and come to a halt.

McCoy realized that it was expensive and wasteful to stop the train every time it needed lubrication. Therefore in 1872, he invented the oil lubricator, a device that oiled the engine parts while the train was under power.

Soon McCoy was designing lubricators for all kinds of machines. People insisted that their machines have McCoy lubricators: they wanted "the real McCoy."

6) **Don't do it just for the financial rewards.** You should be motivated by the gamesmanship. It may sound trite, but people who do things just for the money usually come up shortchanged.

7) **Learn to take rejection.** Don't be turned off by the word "no" because you'll hear it often, as in: "No, we're not looking for that at this time." "No, you'll have to do better than that for us to consider it." "No, your idea isn't original."

Rejection can be positive if it is turned into constructive growth. Don't let it shake your confidence. My experience is that products get better the more times they are presented. Remember, the finest steel goes through the hottest fire.

I like to think of rejection as a rehearsal before the big event. It is the shakedown period similar to the practice of taking a play out of town before it opens on Broadway.

I have rarely licensed a product to the first manufacturer that saw it. And for every product I've licensed, there are many more that never made it off the drawing board, and probably should not have.

It comes down to this. Rejection of your ideas is part of the licensing game. So, if you are going to live by the crystal ball, learn to eat glass.

8) **Believe in yourself.** One of the first steps toward success is learning to detect and follow that gleam of light that flashes across the mind from within. We tend to dismiss without notice our own thoughts, because they are ours. In every work of genius we recognize our own rejected thoughts; they come back at us with a certain alienated majesty.

It is critical that you learn to abide by your own spontaneous impression. Permit nothing to affect the integrity of your own mind.

If you stand for something, you will always find some people for you and some against you. If you stand for nothing, you will find nobody against you, and nobody for you.

Remember the advice of Winston Churchill. "Never, never, never, never, never give in—except to dictates of conscience and duty."

9) **Sell yourself before you sell your ideas.** Be concerned about how you are perceived. You may be capable of dreaming up many ideas for a company to consider, but if you cannot command the respect and attention of the corporate executives, your product will never get off the mark, and you may not be invited back for an encore.

Be smart. Ideas come, ideas go. Know how much to push. Know when to disappear. Where elephants fight, grass does not grow. Don't wear out your welcome. You cannot put a dollar value on access to a corporate executive's valuable space and time.

10) **If it ain't on the page, it ain't on the stage.** Document everything. And follow up. Follow up. Follow up.

Let me remind you that the road to success is always under construction and there are NO short cuts. Short cuts tend to do just what the term implies, give you the short end. There is simply no way to avoid the ache and pain of hard work.

Longfellow put it well when he wrote in *The Ladder of St. Augustine,* "The heights by great men reached and kept, were not attained by sudden flight, but they, while their companions slept, were toiling upward in the night."

In closing this chapter, a chapter designed to inspire and give you confidence, remember to be positive. Trust yourself. Do not be deterred by poor odds. And realize that what you can accomplish is truly amazing.

Basic Types of Protection

Now that you have, as Abraham Lincoln once said, "added the fuel of interest to the fire of genius," here is some basic information you should know about protecting your invention before you go any further.

- **Copyright.** A copyright is a form of protection provided to the creators of "original works of authorship" including literary, dramatic, musical, artistic, and certain other intellectual works. See Chapter 10.

- **Design Patent.** A design patent is granted for any new, original, and ornamental design for an article of manufacture. A design patent only protects the appearance of an article and not its structure or utilitarian features. See Chapter 7.

- **Disclosure Document** Program. If you are not ready or do not care to apply for a patent, but want to officially evidence and register the conception date of an invention, the Patent and Trademark Office offers the Disclosure Document Program. See Chapter 4.

- **Plant Patent.** A plant patent is granted to anyone who has invented or discovered and asexually reproduced any distinct and new variety of plant, including cultivated spores, mutants, hybrids, and newly found seedlings, other than a tuber-propagated plant or a plant found in an uncultivated state. See Chapter 7.

- **Provisional Patent Application.** Good for only one year, a provisional application is less expensive than a regular patent application and a quick and simple way to establish date of invention. See Chapter 7.

- **Trademark.** A trademark is either a word, phrase, symbol or design or combination thereof, which identifies and distinguishes the source of the goods or services of one party from those of others. See Chapter 9.

- **Utility Patent.** A utility patent is granted for a new, useful, and non-obvious process, machine, manufactured article, composition, or an improvement in any of the above. See Chapter 7.

Inventors Beware

INVENTION MARKETING COMPANIES

This Is Only a Test

Empirical observation is always in order. In the name of science and boy scouts everywhere, I tested one of these invention marketing companies. After hearing a toll-free number on late-night radio, I gave the company a call.

My idea was for a device that kept individual strands of spaghetti fresh. The concept was to insert spaghetti strands into clear plastic containers capped off at both ends.

"Terrific. Lots of potential," I was told on the telephone. All I had to do was pay $400 for an initial product assessment.

Never play cards with a man named Doc. Never eat at a place called Mom's. And never hire an invention marketing company that advertises.

Every year the dreams of some 25,000 inventors are dashed by invention marketing companies that reportedly rip them off to the tune of up to $10,000 each, according to U.S. Senator Joe Lieberman (D-CT), a former Connecticut attorney general, who chaired a hearing on invention marketing scams on September 2, 1994, in Room 342 of the Dirksen Senate Office Building on Capitol Hill.

"Necessity may be the mother of invention, but some of these marketing companies are nothing more than deadbeat dads," Lieberman told those attending the Government Affairs Subcommittee on Regulation and Government Information. "They praise all inventions, even those that stand no chance of being brought to market. They paint a rosy picture of huge profits, then do little or nothing to make that dream come true. They say they will make their money from royalties off the sale of the invention, when in reality their profits come from the inventor's up-front fee. And just when the inventor thinks the company is going to get rolling on their behalf, he or she has just been rolled, and the company has moved on to the next victim."

Although 19 states already have legislation to protect inventors (see page 21), Senator Lieberman intends to introduce legislation at the Federal level.

"Inventors are paying invention marketers thousands of dollars in up-front fees for help in licensing their products, and are getting nothing for their investment," according to Bruce Gjovig, director of the Center for Innovation and Business Development at the University of North Dakota.

The Wall Street Journal reported in its February 11, 1994 edition that Robert Lougher, founder of Inventors Awareness Group, Inc., and a former employee of American Inventors Corp.(AIC) in Westfield, Massachusetts, "quit in disgust after concluding that the invention-promotion concern was taking in millions of dollars from unsuspecting inventors—but rarely, if ever, getting its customers' ideas to market."

Lougher testified before the Government Affairs Subcommittee on Regulation and Government Information, that as a marketing director for AIC ". . . I would ultimately witness widespread fraud, listen to horror stories from people who lost their homes, their cars and life savings and sink to the lowest moral depths a human being can sink to."

According to Margaret Leonard, an attorney in the Enforcement Division, Office of Compliance, of the District of Columbia Department of Consumer and Regulatory Affairs, as of September 1, 1994, there were approximately 140 complaints registered with her office against invention marketing firms. Complainants hailed from 35 states and Canada. "The complaints suggest a thriving industry of misrepresentation and systematic exploitation of consumers nationwide," Leonard reports.

The District of Columbia government gets involved because many of these invention marketing firms operate within D.C.'s borders because they like the sound of a national capital address. Leonard says that as a result of her department's efforts, approximately 80 consumers have accepted settlement offers totaling about $200,000. These people paid out some $350,000 to these companies. She indicates that there are approximately 60 unresolved consumer complaints totaling close to $300,000. "The Department is committed to its pursuit of redress for inventors who file complaints against these companies," says Leonard. "Unfortunately, the Department receives new complaints and learns of new companies on a regular basis."

Michael K. Kirk, deputy assistant secretary of commerce and deputy commissioner of patents and trademarks, told the Government Affairs Subcommittee on Regulation and Government Information, "We strongly believe that the problem of unscrupulous invention development organizations is persistent and growing. It is a problem that threatens the independent inventors that are directly affected by the unscrupulous invention marketing organizations. Further, it tarnishes the reputation of the many legitimate organizations dedicated to assisting inventors, and the integrity of the system for the protection and therefore the encouragement of inventions. Finally, it tarnishes the reputation of the USPTO [United States Patent and Trademark Office]."

"Based upon its investigations, the Commission [Federal Trade Commission, or FTC] estimates that over the years, tens of thousands of independent inventors have been victimized by, and have lost tens of thousands of dollars to, ineffective or dishonest companies and individuals offering invention promotion services," says Phoebe D. Morse, director of the Boston Regional Office of the FTC.

Who Are These Parasites?

"About 80 percent of all people claiming to help inventors build a business, market their product, or raise capital are con-men, beggars, thieves, or incompetents," cautions Professor Mark A. Spikell, co-founder of the Entrepreneurship Center in the School of Business Administration at George Mason University.

Alan A. Tratner, president of Inventor's Workshop International (IWIEF), describes most invention marketing companies as cancers on the inventing community, a disease that needs to be eradicated immediately. "Too often inventors lose large amounts of money and are derailed by the unfulfilled promises and the come-ons of these companies," he says.

In a legal brief sent by an unidentified person to *The Lightbulb*, a journal for inventors, an attorney describes invention marketing companies as putting out

Learning the Hard Way

Dave Thomas created an improved joystick for video games. His wife, Susan, saw an ad in *USA Today*: "Have an invention? Need help?" They responded to the ad, and the company, located in Boston, was excited about promoting the device. There was just one hitch: it would take $10,000 to get it off the ground.

Then Susan's brother-in-law, who believed in the invention, was killed in an auto accident. His widow provided the needed $10,000 from the insurance settlement.

Dave and Susan flew to Boston. The "account representative" wined and dined them. They signed a contract and handed over a $10,000 cashier's check.

And now? The account representative has disappeared. The Massachusetts Attorney General's Office doesn't have any record of the company.

And Dave and Susan are $10,000 wiser.

"an assembly line of ambiguity, little truths, big lies and contracts with hidden pockets of deceit!"

According to a report by the Minnesota Attorney General, only 48 of the 9,184 inventors who contracted with one firm got any money from using that company's service, and just 14 of them netted more money than they paid the company. Another study of two invention marketing companies by the Federal Trade Commission indicated that only 18 of over 70,000 inventors were helped to make a profit of $1 or more.

How to Tell the Pros from the Cons

Separating the legitimate invention marketing companies and agents from the bad ones is not always easy. The image the fly-by-nights portray through false and misleading advertising in legitimate media (such as *USA Today*, *Entrepreneur*, *Popular Mechanics*, *Popular Science*, CNN, etc.) is compelling. Messages are tagged with toll free numbers. Paid actors deliver confidence-building, call-to-action TV pitches. Slick four-color brochures picture well-known inventions that the invention marketer had absolutely nothing to do with (like the tricycle, ball point pen, zipper, teddy bear, or radio). Often in small print it will say: "These products are not intended to represent success for inventors who have worked with our firm."

Does advertising ever show evidence of the company's own product placement? Rarely. And, if it does, you will not recognize the product, and inventor testimonials—if any—will be signed with a name and state only. Nothing is offered that would help you contact the inventor for a reference.

There is no way any one company or individual could have enough meaningful contacts in several industries, or expertise in the fields of invention, that would be needed to find a home for the numbers of products that come in as a result of shotgun, mass-media advertising.

I specialize in the toy industry. It takes every bit of energy, and most of my waking hours, to keep up with the inter- and intra-company movements of R&D and marketing executives and to stay ahead of what's in and what's out in terms of product trends. There is no way that I could ever track a multitude of industries simultaneously. No one could. The business of invention licensing depends too much on personal relationships, not unlike all businesses. Success is a combination of know-how and know-who.

Never Pay an Up-Front Fee

Reputable invention marketing services will not require any up-front fees. You invested your ingenuity, time, and money to create and protect your invention. The marketer is obligated, in turn, to invest his or her ingenuity, time, and money to find it a home. What is more fair than that? The moment you pay for services, the carrot is removed; there is no risk for the marketer. With nothing to lose and your deposit in hand, the broker has little reason to display incentive.

"I frequently hear from sadder but wiser inventors who have spent thousands of dollars 'up front' to have some company market their invention," reports Ray Watts, former editor of the *Inventor Assistance Program News*.

A reputable firm will be able to demonstrate a track record of successful product placement and a list of satisfied clients. A reputable company will crow about its accomplishments and urge you to call its inventor clients and industry references.

Direct Mail Tactics

Direct mail pieces run the gamut from inexpensive post cards to envelopes bulging with slick brochures and official looking paperwork. But whether the marketer has spent fifty cents or two dollars per mailer, it does not matter: the stuff isn't worth the paper it's printed on.

Postcards

An inexpensive way to prospect for inventors, postcards from invention marketing companies usually go something like this:

IMPORTANT NOTICE

We have located six companies that produce, market, or sell products in a field to which your invention might apply. You may wish to contact these manufacturers if you are interested in one or more of the following:

A. Licensing your patent.

B. Finding a company to produce your invention.

C. Securing marketing, distribution, and/or sales help.

D. Hiring design or technical support.

Do they hit the hot buttons or what? Then comes the kicker. You are asked to remit a fee of $75 plus $1 for each name.

Once they get your money, all they do is pull names and addresses of appropriate manufacturers from the *Thomas Register* (a directory of manufacturers, products, and services) and send them to you. This satisfies their legal obligation. After you have paid for this service, you will be solicited about additional services.

Fat Envelopes

Other direct mail offers come in envelopes stuffed with all kinds of slick flyers and official-looking confidential disclosure agreements with diploma-like borders. Rarely will you be able to make out the signature of the "authorized agent" who signs off on the "Dear Inventor" cover letter. Nor, as I have already explained, will you recognize any of the products they claim to have commercialized, if any are shown at all.

Most invention marketing companies wrap themselves in Old Glory and Uncle Sam. Names are made to sound government-sanctioned with words like American, Federal, and National. Logos incorporate the bald eagle or another sym-

They [invention marketers] scare the living daylights out of me," says James Kubiatowicz, former director of product development/toys at Spearhead Industries. "They're leeches. They prey on the novices." Then pulling open his desk drawer, he adds, "I've got thirty-two post cards from these guys and I am saving the stamps. One of these days I'll steam them off and put them into a retirement fund.

"Anybody who is breathing and has a dollar can get strokes from these companies. They work on a person's ego."

bol of government. And often they use addresses and phone numbers that imply a Washington, D.C. presence. However, the addresses are nothing more than mail drops that forward correspondence on to the company's office elsewhere, and the phone numbers are routed through D.C. but answered in a boiler room somewhere else in the United States.

New Twist to an Old Scheme

Here is a novel way of getting money from inventors. A letter offers the inventor a free booklet on how to sell inventions. It invites a visit to the company's offices "...just to chat and to see what we look like." It is made clear that there is no obligation whatsoever.

The booklet is given away for free because, the letter states, "... inventions are our business." A visit is encouraged because "some inventors are reluctant to entrust their invention to people they have never seen and I (a company executive) cannot blame them for being cautious." A prepaid response card is enclosed.

The company goes on to build your confidence by separating itself from invention marketing companies. It claims not to neglect inventors after they pay a fee, or steal or mishandle their invention. This company sticks to what it does best, "preparing invention descriptions and compiling lists of manufacturers."

An Employment Agreement is enclosed through which you are asked to employ the company's "technique" in writing descriptive material, preparing folios, and selecting manufacturers that it "believes" may "potentially" be interested in your invention. It asks for no rights to your invention. It does not seek to legally bind you or your invention in any way. The descriptive materials it sells you become your property.

The Agreement states that you understand that the company does "no developing, promoting, or brokering." Its services are "strictly" to prepare invention descriptions and compile lists of manufacturers. It further claims to offer no "evaluation of the merits, practicability, feasibility, potential salability" of the invention.

What does it cost to have this company do a write-up of your invention and provide names of potential manufacturers? $589.50.

While I can see nothing dishonest in the service offered, the question you must ask yourself is, what is the value of such a service? In my opinion, this is a lot of money to pay for a "technique" in writing descriptive material, preparing folios and selecting manufacturers that *might* be interested in your invention.

How Do They Get Your Name?

Anyone who has been awarded a patent over the past twenty-five or thirty years has no doubt received an unsolicited letter that began with the fateful words, *"Dear Inventor: A number of manufacturers have invited us to send them descriptions of your invention...."*

I can always tell when a patent of mine is about to issue. Mailers from invention marketing companies flood into my post office box. Often three or four will come from the same company in one day. They are a harbinger of a new patent much as the robin is of spring.

When a patent issues, notice of it is automatically carried in the *Official Gazette* of the U.S. Patent and Trademark Office. This publication has come out weekly since 1872. It is mailed to subscribers and put on sale every Tuesday. The *Official Gazette* tells its readers:

- the name, city, and state of residence of the applicant, with the mailing address in the case of unassigned patents
- the same information for the assignee, if any
- the filing date
- the serial number of the application
- the patent number
- the title of the invention
- the number of claims
- the U.S. classification by class and subclass
- a selected figure of the drawing, if any, except in the case of a plant patent
- a claim or claims
- international classification
- U.S. patent application data, if any
- foreign priority application data, if any

In the case of a reissue patent, they publish the additional data of the number and date of the original patent and original application.

An annual subscription to the *Official Gazette* ($516 domestic; $687 priority class), is a fairly cheap way for invention marketing companies to obtain a qualified mailing list.

How to Protect Yourself

Levy's Rule of Thumb

My formula is simple. NEVER deal with invention marketing firms or agents that reach you through TV, radio, newspaper and magazine ads, or with unsolicited direct mail appeals. It is as simple as that. No ifs, ands, or buts.

As a general rule, reputable invention marketing companies:

- specialize in one industry where they are well known
- are eager to regale you with their success stories
- willingly introduce you to satisfied clients
- do not take up-front money for services

AIF's Litmus Test

Amateur inventors are full of questions, not the least of which is, "How can I find somebody to help me with my invention who won't rip me off?," says Joanne M. Hayes, owner, publisher, and editor of *Inventors' Digest*.

"The invention development field is loaded with con-men, rip-off outfits, and downright crooks," warns Hayes. With this in mind, she has published the following checklist written by the Affiliated Inventors Foundation (AIF), reprinted here with permission and gratitude.

AIF says that these questions should help you to identify questionable practices and typify activities that are less than honest. The correct response to each question is in boldface type. "If you check any of the non-bold responses, be warned that there is a strong possibility you are dealing with crooks," Hayes advises.

How to Rate An Invention Marketing Company

About Recording Inventions:

Do they imply that the Disclosure Document Program (DDP) of the U.S. Patent Office offers protection or that it is a viable substitute for a U.S. patent (the Patent Office says it is not)?

yes   **no**

About Evaluations:

Do they offer a meaningful evaluation of inventions at a reasonable price (less than $200)?

yes   no

Does their evaluation report point out weaknesses as well as strengths of your invention?

yes   no

Does the evaluation report make specific recommendations concerning such important areas as patenting, model making, marketing, and protection?

yes   no

About Patent Searches:

Do they recommend a patent search before recommending expensive marketing services?

yes   no

If they offer to perform a patent search, do they state specifically that the search will include an opinion of patentability by a registered patent attorney (or registered agent)?

yes   no

If you purchased a patent search, did they tell you the name of the individual responsible for the search and the name and registration number of the individual providing the opinion of patentability?

yes   no

About Marketing Services:

Do they ask for exorbitant front-end cash fees running to hundreds or thousands of dollars for their "best efforts" to sell or present your invention to industry?

yes   **no**

Do they tell you that in most cases to sell or license the rights to your invention you need a working model?

yes   no

Do they tell you anything in conversation, such as the "company is all lined up to buy your invention," which does not end up in the written contract?

yes   **no**

If they think your invention is so promising, are they willing to offer their marketing services on a no-cash, straight-commission basis?

yes   no

Do they suggest trying to market your potentially patentable invention before it reaches the "patent pending" stage?

yes   **no**

About Patent Applications:

Do they recommend filing for a patent application before beginning marketing efforts?

yes   no

Do they quote a firm, fixed price for preparation and filing a patent application (except for patent drawings, which are priced by the sheet)?

yes   no

Do they tell you about the patent application costs after filing, such as the almost-always required amendment(s), the final issue fee, maintenance fees, etc.?

yes   no

About References:

Are they willing to provide you with names and addresses of inventors who have used their services?

yes   no

Is the company willing to provide you with a copy of its latest financial statement indicating its net profit?

yes   no

REMEMBER: If you checked even one non-bold response, you should investigate the company in greater detail before proceeding!

Deceptive Trade Practices

The Federal Trade Commission (FTC) has received many complaints about invention marketing scams, so many that, based upon a case decided in 1978, it has determined that the following practices used in the advertising and marketing of idea or invention marketing services are unfair or deceptive trade practices and are unlawful under Section 5(a)(1) of the FTC Act:

1. For a seller of idea or invention promotion or development services to misrepresent, directly or indirectly, that potential purchasers will be provided with evaluations or appraisals of the patentability, merit, or marketability of ideas or inventions.

2. To represent, directly or indirectly, that the seller of idea or invention promotion or development services, or its officers, agents, representatives, or

Information, Please

Before you sign on the dotted line with an invention marketing firm, find out all the information you can from the FTC or your state attorney general's office. Don't be impressed by a Better Business Bureau recommendation alone.

If the government agency says it is unable to release information on a given company, consider making a request under the Freedom of Information Act or a state's sunshine laws.

employees are registered patent attorneys or patent agents, or are qualified to practice before the U.S. Patent and Trademark Office, unless such is a fact.

3. To misrepresent, directly or indirectly, the scope, nature, or quality of the services performed to develop or refine ideas or inventions.

4. To misrepresent, directly or indirectly, the scope, nature, or quality of the services performed to introduce or promote ideas or inventions to industry.

5. To represent, directly or indirectly, that a seller of idea or invention development service has special access to manufacturers or has been retained to locate new product ideas, unless such is a fact.

6. To misrepresent, directly or indirectly, that a person, partnership, corporation, government agency, or other entity endorses or uses the services of a seller or provider of services.

7. For sellers of idea or invention promotion or development services to fail to disclose, when price information is provided to potential purchasers, all significant fees or charges that may be incurred by purchasers in connection with such services.

8. To misrepresent, directly or indirectly, the background, qualifications, experience, or expertise of a seller or provider of services.

9. For a seller of idea or invention promotion or development services to induce through misleading or deceptive representations the purchase of services that have little or no inherent value, or to offer to provide services that grossly exceed the value of the services actually provided. It is also an unfair or deceptive act or practice to retain money from the sale of such services.

If you want to find out whether the FTC has any ongoing investigation into a particular invention marketing company, use a Freedom of Information Act (FOIA) request. The FTC will not divulge any particulars about an active investigation, legal or otherwise, or even about complaints received; however, if a consent order has been issued against the invention marketing company, you'll find out.

You may have to read between the lines. If under an FOIA request you ask for copies of any and all FOIA requests for information about a specific company, you just may get in return copies of requests sent to the FTC by law firms representing said company. The law firm would not be making a request about its client unless something was amiss.

For example, I once made an FOIA request to find out whether anyone else was interested in a specific invention marketing firm. The FTC sent me copies of requests, although denied, from a television news department, and a law firm. The fact that the FTC had denied their requests told me that the invention marketing firm was being looked at for some reason.

FTC Guidelines

In January, 1994, the FTC issued guidelines to help inventors identify legitimate companies. The following guidelines were developed by the commission's Boston Regional Office. I have annotated them as I felt appropriate.

❑ Early in your discussions with a promotion firm, ask what the total cost of its services will be, i.e., how expensive could it get when all is said and done? Any hesitation on the part of the salesperson should be a warning to you.

It is my experience that legitimate invention brokers do not charge up-front fees for their work. Instead they take a percentage of future advances and royalties. Percentages may run anywhere from one-third to one-half of future earnings, depending upon the person or company's track record for product placement.

One invention marketing company I received a flyer from offers packages ranging from Option #1: $150 per hour plus expenses, to Option #4: $6,950 plus a 25 percent royalty/commission. This may be the outfit that stung Lorraine Leiner, a medical records technician from Hartford, Connecticut.

Ms. Leiner told the Government Affairs Subcommittee on Regulation and Government Information that she sent a check in the sum of $4,950 according to Plan C of an agreement. "Under that plan, I was to get 35 percent of the royalties from my invention. The company would get 65 percent. This was the least expensive option. I took the bait and have been kicking myself ever since for being so trusting."

❑ Be careful of an invention promotion firm that offers to review and evaluate your invention but refuses to disclose details concerning its criteria, system of review, and qualifications of company evaluators. Without this information, you cannot assess the competence of the firm or make meaningful comparisons with other firms. Reputable firms should provide you with an objective evaluation of the merit, technical feasibility, and commercial viability of your invention.

"Pure and simple garbage," is how Fred L. Hart, lead invention coordinator at the U.S. Department of Energy, describes the work of most invention marketing companies. Once Fred showed me two different invention assessment reports that a Pittsburgh invention marketing firm had prepared for two different people about their inventions, a frost pump and a magnetic engine (perpetual motion machine). Both reports were exactly thirty-four pages long. Both were identical, word-for-word, save for the name of the invention and the product descriptions.

The inventors had paid the company for the same boilerplate review. A standard text had been developed and was being used for every product submission. In the case of the perpetual motion machine, Hart said the firm had written and submitted three separate but identical reports on behalf of three different inventors who felt they had found the way around gas engines. The reports were all signed by the same individual who tagged his name with the initials P.E. (Professional Engineer).

❑ Require the firm to check on existing patents. Because unscrupulous firms are willing to promote virtually any idea or invention with no regard for its patentability, they may unwittingly promote an idea for which someone already has been awarded a valid, unexpired patent. This could mean that even if the promotional efforts of your invention are successful, you may find yourself the target of a patent infringement suit.

If no valid, unexpired patent exists for your invention, seek advice from a patent professional such as a registered patent attorney before authorizing the public disclosure of your idea.

❑ Be wary of an invention promotion firm that will not disclose its success and rejection rates. Success rates show the number of clients who made more money from their invention than they paid the invention promotion company. Rejection rates reflect the percentage of all ideas or inventions that were found unacceptable by the invention promotion company. Check with your state and local consumer protection officials to learn if invention promotion firms are required to disclose their success and rejection rates in your area.

Talat I. Abuhamada, an inventor from Germantown, Maryland, sought assistance from American Inventors Corporation (AIC) in Westfield, Massachusetts, for a toy he had invented. AIC asked for $349 up front. However, he was told, if the opinion of the patent attorney/agent was unfavorable, a full refund would be sent to him.

Before paying the $349, Talat was smart enough to write and ask AIC for some information, such as the names and telephone numbers of satisfied inventor clients with whom he could talk. AIC's Scott Favreau answered by letter: "We cannot disclose the names and telephone numbers of any of our clients as this would be regarded as a breach of confidentiality."

Joanne Hayes, editor of *Inventors' Digest* sums it up best when she says, "The inventions are confidential—not the clients! If they've been in business several years, they should be able to give you many names (50, 100, 150). Don't just settle for two or three."

Talat also asked for the names of AIC's most successful products in hopes that he might recognize one of them. This question went unanswered. And as for the number of ideas the company's patent lawyers and/or agents found "unfavorable" last year, and how much money was refunded to the inventors who submitted such ideas, Favreau wrote: "I cannot disclose the number of unfavorable opinions, nor the dollar amount, resulting from the patent search."

In the *Wall Street Journal* piece referred to earlier in this chapter, Ron Boulerice, AIC's owner, acknowledged that his customers' success rate is "less than two percent," but he would not specify how much less. When the paper asked Boulerice to describe some inventions AIC has commercialized, he mentioned a stomach exerciser and a device that holds a tube during stomach operations.

❑ Be wary of a firm that claims to have special access to manufacturers looking for new products, but refuses to document such claims. Legitimate invention promotion firms substantiate their claims, which you can check.

❑ Be skeptical of claims and assurances that your invention will make money. No one can guarantee the success of your invention.

❑ Avoid being taken in solely on a firm's promotional brochures and affiliations with impressive-sounding organizations.

❑ Beware of high-pressure sales tactics.

❑ Investigate the company before making any commitments. Call the Better Business Bureau, local consumer protection agency, and the

The Inventor's Desktop Companion

Attorney General in your state and the state in which the company is located to learn if they know of any unresolved consumer complaints about the firm.

Find out if the FTC has ever taken any action against the company. According to Phoebe D. Morse of the FTC's Boston Regional Office, the Commission does not need a complaint to launch an investigation. See page 20 for information on contacting the FTC.

Be alert. The FTC may close down an invention marketing scam in one state, but nothing can prevent the same people from opening for business the next day in another state, under a new name and with new frontmen. "Chasing these people is like hunting gophers—they disappear down a hole and pop up somewhere else," cautions Gerald G. Udell, a well-known inventor advocate and professor of marketing at Southwest Missouri State University.

❏ Make sure your contract contains all agreed-upon terms, written and verbal, before you sign. Have the agreement reviewed by an attorney.

❏ If you do not get satisfactory answers to all of your questions, do not sign the contract. Once you pay your money to a dishonest company, it is unlikely you will ever get it back.

The FTC Takes Action

Let's take a look at what happened to one company the FTC went after in April of 1993 through the U.S. District Court for the Western District of Pennsylvania. This is a wonderful example of the kinds of misrepresentations invention marketing companies make, and what the Federal government can do about it if it gets involved.

Within ten months of the FTC's complaint, Invention Submission Corporation of Pittsburgh agreed to pay $1.2 million for consumer redress as part of a settlement. The FTC charged that this company had misrepresented the nature, quality, and success rate of the invention promotion services it sold.

The FTC alleged that despite the company's representations to the contrary, virtually none of its customers earned more from their inventions than they paid for promotion services. In addition to the redress payment, the settlement required the company to provide to prospective clients, on initial contact, an affirmative disclosure of its success rate, and to give consumers a right to cancel their contracts and obtain refunds.

The FTC's complaint detailing its allegations also named Western Invention Submission Corp., Intromark, Inc., and the parent company of all three, Technosystems Consolidated Corp., as well as Martin S. Berger, who is listed as sole officer and director of the firms (collectively known as ISC). All of the defendants are located in Pittsburgh.

According to the FTC complaint, ISC sells a variety of invention promotion services to individual inventors in up to three stages. First, the company prepares a "Basic Information Package" concerning the idea or product, for $395 to $595. In the second stage, the complaint states that the defendants offer to provide certain promotional services under a "Submission Agreement," at a cost ranging from $3,450 to $4,890. Third, ISC provides any leads it may receive on a client's idea to

Intromark, Inc., which allegedly attempts to negotiate a license for the idea or invention. According to the FTC complaint, clients agree to share with ISC a percentage of any payments received as a result of the company's services.

The FTC charged that ISC made numerous misrepresentations about the success rate and financial gains achieved by its customers. The FTC charged also that ISC incorrectly stated that it would evaluate or appraise the merit or marketability of its client's ideas or inventions, and that it has specialized, valuable access to manufacturers. In addition, the FTC charged that ISC misrepresented its background, qualifications, experience, and expertise as an invention promoter.

In another case, the FTC charged in 1991 that the invention promotion company American Idea Management (AIM) of Stoneham, Massachusetts, its two successor corporations, and three individuals misrepresented the nature, quality, and success rate of invention promotion services they had sold to inventors. Proposed consent decrees to settle the charges would prohibit future misrepresentations. In addition, the defendants would pay $570,000 plus interest to provide partial refunds to consumers injured by the defendants' allegedly deceptive activities.

This case was against AIM; Idea Management and Patent Assistance Corporation, also of Stoneham; Technology Licensing Consultants, Inc. of Pittsburgh; and corporate officers or owners: Suzanne Kameese of Lynfield, Massachusetts, and Anita and Lowell French, both of Pittsburgh.

These companies had operated since 1984, according to the FTC complaint, selling a variety of invention promotion services to individual inventors. The defendants solicited inventors to submit ideas or products for a free "initial review." They then offered the inventor—for a fee ranging from $450 to $595—a written "research report" that often included a patent novelty search and a written opinion of patentability prepared by a registered patent attorney. If the "research report" was favorable, the defendants then offered marketing, promotion, and licensing services for an additional fee ranging from $3,500 to $9,750.

What To Do If You Are a Victim

If you think that you have been victimized by a fraudulent invention marketing company, first contact the firm in writing and attempt to get your money back. If you are unsuccessful, report the problem to the FTC, your local consumer protection agency, the Better Business Bureau, the Attorney General (Consumer Protection Division) in your state and the state where the company is located, and any media you think might be interested in following up on your story.

Complaints to the FTC

You can report problems to the Federal Trade Commission by writing to: Correspondence Branch, Federal Trade Commission, 6th and Pennsylvania Avenue NW, Washington, D.C. 20580. Or call the Complaint and Inquiry Branch at: (202) 326-2222. The TDD number is: (202) 326-2502.

If you send in a written complaint, I recommend sending a photocopy to the person spearheading the FTC's work to stop invention marketing scams: Phoebe Morse, director of the FTC's Boston office. She may be reached by telephone at (617) 565-7240 or by writing to the Federal Trade Commission, 10 Causeway Street, Suite 1184, Boston, MA 02222-1073.

For good measure, bring your U.S. senator or representative into the loop as well by sending him or her a copy of FTC correspondence. This will assure that you get timely action. You should know beforehand that the FTC generally does not intervene in individual disputes. However, the information you provide may indicate to them a pattern of possible law violations.

If you wish to visit or contact an FTC Regional Office, here is a complete list of them.

1718 Peachtree Street NW, Suite 1000
Atlanta, Georgia 30367
(404) 347-4836

1405 Curtis Street, Suite 2900
Denver, Colorado 80202-2393
(303) 844-2271

101 Merrimac Street, Suite 810
Boston, Massachusetts 02114-4719
(617) 424-5960

11000 Wilshire Boulevard, Suite 13209
Los Angeles, California 90024
(310) 575-7575

55 East Monroe Street, Suite 1437
Chicago, Illinois 60603
(312) 353-4423

150 William Street, Suite 1300
New York, New York 10038
(212) 264-1207

668 Euclid Avenue, Suite 520-A
Cleveland, Ohio 44114
(216) 522-4207

901 Market Street, Suite 570
San Francisco, California 94103
(415) 744-7920

100 N. Central Expressway, Suite 500
Dallas, Texas 75201
(214) 767-5501

2806 Federal Bldg.
915 Second Ave.
Seattle, Washington 98174
(206) 220-6363

Complaints to Attorneys General

Every state Office of Attorney General has a Consumer Fraud Division or a department for handling such complaints. In states that have legislation on the books to protect inventors from front-money frauds, it will be even easier for you to get action. They understand the problem.

A Mere 2%

Few inventions, patented or otherwise, are ever commercialized. According to experts used in FTC cases, an invention promotion firm that does not reject most of the inventions it reviews may be unduly optimistic, if not dishonest, in its evaluations.

"A red flag should be raised if an invention service accepts and is positive about all ideas because less than two percent of all new ideas are commercially and technically viable," warns Bruce Gjovig, founder and director of the Center for Innovation and Business Development at the University of North Dakota.

More and more states are passing legislation that favors inventors. At this writing, there are 19 states: California, Connecticut, Illinois, Iowa, Kansas, Massachusetts, Minnesota, Nebraska, North Carolina, North Dakota, Ohio, Oklahoma, South Dakota, Tennessee, Texas, Utah, Virginia, Washington, and Wisconsin.

For example, under a tough Minnesota law, invention marketing companies that operate in the state must, among other things, disclose to prospective customers the company's success rates.

Penny Becker, executive director of the Minnesota Inventors Congress, stated in a *Wall Street Journal* article that Western Invention Submission reported that of 7,651 customers who had signed contracts, "zero have received money in excess of fees paid."

Unfortunately, most inventors who get ripped off by the paracreative slugs who operate under the guise of invention marketing services do not even realize that their state has protective legislation.

Here are some highlights from the laws of Minnesota and Virginia. I would encourage you to write to any of the aforementioned states for a complete copy of its legislation.

If your state does not have protective legislation for inventors, get the word to the appropriate elected officials. Every state should have some sort of regulation on the books that sets fair standards for invention marketing companies.

Minnesota (Invention Services 325A.02)

1. Notwithstanding any contractual provision to the contrary, inventors have the unconditional right to cancel a contract for invention development services for any reason at any time before midnight of the third business day following the date the inventor gets a fully executed copy.

2. A contract for invention development services shall be set in no less than 10-point type.

3. An invention developer who is not a lawyer may not give you legal advice with respect to patents, copyrights, or trademarks.

4. The invention marketer must tell you:

    a. the total number of customers who have contracted with him up to the last thirty days; and

    b. the number of customers who have received, by virtue of the invention marketer's performance, an amount of money in excess of the amount of money paid by such customers to the invention marketer pursuant to a contract for invention development services.

5. The contract shall state the expected date of completion of invention marketing services.

6. Every invention marketer rendering invention development services must maintain a bond issued by a surety admitted to do business in the state, and equal to either ten percent of the marketer's gross income from the invention development business during the preceding fiscal year, or $25,000, whichever is larger.

1. No invention developer may acquire any interest, partial or whole, in the title to the inventor's invention or patent rights, unless the invention developer contracts to manufacture the invention and acquires such interest for this purpose at or about the time the contract for manufacture is executed.

2. The developer must tell you if they intend to spend more for their services than the cash fee you will have to pay.

3. The Attorney General has the mandate to enforce the provisions of this chapter, and recover civil penalties.

Help from the News Media

If the invention marketing company is national, you may wish to tell your story to reporters who cover the consumer beat at network level news-gathering operations, national daily newspapers, weekly and/or specialized magazines, and radio programs. You may wish to begin with local media just to get the story in play.

Small Claims Court

Depending upon the amount of money owed, Small Claims Court may be an option if the invention marketing company is local.

Meanwhile on Capitol Hill

As mentioned at the beginning of this chapter, Senator Joe Lieberman intends to use the information gathered at his September 1994 hearing, together with any comments that other members of the public wish to submit for the record, in his preparation of federal legislation.

To send your comments about the subject of invention marketing scams to Senator Lieberman, direct your correspondence to him care of the United States Senate, Washington, D.C. 20510-4601. Telephone: (202) 224-4041. Fax: (202) 224-9750.

You should also encourage your U.S. Representative to support the Lieberman bill when it comes out of Committee. It will propose that:

- all invention marketing companies must register with the U.S. Patent and Trademark Office (PTO)

- invention marketing companies must tell prospective clients how many invention submissions they reject, and how many actually make a profit

- the PTO must maintain a complaint log so that consumers could check to see if the company or individual they are dealing with has any outstanding complaints, and law enforcers could keep better track of troublesome firms

- the power of consumers to sue deceptive marketing companies in court must be expanded

How to Find an Honest Broker

Here are two methods I endorse that, while not guaranteed, put the odds more in your favor that you will not be ripped off.

Build a Box

Carl E. Haskett, inventor of the Haskett SunClock, an elaborate sundial that is as precise as a watch, is a former consultant at Booz, Allen, Hamilton. He says that in dealing with consultants the inventor must "build a box so tight around the problem that there are no holes in corners." A University of Oklahoma graduate with a degree in Engineering Physics, Haskett says this means the task given the consultant must be defined so carefully that the consultant cannot expand it without absolute permission.

Inventor Groups

Call an inventor organization, preferably a local one whose members may be able to recommend an agent. I think you will get a more complete picture of the situation and gain more confidence in the agent if you sit down with his or her satisfied customer.

Industry Referrals

Ask the company to whom you would like to license your invention to recommend an agent. While not all companies are comfortable recommending invention brokers, many will. For example, if you approach Milton Bradley, the world's largest game manufacturer, and ask for the name of a broker with whom the company does business, it will send you a list of names. You must, of course, talk to the agents and strike your own level of confidence and a fair deal; nonetheless, you know up-front that the door at Milton Bradley is open to them. Further, anyone Milton Bradley recommends most likely has excellent relationships with other game manufacturers too.

Invention Consultants

Do not confuse invention marketing firms with legitimate invention marketing consultants. I strongly believe in paying consultants whenever their expertise can contribute to the progress and success of a project.

What is consulting? I like the definition offered by England's Institute of Management Consultants. It defines consulting as: "The service provided by an independent and qualified person or persons in identifying and investigating problems concerned with policy, organization, procedures and methods; recommending appropriate action and helping to implement these recommendations."

There are as many types of consultants as there are problems to solve. These experts can bring new techniques and approaches to bear on an inventor's work. This contribution can range from helping to bridge a technological gap to the special knowledge and talent required to successfully license or market a particular innovation.

Advisors can provide impartial points of view by seeing challenges in a fresh light. They operate outside existing frameworks and free from existing beliefs, politics, problems, and procedures inherent in many organizations or situations.

Most consultants operate on the basis of an hourly rate plus expenses. Inventors, however, by the nature of their work, are often able to make equity deals whereby in return for their advice, consultants are given participation in any profits said invention might generate. Inventors should think long and hard before doing something like this because it is often less expensive to risk the cash and hold all the points possible in-house.

Do not think that consultants have all the answers. They do not. Consulting is very hard work and not everything can be solved as quickly as one would like. Do not look for miracle solutions.

Shop around. Get references on any consultant or research organization you are considering. Don't be impressed by a consultant's or organization's professional association alone (such as a university affiliation). Their success rate in fields related to yours is what matters. How much can they do with a single phone call? Results are what you want, not just paper reports.

A very good source for consultant leads is Gale Research Inc.'s *Consultants and Consulting Organizations Directory*. To find a copy, consult your local library.

The Pictionary Story

One example of an inventor who took in consulting partners to make his project materialize is Robert Angel, inventor of the popular game Pictionary. Angel knew that he had a terrific concept, but needed some graphic support. He asked artist Gary Everson to help him design the game board in return for points in the venture. Everson agreed. The then 26-year-old inventor also needed assistance in making business decisions, so he gave points as well to an accountant named Terry Langston.

Pictionary went on to be the bestselling game of 1987, grossing more than $52 million at retail. In 1988 the product's sales soared to an astounding $120 million. To date more than 14 million units of the quick draw game have been sold, and the three partners have become wealthy, receiving royalties on every game sold.

Wisconsin Innovation Service Center

One of the better and most reputable places to go for help in assessing the marketability of your invention is the Wisconsin Innovation Service Center (WISC), a cooperative effort between the University of Wisconsin-Whitewater Small Business Development Center, UW-Extension, and the Small Business Administration.

Since it opened its doors in 1980, WISC has evaluated thousands of new product ideas at the request of inventors and small businesses. It has many success stories.

For information, write: Wisconsin Innovation Service Center, 402 McCutchan Hall, University of Wisconsin-Whitewater, Whitewater, Wisconsin 53190. Telephone: (414) 472-1365.

A Happy Ending

"I actually made my consultant a partner," says Richard Tweddell III, inventor of *Vegi*Forms, a device that press-molds vegetables, while still on the vine, into the likenesses of human faces. A professional toy designer at Kenner Products in Cincinnati, Ohio, Tweddell credits his consultant with moving his company from ground zero. "He showed me how to license the product, he reviews all of our licensing agreements and he found companies that were interested in taking the rights to [*Vegi*Forms]."

The Patent & Trademark Office

> *Invention is the fruit of man's brain. Industries grow in proportion to invention. Therefore, the government must aid progress by fostering the inventive genius of its citizens.*
> —DANIEL WEBSTER

To find out more about the history of the PTO, consult their booklet entitled *The Story of the U.S. Patent and Trademark Office*. In that publication, they summarize the significance of the patent process:

"The patent system is one of the strongest bulwarks of democratic government today. It offers the same protection, the same opportunity, the same hope of reward to every individual. For 198 years it has recognized, as it will continue to recognize, the inherent right of an inventor to his government's protection. The American patent system plays no favorites. It is as democratic as the Constitution which begot it."

Once upon a Time . . .

On April 10, 1790, President George Washington signed the bill that laid the foundation of the modern American patent system. This was the first time in history that the innate right of an inventor to profit from his or her invention was recognized by law. At that time, fees for a patent were between $4 and $5, and patents remained in effect for 14 years.

Previously, privileges granted to an inventor were dependent upon the prerogative of a monarch or upon a special act of a legislature or court. The first patent on this continent was granted by the Massachusetts General Court to Samuel Winslow in 1641 for a novel method of making salt. The first patent on machinery was granted by the same court to Joseph Jenkes in 1646; he had invented a mill for manufacturing scythes.

The responsibility for granting patents in 1790 was shared by a board consisting of the secretary of state, the secretary of war, and the attorney general. Board members had the power to issue a patent. However, the responsibility for the administration of the patent laws was given to the Department of State.

As secretary of state, Thomas Jefferson was not only the first supervisor of the Patent Office, but also in effect its first patent examiner. No one could have been more ably fitted to administer the first American patent laws.

A mathematician, astronomer, architect, and student of languages, he was undoubtedly the most accomplished and versatile person in public life at the time. Not only did Jefferson appreciate the value of patents from an intellectual standpoint, but as an inventor himself he had firsthand knowledge of the problems faced by creative people. Although he never filed for a patent himself, Jefferson invented a number of things, one of which—an improvement in the mold board of the plough—had a significant effect on the agricultural development of this country and earned him a decoration from the French Institute. He also invented a revolving chair which his enemies accused him of designing "so as to look all ways at once," a folding chair or stool which could be used as a stick, a machine for treating hemp, and a pedometer.

In 1828 an Englishwoman who visited America wrote, "The Patent Office is a curious record of fertility of the mind of man when left to his own resources."

In 1849, the Act of March 3 transferred the Patent Office from the State Department to the newly created Department of the Interior. On July 8, 1870, the commissioner of patents was given jurisdiction to register trademarks and the power to make rules and regulations concerning them. On April 1, 1926, by executive order, the Patent Office was transferred to the Department of Commerce. Then on January 2, 1975, Public Law 93-596 changed the name of the Patent Office to the Patent and Trademark Office.

Since its inception, the patent system has encouraged the genius of millions of inventors. It has protected these creative individuals by allowing them an opportunity to profit from their labors, and has benefited society by systematically recording new inventions and releasing them to the public once the inventors' limited rights have expired.

The Patent and Trademark Office has recorded and protected Morse's telegraph, McCormack's reaper, Bell's telephone, and Edison's incandescent lamp. It has fostered the genius of Goodyear and Westinghouse, of Whitney and the Wright Brothers, of Mergenthaler and Ives, of Baekeland and Hall.

Under the patent system, American industry has flourished. New products have been invented, new uses for old ones discovered, and employment given to millions. And, most important, under the patent system a small, struggling nation has grown into the greatest industrial power on earth.

The Mandate of the PTO

The PTO administers the laws relating to patents and trademarks to promote industrial and technological progress in the United States and to strengthen the national economy. It develops and advises the Secretary of Commerce and the President on intellectual property policy, including copyright matters. Also, in cooperation with the International Trade Administration, the PTO advises the Department of State and other agencies of the U.S. government, such as the United States Trade Representative, on trade-related aspects of intellectual property.

What Does It Cost to Run the PTO?

The PTO no longer receives federal funding. Total revenues in fiscal year 1994 were $544 million, against expenses of $485 million.

How Many People Work at the PTO?

As of this writing, the Patent and Trademark Office has over 5,000 employees, of whom about half are examiners and others with technical and legal training.

How Much Work Do They Handle?

More than 5 million patents have been issued since the Patent Office was established. Each year, the PTO receives over five million pieces of mail. In 1994, the PTO received 201,554 new patent applications.

A Unique Agency

The PTO is one of the most unusual branches of the U.S. Government. Its examining staff of over 1,800 is trained in all branches of science and examines thoroughly every application to determine whether a patent may be granted—a task, in these days, involving the most exhaustive research.

Not only must the examiners search United States and foreign patents to learn if a similar patent has been issued, but they must study scientific books and publications to discover whether the idea has ever been described.

Top 20 Corporations

According to a list compiled by Intellectual Property Owners, a non-profit association of patent, trademark, copyright, and trade secret owners, the following were the top 20 corporations receiving U.S. patents in 1994:

IBM Corp.	1298
Canon K.K.	1096
Hitachi, Ltd.	976
Mitsubishi Denki K.K.	972
General Electric Co.	970
Toshiba Corp.	968
NEC Corp.	897
Eastman Kodak Co.	888
Motorola, Inc.	837
Matsushita Electric Industrial Co., Ltd.	771
Sony Corp.	656
Xerox Corp.	611
AT&T Corp.	595
Fujitsu, Ltd.	592
Fuji Photo Film Co., Ltd.	545
Minnesota Mining & Manufacturing Co.	519
E. I. Du Pont de Nemours & Co.	486
Texas Instruments, Inc.	479
Sharp K.K.	454
Hewlett-Packard Co.	428

Each week, 65,000 pages of new documents are added to the estimated 250 million pages already on file at the PTO.

In mid April, 1994, Bruce A. Lehman, Assistant Secretary of Commerce and Commissioner of Patents and Trademarks, released his annual report for fiscal year 1994. Here are some highlights of PTO activity during that year:

Patents

- The PTO issued 113,268 patents. Fifty-seven percent of the patents went to U.S. inventors—the highest percentage of patents awarded to U.S. inventors in the past nine years.
- It processed a record number of 201,554 patent applications, of which 186,123 were for utility patents.

Trademarks

- The PTO registered 68,853 trademarks, a reduction in registrations from 1993, when it was 86,122.
- It processed a record number of 155,376 trademark applications.

General Information and Correspondence

The PTO is a huge bureaucracy. Without a map and a compass, it can be frustrating or near impossible for a beginner (especially one living outside the Washington D.C. area and working long-distance). To help you quickly reach the most appropriate person for your needs, see Appendix 2 for the current PTO telephone and mailbox directory.

All business with the Patent and Trademark Office should be transacted by writing to: Commissioner of Patents and Trademarks, Washington, D.C. 20231.

The principal location of the PTO (in other words, where the Commissioner's office is situated) is Crystal Plaza 3, 2021 Jefferson Davis Highway, Arlington, Virginia. Free PTO shuttle buses run between sites from 6:30 a.m. to 6:30 p.m.

How to Correspond Efficiently with the PTO

Separate letters (but not necessarily in separate envelopes) should be written in relation to each distinct subject of inquiry, such as assignments, payments, orders for printed copies of patents, orders for copies of records, and requests for other services. None of these should be included with letters responding to PTO actions in applications.

If your letter concerns a patent application, include the serial number, filing date, and Group Art Unit number. When a letter concerns a patent, include the name of the patentee, the title of the invention, the patent number, and date of issue.

If ordering a copy of an assignment, provide the book and page or reel and frame record, as well as the name of the inventor; otherwise, the PTO will assess an additional charge to cover the time consumed in searching for the assignment.

The PTO will not send or show you applications for patents. These applications are not open to the public, and no information concerning them is released except on written authority of the applicant, assignee, or designated attorney, or when necessary to the conduct of the PTO's business. You can write for and receive, however, records of any decisions, the records of assignments other than those relating to assignment of patent applications, books, or other records and papers in the PTO that are open to the public.

The PTO will not respond to inquiries concerning the novelty and patentability of an invention in advance of the filing of an application; give advice as to possible infringement of a patent; advise on the propriety of filing an application; respond to inquiries as to whether or to whom any alleged invention has been patented; or act as an expounder of the patent law or as a counselor for individuals, except in deciding questions arising before it in regularly filed cases.

Keep copies of everything you send the PTO. Don't forget to put your name and return address on all papers and the envelope.

In Case of Emergency . . .

The PTO has established a contingency plan for filing any paper or paying any fee in the Office in the event of an emergency caused by any major interruption in the country's mail service.

Upon the determination by the commissioner of patents and trademarks that such an emergency exists, the commissioner will cause to be printed a notice of the plan in the *Wall Street Journal* and make it available by telephone at (703) 305-9723. Also, certain publications, patent bar groups, and other organizations closely associated with the patent system will be notified. Termination of the emergency program will be similarly announced. Where the postal emergency is not nationwide, the commissioner will designate the areas of the country in which the procedures outlined will be in effect.

The plan calls for the U.S. Department of Commerce district offices to be designated as emergency receiving stations for filing papers and paying fees in the PTO.

Disclosure Document Program

Buried Treasures

Deep inside a mountain in Pennsylvania, tunneled more than 200 feet beneath the earth's surface, stands an impregnable fortress. Armed guards watch over the three-foot-thick, three-ton gate at the entrance of the former mine. Inside temperature is kept at a constant 56 degrees. Located 55 miles north of Pittsburgh, the cavernous National Underground Storage (NUS) area is where the PTO holds important documents safe from fire, earthquakes, or even nuclear war.

When you have something hot, burn it. When it gets cold, sell it for ice.

The First Step

If you are not quite ready, or do not wish to apply for a patent yet, but want to officially register the conception date of your invention, the PTO offers the Disclosure Document Program. For a fee of $10, the PTO will preserve your idea on file for a period of two years. This inexpensive recognition will strengthen your case if any conflict arises as to the date of the conception of your idea, but it is not meant to replace your inventor's notebook or an actual patent.

Disclosure Document Program Requirements

The requirements are simple. Send the PTO a paper disclosing the invention. Although there are no stipulations as to content, and claims are not required, the benefits afforded by the Disclosure Document Program will depend directly upon the adequacy of your disclosure. Therefore, it is strongly recommended that the document contain a clear and complete explanation of the manner and process of making and using the invention in sufficient detail to enable a person having ordinary knowledge in the field of the invention to make and use the invention. When the nature of the invention permits, a drawing or sketch should be included. The use or utility of the invention should be described, especially in chemical inventions.

This disclosure is limited to written matter or drawings on heavy paper, such as linen or plastic drafting material, having dimensions or being folded to dimensions not to exceed 8½ x 13 inches (21.6 by 33 cm.). Photographs are also acceptable. Number each page, and make sure that the text and drawings are of such quality as to permit reproduction.

Disclosure Document Request and Fee

In addition to the fee, the Disclosure Document must be accompanied by a stamped, self-addressed envelope (SASE) and a separate paper in duplicate, signed by you as the inventor. These papers will be stamped by the PTO upon receipt, and the duplicate request will be returned in the SASE together with a notice indicating that the Disclosure Document does not provide patent protection and that a patent application should be diligently filed if patent protection is desired.

The $10 fee must accompany the Disclosure Document when it is submitted to the PTO. A check or money order must be made payable to "Commissioner of Patents and Trademarks." Mail with the Disclosure Document to: Commissioner of Patents and Trademarks, Box DD, Washington, D.C. 20231.

Fees are subject to change annually. Fees may be confirmed by calling the Public Service Branch at: (703) 308-4357.

Your request may take the following form:

Date: 1/1/96

To: Commissioner of Patents & Trademarks

From: Jane Q. Inventor

Dear Commissioner:

The undersigned being the inventor of the disclosed invention, request that the enclosed papers be accepted under the Disclosure Document Program, and that they be preserved for a period of two years. A check in the sum of $10 is attached hereto.

My invention is (insert description).

There are (insert number) different approaches I intend to take. They are illustrated on the enclosed (insert number) sheets.

Thank you very much,

Jane Q. Inventor

456 Maple Road

Midtown, Michigan 48010

Tel: (810) 555-1234

Be advised: the two-year retention period should not be considered to be a "grace period" during which you can wait to file a patent application without possible loss of benefits. It must be recognized that in establishing priority of invention an affidavit or testimony referring to the Disclosure Document must usually also establish diligence in completing the invention or in filing the patent application since the filing of the Disclosure Document.

Also be reminded that any public use or sale in the U.S. or publication of the invention anywhere in the world more than one year prior to the filing of your patent application on the invention disclosed will prohibit the granting of a patent on it.

The Disclosure Document is not a patent application, and the date of its receipt in the Patent and Trademark Office will not become the effective filing date of any patent application subsequently filed. It will be retained for two years and then be destroyed unless referred to in a separate letter in a related application within two years.

The program does not diminish the value of the conventional witnessed and notarized records as evidence of conception of your invention, but it should provide a more credible form of evidence than that provided by the popular practice of mailing a disclosure to oneself or another person by registered mail.

Inventor's Notebook

In addition to the formal PTO Disclosure Document, you should maintain an inventor's notebook in which to record your ideas through words and sketches and the dates of their inception. Some inventors have their entries witnessed and signed by someone else. This documentation could prove very valuable if you ever have to prove in court the date of an invention.

I have designed a patent worksheet upon which to record my ideas. I keep these sheets and other notes in a three-ring binder. Feel free to use it as is or modify it as required. This worksheet can be found on page 121.

CHAPTER

5

It All Begins with a Search

Never bring up the artillery until you bring up the ammunition.

How Novel Is Your Idea?

There you sit eating dinner when, without warning, the Mother of All Ideas makes its way through your mind and into the fore of your brain. You leap up, write it down. Then just as you re-approach Alpha Level, you ask yourself the sobering question that always follows the Mother of All Ideas: *has the concept already been patented by someone else?*

You are not alone. This is the same question every inventor asks after feeling the kind of exhilaration that only the Mother of All Ideas can cause.

The Search

The answer to this question can only be found through a deep patent search. The search will tell you if your idea has been patented already, and if so, if the patent is still in force. The PTO's cataloging system is pretty complex, and the number of patents you may need to look through can be staggering. The PTO has issued about 5.2 million patents. At the Office's Public Search Room in Crystal City, Virginia, there are 27 million references on file, according to Oscar Mastin, a PTO public affairs officer.

You cannot avoid doing a search. The examiner will do one anyway and if your application is rejected based upon prior art (patents that have previously issued), you'll have lost the application cost not to mention the significant time and energy you invested.

Further, even if none of the earlier patents show all the details of the invention, they may point out important features or better ways of doing the invention. If this is the case, you may not want to get patent protection on an invention that could encounter commercial difficulty.

In the event nothing is found to prevent or delay your application, the information gathered by a search will prove helpful, acquainting you with the details of patents related to your invention.

How to Conduct a Patent Search

You may approach the search in three ways. You could hire a patent attorney, directly engage the services of a professional patent search firm, or do it yourself. Let's look at all these options.

Eureka!

The PTO's research collection is immense. More than 30 million documents are on hand, including over five million U.S. patents, nine million cross-references, and 16 million foreign patents. The Scientific Library maintains a collection of 120,000 volumes and provides access to commercial databases.

Patent Attorney–Directed Search

Going through a lawyer will cost the least amount of time and the most money. Good attorneys do not conduct their own searches; they do not have the time. They are too busy drafting claims, going to court, and fulfilling other lawyerly duties. This is no different than dentists who no longer find it cost-effective to clean teeth, or physicians who do not give injections because their time is better spent on services that can generate greater income.

Patent attorneys have professional researchers do this work. You hire the attorney. The attorney hires the searcher. Then the attorney adds a handling fee to the search bill, sometimes as much as several hundred percent. To save this extra expense, many inventors do the search themselves or hire their own researcher.

But be advised: some patent attorneys will not render an opinion on a search conducted by anyone other than their own searcher. I have always told my lawyers that if they would not accept the work of my search firm, or searches done by myself, I would go elsewhere where such work would be acceptable. I figure I am paying the bills and if I am willing to take the risk, the lawyer should not have a problem.

I do not go through a patent attorney because I see no reason to pay for a legal opinion when one may not be needed. If search results show no prior art in my field of invention, I do not need an attorney to tell me that the coast is clear. Conversely, if a search reveals prior art that is spot on my invention, I do not need an attorney to tell me that my idea has been done before.

If you hire a lawyer, get a quote in advance. The fee will be based on how all-encompassing you want the search to be.

Direct Hire Professional Search

If you want to save the lawyer's mark-up, consider going directly to a patent search firm. Searchers are best found through inventor grapevines, inventor associations or university intellectual property departments.

You can also check the yellow pages under "Patent Searchers." But be careful not to fall into a trap set by some disreputable invention marketing organizations where they list themselves in the phone book under "Patent Searchers" with a toll-free number. This is another way they hook unsuspecting inventors into service contracts. Get all the facts up-front.

Some reputable search companies will ask for money up-front if you are not known to them. This is understandable. Just make sure that you are told the cost of the search beforehand, and get references if you are working long distance.

Ask for a rate sheet, then get on the phone and discuss the exact services you need. There are different charges for different assignments (for example, searching utility patents versus design patents). Electronic, chemical, biological, botanical, and medical searches are often more expensive. And, in most cases, there will be incidental charges for copies, phone and fax, online fees, and shipping and handling of your materials. This is all standard.

Most important, ask if your searcher is experienced in your field of invention. For example, if you are searching a plant patent, don't hire someone skilled in mechanical devices.

Unless you have a history with a search firm, ask for a letter of non-disclosure before you sign on.

The cost to search a utility patent here in the D.C. area runs between $150 and $250. Once the search has been completed, you may want to obtain an opinion as to the patentability of your invention; if so, add the cost of your lawyer. For more information on hiring patent attorneys, see Chapter 6.

My company uses Greentree Information Services (GIS), located in Bethesda, Maryland; the telephone number is (301) 469-0902. Its president is George Harvill. You may wish to contact GIS as part of your comparative shopping. There is no one better or more honest in the business than the people who own and operate Greentree. The average patent search by GIS starts around $200.

The Do-It-Yourself Patent Search

Several methods are available to you should you decide to do the patent search yourself.

The PTO operates a Patent Public Search Room located in Arlington, Virginia. Here, every U.S. patent granted since 1839 (over five million and growing) may be searched and examined. Many inventors like to make at least one pilgrimage to the Patent Public Search Room. It is located less than five minutes from National Airport by taxi. Metro Rail serves it off the Blue and Yellow Lines, Crystal City Station. There are several hotels within walking distance.

Upon your arrival the PTO will issue you, at no cost, a non-transferable user pass for the day. It is wise to double check the hours of operation by calling (703) 308-0595. Depending where in the facility you want to search, the Patent Public Search Room is typically open from 8:00 a.m. to 8:00 p.m.

The Patent Public Search Room is really something to behold. You can touch and feel the original documents, including everything from Abraham Lincoln's 1849 patent (No. 6,469) for a device to buoy vessels over shoals, to Auguste Bartholdi's design patent on a statue entitled "Liberty Enlightening the World," better known as the Statue of Liberty.

There are a number of computerized PTO search systems available to the public. The Automated Patent System (APS) is online in the Patent Public Search Room at Crystal City and, as of this printing, at fourteen Patent and Trademark Depository Libraries (PTDLs) nationwide. Call your closest PTDL to see if it has the service. The list of PTDLs starts on page 39.

While the APS system is excellent, you will want to plan your search and know how it works before you log on. On occasion, there are free—by appointment—classes conducted by PTDL staffers.

The Classified Search Image Retrieval system (CSIR) is currently located at the PTO in Crystal City. There are ten workstations open to the public; the cost is $50 per hour. To make reservations to use the machine, call (703) 308-6001.

Questions about these systems may be directed to: Edith Wikniss, Manager, PTO Public Search Room, (703) 308-0595; the PTO's Public Service Center at (703) 308-4357; or your local PTDL librarian.

According to the *Inventor-Assistance Program News,* a publication of the States Inventors Initiative, patent searches can now be conducted via the Internet. The Electronic Data Systems (EDS) Shadow Patent Office (SPO) has opened an online patent search facility. SPO maintains a full-text patent database containing more than 1.7 million utility patents issued from January 1, 1972 to

Inventors, Unite!

Inventor-Assistance Program News is the main information transfer mechanism of the States Inventors Initiative, which is conducted by the staff of the Battelle Pacific Northwest Laboratories (PNL), located in Richland, Washington. PNL is operated for the Department of Energy (DOE) by Battelle Memorial Institute. The purpose of the States Inventors Initiative is to encourage inventor organizations, and any associations that assist inventors. It shares information among and with these organizations.

Available free of charge to qualified inventor associations, the *Inventor-Assistance Program News* is published periodically as a limited edition publication. Electronic copies of this newsletter are available upon request via e-mail (ipn_sii@pnl.gov), or via anonymous FTP at ftp.pnl.gov in the /pub/sii directory.

For information and to see if you qualify, contact Robin Conger, Battelle Pacific Northwest Laboratories, P.O. Box 999, K8-11, Richland, Washington 99352. The phone number is (509) 372-4328; the fax number is (509) 372-4369.

the present. It is updated weekly with all the new patents issued by the Patent and Trademark Office.

To make use of these services, you need to register with the EDS Shadow Patent Office and have access to electronic mail. Simply send an e-mail message of fifty words or less containing a description of the invention you are working with, and SPO's proprietary Concept Search Engine will send you the first page only of the 25 patents closest to your description. The cost of this service is $74. SPO also offers many other types of online searches.

For more information, send an e-mail message to: spo_patent@spo.eds.com; or call them at 1-800-258-6739.

Whether you do a manual or a computer search, make sure to look through *every possible* class and subclass that the patent office personnel suggest or that you feel are pertinent, and then some.

Same Day Search Service

There are many patent search services that now offer same-day service via fax. For a price, some will give service within the hour. These services have people on duty at the PTO's Patent Public Search Room in Crystal City, Virginia. Typically, they will make copies of the original patents and then rush them to a nearby fax machine.

This is not an inexpensive service. Faxpat, a company located in Arlington, for example, charges the following rates. I include this specific data as a guideline and not an endorsement. I encourage you to compare prices from several services before engaging one. For more information, fax Faxpat at 1-800-666-1233.

U.S. Patents

Sent same day:	$1/page, $9 minimum each
Sent next day:	$9 each
Sent in 3-5 days:	$6 each

Foreign Patents and Literature

Same day, next day, etc.:	$1/page, $12 minimum each

Trademarks

Sent same or next day:	$7 each
Sent in 3-5 days:	$4 each

U.S. Patent or Trademark Files

Sent same day:	$1/page, $45 minimum each

Cited references are processed as same-day patents unless instructed otherwise.

Files may be adhesive bound with clear and plastic covers, or hole-punched and metal clipped. However, files will be hole-punched unless adhesive binding is requested.

The faxing rate is equal to the document costs plus $3/page, $24 minimum each. The pages are faxed within one hour.

Patent Search Steps

No matter where you decide to conduct a search, certain steps must be taken in any patent search. Here is a brief guide to manual searches for U.S. patents.

Step 1: If you know the patent number, go to the *Official Gazette* to read a summary of the patent. This publication is available at any patent search facility and in many public library reference rooms.

Step 2: If you know the patentee or assignee, look at the *Patent/Assignee Index* to locate the patent number. This is available at any of the patent search facilities. In Crystal City, it is on microfiche and in card catalogues.

Step 3: If you know the subject, start with the *Index to the U.S. Patent Classification System.*

Step 4: Once you have jotted down the class(es) and subclass(es) out of the *Index,* refer to the *Manual of Classification* and check this information in relation to the hierarchy to see if it is close to what you need. The *Manual of Classification* is available at all patent search facilities.

Step 5: Using the class/subclass numbers you have found, look at the U.S. Patent Classification Subclass and Numeric Listing and copy the patent numbers of patents assigned to the selected class/subclass. If you are at the Crystal City facility, take the class/subclass numbers into the stacks of patents and begin "pulling shoes." To pull shoes is to physically remove patent groupings from the open shelves.

Step 6: Then, using the *Official Gazette* again, look at the summaries of those patents. At the Crystal City facility, you will not have to go back to this publication since the actual patents are there.

Step 7: Upon locating the relevant patents, examine the complete patent in person or on microfilm—depending upon where you conduct the search. Then make copies of the relevant ones.

Patent Classification System

Patents are arranged according to a classification system of more than 400 classes and 115,000 subclasses. The *Index to the U.S. Patent Classification System* is an alphabetical list of the subject headings referring to specific classes and subclasses of the U.S. patent classification system. The Classifications are intended as an initial means of entry into the PTO's classification system and should be particularly useful to those who lack experience in using the classification system or who are unfamiliar with the technology under consideration.

The Classifications are to searching a patent what the card catalog is to looking for a library book. It is the only way to discover what exists in the field of prior art. The Classifications are a star to steer by, without which no meaningful patent search can be completed. Before you begin your search, use the Classifications to plot your direction. First, look for the term that you feel best represents your invention. If a match cannot be found, look for terms of approximately the same meaning, for example, describing a similar function, effect, or applications. By doing some homework before you begin searching, such as familiarizing yourself with the *Index* and locating the class and subclass numbers for terms that pertain to your invention, you'll save time.

Once you have recorded the identifying numbers of possibly pertinent classes and subclasses, refer to the *Manual of Classification,* a loose-leaf PTO volume

The *Manual of Classification* and the *Index to the U.S. Patent Classification System* may be purchased from the Superintendent of Documents, U.S. Government Printing Office (USGPO), Mail List Section, Washington D.C. 20402. The telephone number is (202) 783-3238.

listing the numbers and descriptive titles of more than 300 classes and 95,000 subclasses used in the subject classification of patents, with an index to classifications. This manual is also available at the Patent and Trademark Depository Libraries.

The Classifications are arranged with subheadings that can extend to four levels of indentation. A complete reading of a subheading includes the title of the most adjacent higher heading, and so on until there are no higher headings. Some headings will reference other related or preferred entries.

New classes and subclasses are continuously based upon breaking developments in science and technology. Old classes and subclasses are rendered obsolete by technological advance. In fact, if you have suggestions for future revisions of the classifications, or if you find omissions or errors, you are encouraged to alert the PTO. Send your suggestions to Editor, U.S. Patent Classification, Office of Documentation, U.S. Patent and Trademark Office, Washington, D.C. 20231.

Nearby the Patent Public Search Room is the Scientific Library of the Patent and Trademark Office. The Scientific Library makes publicly available over 120,000 volumes of scientific and technical books in various languages, about 90,000 bound volumes of periodicals devoted to science and technology, the official journals of 77 foreign patent organizations, and over 12 million foreign patents. The hours are from 8:45 a.m. to 4:45 p.m.

Patent and Trademark Depository Libraries

I think that every inventor should do at least one hands-on patent search to fully understand and appreciate the process. Obviously, not everyone can visit the Patent and Trademark Office's Public Search Room in Arlington, Virginia. If you cannot make the trip to Washington, D.C., you may inspect copies of the patents at a Patent and Trademark Depository Library (PTDL), a nationwide network of prestigious academic, research, and public libraries. According to Carole A. Shores, Director, Center for Patent and Trademark Information, it is the goal of the PTO to have PTDLs in each state. "Our mission is a very simple one," explains Shores. "It is to bring more information and help to all the people out there that need it and can't afford to pay big money to get it. What makes our program special is that we really listen to their needs and when they bring requests to us we try very hard to give them what they want."

The PTDLs continue to be one of the PTO's most effective mechanisms for publicly disseminating patent information. PTDLs receive current issues of U.S. patents and maintain collections of patents issued earlier. The scope of these collections varies from library to library, ranging from patents of only recent years to all or most of the patents issued since 1790.

The patent collections in the PTDLs are open to the general public and I have always found the librarians very willing to take the time to help newcomers gain effective access to the information contained in patents. In addition to the patents, PTDLs usually have all the publications of the U.S. Patent Classification System, including the *Manual of Classification, Index to the U.S. Patent Classification System, Classifications Definitions, Official Gazette of the United States Patent and Trademark Office.*

The following is a list of Patent and Trademark Depository Libraries that make available their patent collections free of charge to the public. Because of variations among the PTDLs in their hours of service and the scope of their patent collections, anyone contemplating use of a particular library is well-advised to contact the library in advance about its collection and hours, to avoid possible inconveniences.

Alabama

Auburn University Libraries	(205) 844-1747
Birmingham Public Library	(205) 226-3620

Alaska

Anchorage: Z. J. Loussac Public Library	(907) 562-7323

Arizona

Tempe: Noble Library, Arizona State University	(602) 965-7010

Arkansas

Little Rock: Arkansas State Library	(501) 682-2053

California

Los Angeles Public Library	(213) 228-7220
Sacramento: California State Library	(916) 654-0069
San Diego Public Library	(619) 236-5813
San Francisco Public Library	(415) 557-4488
Sunnyvale Patent Clearinghouse	(408) 730-7290

Colorado

Denver Public Library	(303) 640-8847

Connecticut

New Haven: Science Park Library	(203) 786-5447

Delaware

Newark: University of Delaware Library	(302) 831-2965

District of Columbia

Howard University Libraries	(202) 806-7252

Florida

Fort Lauderdale: Broward County Main Library	(305) 357-7444
Miami-Dade Public Library	(305) 375-2665
Orlando: University of Central Florida Libraries	(407) 823-2562
Tampa Campus Library, University of South Florida	(813) 974-2726

Georgia

Atlanta: Price Gilbert Memorial Library, Georgia Institute of Technology	(404) 894-4508

Hawaii

Honolulu: Hawaii State Public Library System	(808) 586-3477

Idaho

Moscow: University of Idaho Library	(208) 885-6235

Illinois

Chicago Public Library	(312) 747-4450
Springfield: Illinois State Library	(217) 782-5659

Indiana

Indianapolis-Marion County Public Library	(317) 269-1741
West Lafayette: Siegesmund Engineering Library, Purdue University	(317) 494-2873

Iowa

Des Moines: State Library of Iowa	(515) 281-4118

Kansas

Wichita: Ablah Library, Wichita State Library	(316) 689-3155

Kentucky

Louisville Free Public Library	(502) 574-1611

Louisana

Baton Rouge: Troy H. Middleton Library, Louisana State University	(504) 388-2570

Maine

Orono: Raymond H. Fogler Library, University of Maine	(207) 581-1691

Maryland

College Park: Engineering and Physical Sciences Library, University of Maryland	(301) 405-9157

Massachusetts

Amherst: Physical Sciences Library, University of Massachusetts	(413) 545-1370
Boston Public Library	(617) 536-5400, Ext. 265

Michigan

Ann Arbor: Engineering Library, University of Michigan	(313) 764-5298
Big Rapids: Abigail S. Timme Library, Ferris State University	(616) 592-3602
Detroit Public Library	(313) 833-1450

Minnesota

Minneapolis Public Library and Information Center	(612) 372-6570

Mississippi

Jackson: Mississippi Library Commission	(601) 359-1036

Missouri

Kansas City: Linda Hall Library	(816) 363-4600
St. Louis Public Library	(314) 241-2288, Ext. 390

Montana

Butte: Montana College of Mineral
    Science and Technology Library   (406) 496-4281

Nebraska

Lincoln: Engineering Library,
    University of Nebraska-Lincoln   (402) 472-3411

Nevada

Reno: University of Nevada-Reno Library   (702) 784-6579

New Hampshire

Durham: University of New Hampshire Library   (603) 862-1777

New Jersey

Newark Public Library   (201) 733-7782

Piscataway: Library of Science and Medicine,
    Rutgers University   (908) 445-2895

New Mexico

Albuquerque: University of New Mexico General Library   (505) 277-4412

New York

Albany: New York State Library   (518) 474-5355

Buffalo and Erie Country Public Library   (716) 858-7101

New York Public Library (The Research Libraries)   (212) 930-0917

North Carolina

Raleigh: D.H. Hill Library,
    North Carolina State University   (919) 515-3280

North Dakota

Grand Forks: Chester Fritz Library,
    University of North Dakota   (701) 777-4888

Ohio

Cincinnati and Hamilton County, Public Library of   (513) 369-6936

Cleveland Public Library   (216) 623-2870

Columbus: Ohio State University Libraries   (614) 292-6175

Toledo/Lucas County Public Library   (419) 259-5212

Oklahoma

Stillwater: Oklahoma State University
    Center for International Trade Development   (405) 744-7086

Oregon

Salem: Oregon State Library   (503) 378-4239

Pennsylvania

Philadelphia, The Free Library of   (215) 686-5331

Pittsburgh, The Carnegie Library of   (412) 622-3138

University Park: Pattee Library, Pennsylvania State University   (814) 865-4861

Don Coster, President of the Nevada Inventor's Association, reports that the PTDL in Reno is "run by a person who trains the staff to the point that you cannot tell who's in charge. So, therefore, anytime you walk in you get help, and the help is for the guy that doesn't know what to do or who to write to or who to talk to. I think without those Patent and Trademark Depository Libraries we'd [independent inventors] be floating in limbo."

Rhode Island

Providence Public Library (401) 455-8027

South Carolina

Charleston: Medical University of
    South Carolina Library (803) 792-2372

Clemson: R. M. Cooper Library,
    Clemson University (803) 656-3024

South Dakota

Rapid City: Devereaux Library,
    South Dakota School of
    Mines and Technology (605) 394-2418

Tennessee

Memphis & Shelby County Public Library and
    Information Center (901) 725-8877

Nashville: Stevenson Science Library,
    Vanderbilt University (615) 322-2775

Texas

Austin: McKinney Engineering Library,
    University of Texas at Austin (512) 495-4500

College Station: Sterling C. Evans Library,
    Texas A & M University (409) 845-3826

Dallas Public Library (214) 670-1468

Houston: Fondren Library, Rice University (713) 527-8101, Ext. 2587

Utah

Salt Lake City: Marriott Library,
    University of Utah (801) 581-8394

Virginia

Richmond: James Branch Cabell Library,
    Virginia Commonwealth University (804) 828-1104

Washington

Seattle: Engineering Library,
    University of Washington (206) 543-0740

West Virginia

Morgantown: Evansdale Library,
    West Virginia University (304) 293-2510

Wisconsin

Madison: Kurt F. Wendt Library,
    University of Wisconsin-Madison (608) 262-6845

Milwaukee Public Library (414) 286-3247

Wyoming

Casper: Natrona County Public Library (307) 237-4935

PTDL Reference Material Inventory

As mentioned above, a PTDL branch may have a specialized collection, but each should have the following as a basic part of its inventory.

Patents

Index to the U.S. Patent Classification System

Classification Definitions (microfiche)

Official Gazette-Patents (OG)

Index of Patents, Part I and Part II (comes with Patent OGs)

Patentee/Assignee Index (microfiche)

Basic Facts about Patents

General Information concerning Patents

Attorneys and Agents Registered to Practice before the U.S. Patent and Trademark Office

Manual of Patent Examining Procedure Concordance, United States Classification to International Patent Classification

National Inventors Hall of Fame

Trademarks

Official Gazette-Trademarks (OG)

Index of Trademarks (comes with Trademark OGs)

Trademark Manual of Examining Procedure

Basic Facts about Trademarks

Miscellaneous

Directory of Patent and Trademark Depository Libraries

Consolidated Listing of Official Gazette Notices—Re: Patent and Trademark Office Practices and Procedures

Code of Federal Regulations, Title 37; Patents, Trademarks, and Copyrights

Patents and Trademarks Style Manual

Annual Report of the Commissioner of Patents and Trademarks

Story of the Patent and Trademark Office

Current Subscriptions Provided with Statutory Fee

Utility Patents, microfilm

Design Patents, microfilm

Plant Patents

Statutory Invention Registrations, microfilm

Changes to Patents, microfilm (Certificates of Correction Disclaimers, Reissues, Reexamination Certificates)

A Search Light

An illuminating source of basic information on databases and their services is the Gale Directory of Databases. Volume One covers online databases, and Volume Two covers CD-ROM, Diskette, Magnetic Tape, Handheld, and Batch Access Database Products.

There are over 60 online and approximately 20 microcomputer-based CD-ROM patent databases available commercially.

Online Searching Can Be Pricey

The cost of searching patents via databases varies with each database and online service. First you must open an account with the service provider. For a nominal yearly subscription fee, you'll be given a user ID and a password. Certain databases charge for connect time ($100 to $300 per hour). There are display or print charges ($.40 to $1.50 per record, although complete record displays can run to $20/record). Then there are the telecommunication charges ($8 to $15 per hour).

Other Materials Provided to PTDLs by the PTO

PTDL Mailing List, Representatives

PTDL Mailing List, Directors

Patent and Trademark Office Fee Schedule

U.S. Trademark Law; Rules of Practice, Forms & Federal Statues, U.S. Trademark Association

Trademark Examination Guides

Trademark Classification of Services

Trademark Classification of Goods

"A Trademark Is Not a Patent or a Copyright" (USTA reprint)

"How to Use a Trademark Properly" (USTA reprint)

"A Guide to the Care of Trademarks" (USTA reprint)

PTO Information Contacts (telephone directory)

IEEE Patents and Patenting for Engineers and Scientists (reprint)

CASSIS brochures

CASSIS User's Guide (card)

CASSIS User's Manual

CASSIS Location Manager's Manual

CASSIS search examples

CASSIS tips

Disclosure Document Program brochures

Sample PTDL/PTO Conference Agenda

Sample List of Conference Attendees

ADLIBS

Handbook on the Use of the U.S. Patent Classification System for Patent and Trademark Depository Libraries

Even though I live near enough to the Alexandria, Virginia, Public Search Room, often I opt to do my work at the University of Maryland's PTDL, located within its Engineering and Physical Sciences Library. Crowded with books and not much larger than a small meeting room, the library offers only a couple of chairs. However, here I can do my search work and have access to a very extensive collection of technical publications outside the PTDL that I enjoy browsing through for ideas and technologies. Such "extras" are not available at the main government facility, and I consider them a bonus.

Online Patent Searches

As is evident from the list above, all PTDLs have in their collection the Classification and Search and Support Information System (CASSIS), an online computer database.

CASSIS brings you all the data available at the Patent and Trademark Search Room. And it can streamline the manual search procedure by providing electronic access to basic patent search tools. CASSIS can help you define a "field of search" and identify the patents in that field. It provides classifications of a patent, supplies patents in a classification, displays patent titles and/or company names, structures classification titles, finds key words in classification titles and patent abstracts, and much more.

Other Databases

Databases are available that savvy searches use to augment their work. With access through host systems such as ORBIT and DIALOG, these files include, for example, Derwent's World Patent Index covering 33 countries and technology back to 1963. The information provided by Derwent is in the form of abstracts and bibliographic background. Some databases just supply bibliographic data. If you decide to use one, find out about the scope of its information.

Computer-aided searches are not by any means the final word. The best they can do is generate abstracts of patents. The searcher must ultimately retrieve copies of the full patents and study them.

More Help—If You Have the Time

If you want help after referring to the *Index to the U.S. Patent Classification System,* you may write for free assistance to the Commissioner of Patents and Trademarks. Your letter might look something like this:

---

1 January 1996

Commissioner of Patents and Trademarks

Att: Patent Search Division Washington, D.C. 20231

Please let me know what subject area or class(es) and subclass(es) cover my idea. The enclosed sketches on the back of this sheet, with each part labeled, show my intended invention.

My idea has certain features of structure, mode of operation, and intended uses which I have defined below.

I understand that there is no charge for this information.

Thank you for your help.

1. Features of structure, or how it is constructed.

2. Mode of operation, or how it works.

3. Intended uses, or purpose of idea.

4. Rough sketches of idea, viewed from all sides, with labels to identify each part. [If necessary use extra sheets of paper, either plain or lined. These sketches may be made in pencil and need not be drawn to scale.]

If you live in a city with a Patent and Trademark Depository Library (PTDL), a librarian there can usually assist you in finding a list of those who make a living searching patents.

Don't be put off if a librarian is reluctant to give a specific recommendation; they are not encouraged to personally recommend any patent searcher or search organization.

How to Order Copies of Searched Patents (Prior Art)

You may order from the Patent and Trademark Office copies of original patents or cross-referenced patents contained in subclasses comprising the field of search. Mail your request to: Patent and Trademark Office, Box 9, Washington, D.C. 20231.

Payment may be made by check, coupons, or money. Expect a wait of up to four weeks when ordering copies of patents from the PTO.

For the convenience of attorneys, agents, and the general public in paying any fees due, deposit accounts may be established in the PTO with a minimum deposit of $50. For information on this service, call (703) 308-0902.

What to Do with Your Search Results

Study the results of the patent research. You may be out of luck if a previously patented invention is very similar to yours; your invention may even be infringing on another invention. On the other hand, one or more patents may describe inventions that are intended for the same purpose as yours, but are significantly different in various ways. Look these over and decide whether it is worthwhile to proceed. Consult a patent attorney if you have any doubt.

If the features that make your invention different from the prior art provide important advantages, you should discuss the situation with your attorney to determine whether a fair chance exists of obtaining a patent covering these features.

I have found from experience that a good patent attorney can often get some claim to issue, albeit not always a strong one. A patent for patent's sake is usually possible. Whether it will be worth the paper it is printed on is another matter. Do not make this decision lightly, because the patent process is not cheap. The average utility patent will cost about $2,500.

PTO Patent References: Quick Look-Up

CASSIS (Classification and Search Support Information System) Patent Mode (PM)

An online database used to find the one original class and subclass for each patent and all official and unofficial cross-reference class(es)/subclass(es) for that patent.

U.S. Patent Classification—Numeric Listing

Microfilm publication that lists all U.S. Patents in ascending numerical sequence, giving each patent's original class and subclass and all official and unofficial cross-reference class(es)/subclass(es), current as of the date of the publication. Updated every six months. 22 reels.

Patentee/Assignee Index

Microfiche publication that lists in alphabetical order five to six year's worth of patentee's and assignees at the time each patent was issued with corresponding patent number. Updated quarterly. More than 250 microfiche.

Index of Patents, Part I—List of Patentees

Annual publication that lists in alphabetical order all patentees and assignees at the time each patent was issued with corresponding patent number. Approximately 2,000 pages per year. 1926 to present. Predated by Annual Report of the Commissioner of Patents, 1790–1925.

"List of Patentees"

In the back of each weekly issue of the *Official Gazette* is an alphabetical list of patentees and assignees for the week with corresponding patent number.

Index to the U.S. Patent Classification

Alphabetical list of approximately 63,000 common, informal subject headings or terms that refer to specific class(es)/subclass(es) in the classification system used to categorize patents. It is intended as a means for initial entry into the classifications system. Approximately 240 pages.

CASSIS Index Mode (IM)

An online database used to find class(es)/subclass(es) corresponding to common, informal terms and phrases in the *Index to the U.S. Patent Classification System* by using a given word or word stem (or various logical combinations of words and word stems). Once the words are identified, the corresponding class(es)/subclass(es) may be obtained with the List Index (LI) command, and with the List Detail (LD), or entry number, command which provides sub-entries for a specified index entry number from the (LI) listing.

CASSIS Word Mode (WM)

An online database used to find class(es)/subclass(es), the titles of which include a given word or word stem (or various logical combinations of words and word stems). The file comprises the words in the titles of the *Manual of Classification*. Once the words are identified, the corresponding class(es)/subclass(es) may be obtained with the "List" (L) command.

CASSIS Patent Abstract Mode (PA)

An online database used to identify patents, the abstracts of which include a given word or word stem (or various logical combinations of words or word stems). The file comprises the words in the abstracts from the most recent one to two years' worth of patents. Once the patents are identified, the system will locate all class(es)/subclass(es) under which the patents are filed using the "Get Classifications" (GC) command. These class(es)/subclass(es) may then be obtained by using the "List Classifications" (LC) command, which lists the class(es)/subclass(es) in ascending number order with the number of patents in each; or they may be obtained by using the "Rank Classifications" (RC) command which lists the class(es)/subclass(es) in descending order by number of patents in each.

Manual of Classification

Presents listings of the approximately 390 main groupings (classes) and approximately 113,000 smaller groupings (subclasses) into which patented subject matter is classified. Each subclass has a short, descriptive title often arranged in a specific hierarchical order designed by dots for indentation levels. About 1,400 pages.

CASSIS Title Mode (TM) Page Submode (P)

An online database used to examine the titles of the various class(es)/subclass(es) exactly as they are displayed on a page of the *Manual of Classification*. The level of a subclass in the hierarchy is designated by the number of dots (called the indent level) preceding the subclass title.

CASSIS Title Mode (TM) Full-title Submode (F)

An online database used to examine the full title of a single class/subclass, which includes the titles of the higher subclass(es) and class.

CASSIS Title Mode (TM) Coordinate Submode (C)

An online database used to examine the hierarchical relationship of a given class/subclass to the next higher level of classification (the "parent") and up to eight of the subordinate subclasses (the "children").

Classification Definitions

Detailed definitions for each class and official subclass included in the *Manual of Classification*. The definitions indicate the subject matter to be found in or excluded from a subclass; they limit or expand in precise manner the meaning intended for each subclass title; they serve as a guide to users of the *Manual* to refer to the same subclass for patents on a particular technology by eliminating, as much as possible, subjective and varying interpretations of the meanings of subclass titles. The definitions provide further guidance through "search notes," which illustrate the kinds of information that can be found in a subclass and direct the searcher to other related subclasses that may contain relevant information. About 10,000 pages. *Classification Definitions* is also available on microfiche. About 360 fiche.

CASSIS Classification Mode (CM)

An online database used to list all of the patent numbers that have been assigned to a particular class/subclass as an original reference, or an official or unofficial cross-reference.

U.S. Patent Classification—Subclass Listing

A microfilm publication that lists all patents classified in each class/subclass and digest of the U.S. Patent Classification System. Within each subclass, the patents are listed in numerical sequence and categorized by an original reference, or an official or unofficial cross-reference. Updated every six months. 11 reels.

"Classifications of Patents"

In the back of each weekly issue of the *Official Gazette* is a list of all the patent numbers assigned to each class/subclass for that week.

Official Gazette of the U.S. Patent and Trademark Office (Patent Section)

Weekly publication that presents an entry according to orginal class/subclass designation for each of the approximately 1,500 patents issued. Each entry includes certain bibliographic information and a brief summary or abstract in the form of one or more representative patent claims (depending on date), together with (where appropriate) a reduced representative drawing. About 300 pages per week.

CDR File

Annual, cumulative, computer-produced book listing patent numbers with corresponding reel and frame of the CDR microfilm for Corrections, Disclaimers, Reissues, and reexamination certificates associated with the original patent. Issued

Milwaukee native James Pinske, inventor of the B'zarts flying toy, found his nearby PTDL a great resource. "If I got into a snag, I just went to a librarian and got complete information." Pinske adds that the independent inventor should feel very comfortable using the PTDL. He says, "Once you know what you're going after, the librarians will give unqualified support."

since 1973; earlier CDR material was filmed in place with the original patent. About 235 pages and four reels per year.

In Conclusion . . .

When determining which search method is best for you, consider the following: how much money can you spare, how much time can you spare, and how well could you do the job yourself? Finding out early that the Mother of All Ideas is old news can save you a lot in developmental and legal costs, not to mention time. Conversely, you just might find out that your idea is the next sliced bread.

CHAPTER

6

Do You Need a Patent Attorney?

Tips for Dealing with Patent Attorneys

- Make sure your patent attorney is registered by the PTO.

- Make sure your patent attorney is a specialist in your field.

- Ask lots of questions.

- A bigger law firm is not always better.

- Shop around and get estimates before retaining a lawyer.

- The more information you can give an attorney, the less it will cost you and the better your patent is apt to be.

- Find out in advance possible hidden charges for such things as faxes, xeroxes, and postage.

If you think it is expensive to use a patent attorney, see what it costs you when you do not.

While it is perfectly legal for you to prepare and conduct your own proceedings at the PTO, unless you really know what you are doing, I would recommend that you retain patent counsel to handle utility patents. On the other hand, design patents can be handled by yourself (*pro se*) by following simple instructions (see pages 73–74).

I see it like this. You can do your own income tax too. It's not illegal. And for people who do not make a lot of money, this may be the most cost-effective way to deal with taxes. But people who have (or hope to have) substantial estates hire CPAs. They don't calculate their income taxes themselves with a do-it-yourself, self-help book. There is too much at stake. Wealthy people use CPAs to level the playing field and give themselves every benefit of the law.

In this same way, smart inventors use experienced patent counsel to assure that they obtain the strongest patent protection available on their inventions. There is too much at stake. Smart inventors do not rely on patent-it-yourself books.

This chapter is not written to encourage you to do utility patents on your own. It is designed to give you an overview of the patent process and show you some ways to save money along the way. But I can tell you without equivocation that time after time, in my experience, the value of my patents has been enhanced by the contributions of a savvy patent attorney. The fact is, anyone can get a patent on almost anything. What a good patent attorney will do is assure that the patent awarded adequately protects your invention.

I do not use my patent counsel to negotiate licensing agreements. See page 236 for my thoughts on lawyers and contract negotiations.

Here are some things to consider when hiring patent counsel. Read them carefully. They may save you time and money.

- *Make sure your patent attorney is registered by the PTO.* The Patent & Trademark Office keeps a register of more than 17 thousand attorneys and agents. To be listed, a person must comply with the regulations prescribed by the PTO, which require proof that the person is of good moral character and of good repute and that he or she has the legal, scientific, and technical qualifications necessary to give inventors valuable service. Certain qualifications must be demonstrated by the passing of an exam. Those admitted must have a college degree in engineering or science or the equivalent of such a degree.

A listing of patent lawyers and agents registered by the PTO, organized state-by-state, city-by-city, may be found at your public library in the second edition of *The Inventing & Patenting Sourcebook* (Gale Research). This listing can also be found at any of the Patent & Trademark Depository Libraries nationwide in the publication *Patent Attorneys and Agents Registered to Practice before the U.S. Patent and Trademark Office,* a current copy of which can also be purchased for $31 from the U.S. Government Printing Office, Superintendent of Documents, Washington, D.C. 20402. For information, call: (202) 512-1800; send faxes to (202) 512-2250.

- *Make sure your patent attorney is a specialist in your field.* Just as you would not hire a dermatologist to do heart surgery, you would not want an electronic patent specification written by an attorney whose specialty is mechanical engineering. Get proof of a lawyer's expertise in your particular field of invention. Interview more than one candidate for the work.

- *Ask lots of questions.* Does the lawyer draft the patent or is it done by an assistant? How long will it take to complete the job? Lawyers like to have clients pick up their overhead on top of paying their fees. Ask how much the lawyer charges for photocopies. If it is too much, negotiate a better price. A lawyer once tried to charge me 50 cents per page. I know a law firm in D.C. that charges 75 cents per page. This is crazy. You can buy copy service for less than 5 cents per page. How much mark-up, if any, is fair? Is there a cost for local incoming and outgoing faxes? I refuse to pay for local faxes. How are telephone consults billed? If the firm is inflexible, go elsewhere. I refuse to pay separately for a firm's overhead.

  It's a buyer's market. It pays to shop around. Attorneys are able to give close estimates of expected charges once they understand the scope of work. Keep the attitude that you are selling money versus buying services.

  Once you have an acceptable estimate, ask your attorney to agree in writing to that price and cap it off. Otherwise, you may find yourself mired in high fees. Make a package deal whenever you can.

  Don't embarrass yourself by insisting that a lawyer sign a confidentiality agreement before you'll disclose your invention. Lawyers do not steal ideas.

- *Bigger is not always better.* Chances are at a large firm, you will be considered small potatoes. Big firms need hefty corporate retainers and litigation to make ends meet. Independent inventors on limited budgets are not going to do it. If you go to a large law firm, therefore, don't be surprised if your work is drafted by someone other than the lawyer you meet. I have experienced situations where the so-called "rainmaker" brings in the client, but the work is done by a less experienced lawyer that you don't meet.

  If you use a large firm, and several lawyers show up for a meeting, ask who they are, their purpose, and if you are paying for them. If you are, be very sure they are needed.

- *It all begins with a search.* Before you start drafting claims, a good patent attorney will rightfully insist upon a search. Be aware, however, law firms do not conduct searches themselves. They hire a search service to do the leg work and/or computer search. Law firms add to the search fees anywhere from 40 percent to 100 percent or more depending upon the firm and city.

  If you decide to allow a lawyer to have the search done, find out first what it will cost and how much search time you are buying.

  And remember, if you want to save the lawyer's mark-up, you can do what I do and hire a patent searcher yourself.

Multi-Million Dollar Bail

In the early 1920s, R. A. Watkins, the owner of a small printing plant in Illinois, was approached by a man who wanted to sell him the rights to a homemade device made of waxed cardboard and tissue on which messages could be printed and then easily erased by lifting up on the tissue. Watkins wanted to think about the proposition overnight, and told the man to return the next day.

In the middle of the night, Watkins's phone rang; it was the inventor calling from jail. The man said that if Watkins would bail him out, he could have the device. Watkins agreed and went on to acquire the U.S. patent rights as well as the foreign rights to the device, which he called the Magic Slate.

Tripping Out

Are you confused about certain TRIPs (**T**rade **R**elated Aspects of **I**ntellectual **P**roperty Rights)? If so, there is information available from the following organizations.

- *PTO:* Charles Van Horn, former PTO TRIPs maven, narrates a one-hour video explaining the changes in PTO regulations. To order, call the Office of Public Affairs at (703) 305-8341.

- *American Intellectual Property Law Association:* $25 buys a video entitled "Twenty Year Patent Term Videotape." It is packaged with a printed copy of Title V of the GATT. To order, write to: AIPLA, 2001 Jefferson Davis Highway, Suite 203, Arlington, Virginia 22202.

- *American Bar Association's Section on Intellectual Property Law:* If you don't have a VCR, you can read about TRIPs in ABA's publication *Patent Aspects of GATT.* The cost is $65; to order, call (202) 783-5070.

- *Intellectual Property Owners:* IPO has packaged a copy of TRIPs legislation and statements of administrative action concerning the GATT bill. Members get it for free; non-members pay $25. To order, call (202) 466-2396.

Some lawyers will insist upon having their own service conduct patent searches, saying that they cannot guarantee the work of anyone they don't know. Since no one can guarantee anything in searches, I have always, in such cases, stuck with my searcher and hired a more flexible lawyer. In one instance, my search firm began doing work for my law firm once the quality of its work was seen.

You could conduct the search yourself, if you live near one of the Patent and Trademark Depository Libraries (PTDLs). Librarians can be very helpful. You may also find professional patent searchers working at the PTDLs. See Chapter 5 for information on PTDLs and conducting a patent search.

- *Don't pay more than you have to for patent drawings.* If you rely on a patent attorney for drawings, understand that he or she will not put pen to paper. Lawyers hire professional patent draftsmen, just as they hire professional searchers, and typically add a premium of no less than 40 percent to the fees.

I always have our patent drawings done by a professional of my choice. You can find the fair market price by calling to get quotes. Look in the yellow pages, ask other inventors for recommendations, or speak with the librarian at the nearest PTDL.

- *Be prepared to pay for utility patents.* Lawyers usually charge $2,500 to $5,000, depending upon the complexity of invention and number of claims. The more work the lawyer must do, the more you pay. On the other hand, the more you can write up about the features of your invention for the attorney, the less you'll have to pay because this will cut the time he or she will have to spend on such elements as specifications and claims.

- *Use an alternate route for design patents.* I never use an attorney for design patents. I have the search done by Greentree and then, if the field looks open, I prepare the specification myself using guidelines provided by the PTO. An example may be found on page 74.

A Legal Scam

Some individuals and organizations that are not registered advertise their services in the fields of patent searching and invention marketing and development. Such individuals and organizations **cannot** represent inventors before the PTO, and, since they are not subject to PTO discipline, the Office cannot assist you in dealing with them.

While calling to acquire information on organizations that I felt might be helpful to inventors, I came across a nifty scam used by some lawyers to attract patent, trademark, and copyright clients. What appeared in telephone directories and source material to be professional councils specific to patents, trademarks, and copyrights were actually law offices. In one case, the phone numbers for three different councils led to the same law firm. The person listed as president of the three respective organizations was actually a senior partner in the law firm.

National Council of Intellectual Property Law Associations

Patent attorneys who are active in the National Council of Intellectual Property Law Associations (NCIPLA) should be particularly up-to-date on PTO matters. Established more than 35 years ago, the NCIPLA consists of some 40 local and regional patent law associations.

For more information on NCIPLA, contact your nearest member association, or write to: NCIPLA, Crystal Park 1, Suite 208B, 2011 Crystal Drive, Arlington, Virginia 22202. The telephone number is (703) 305-834

The following list comprises the current officers of the NCIPLA. This is by no means an exhaustive list of attorneys registered to practice before the PTO; that list numbers more than 17,000 and can be obtained at your local Patent and Trademark Depository Library. Not all of the attorneys listed here are in private practice and available to consult with you on your invention; regardless, it may be worth a call to them to get information on their local NCIPLA chapter.

CALIFORNIA

William H. Benz
Irell & Manella
545 Middlefield, Suite 200
Menlo Park, CA 94025

David A. Dillard
Christie Parker and Hale
350 W. Colorado Blvd., Suite 500
Pasadena, CA 91105

James E. Hawes
Hawes & Fischer
660 Newport Center Dr., #460
Newport Beach, CA 92660

Virginia H. Meyer
McCubbrey Bartels Meyer & Ward
One Post St., Suite 2700
San Francisco, CA 94104-5231

Edward G. Poplawski
Pretty Schroeder Bruegemann & Clark
444 Flower St., Suite 2000
Los Angeles, CA 90071

Edith A. Rice
Raychem Corporation
300 Constitution Dr.
Menlo Park, CA 94025

William C. Rooklidge, President
Orange County Patent Law Association
c/o Knobbe, Martens et al.
620 Newport Center Dr., 16th Floor
Newport Beach, CA 92660

Thomas F. Smegal, Jr.
Graham & James
One Maritime Plaza, #300
San Francisco, CA 94111

Neil A. Smith
Limbaugh & Limbaugh
2001 Ferry Building
San Francisco, CA 94111

COLORADO

William W. Cochran II
Hewlett Packard Co.
3404 E. Harmony Rd.
Fort Collins, CO 80525

Bill O'Meara
Klass, Law, O'Meara & Malkin
1999 Broadway, Suite 225
Denver, CO 80202

CONNECTICUT

Ted Carvis
St. Onge Steward Johnson & Reens
986 Bedford St.
Stamford, CT 06905

James R. Frederick
Parmelee, Bollinger & Bramblatt
4670 Summer St.
Stamford, CT 06901

Hall of Fame

As of July 1995, the National Inventors Hall of Fame will be housed in an all-new facility. For information, write to Inventure Place, 221 South Broadway, Akron, Ohio 44308; or call (216) 762-4463.

Applications for nomination may be obtained by writing to the National Inventors Hall of Fame Foundation, Inc., P.O. Box 1553, Akron, Ohio 44309-1553.

Leonard B. Mackey
700 Strawberry Ln., Apt. 5-3E
Stamford, CT 06902

Sandra M. Nolan
Bristol-Myers Squibb Co.
5 Research Parkway
P.O. Box 5100
Wallingford, CT 06492-7660

Mary R. Norris
Wiggin & Dana
One Century Tower
New Haven, CT 06508-1832

DELAWARE

David W. Westphal
DuPont Legal Department
Barley Mill Plaza, P17-2288
P.O. Box 80017
Wilmington, DE 19880

DISTRICT OF COLUMBIA

Joseph A. Degrandi
Beveridge Degrandi Weilacher & Young
1850 M St. NW, Suite 800
Washington, D.C. 20036

Carol Einaudi
Finnegan, Henderson et al.
1300 L St. NW, Suite 700
Washington, D.C. 20005-3315

Michael A. Grow
Vorys, Sater, Seymour & Pease
1828 L St. NW
Washington, D.C. 20036

David W. Hill
Finnegan, Henderson et al.
1300 L St. NW, Suite 700
Washington, D.C. 20005-3315

Ann Kornbau
Browdy & Neimark
419 7th St., NW, Suite 300
Washington, D.C. 20004

James H. Laughlin, Jr.
Lane & Mittendorf
919 18th St. NW, Suite 800
Washington, D.C. 20006

James F. McKeown
Evenson, Wands, Edwards et al.
1200 G St. NW, Suite 700
Washington, D.C. 20005

Thomas L. Peterson
National IPLA
Banner, Birch et al.
1001 G St. NW
Washington, D.C. 20001

H. C. Wamsley
1255 Twenty-Third St. NW
Washington, D.C. 20037

John Whelan
Fitzpatrick, Cella & Harper
1001 Pennsylvania Ave. NW, Suite 650
Washington, D.C. 20004

FLORIDA

Richard M. Saccocio
Dominick & Saccocio
6175 NW 153 St., Suite 225
Miami Lakes, FL 33014

GEORGIA

Todd Deveau
Deveau Cotton & Marquis
2 Midtown Plaza, Suite 1400
1360 Peachtree St. NE
Atlanta, GA 30309-3999

Joel Goldman
Troutman Sanders
600 Peachtree St. NE, Suite 5200
Atlanta, GA 30308

Joyce B. Klemmer
Smith Gambrell & Russell
1230 Peachtree St. NE, Suite 3100
Atlanta, GA 30309-3592

ILLINOIS

John J. Chrystal
Ladas & Parry
224 S. Michigan Ave.
Chicago, IL 60604

Charles A. Laff
401 N. Michigan Ave., Suite 1700
Chicago, IL 60611

Timothy T. Patula
Patula & Associates
1116 S. Michigan Ave., 14th Floor
Chicago, IL 60603

Thomas F. Peterson
Ladas & Parry
224 S. Michigan Ave.
Chicago, IL 60604

INDIANA

Dan Boots
William Brink & Olds
1 Indiana Square, Suite 2425
Indianapolis, IN 46204

George T. Dodd
2400 Fort Wayne National Bank Bldg.
Fort Wayne, IN 46802-2387

IOWA

Kent Herink
Davis Hockenberg et al.
2300 Financial Center
Des Moines, IA

KENTUCKY

J. Ralph King
King & Schickli
3070 Harrodsburg Rd., Suite 210
Lexington, KY 40503

Important!

A caveat: Do not rely on your patent attorney for prototyping advice, manufacturing processes, or insights into the day-to-day complexities of marketing.

MARYLAND

Ed Cabic
W.R. Grace & Company
7379 Route 32
Columbia, MD 21044-4098

Maurice H. Klitzman
9811 Inglemere Dr.
Bethesda, MD 10817

Frank L. Neuhauser
10030 Chapel Rd.
Potomac, MD 20854

MASSACHUSETTS

Jason Mirabito
Wolf Greenfield & Facks
600 Atlantic Ave.
Boston, MA 02210

David J. Powsner
Lahive & Cockfield
60 State St., Suite 510
Boston, MA 02109

MICHIGAN

Cary W. Brooks
General Motors Corp. Pat. Sec.
3031 West Grand Blvd.
P.O. Box 333
Detroit, Michigan 48232

Gerald P. Dundas
Vickers Incorporated
5445 Corporate Dr.
P.O. Box 302
Troy, MI 48007-0301

Bill Honacker
Howard & Howard
1400 N. Woodward Ave., Suite 250
Bloomfield Hills, MI 48304

**Gary L. Newtson,
President**
American IPLA
c/o Harness, Dickey et al.
5445 Corporate Dr., Suite 400
Troy, MI 48098

Robert M. O'Keefe
Dow Chemical Company
P.O. Box 1967
Midland, MI 48641-1967

Steven L. Permut
Reising Ethington Barnard et al.
P.O. Box 4390
Troy, MI 48099

Paula Ruth
Dow Chemical Company
P.O. Box 1967
Midland, MI 48641-1967

Sidner B. Williams, Jr.
Upjohn Company
301 Henrietta
Kalamazoo, MI 49001

MINNESOTA

Conrad Hansen
Moore & Hansen
3000 N. West Center
Minneapolis, MN 55402

Michelle M. Michel
Merchant & Gould
3100 N. West Center
Minneapolis, MN 55402

MISSOURI

Annette P. Heller
Heller & Kelper
721 Emerson Rd., Suite 569
St. Louis, MO 63141-6709

John W. Kelper III
Heller & Kelper
721 Emerson Rd., Suite 569
St. Louis, MO 63141-6709

NEW JERSEY

Albert Gazzola
Weingram & Zall
P.O. Box 927
Maywood, NJ 07607

Charles P. Kennedy
Lerner David et al.
600 S. Avenue West
Westfield, NJ 07090

Stephen I. Miller
101 Wood Ave.
Iseline, NJ 08830-0770

M. Andrea Ryan
Warner-Lambert Co.
201 Tabor Rd.
Morris Plains, NJ 07950

NEW YORK

Charles P. Baker
Fitzpatrick Cella Harper & Scinto
277 Park Ave.
New York, NY 10172

William J. Gilbreth
1251 Avenue of the Americas, 49th Floor
New York, NY 10020

H. Walter Haeussler
Cornell Research Foundation Inc.
20 Thornwood Dr., Suite 105
Ithaca, NY 14850

James E. McGinness
General Electric—Corporate R&D
Bldg K-1, Rm. 3A64
P.O. Box 8
Schenectady, NY 12301

Paul F. Morgan
Xerox Corporation
Xerox Square—020
Rochester, NY 14644

Sam Pace
Bausch & Lomb
P.O. Box 54
Rochester, NY 14601-0054

James Riesenfeld Esq.
Johnson & Johnson
One Johnson & Johnson Plaza
New Brunswick, NY 08933-7003

David J. Rosenblum
240 E. 83rd St.
New York, NY 10028

James C. Simmons
Hodgson Russ Andrews Woods et al.
1800 One M&T Plaza
Buffalo, NY 14201

K. McNeill Taylor, Jr.
Corning Incorporated
SP-FR-212
Corning, NY 14831

William A. Teoli
GE Company—Corporate R&D
Bldg. K1 Rm 3A70
P.O. Box 8
Schenectady, NY 12301

Kathleen Terry
516 Capen Hall
SUNY Buffalo
Buffalo, NY 14260-1661

NORTH CAROLINA

Howard A. MacCord, Jr.
Rhodes Coats & Bennett
P.O. Box 2974
Greensboro, NC 27401

Ken Seaman
IBM Corp.
Dept. 18A, Bldg. 204-3
Charlotte, NC 28257

OHIO

James A. Baker
Cleveland IPLA
17325 Euclid Ave.
Cleveland, OH 44112

Thomas J. Burger
Wood, Herron & Evans
2700 Carew Tower
Cincinnati, OH 45202

Thomas E. Fisher
Watts Hoffman Fisher et al.
100 Erieview Plaza, Suite 2850
Cleveland, OH 44114

William A. Heidrich
Cincinnati IPLA
Quantum Chemical Co.
11500 Northlake Dr.
Cincinnati, OH 45249

Ralph Jocke
Walker & Jocke
231 S. Broadway
Medina, OH 44256

Roger A. Johnston
Eaton Corporation
Eaton Center
Cleveland, OH 44114

Mark P. Levy
Thompson, Hine & Flory
2000 Courthouse Pl. NE
P.O. Box 8001
Dayton, OH 45401-8801

David D. Murray
William Brinks Olds et al.
1130 Edison Plaza
Toledo, OH 43604-1537

Phillip J. Pollick
Watkins Dunbar & Pollick
2941 Kenny Rd., Suite 260
Columbus, OH 43221

David C. Purdue
2735 North Holland-Sylvania Rd., Suite B-2
Toledo, OH 43615-1844

Dennis Raincar
Borden Inc.
180 E. Broad St.
Columbus, OH 43215

Charles F. Schroeder
608 Madison Ave., Suite 819
Toledo, OH 43604-1140

Charles N. Shane
Mead Corporation
Courthouse Plaza NE
Dayton, OH 45463

David Thomas
Bridgestone/Firestone Inc.
1200 Firestone Pkwy
Akron, OH 44317

Leonard Williamson
The Procter & Gamble Company
11520 Reed Hartman Hwy.
Cincinnati, OH 45241

William S. Wyler
Schwartz Manes & Ruby
2900 Carew Tower
441 Vine St.
Cincinnati, OH 45202-3090

OKLAHOMA

Lynda S. Jolly
Phillips Petroleum Company
232 Patent Library Bldg.
Research Center
Bartlesville, OK 74004

David W. Westphal
Conco Incorporated
P.O. Box 1267
Ponca City, OK 74602-1267

OREGON

Alan T. McCollom
650 American Bank Bldg.
621 SW Morrison St.
Portland, OR 97205

PENNSYLVANIA

Richard L. Byrne
700 Koppers Bldg.
436 7th Ave.
Pittsburgh, PA 15272

John C. Dorfman
Dann Dorfman Herrel & Skillman
1601 Market St.
Philadelphia, PA 19103-2306

Michael D. Fox
Gefsky & Lehman
2301 One PPG Place
Pittsburgh, PA 15222

John W. Jordan IV
Grisby Gaca & Davies
One Gateway Center, 10th Floor
Pittsburgh, PA 15222

Lodge Complaints Here

If you wish to lodge a formal complaint about a registered patent attorney or agent, write to the PTO's Office of Enrollment at P.O. Box OED, U.S. Patent and Trademark Office, Washington, D.C. 20231. You may also call the Office of Enrollment at (703) 308-5285, or (703) 308-5316.

Alan Ratner
500 N. Gulph Rd.
P.O. Box 980
Valley Forge, PA 19482

Kenneth J. Stachel
PPG Industries Incorporated
Pittsburgh, PA 15272

Frederick B. Ziesenheim
Webb Burden Ziesenheim & Webb
436 7th Ave.
700 Koppers Bldg.
Pittsburgh, PA 15219-1818

SOUTH CAROLINA

Richard M. Moose
Dority & Manning
700 E. North St., Suite 15
Greenville, SC 29601

TEXAS

Margaret A. Boulware
Vaden Eickenroht & Boulware
One Riverway #110
Houston, TX 77056

Robert M. Chiavillo, Jr.
Baker & Botts
2001 Ross Ave.
800 Tramell Crow Ctr.
Dallas, TX 75201

Russ Culbertson
Shafer & Culbertson
1250 Capital of Texas Hwy.
Bldg. One, Suite 105
Austin, TX 78746

John R. Kirk
Jenkins & Gilchrist
1100 Louisiana, Suite 1800
Houston, TX 77002

William L. LaFuze
Vinson & Elkins
1001 Fannin 2500—1st City Tower
Houston, TX 77002-6760

David L. McCombs
Haynes & Boone
3100 Nationsbank Plaza, 901M
Dallas, TX 75202-3789

Patricia C. Ohlendorf
Executive Vice President & Provost Office
University of Texas
Main Bldg., Rm. 201
Austin, TX 78712-1111

Jeffrey W. Tayon
Conley Rose & Tayon
1850 Texas Commerce Tower
600 Travis
Houston, TX 77002

Frank S. Vaden III
Vaden Eickenroht Thompson et al.
One Riverway, Suite 1100
Houston, TX 77056-1982

UTAH

Charles L. Roberts
Madsen & Metcalf
36 S. State St.,
1300 Beneficial Life Tower
Salt Lake City, UT

VIRGINIA

Michael W. Blommer
American IPLA
2001 Jefferson Davis Hwy., Suite 20
Arlington, VA 22202

Isaac Fleischmann
3161 Ravenwood Dr.
Falls Church, VA 22044

Jeffrey L. Forman
IBM Corp.
1755 S. Jefferson Davis Hwy, Suite 605
Arlington, VA 22202

WASHINGTON

John C. Hammar
Boeing Defense & Space Group
P.O. Box 3999, M/S 80-PA
Seattle, WA 98124-2499

WISCONSIN

Gary A. Essmann
Andrus Sceales Starke & Sawall
100 E. Wisconsin Ave.
Milwaukee, WI 53202

Andrew S. McConnell
Andrus Sceales Starke & Sawall
100 E. Wisconsin Ave., Suite 1100
Milwaukee, WI 53202

CHAPTER 7

How to Apply for a U.S. Patent

An invasion of armies can be resisted, but not an idea whose time has come.

—VICTOR HUGO

GATT Brings Major Changes to U.S. Patent Law

On December 7, 1994, President Bill Clinton signed into law the long-debated Uruguay Round Agreement Act, better known as the GATT (General Agreement on Tariffs and Trade) bill. In it are several intellectual property provisions, specifically major changes in U.S. patent law, that are of importance to inventors. They are as follows.

❏ The 20-year patent term measured from the date of filing the first application replaces the 17-year term measured from the date the patent is granted. The 17-year term has been a part of U.S. law since 1861.

A term measured from the date of filing means that the longer it takes a patent to issue, the shorter its effective term. In other words, if a patent takes 5 years to issue, under the proposed 20-year scheme the inventor would only get 15 years to commercialize and earn royalties on the invention before the patent expires. Under the 17-year system, the length of time it takes a patent application to get through the PTO does not affect the life of a patent; the clock does not begin ticking until a patent issues.

Thinking about a worst-case scenario, under the proposed 20-year plan, a competitor could file meritless interferences to delay the issuance of a patent, thereby reducing the patent's life when it does issue. Further, according to inventor Paul Heckel, a founder of Intellectual Property Creators (IPC), a patent examiner could "use the threat of a shorter term to get inventors to accept narrower claims than are deserved."

Who wins by having a 20-year term? It is favored by large multinational corporations because by cutting the term of patent protection on new technologies the companies will reduce potential royalty payments on the inventions they wish to exploit. IPC's Heckel sees the battle over patent terms as the one that will "likely be the decisive battle of this war for the soul of American innovation."

Herbert C. Wamsley, executive director of the Intellectual Property Owners Association (IPO), recommends that patent applicants and their attorneys become

Explanation Pending

What does "patent pending" signify? These words are put on a product by a manufacturer to inform the public (and competition) that an application for patent on that item has been filed at the PTO. It is basically a warning to competitors to stay away.

The patent pending designation is not used lightly; the law imposes a fine on those who use these words falsely to deceive.

familiar with the new patent term immediately. Wamsley cautions, "The provisions for phasing in the 20-year term are exceedingly complex."

The PTO has set up a toll-free information number should you have any questions about the GATT: 1-800-PTO-2224.

❑ Beginning in June of 1995, a new option for inventors is the filing of a provisional application for a fee of $75 by a small entity applicant or $150 by a large applicant. Provisional applications will not have to show patent claims, but they must contain a description of the invention that satisfies the existing disclosure requirements of Patent Code, Section 112. Provisional applications will not be examined by the Patent and Trademark Office (PTO). They must be followed by a regular application within one year.

A provisional application has two main purposes according to Jennifer Bahr, a primary examiner at the PTO. It establishes a priority date for the date of invention and an offensive date if a patent issues on an application which claims priority back to the provisional application. In other words, the effective date of the patent would be the date of the provisional application.

When you file a provisional application, the PTO clock does not start running for a year, while at the same time establishing a priority or offensive date. In other words, you can buy yourself a year. This provisional application does not replace the regular patent application.

The PTO requires that all provisional applications be accompanied by a special cover sheet which will specifically identify them as such. As this book goes to press, the PTO has not yet generated this cover sheet. You can obtain a copy of this form by calling (703) 308-HELP or by faxing (703) 305-7786.

❑ Patent interference proceedings in the PTO will change in that activities in a GATT country become admissible to prove the date of invention. This means someone from a foreign country can bring an interference action against a U.S. patent holder. For example, a toy inventor in Japan could say that he invented something before I did and the PTO would accept his request for an interference proceeding.

❑ The definition of patent rights, that is, the rights to exclude others from making, using, and selling, will be augmented to encompass rights to exclude others from importing or offering for sale.

❑ Proceedings at the U.S. International Trade Commission to block infringing imported products (Section 337 of the Tariff Act of 1930) will change to respond to charges that the proceedings discriminate against foreign nationals.

IPO Booklet on GATT

The Intellectual Property Owners Association has written a 120-page booklet entitled, *GATT Implementing Legislation*. The booklet includes intellectual property-related excerpts from the GATT implementing bill.

Single copies are available to members free-of-charge. The price for non-members is $25. To order call: (202) 466-2396.

Patentability of Computer Programs

Can computer programs be patented? The answer is not simple. Under certain tests the PTO will award patent protection to a piece of software. The former Court of Customs and Patent Appeals (CCPA), known today as the Court of Appeals for the Federal Circuit (CAFC), has held that computer processes are statutory unless they fall within a judicially determined exception.

The original cases that went to the U.S. Supreme Court from the CCPA provided guidance to the PTO as to the patentability of computer-related inventions and software. However, there is disagreement between the PTO and some patent attorneys who practice before the PTO on the interpretation of the cited court cases.

As if this situation was not murky enough, in recent years there have been a spate of cases which are prompting the PTO to review its guidelines. As of this writing, in March of 1995, two cases pending before the CAFC (*In re Lowry* and *Ex parte Beauregard)* could go to the U.S. Supreme Court.

I do not develop computer software. I have no experience in the area of patenting computer software. But since the topic is of interest to many of my readers, I interviewed Gerald Goldberg, director of Group 2400 at the PTO, and an expert on computer programs and mathematical algorithms as patentable subject matter.

I strongly recommend that you seek the advice of patent counsel with experience in this highly specialized field. If after reading this section you have questions, as well you may since everything concerning the patentability of computer software is in a state of flux, and you want the latest report on the aforementioned cases *vis-a-vis* the PTO, you may wish to contact Stephen G. Kunin, Deputy Assistant Commissioner for Patent Policy at (703) 305-8850. Mr. Kunin is a career official who will ultimately issue the PTO policy guidelines.

If you would like to contact Mr. Gerald Goldberg, he may be reached by mail at PTO, Washington, D.C. 20231. His telephone number is (703) 308-5443; the fax number is: (703) 308-5360. And his Internet address is goldberg@uspto.gov. I found him to be extremely responsive, cooperative, and patient, in addition to being up-to-date on all the issues.

Inventions may be patented only if they fall within one of the four statutory classes of subject matter of 35 U.S. Code § 101: "process, machine, manufacture, or composition of matter." No patent is available for a discovery, however useful, novel, and non-obvious, unless it falls within one of those express categories. Subject matter that does not fall within one of these statutory classes is said to be "non-statutory" or to be "unpatentable subject matter."

Still today, mathematical algorithms, the basis for computer programs, *per se* are not a statutory "process" under 35 U.S. Code § 101. The courts have defined a mathematical algorithm as "a procedure for solving a given type of mathematical problem." They determined that a mathematical algorithm, or mathematical formula, like a law of nature, cannot be the subject of a patent. However, there is currently an exception.

The exception applies only to **mathematical** algorithms; any process is an "algorithm" in the sense that it is a step-by-step procedure to arrive at a given result.

As Justice Stone of the U.S. Supreme Court explained in *Mackay Radio & Telegraph Co. v. Radio Corp. of America,* 306 U.S. 86, 94 (1939): "While a scientific truth, or the mathematical expression of it, is not a patentable invention, a novel and useful structure created with the aid of knowledge of scientific truth may be."

The U.S. Supreme Court recognizes that mathematical algorithms are "the basic tools of scientific and technological work" *(Benson,* 409 U.S. at 67, 175 USPQ at 675) and should not be the subject of exclusive rights, whereas technological application of scientific principles and mathematical algorithms furthers the constitutional purpose of promoting "the Progress of . . . Useful arts" (U.S. Constitution). It is also recognized that mathematical algorithms may be the most precise way to describe the invention.

When claims involve mathematical algorithms, the goal of the patent examiner in deciding whether to grant a patent is to answer the question: "What did the applicant invent?" If the claimed invention is a mathematical algorithm, it is improper subject matter for patent protection, whereas if the claimed invention is an application of the algorithm, § 101 will not bar the grant of a patent.

According to the Court of Customs and Patent Appeals (CCPA): "If it appears that the mathematical algorithm is implemented in a specific manner to define structural relationships between the physical elements of the claim (in apparatus claims) or to refine or limit claim steps (in process claims), the claim being otherwise statutory, the claim passes muster under § 101. If, however, the mathematical algorithm is merely presented and solved by the claimed invention, as was the case in *Benson* and *Flook* (437 U.S. at 585 n.1, 198 USPQ at 195 n. 1), and is not applied in any manner to physical elements or process steps, no amount of post-solution activity will render the claim statutory; nor is it saved by a preamble merely reciting the field of use of the mathematical algorithm."

Mr. Goldberg summarizes the issue in this way. As a stand-alone invention, a software program *per se* may not be patentable, but in view of new court decisions, the PTO interpretation of existing case law may change.

If you don't ask the question, the answer is an automatic no. If you have a computer software program that you wish to patent, it just may be possible. Consult a competent patent lawyer, one who stays up to date with the case law. Or watch for news about this in the *Official Gazette* or the *Manual of Patent Examining Procedure.*

Treaties and Foreign Patents

The rights granted by a U.S. patent extend only throughout the territory of the United States and have no effect in a foreign country. Therefore, to receive patent protection in other countries you'll have to make a separate application in each of the other countries or in regional patent offices. Almost every country has its own patent laws.

The laws in many countries differ from our own. In most foreign countries, publication of the invention before the date of the application will bar the right to a patent. Most foreign countries require maintenance fees and that the patented invention be manufactured in that country within a certain period, usually about three years. If no manufacturing occurs within that period, the patent may be subject to the grant of compulsory licenses to any person who may apply for a license.

The Paris Convention

The Paris Convention for the Protection of Industrial Property is a treaty relating to patents which is followed by 93 countries, including the United States. It provides that each country guarantee to the citizens of the other countries the same rights in patent and trademark matters that it gives to its own citizens. The Paris Convention is administered by the World Intellectual Property Organization (WIPO) in Geneva, Switzerland.

The treaty also provides for the right of priority in the case of patents, design patents, and trademarks. This right means that on the basis of a regular first application filed in one of the member countries, the applicant may, within a certain period of time, apply for protection in all of the other member countries. These later applications will then be regarded as if they had been filed on the same day as the first application. Thus, these later applications will have priority over applications for the same invention that may have been filed during the same period by other persons. Moreover, these later applications, being based on the first application, will not be invalidated by any acts accomplished in the interval, such as, for example, publication or exploitation of the invention, sale of copies of the design, or use of the trademark.

The timeframe allowed for subsequent applications in other member countries is 12 months in the case of utility patents and 6 months in the case of design patents and trademarks.

Patent Cooperation Treaty

Negotiated at a diplomatic conference in Washington, D.C., in June of 1970, the Patent Cooperation Treaty (PCT) came into force on January 24, 1978, and is presently adhered to by the 53 countries listed alphabetically below.

Australia	Denmark
Austria	Finland
Barbados	France
Belgium	Gabon
Benin	Germany
Brazil	Greece
Bulgaria	Guinea
Burkina Faso	Hungary
Cameroon	Ireland
Canada	Italy
Central African Republic	Ivory Coast
Chad	Japan
Congo	Korea (North)
Czech Republic	Korea (South)

Professor Mark Spikell of George Mason University offers these pearls of wisdom to inventors:

- Your invention must have something that will prevent a competitor from producing it once you have established a market.

- You must have a very sound and conservative business plan, plus a team of expert management.

- You must be sufficiently financed.

- You must be willing to trade your personal involvement in the business for money. If a reasonable offer is presented to you to sell out your interest, you must be of a mind not to fight it.

Top 20 Foreign Patents

The following is a list of the twenty foreign countries whose residents received the most U.S. patents in fiscal year 1993, including utility, design, plant, and reissue patents.

Japan	22,942
Germany	7,172
France	3,165
United Kingdom	2,463
Canada	2,198
China (Taiwan)	1,453
Italy	1,452
Switzerland	1,193
Netherlands	961
Korea, Rep. of	789
Sweden	743
Australia	433
Hungary	358
Belgium	351
Finland	328
Austria	320
Dominican Republic	288
Hong Kong	174
Spain	160
Norway	120

Liechtenstein
Luxembourg
Madagascar
Malawi
Mali
Mauritania
Monaco
Mongolia
Netherlands
New Zealand
Norway
Poland
Portugal

Romania
Russia
Senegal
Slovakia
Spain
Sri Lanka
Sudan
Sweden
Switzerland
Togo
United Kingdom
United States

Under U.S. law it is necessary, in the case of inventions made in that country, to obtain a license from the Commissioner of Patents and Trademarks before applying for a patent in a foreign country. Such a license is required if the foreign application is to be filed before an application is filed in the United States or before the expiration of six months from the filing of an application in the United States.

If the invention has been ordered to be kept secret, the consent to the filing abroad must be obtained from the Commissioner of Patents and Trademarks during the period the secret is in effect.

Foreign Applications for U.S. Patents

Any person of any nationality may make application for a U.S. patent so long as that person is the inventor of record. The inventor must sign the same oath and declaration (with certain exceptions).

An application for a patent filed in the United States by any person who has in the past regularly filed an application for a patent for the same invention in a foreign country (which affords similar privileges to U.S. citizens) shall have the same force and effect for the purpose of overcoming intervening acts of others. The requirement is that it be filed in the United States on the date on which the application for a patent (for the same invention) was first filed in the foreign country, provided that the application in the United States is filed within 12 months (6 months in the case of a design patent) from the earliest date on which any such foreign application was filed. A copy of the foreign application certified by the patent office of the country in which it was filed is required to secure this right of priority.

If any application for patent has been filed in any foreign country prior to application in the United States, the applicant must, in the oath or declaration accompanying the application, state the country in which the earliest such application has been filed, giving the date of filing the application. All applications filed more than a year before the filing in the United States must also be recited in the oath or declaration.

An oath or declaration must be made with respect to each and every application. When the applicant is in a foreign country, the oath or affirmation may be before any diplomatic or consular officer of the United States. It may also be made before any officer having an official seal and authorized to administer oaths in the foreign country, whose authority shall be proved by a certificate of a U.S. diplomatic or consular officer. In all cases, the oath is to be attested by the proper official seal of the officer before whom the oath is made.

When the oath is taken before an officer in the country foreign to the United States, all the application papers (except the drawings) must be attached together and a ribbon passed one or more times through all the sheets. The ends of the ribbons are to be brought together under the seal before the latter is affixed and impressed, or each sheet must be impressed with the official seal of the officer before whom the oath was taken. If the application is filed by the legal representative (executor, administrator, etc.) of a deceased inventor, the legal representative must make the oath or declaration.

When a declaration is used, the ribboning procedure is not necessary, nor is it necessary to appear before an official in connection with the making of a declaration.

Your Patent Application

If the results of your patent search are positive, a patent application may be prepared and filed in the PTO. The patent search is explained in Chapter 5.

You should expect, from submission of your application to issuance, a minimum timeframe of a year and a half. The PTO's goal is to issue patents within 18 months. In fiscal year 1994, the overall pendancy rate for utility, plant, and reissue patent applications was 19 months. In 1993, it was slightly higher, at 19.5 months.

The majority of patents fall into one of two categories: utility patents or design patents (the third category being plant patents). The difference between a utility patent and a design patent is explained by Michael K. Kirk, Deputy Assistant Secretary of Commerce and Deputy Commissioner of Patents and Trademarks, as follows:

> To understand the difference between a utility and a design patent, let's take the example of a flower pot having a novel self-watering mechanism.
>
> A utility patent protects the invention's structural or functional features, for example the self-watering mechanism. A utility patent must describe the invention so that one skilled in the relevant technology can make use of it. In the example given, the application must include a written description of the watering mechanism, including drawings, if necessary. The utility patent application must also include one or more claims which define in words that which the applicant considers to be the invention.
>
> A design patent protects the ornamental design for an article of manufacture—in a word, the way a product "looks." The protection is limited to the design shown in the drawings that comprise the application. For example, a design patent application may be drafted and filed for a flower pot having a floral design added to its surface. Assuming the application issues as a design patent, anyone is permitted to make the same flower pot, provided it does not have the same floral design.

Indeed, such a design patent would not prevent anyone from making pots having the inventor's automatic watering mechanism. A design patent is easier (thus, less costly) to draft than a utility application, and, because of its limited scope of protection, relatively easy to get. It is also less valuable commercially because of its limited scope.

Small Entity Status Reduces Fees by Half

On August 27, 1982, Public Law 97-247 provided that funds would be made available to the PTO to reduce by 50 percent the payment of fees by independent inventors, small business concerns, and nonprofit organizations.

The reduced fees include those for patent application, extension of time, revival, appeal, patent issues, statutory disclaimer, and maintenance on patents based on applications filed on or after August 27, 1982.

Fees that are not reduced include petition and processing, (other than revival), document supply, certificate of correction, request for re-examination, international application fees, and certain maintenance fees.

What is an Independent Inventor?

The PTO considers an inventor as independent if the inventor (1) has not assigned, granted, conveyed, or licensed, and (2) is under no obligation under contract or law to assign, grant, convey, or license any rights in the invention to any person who could not likewise be classified as an independent inventor if that person had made the invention, or to any concern which would not qualify as a small concern or a nonprofit organization.

What is a Small Business Concern?

The PTO defines a small business as one whose number of employees, including those of its affiliates, does not exceed 500 persons. The definition also requires a small business, for this purpose, to be one which has not assigned, granted, conveyed, or licensed, and is under no obligation under contract or law to assign, grant, convey, or license, any rights in the invention to any person who could not be classified as an independent inventor if that person had made the invention, or to any concern which would not qualify as a small business concern or a nonprofit organization.

What is a Nonprofit Organization?

To be recognized as a nonprofit organization it must be so accredited by a nationally recognized accrediting agency or association, or of the type described in Section 501(c)(3) of the IRS Code of 1954-26 U.S.C. 501(c)(3), and which is exempt from taxation under 26 U.S.C. 501(a).

Facsimile Transmissions to the PTO

Since November 1, 1988, certain papers to be filed in national patent applications and re-examination proceedings for consideration by the Office of the Assistant Commissioner for Patents, the Office of the Deputy Assistant Commissioner for Patents, and the Patent Examining Groups (Patent Examining Corps) have been allowed to be submitted to the PTO by facsimile transmission.

Examples of what you can send in by fax include: amendments, responses to restriction requirements, requests for reconsideration before an examiner, petitions, powers of attorney, notices of appeal, and appeal briefs.

Among those documents that the PTO **will not** accept by fax are: new or continuing patent applications of any type, assignments, issue fee payments, maintenance fee payments, declarations or oaths under 37 CFR 1.63 or 1.67, formal drawings, and all papers relating to international patent applications. Also excluded are papers to be filed in applications that are subject to a secrecy order under 37 CFR 5.1-5.8, and directly related to the secrecy order content of the application.

Your facsimile submissions may include a certificate for each paper stating the date of transmission. A copy of the facsimile submission with a certificate faxed therewith will be evidence of transmission of the paper should the original be misplaced. The person signing the certificate should have a reasonable basis to expect that the paper would be transmitted on the date indicated. An example of a preferred certificate is the following:

Certification of Facsimile Transmission

I hereby certify that this paper is being facsimile transmitted to the Patent and Trademark Office on the date shown below.

_____

Type or print name of person signing certification

_____

Signature   Date

When possible, the certification should appear on a portion of the paper being transmitted. If the certification is presented on a separate paper, it must identify the application to which it relates, and the type of paper being transmitted (e.g., amendment, notice of appeal, etc.).

In the event that the facsimile submission is misplaced or lost in the PTO, the submission will be considered filed as of the date of the transmission, if the party who transmitted the paper:

1. Informs the PTO of the previous facsimile transmission promptly after becoming aware that the submission has been misplaced or lost;

2. Supplies another copy of the previously transmitted submission with the Certification of the Transmission; and

3. Supplies a copy of the sending unit's report confirming transmission of the submission. In the event that a copy of the report is not available, the party who transmitted the paper may file a declaration under 37 CFR 1.68, which attests on a personal knowledge basis or to the satisfaction of the Commissioner to the previous timely transmission.

Just the Fax, Ma'am

Did you know that you can now send a fax directly to the PTO's front offices?

- Assistant Commissioner for Patents: (703) 305-8825
- Assistant Commissioner for Trademarks: (703) 308-7220

For all the fax numbers at the PTO that you could possibly need, please refer to the PTO phone directory in Appendix 2.

If all criteria above cannot be met, the PTO will require you to submit a verified showing of facts. Such a showing must indicate to the satisfaction of the Commissioner the date the PTO received the submission. PTO fax numbers may be found in Appendix 2.

Utility Patents

At this point let me remind you that utility patent applications are complicated legal documents to draft. Accordingly, I strongly recommend that you hire the services of competent legal counsel—someone admitted to practice before the PTO—to draft, file and prosecute your applications. See Chapter 6 for information on how to hire legal counsel. A sample utility patent is shown at the end of this chapter, on pages 87-94.

The following information is included to help you file a utility patent application, should you opt to do it yourself. Even if you decide to use a lawyer, this information will familiarize you with the process.

Your complete utility patent application should contain the information listed below and should be arranged in the following order.

1. *A letter of transmittal.* A transmittal letter should be filed with every patent application to instruct the PTO on the services you desire in the processing of your application. In this cover letter, inform the commissioner of your name, address, and telephone number, the type of application, the title of your invention, and the contents of the application. A sample transmittal letter is shown on page 101.

2. *Specification.* The specification must include a written description of the invention and of the manner and process of making and using it, and is required to be in such full, clear, concise, and exact terms as to enable any person skilled in the art or science to which the invention pertains, or with which it is most nearly connected, to make and use it.

   The specification must set forth the precise invention for which you are seeking patent protection, in such a way as to distinguish it from other inventions and from what is old. It must describe completely a specific embodiment of the process, machine, manufacture, composition of matter or improvement invented, and must explain the method of operation or principle whenever applicable. Write up the best embodiment you contemplate for your invention.

   In the case of an improvement to an invention, the specification must particularly point out the part(s) of the process, machine, manufacture, or composition of matter to which your improvement relates, and the description should be confined to the specific improvement and to such parts as necessarily cooperate with it or as may be necessary to complete understanding or description of it.

   The pages of the specification, including claims and abstract, should be numbered consecutively, starting with page one. The page numbers should be centrally located above, or preferably, below the text.

   a. *Title.* The title of the invention, which should be as short and specific as possible (best if it does not exceed 280 typewritten spaces), should appear as the heading on the first page of the specification, if it does not otherwise appear at the beginning of the application.

b. *Cross-Reference to related applications, if any.*

c. *Reference to a microfiche appendix, for computer program listings, if any.* The total number of microfiche and total number of frames should be specified.

d. *Background of the Invention.* The specification should set forth the background of the invention in two parts:

- Field of the Invention: This section should include a statement of the field of endeavor to which the invention pertains. This section may also include a paraphrasing of the applicable U.S. patent classification definitions or the subject matter of the claimed invention. This section may also be titled "Technical Field."
- Description of the related art (or prior art): This section should contain a description of information known to the applicant, including references to specific documents, which are related to the applicant's invention. This section should also contain, if applicable, references to specific art-related problems involved in the prior art which are solved by the applicant's invention.

e. *Summary of the Invention.* This section should present the substance or general idea of the claimed invention in summarized form. The summary may point out the advantages of the invention or how it solves previously existing problems, preferably those problems identified in the Background of the Invention. A statement of the object of the invention may also be included. The summary should precede the detailed description.

f. *Abstract of the Disclosure.* The purpose of the abstract is to enable the PTO and the public to determine quickly from a cursory inspection the nature and gist of the technical disclosures of the invention. The abstract points out what is new in the art to which the invention pertains. It should be in narrative form and generally limited to a single paragraph on a separate page.

g. *Claims*

- The specification must conclude with a claim particularly pointing out and distinctly claiming the subject matter which you regard as your invention.
- More than one claim may be presented provided the claims differ substantially from each other and are not unduly multiplied.
- One or more claims may be presented in dependent form, referring back to and further limiting another claim or claims in the same application. Any dependent claim which refers to more than one other claim ("multiple dependent claim") shall refer to such other claims in the alternative only. A multiple dependent claim shall not serve as a basis for any other multiple dependent claim. For fee calculation purposes, a multiple dependent claim will be considered to be that number of claims to which direct reference is made therein. For fee calculation purposes, also, any claim depending from a multiple dependent claim will be considered that number of claims to which direct reference is made in that multiple dependent claim.
- The claim or claims must conform to the invention as set forth in the remainder of the specification and the terms and phrases used in the

Specification Checklist

1. Title of the invention

2. Cross-references to related applications, if any

3. Reference to a microfiche appendix, for computer program listings, if any

4. Background of the invention

5. Summary of the invention

6. Abstract of the disclosure

7. Claims

Utility Patent Application Checklist

1. Transmittal letter

2. Specification

3. Drawings

4. Oath or Declaration

5. Fee payments

claims must find clear support or antecedent basis in the description so that the meaning of the terms in the claims may be ascertainable for reference to the description.

3. *Drawings.* A patent application is required to contain drawings if drawings are needed for the understanding of the subject matter sought to be patented. The drawings must show every feature of the invention as specified in the claims. Omission of drawings will cause an application to be considered incomplete. An application for a design patent must contain at least one drawing.

When there are drawings, there must be a brief description of the several views of the drawings, and the detailed description of the invention must refer to the different views by specifying the numbers of the figures and to the different parts by use of reference letters or numerals (preferably the latter). For specific guidelines to drawings, refer to the section in this chapter entitled "Patent Drawings."

4. *Oath or Declaration.* This document must be signed by all of the actual inventors. An oath may be administered by any person within the United States, or by a diplomatic or consular officer of a foreign country, who is authorized by the United States to administer oaths. A declaration does not require any witness or person to administer or verify its signing. Thus, use of a declaration is preferable.

The document must identify the application to which it is directed. It must give the name, city, and either state or country of residence, country of citizenship, and post office address of each inventor and it must state whether the inventor is a sole or joint inventor of the invention claimed. Additionally, designation of a correspondence address is needed on the oath or declaration. Providing a correspondence address will help to ensure prompt delivery of all notices, official letters, and other communications.

5. *Fee payment.* And last but not least, don't forget to put in your check. Fees for patent applications are subject to change and should **always** be double-checked before filing. The current fees may be found on pages 122–25. Having made improper payments in the past, which caused delays and penalties, I now call several times before writing my check to the PTO. For the most up-to-date fees, call (703) 308-4357.

Please remember that two sets of fees exist, one for a small entity and one for other than a small entity. If you qualify as a small entity for patent fee purposes, you should complete one of the forms verifying small entity status in order to receive the benefit of reduced fees. These forms can be found starting on page 106.

If you find that you need additional guidance filing your application, hire a patent attorney, or call the PTO's Special Processing Branch at (703) 308-1202.

Models Not Generally Required

Now that you know what *is* required for your patent application, here is something that generally is *not* required. Models were once required in all cases admitting a model, as part of the application, and these models became part of the record of the patent. Such models are no longer generally required (the description of the invention in the specification, and the drawings, must be sufficiently full and complete and capable of being understood to disclose the invention without the aid of a model) and will not be accepted unless specifically called for by the examiner.

When the invention relates to a composition of matter, the applicant may be asked to furnish specimens of the composition, or of its ingredients or intermediates, for the purpose of inspection or experiment.

Design Patents

If you want an inexpensive patent that will give you little actual protection but will still meet the requirement that a patent issue, maybe a design patent is for you. It appears to be popular with lots of folks: more than six thousand new designs were patented last year.

Inventors get design patents on just about anything, for example, baby bibs, sweatbands, tissue box holders, dishes, game boards, vending machines, telephones, pencils and pens, and even internal combustion engines.

Rubbermaid has one for a cereal container. Totes protects its umbrella handle designs. The Parker Pen Company takes them out for writing instruments. And Ford Motor Company has them on car parts such as automobile quarter panels.

For little expense and effort, a design patent is another way to stake out a claim. It permits you to legally post a no trespass sign in the form of Patent Pending, Patent Applied For, or Patent No. 1,234,567.

If you've invented any new, original, and ornamental designs for an article of manufacture, a design patent may be appropriate. A design patent protects only the appearance of an article and not its structure or utilitarian features. The proceedings relating to granting of design patents are the same as those relating to other patents with few differences.

In design cases as in "mechanical" cases, novelty and nonobviousness are necessary prerequisites to the granting of a patent. In the case of designs, the inventive novelty resides in the shape or configuration or ornamentation as determining the appearance or visual aspect of the object or article of manufacture in contradistinction to the structure of a machine, article of manufacture, or the constitution of a composition of matter. Simply put, it is the appearance of the object that creates an impression upon the mind of the observer.

Do You Need A Patent Attorney for Design Patents?

This is, of course, a personal decision. I do not use a patent attorney's services for design patent applications. Unlike the complicated business of utility patents, the design patent application process is very easy and uncomplicated. I have found that paying a lawyer to do my design patents is like tossing money out the window because the application form is simple, esoteric language is not required to draft the claims, and searches are actually so pleasant an experience that I often do them myself. A design patent search doesn't require much reading. You're just looking at a lot of line drawings, many of which are fascinating.

I spent $500 learning that lawyers were not necessary in such matters. At the time I had been using a fine law firm for utility patents and I naturally released our first design patent to them. I knew no better.

My lawyer said that he would "write up the specifications." I was asked to provide his draftsman with a prototype of the design. It was a tricycle with a main-

Designing Inventors

The range of ornamental appearances that have been patented during the more than 150 years design patents have existed is very impressive. Over 270,000 designs have received design patent protection since the first one was granted to George Bruce for "Printing Type" on November 9, 1842.

frame shaped like a toothpaste tube. It ultimately became Proctor & Gamble's Crest Fluorider. When the design patent arrived, I saw for the first time the specifications. As you will see from the sample specification I have provided below, design patents are obvious cash cows for patent attorneys.

Design Patent Title

The title is of great importance in a design patent application. It serves to identify the article in which the design is embodied and which is shown in the drawing by a name generally used by the public. The title should be to a specific definite article. Thus a stove would be called a "Stove" and not "Heating Device." The same title is used in the preamble to the specification, in the description of the drawing, and in the claim.

To allow latitude of construction it is permissible to add to the title "or similar article." The title must be in the singular.

Design Patent Specification

To create your own design patent specification document, all you have to do is use the below example and fill in the appropriate blanks. This is a sample only and, unlike other forms in this book, should not be returned to the PTO. The petition, specification, and claim must be typed on legal-sized paper.

To the Commissioner of Patents and Trademarks:

Your petitioner, _____, a citizen of _____, and a resident of _____, prays that letters patent may be granted for the design for a _____ (insert title of design), as set forth in the following specification.

Figure 1 is a _____ (insert front, side, rear, or appropriate description) view of a _____ (insert title of design) showing my (our) new design; Figure 2 is a _____ thereof.

I (We) Claim:

The ornamental design for a _____ (insert title of design) as shown and described.

What to Send the PTO

The design patent application must include the following:

1. specification and single claim

2. drawing

3. declaration

4. verified statement of small entity, if applicable

5. a filing fee submitted in the form of a check or money order made out to the Commissioner of Patents and Trademarks

Plant Patents

The law also provides for the granting of a patent to anyone who has invented or discovered and asexually reproduced any distinct and new variety of plant, including cultivated sports, mutants, hybrids, and newly found seedlings, other than a tuber-propagated plant or a plant found in an uncultivated state.

Asexually propagated plants are those that are reproduced by means other than from seeds, such as by the rooting of cuttings, by layering, budding, grafting, inarching, and so on.

With reference to tuber-propagated plants, for which a plant patent cannot be obtained, the term "tuber" is used in its narrow horticultural sense as meaning a short, thickened portion of an underground branch. The only plants covered by the term "tuber-propagated" are the Irish potato and the Jerusalem artichoke.

Elements of a Plant Application

An application for a plant patent consists of the same parts as other applications. As of this printing, a plant patent has a term of 17 years.

The application papers for a plant patent and any responsive papers pursuant to the prosecution must be filed in duplicate but only one need be signed (in the case of the application papers the original should be signed); the second copy may be a legible copy of the original. The reason for providing an original and duplicate file is that the duplicate file is sent to the Agricultural Research Service, Department of Agriculture, for an advisory report on the plant variety.

Plant Specification and Claim

The specification should include a complete detailed description of the plant and the characteristics thereof that distinguish it over related known varieties, and its antecedents, expressed in botanical terms in the general form followed in standard botanical textbooks or publications dealing with the varieties of the kind of plant involved (evergreen tree, dahlia plant, rose plant, apple tree, etc.), rather than a mere broad nonbotanical characterization such as commonly found in nursery or seed catalogs. The specification should also include the origin or parentage of the plant variety for which you are seeking a patent and must particularly point out where and in what manner the variety of plant has been asexually reproduced. Where color is a distinctive feature of the plant the color should be positively identified in the specification by reference to a designated color as given by a recognized color dictionary. Where the plant variety originated as a newly found seedling, the specification must fully describe the conditions (cultivation, environment, etc.) under which the seedling was found growing to establish that it was not found in an uncultivated state.

A plant patent is granted on the entire plant. It therefore follows that only one claim is necessary and only one is permitted.

Plant Oath or Declaration

The oath or declaration required of the applicant in addition to the statements required for other applications must include the statement that the applicant has asexually reproduced a new plant variety. A declaration for Plant Patent Application (35 U.S.C. 161) may be found on page 116.

Plant Patent Drawings

Plant patent drawings are not mechanical drawings and should be artistically and competently executed. The drawing must disclose all the distinctive characteristics of the plant capable of visual representation. When color is a distinguishing characteristic of the new variety, the drawing must be in color. Two duplicate copies of color drawings must be submitted. Color drawings may be made either in permanent water color or oil, or in lieu thereof may be photographs made by color photography or properly colored on sensitized paper. The paper in any case must correspond in size, weight, and quality to the paper required for other drawings. Mounted photographs are acceptable.

Specimens

Specimens of the plant variety, its flower or fruit, should not be submitted unless specifically called for by the examiner.

Fees and Correspondence

The filing fee on each plant application and the issue fee can be found in the fee schedule on pages 122-125. For a qualifying small entity, filing and issue fees are reduced by half.

All inquiries relating to plant patents and pending plant patent applications should be directed to the Patent and Trademark Office and not to the Department of Agriculture.

The Plant Variety Protection Act

The Plant Variety Protection Act (Public Law 91-577), approved December 24, 1970, provides for a system of protection for sexually reproduced varieties, for which protection was not previously provided, under the administration of a Plant Variety Protection Office within the Department of Agriculture. Requests for information regarding the protection of sexually reproduced varieties should be addressed to Commissioner, Plant Variety Protection Office, Agricultural Marketing Service, National Agricultural Library Building, Room 500, 10301 Baltimore Boulevard, Beltsville, Maryland 20705-2351.

Patent Drawings

Unless you are capable of doing this work yourself, hire a competent, experienced patent draftsman to make your drawings. The requirements for drawings are strictly enforced. Professional draftsmen will stand behind their work and guarantee revisions if requested by the PTO due to inconsistencies in the drawings. I can tell you from firsthand experience that many times my professional draftsman has had to redo drawings on his own time because of PTO objections. Even the best draftsmen have their work rejected from time to time by PTO examiners.

Because the design patent is granted for the appearance of the article, the drawing in the design patent is more critical than the drawing in a utility patent. The design drawing is the disclosure of the claimed design, whereas the utility drawing is intended to provide only an exemplary illustration of some aspects of the mechanism described in the specification and claims. The design drawing must therefore satisfy the disclosure requirements of 35 USC 112, first and second paragraphs, which are met by the specification of the utility patent. The design drawings must also meet the technical requirements of 37 CFR 1.84 (Standards for Drawings), although not all of the drafting procedures included in this rule are appropriate or permitted in a design drawing. Some of these exclusions are identified in the following paragraph.

Requirements for Contents of Design Drawings

To meet the disclosure requirements of 35 USC 112, the drawing must include a sufficient number of views to constitute a complete disclosure of the appearance of the article. These views must show the claimed design in solid lines, and these views must be consistent in all details of the design. These views must be appropriately shaded to show clearly the character and/or contour of the surfaces represented. This shading is of particular importance in the showing of three-dimensional articles, where it is necessary to delineate plane, concave, convex, raised and/or depressed surfaces of the article, and to distinguish open and solid surfaces. Transparent and translucent surfaces should be indicated by oblique line shading as shown in design patent Des. 286,841 on page 96 and on the "Symbols for Draftsmen" (see below). Broken lines may not be used to indicate unimportant or immaterial features of the design; there are no features of a claimed design that are unimportant or immaterial. (This issue was determined by court decisions: *In re Blum,* 153 USPQ 177 CCPA 1967; *In re Zahn* 204 USPQ 988, CCPA 1980.) Broken lines may not be used to show movement or alternate positions in a design drawing, or to show hidden structure through opaque surfaces. Broken lines may be included to show environmental structure. Only the claimed design may be shown in solid lines in the drawing; unclaimed subject may only appear in broken lines. Legends and reference characters identifying portions of the design may not be used in a design drawing, although they are permitted in a utility drawing.

Color Drawings

Color drawings are permitted on rare occasions in utility applications as the only practical medium to disclose the subject matter sought to be patented. Color drawings are not permitted for any reason in design patent applications. However, color may be illustrated in design patent drawings by means of shading, utilizing specific drafting symbol patterns, examples of which are shown in the "Symbols for Draftsmen," see below. When such shading patterns are used in design patent drawings, special descriptions are required in the specification. Examples of design patent drawings with such shading and associated descriptions are also enclosed.

Photographs

Photographs are not ordinarily accepted in utility and design applications. The PTO will accept photographs in utility and design patent applications only after granting a petition filed under 37 CFR 1.84(b). Photographs submitted in a design application must comply with the disclosure requirements of 35 USC 112, paragraphs 1 and 2, and must show only the design for the article claimed. No environmental structure may be shown. Photographs and ink drawings may not be combined in a single application.

Patent Laws and Rules for Design Patent Drawings

Patent laws and rules applicable to design drawings appear below in the following order:

Top 10 States

The following is a list of the ten American states whose residents were awarded the most U.S. patents in 1994.

California	10,472
New York	5,522
Texas	4,089
New Jersey	3,328
Michigan	3295
Illinois	3,266
Pennsylvania	3,085
Ohio	3,078
Massachusetts	2,669
Florida	2,241

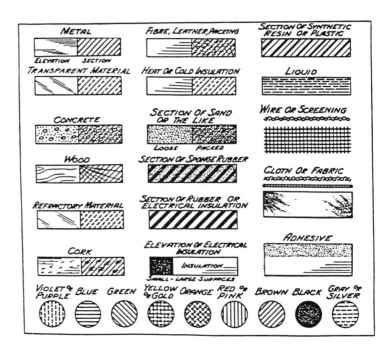

Symbols for Draftsmen

- 35 USC 112, first and second paragraphs
- 37 CFR 1.84
- 37 CFR 1.123
- 37 CFR 1.152

35 USC 112, first and second paragraphs. This specification shall contain a written description of the invention, and of the manner and process of making and using it, in such full, clear, concise, and exact terms as to enable any person skilled in the art to which it pertains, or with which it is most nearly connected, to make and use the same, and shall set forth the best mode contemplated by the inventor of carrying out his invention.

The specification shall include with one or more claims particularly pointing and distinctly claiming the subject matter which the applicant regards as his invention.

37 CFR 1.84. Standards for drawings.

1. *Drawings.* There are two acceptable categories for presenting drawings in utility applications:

   a. *Black ink.* Black and white drawings are normally required. India ink, or its equivalent that secures solid black lines, must be used for drawings, or

   b. *Color.* On rare occasions, color drawings may be necessary as the only practical medium by which to disclose the subject matter sought to be patented in a utility patent application or the subject of a statutory invention registration. The PTO will accept color drawings in utility patent applications and statutory invention registrations only after granting a petition filed under this paragraph explaining why the color drawings are necessary. Any such petition must include the following:

- The appropriate fee set forth in 37 CFR 1.17(h)
- Three sets of color drawings
- The specification must contain the following language as the first paragraph in that portion of the specification relating to the brief description of the drawing: "The file of this patent contains at least one drawing executed in color. Copies of this patent with color drawing(s) will be provided by the Patent and Trademark Office upon request and payment of the necessary fee." If the language is not in the specification, a proposed amendment to insert the language must accompany the petition.

1. *Photographs.*

   a. *Black and white.* Photographs are not ordinarily permitted in utility and design patent applications. However, the PTO will accept photographs in utility and design patent applications only after granting a petition filed under this paragraph which requests that photographs be accepted. Any such petition must include the following:

   - The appropriate fee set forth in 37 CFR 1.17(h)
   - Three sets of photographs. Photographs must either be developed on double weight photographic paper or be permanently mounted on bristol board. The photographs must be of sufficient quality so that all details in the drawing are reproducible in the printed patent.

   b. *Color.* Color photographs wil be accepted in utility patent applications if the conditions for accepting color drawings have been satisfied.

3. *Identification of drawings.* Identifying indicia, if provided, should include the application number or the title of the invention, inventor's name, docket number (if any), and the name and telephone number of a person to call if the PTO is unable to match the drawings to the proper application. This information should be placed on the back of each sheet of drawings a minimum distance of 1.5 cm (5/8 inch) down from the top of the page.

4. *Graphic forms in drawings.* Chemical or mathematical formulas, tables, and waveforms may be submitted as drawings, and are subject to the same requirements as drawings. Each chemical or mathematical formula must be labeled as a separate figure, using brackets when necessary, to show that information is properly integrated. Each group of waveforms must be presented as a single figure, using a common verticle axis with time extending along the horizontal axis. Each individual waveform discussed in the specification must be identifed with a separate letter designation adjacent to the verticle axis.

5. *Type of paper.* Drawings submitted to the PTO must be made on paper which is flexible, strong, white, smooth, nonshiny, and durable. All sheets must be free from cracks, creases, and folds. Only one side of the sheet shall be used for the drawing. Each sheet must be reasonably free from erasures and must be free from alterations, overwritings, and inter-lineations. Photographs must either be developed on double weight photographic paper or be permanently mounted on bristol board.

6. *Size of paper.* All drawing sheets in an application must be the same size. One of the shorter sides of the sheet is regarded as its top. The size of the sheets on which drawings are made must be:

   a. 21.6 cm by 35.6 cm (8½ by 14 inches),

Dutch Treats

As an adolescent, Dutch inventor Cornelius Van Drebbel was apprenticed to an engraver, but he soon developed an interest in alchemy and mechanical inventions.

Drebbel has been credited with constructing the first compound microscope using two sets of convex lenses.

During the early 1620s, Drebbel designed and built his most famous invention, the submarine. Although a similar design had been described some fifty years earlier, Drebbel's is the first known submarine to have been constructed. Consisting mainly of greased leather stretched over a wooden frame, Drebbel's submarine was propelled by oars projecting through the sides and sealed with leather flaps. The vessel was capable of traveling twelve to fifteen feet below the surface, and fresh air was supplied by tubes running to the surface with floats at the top.

Drebbel also invented the first thermostat, which used a column of mercury and a system of floats and levers to hold a steady temperature within a furnace. He later invented an incubator for hatching eggs which used the same principle for temperature regulation.

b.   21.6 cm by 33.1 cm (8½ by 13 inches),

c.   21.6 cm by 27.9 cm (8½ by 11 inches), or

d.   21.0 cm by 29.7 cm (DIN size A4).

7.   *Margins.* The sheets must not contain frames around the sight, i.e., the usable surface. The following margins are required.

   a.   On 21.6 cm by 35.6 cm (8½ by 14 inches) drawing sheets, each sheet must include a top margin of 5.1 cm (2 inches) and bottom and side margins of .64 cm (¼ inch) from the edges, thereby leaving a sight no greater than 20.3 by 29.8 cm (8 by 11¾ inches).

   b.   On 21.6 cm by 33.1 cm (8½ by 13 inches) drawing sheets, each sheet must include a top margin of 2.5 cm (1 inch) and bottom and side margins of .64 cm (¼ inch) from the edges, thereby leaving a sight no greater than 20.3 by 29.8 cm (8 by 11¾ inches).

   c.   On 21.6 cm by 27.9 cm (8½ by 11 inches) drawing sheets, each sheet must include a top margin of 2.5 cm (1 inch) and bottom and side margins of .64 cm (¼ inch) and bottom and side margins of .64 cm (¼ inch) from the edges, thereby leaving a sight no greater than 20.3 by 24.8 cm (8 by 19¾ inches).

   d.   On 21.9 cm by 29.7 cm (DIN size A4) drawing sheets, each sheet must include a top margin of at least 2.5 cm, a left side margin of 2.5 cm, a right side margin of 1.5 cm, and a bottom margin of 1.0 cm, thereby leaving a sight no greater than 17.0 by 26.2 cm.

8.   *Views.* The drawing must contain as many views as necessary to show the invention. The views may be plan, elevation, section, or perspective views. Detail views of portions of elements, on a larger scale if necessary, may also be used. All views of the drawing must be grouped together and arranged on the sheet(s) without wasting space, preferably in an upright position, clearly separated from one another, and must not be included in the sheets containing the specification, claims, or abstract. Views must not be connected by projection lines and must not contain center lines. Waveforms of electrical signals may be connected by dashed lines to show the relative timing of the waveforms.

   a.   *Exploded views.* Exploded views with the separated parts embraced by a bracket to show the relationship or order of assembly of various parts are permissible. When an exploded view is shown in a figure which is on the same sheet as another figure, the exploded view should be placed in brackets.

   b.   *Partial views.* When necessary, a view of a large machine or device in its entirety may be broken into partial views on a single sheet, or extended over several sheets if there is no loss in facility of understanding the view. Partial views drawn on separate sheets must always be capable of being linked edge to edge so that no partial view contains parts of another partial view. A small-scale view should be included showing the whole formed by the partial views and indicating the position of the parts shown. When a portion of a view is enlarged for magnification purposes, the view and the enlarged view must each be labeled as separate views.

- Where views on two or more sheets form, in effect, a single complete view, the views on the several sheets must be so arranged that the complete figure can be assembled without concealing any part of the views appearing on the various sheets.
- A very long view may be divided into several parts placed one above the other on a single sheet. However, the relationship between the different parts must be clear and unambiguous.

c. *Sectional views.* The plan upon which a sectional view is taken should be indicated on the view from which the section is cut by a broken line. The ends of the broken line should be designated by Arabic or Roman numerals corresponding to the view number of the sectional view, and should have arrows to indicate the direction of sight. Hatching must be used to indicate section portions of an object, and must be made by regularly spaced oblique parallel lines spaced sufficiently apart to enable the lines to be distinguished without difficulty. Hatching should not impede the clear reading of the reference characters and lead lines. If it is not possible to place reference characters outside the hatched area, the hatching may be broken off wherever reference characters are inserted. Hatching must be at a substantial angle to the surrounding axes or principal lines, preferably 45 degrees. A cross section must be set out and drawn to show all of the materials as they are shown in the view from which the cross section was taken. The parts in cross section must show proper material(s) by hatching with regularly spaced parallel oblique strokes, the space between strokes being chosen on the basis of the total areas to be hatched. The various parts of a cross section of the same item should be hatched in the same manner and should accurately and graphically indicate the nature of the material(s) as illustrated in the cross section. The hatching of juxtaposed different elements must be angled in a different way. In the case of large areas, hatching may be confined to an edging drawn around the entire inside of the outline of the area to be hatched. Different types of hatching should have different conventional meanings as regards the nature of a material seen in cross section.

d. *Alternate position.* A moved position may be shown by a broken line superimposed upon a suitable view if this can be done without crowding; otherwise, a separate view must be used for this purpose. **Alternate positions may not be shown in broken lines in a design drawing. See section labeled 37 CFR 1.152.**

e. *Modified forms.* Modified forms of construction must be shown in separate views.

9. *Arrangement of views.* One view must not be placed upon another or within the outline of another. All views on the same sheet should stand in the same direction and, if possible, stand so that they can be read with the sheet held in an upright position. If views wider than the width of the sheet are necessary for the clearest illustration of the invention, the sheet may be turned on its side so that the top of the sheet, with the appropriate top margin to be used as the heading space, is on the right hand side. Words must appear in a horizontal, left to right fashion when the page is either upright or turned so that the top becomes the right side, except for graphs utilizing standard scientific convention to denote the axis abscissas (of X) and the axis of ordinates (of Y).

10. *View for "Official Gazette."* One of the views should be suitable for publication in the *Official Gazette* as the illustration of the invention.

11. *Scale.*

   a. The scale to which a drawing is made must be large enough to show the mechanism without crowding when the drawing is reduced in size to two-thirds in reproduction. Views of portions of the mechanism on a larger scale should be used when necessary to show details clearly. Two or more sheets may be used if one does not give sufficient room. The number of sheets should be kept to a minimum.

   b. When approved by the examiner, the scale of the drawing may be graphically represented. Indications such as "actual size" or "scale 1/2" on the drawings, are not permitted, since those lose their meaning with reproduction in a different format.

   c. Elements of the same view must be in proportion to each other, unless a difference in proportion is indispensable for the clarity of the view. Instead of showing elements in different proportion, a supplementary view may be added giving a larger scale illustration of the element of the initial view. The enlarged element shown in the second view should be surrounded by a finely drawn or "dot-dash" circle in the first view indicating its location without obscuring the view.

12. *Character of lines, numbers, and letters.* All drawings must be made by a process which will give them satisfactory reproduction characteristics. Every line, number, and letter must be durable, clean, black (except for color drawings), sufficiently dense and dark, and uniformly thick and well-defined. The weight of all lines and letters must be heavy enough to permit adequate reproduction. This requirement applies to all lines however fine, to shading, and to lines representing cut surfaces in section views. Lines and strokes for different thicknesses may be used in the same drawing where different thicknesses have a different meaning.

13. *Shading.* The use of shading in views is encouraged if it aids in understanding the invention and if it does not reduce legibility. Shading is used to indicate the surface or shape of spherical, cylindrical, and conical elements of an object. Flat parts may also be lightly shaded. Such shading is preferred in the case of parts shown in perspective, but not for cross sections. Spaced lines for shading are preferred. These lines must be thin, as few in number as practicable, and they must contrast with the rest of the drawings. As a substitute for shading, heavy lines on the shade side of objects can be used except where they superimpose on each other or obscure reference characters. Light should come from the upper left corner at an angle of 45 degrees. Surface delineations should preferably be shown by proper shading. Solid black shading areas are not permitted, except when used to represent bar graphs or color.

14. *Symbols.* Graphical drawing symbols may be used for conventional elements when appropriate. The element for which such symbols and labeled representations are used must be adequately identified in the specification. Known devices should be illustrated by symbols which have universally recognized conventional meaning and are generally accepted in the art. Other symbols which are not universally recognized may be used, subject to approval by the

PTO, if they are not likely to be confused with existing conventional symbols and if they are readily identifiable.

15. *Legends.* Suitable descriptive legends may be used, or may be required by the examiner, where necessary for understanding of the drawing, subject to approval by the PTO. They should contain as few words as possible.

16. *Numbers, letters, and reference characters.*

    a.   Reference characters (numerals are preferred), sheet numbers, and view numbers must be plain and legible, and must not be used in association with brackets or inverted commas, or enclosed with outlines, e.g. encircled. They must be oriented in the same direction as the view so as to avoid having to rotate the sheet. Reference characters should be arranged to follow the profile of the object depicted.

    b.   The English alphabet must be used for letters, except where another alphabet is customarily used, such as the Greek alphabet to indicate angles, wavelength, and mathematical formulas.

    c.   Numbers, letters, and reference characters must measure at least .32 cm (⅛ inch) in height. They should not be placed in the drawing so as to interfere with its comprehension. Therefore, they should not cross or mingle with the lines. They should not be placed upon hatched or shaded surfaces. When necessary, such as indicating a surface or cross section, a reference character may be underlined and a blank space may be left in the hatching or shading where the character occurs so that it appears distinct.

    d.   The same part of an invention appearing in more than one view of the drawing must always be designated by the same reference character, and the same reference character must never be used to designate different parts.

    e.   Reference characters not mentioned in the description shall not appear in the drawings. Reference characters mentioned in the description must appear in the drawings.

17. *Lead lines.* Lead lines are those lines between the reference characters and details referred to. Such lines may be straight or curved and should be as short as possible. They must originate in the immediate proximity of the reference character and extend to the feature indicated. Lead lines must not cross each other. Lead lines are required for each reference character except for those which indicate the surface or cross section on which they are placed. Such a reference character must be underlined to make it clear that a lead line has not been left out by mistake. Lead lines must be executed in the same way as lines in the drawing.

18. *Arrows.* Arrows may be used at the ends of lines, provided that their meaning is clear.

    a.   On a lead line, a freestanding arrow to indicate the entire section towards which it points.

    b.   On a lead line an arrow touching a line to indicate the surface shown by the line looking along the direction of the arrow, or

    c.   To show the direction of movement. **Arrows may not be used to show movement in a design drawing.**

A Couple of Sharp Inventors

Men have been shaving off their beards with sharp implements since ancient times. Cave paintings show shells, shark's teeth, and sharpened flint used as razors. But there are two American inventors who changed the face of mankind by revolutionizing shaving habits.

King Camp Gillette invented the first disposable blade in 1901. Gillette created a thin, double-edged blade that was fastened to a special guarded holder and could be thrown away when it got dull. An entire generation was converted to the safety razor when the U.S. government issued Gillette razors to its troops during World War I.

Jacob Schick, a retired U.S. Army colonel and inventor, was inspired to develop a razor that worked without soap or water. After World War I, Schick devoted himself to inventing an electric razor. His wife mortgaged their Connecticut home to finance the venture. The patented design that resulted used a series of slots to hold the hairs while a series of moving blades cut the hair off.

19. *Copyright or Mask Work Notice.* A copyright or mask work notice may appear in the drawing, but must be placed within the sight of the drawing immediately below the figure representing the copyright or mask work material and be limited to letters having a print size of .32 cm to .64 cm (⅛ to ¼ inches) high. The content of the notice must be limited to only those elements provided for by law.

20. Numbering for sheets of drawing. The sheets of drawings should be numbered in consecutive Arabic numerals, starting with 1, within the sight as defined in paragraph seven of this section. These numbers, if present, must be placed in the middle of the top of the sheet, but not in the margin. The numbers can be placed on the right hand side if the drawing extends too close to the middle of the top edge of the usable surface. The drawing sheet numbering must be clear and larger than the numbers used as reference characters to avoid confusion. The number of each sheet should be shown by two Arabic numerals placed on either side of an oblique line, with the first being the sheet number, and the second being the total number of sheets of drawings, with no other markings.

21. *Numbering of views.*

   a. The different views must be numbered in consecutive Arabic numerals, starting with 1, independent of the numbering of the sheets and, if possible, in the order in which they appear on the drawing sheet(s). Partial views intended to form one complete view, on one or several sheets, must be identified by the same number followed by a capital letter. View numbers must be preceded by the abbreviation "Fig." Where only a single view is used in an application to illustrate the claimed invention, it must not be numbered and the abbreviation "Fig." must not appear.

   b. Numbers and letters identifying the views must be simple and clear and must not be used in association with brackets, circles, or inverted commas. The view numbers must be larger than the numbers used for reference characters.

22. *Security markings.* Authorized security markings may be placed on the drawings provided they are outside the sight, preferably centered in the top margin.

23. *Corrections.* Any corrections on drawings submitted to the PTO must be durable and permanent.

24. *Holes.* The drawing sheets may be provided with two holes in the top margin. The holes should be equally spaced from the respective side edges, and their center lines should be spaced 7.0 cm (2½ inches) apart.

37 CFR 1.124 Amendments to the drawing. No change in the drawing may be made except with permission of the PTO. Permissible changes in the construction shown in any drawing may be made only by the submission of a substitute drawing by applicant. A sketch in permanent ink showing proposed changes, to become part of the record, must be filed for approval by the examiner and should be a separate paper.

37 CFR 1.152 Design Drawing. The design must be represented by a drawing that complies with the requirement of 37 CFR 1.84, and must contain a sufficient number of views to constitute a complete disclosure of the appearance of the article. Appropriate surface shading must be used to show the character or contour of the surfaces represented. Solid black surface shading is not permitted except when used to represent color contrast. Broken lines may be used to show visible environmental structure, but may not be used to show hidden planes and surfaces which cannot be seen through opaque materials. Alternate positions of a design component, illustrated by full and broken lines in the same view are not permitted in a design drawing. Photographs and ink drawings must not be combined in one application. Photographs submitted in lieu of ink drawings in design patent applications must comply with 37 CFR 1.84(b) and must not disclose environmental structure but must be limited to the design for the article claimed. Color drawings and color photographs are not permitted in design patent applications.

Applications Filed without Drawings

Not all applications require drawings. It has been a long approved procedure, for example, to accept a process case (in other words, a case having only process or method claims), which is filed without drawings.

Other situations where drawings are usually not considered essential for a filing date are:

1. *Coated articles or products.* If the invention resides only in coating or impregnating a conventional sheet, for example, paper or cloth, or an article of known and conventional character with particular composition.

2. *Articles made from a particular material or composition.* If the invention involves making an article of a particular material or composition, unless significant details of structure or arrangement are involved in the claims.

3. *Laminated structures.* If the invention involves only lamination of sheets (and coatings) of specified material, unless significant details of structure or arrangement (other than the mere order of layers) are involved in the claims.

4. *Articles, apparatus, or systems where sole distinguishing feature is presence of a particular material.* If the invention resides solely in the use of particular material in an otherwise old article, apparatus, or system recited broadly in its claim.

Photographs are not normally considered to be proper drawings. Photographs are acceptable for a filing date and are generally considered to be informal drawings. Photographs are only acceptable where they come within special categories. Photolithographs are never acceptable.

The PTO is willing to accept black and white photographs or photomicrographs (not photolithographs or other reproductions of photographs made using screens) printed on sensitized paper in lieu of India ink drawings, to illustrate inventions which are incapable of being accurately or adequately depicted by India ink drawings restricted to the following categories: crystalline structures, metallurgical microstructures, textile fabrics, grain structures, and ornamental effects.

Who Does the Drawing?

You have several options. You can do it yourself. To this end, standards for drawings are included above to guide your efforts and keep you to the letter of the required standards. I have never attempted to do the drawings for a utility patent, but I have been successful at doing a few uncomplicated design patent drawings.

You could ask that your patent attorney arrange for the drawings. This is all right, but understand that just as in the search business, the lawyer or law firm will be adding a premium of typically no less than 40 percent to the fees charged by the draftsman.

The surest and least expensive way, if you are not a draftsman, is to take bids and hire your own draftsman. Find the fair market price by calling around. Look in the telephone directory or ask at a regional patent library for candidates. Get a guarantee for PTO acceptance from the draftsman if you contract the work yourself.

United States Patent [19]

DeLay, Jr.

[11] Patent Number: 4,557,395

[45] Date of Patent: Dec. 10, 1985

[54] **PORTABLE CONTAINER WITH INTERLOCKING FUNNEL**

[75] Inventor: Victor A. DeLay, Jr., Largo, Fla.

[73] Assignee: E-Z Out Container Corp., Clearwater, Fla.

[21] Appl. No.: 717,439

[22] Filed: Mar. 28, 1985

[51] Int. Cl.⁴ B65D 3/04
[52] U.S. Cl. .... 220/86 R; 220/85 F; 220/1 C; 141/98
[58] Field of Search .... 220/86 R, 85 F, 1 C; 141/98, 331; 220/360

[56]   **References Cited**

U.S. PATENT DOCUMENTS

1,554,589	9/1925	Long	220/1 C X
3,410,438	11/1968	Bartz	220/1 C
4,010,863	3/1977	Ebel	220/1 C
4,149,575	4/1979	Fisher	220/85 F X
4,162,020	7/1979	Kirkland	220/1 C X
4,296,838	10/1981	Cohen	220/1 C X
4,301,841	11/1981	Sandow	220/1 C X

Primary Examiner—Steven M. Pollard
Attorney, Agent, or Firm—Stanley M. Miller

[57]   **ABSTRACT**

A portable, vented container for dirty oil, of the type having a small fill spout and having increased utility when used in conjunction with a funnel. A vent closure member and a funnel securing latch are integral with the funnel so that when the funnel is inverted and positioned in surmounting relation to the container, the vent closure member closes the vent and the securing latch is engaged by a fill spout cap which engagement secures the funnel against movement and hence maintains the vent closure as well. Removal of the fill spout cap releases the funnel, and positioning the funnel into its operative position relative to an automotive oil drain plug separates the vent closure portion of the funnel from the vent. An elongate extension member having a flexible medial portion is further provided.

20 Claims, 12 Drawing Figures

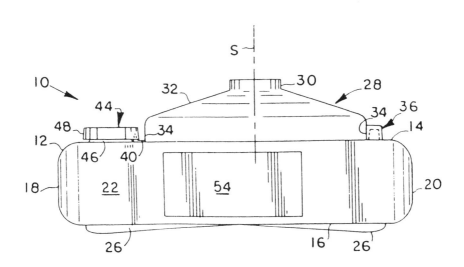

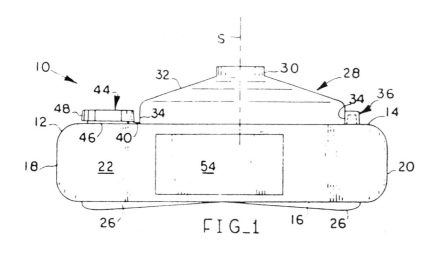

FIG_1

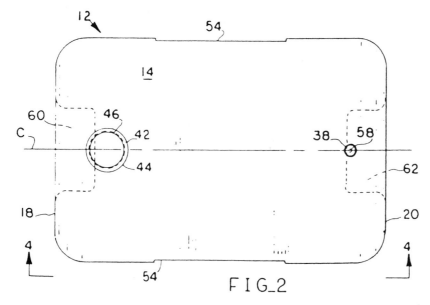

FIG_2

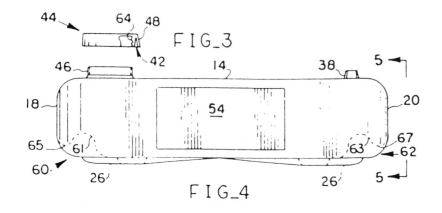

FIG_3

FIG_4

U.S. Patent   Dec. 10, 1985   4,557,395

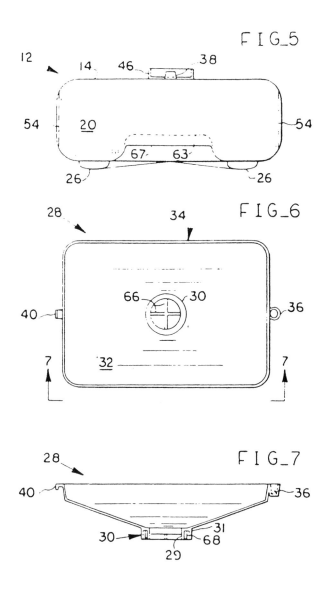

FIG_5

FIG_6

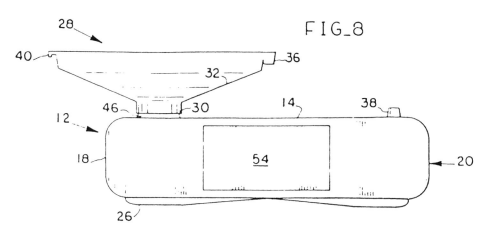

FIG_7

FIG_8

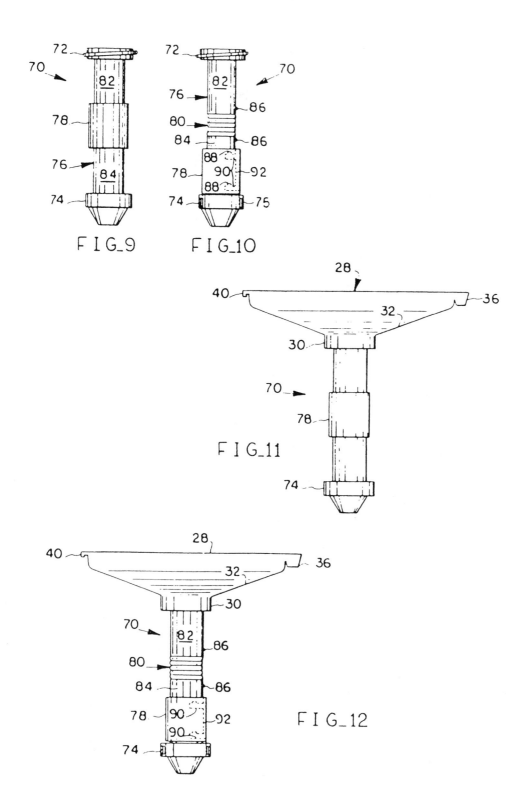

FIG_9

FIG_10

FIG_11

FIG_12

4,557,395

PORTABLE CONTAINER WITH INTERLOCKING FUNNEL

BACKGROUND OF THE INVENTION

1. Field of the Invention

This invention relates generally to containers having small fill spouts, and more particularly this invention relates to a vented container the vent of which is closed when the funnel is stored in latching engagement with the container body.

2. Description of the Prior Art

A thorough description of the prior art in the the field to which this invention pertains may be found in my co-pending application having a filing date of Sept. 14, 1983, Ser. No. 06/531,948. Moreover, the most pertinent prior art is believed to be the container for dirty oil disclosed in said application.

Other patents of interest are: U.S. Pat. Nos. 4,403,692 to Pollacco (1983); 822,854 to Cosgrave (1906); 2,576,154 to Trautvetter (1951); 4,098,393 to Meyers (1978); 4,217,940 to Wheeler and others (1980); and 4,301,841 to Sandow (1981).

Of the known containers, only the container provided by the present inventor and disclosed in the above-identified patent application contains a means whereby the funnel of the container can be conveniently stored when not in use.

Containers having small fill spouts are normally vented to allow the air inside the container to escape as liquid fluids are charged thereinto. Typically, the vent is provided in the form of an upstanding coupling which is provided with a closure member in the form of a cap which may or may not be attached to the coupling itself. Where the cap is attached to the coupling, its loss is safeguarded against but still the user of the container must remember to open and close the vent as needed. Vent caps that are not attached to their couplings are usually lost.

There is a need, therefore, for a vent cap that is safeguarded against loss, and which also opens and closes the vent as needed without requiring the user thereof to remember to open and close such vent.

Another common problem with small-mouthed containers is that the funnels which must be used therewith are often lost. Pollacco solves this problem by permanently securing his funnel to his container. This storage expedient is unsatisfactory because it is important to maintain funnels of the type used to fill automotive crankcases in a substantially clean condition as the introduction of dirt into a crankcase can damage engine parts.

Therefore, there is a need for a funnel storage apparatus capable of storing a funnel in an inverted position when it is not in use. The storage apparatus that is needed would also safeguard against the loss of the funnel.

The art has heretofore developed elongate funnel extension members of the type disclosed by Cosgrave, Trautvetter, and the present inventor, but the same are inflexible and thus inadequate and lacking in utility in certain specific environments.

SUMMARY OF THE INVENTION

The longstanding but heretofore unfulfilled need for a portable container for dirty oil having the desireable features of a self-opening and self-closing vent, a funnel that is storable in an inverted position and which is also secured against loss, is now fulfilled by the invention disclosed hereinafter and summarized as follows.

The container is of parallelepiped form and has finger-receiving recesses formed in its opposite ends, on the underside thereof, which recesses are grasped by an individual when transporting the container.

The top of the container includes a large, imperforate medial portion against which the rim of the funnel is seated when the funnel is in its storage position.

A fill spout of small diameter projects upwardly from the top of the container, and is disposed near the periphery of the container so the medial portion of the container can receive the stored funnel, as aforesaid.

A sleeve member which defines a vent opening projects upwardly from the top of the container as well, but is disposed in longitudinally spaced relation to the fill spout so that it is near the periphery of the container opposite from the fill spout.

The longitudinal axis of symmetry of the container bisects the finger-receiving recesses or handles, the fill spout, the vent-defining sleeve, and the funnel when the latter is in its stored position. In this manner, the container is stable when transported.

The funnel has an integral vent closure member that projects outwardly from the rim of the funnel, in radial relation to the funnel's axis of symmetry. A latch member used to secure the stored funnel against movement is also formed integral to the funnel, extends radially with respect to said axis from the rim thereof, and is positioned in opposition to the vent closure member.

The funnel's size and the amount of space between the fill spout and the vent opening are selected so that when the funnel is inverted and placed in the center of the medial portion of the top wall of the container, and properly rotated about its axis of symmetry, the vent closure member will align with and seal the vent opening and the latch which is opposed to the vent closure member will be positioned in close proximity to the fill spout.

A novel fill spout closure member in the form of a double-walled cap, when brought into screw threaded engagement with the fill spout, will seal the spout and simultaneously overlie the funnel latch to secure the funnel against displacement.

The novel cap's first wall is internally threaded and thus adapted for screw threaded engagement with the externally threaded fill spout. It outer wall defines an annular recess having an open bottom, which recess surrounds the first wall and which receives the funnel latch therewithin. The annular configuration of the recess eliminates any need for aligning the cap with respect to the latch.

In this manner, the act of inverting the funnel and placing it in its storage position on the top wall of the funnel will close the vent if the proper alignment is made. Once the vent has been closed, no further alignment is required as the sealing of the fill spout by the novel cap will also secure the funnel as desired.

Thus, when the funnel is deployed into its operative configuration, the user of the invention need only remove the fill spout cap, as such will release the funnel from its stored position. The act of placing the funnel's spout into the container's fill spout then serves to open the vent.

A funnel extension member having a flexible medial portion is also disclosed hereinafter. A slideably mounted rigid sleeve member serves to delete the flexi-

4,557,395

3

bility function of the extension member when desired when such sleeve member is positioned in registration with the flexible portion of the member. However, the flexibility of the member is restored upon slidingly displacement of the sleeve away from the flexible medial portion.

An important object of this invention, therefore, is to provide a container for dirty oil that includes a funnel as an attachment to the container so that the funnel is not easily misplaced.

Another object is to provide an attachment means that protects the sloping inside walls of the funnel contamination when the funnel is stored.

Another object of this invention is to provide a means whereby the vent of a container can be automatically opened and closed at the time the container's funnel is placed into its operative position and its storage position, respectively.

Other objects will become apparent as this description proceeds.

The invention accordingly comprises the features of construction, combination of elements and arrangement of parts that will be exemplified in the construction hereinafter set forth, and the scope of the invention will be indicated in the claims.

BRIEF DESCRIPTION OF THE DRAWINGS

For a fuller understanding of the nature and objects of the invention, reference should be made to the following detailed description, taken in connection with the accompanying drawings, in which:

FIG. 1 is a side elevational view of the container with the funnel stored in its inverted position thereatop;

FIG. 2 is a top plan view of the container body member;

FIG. 3 is a partially cut away side elevational view of the novel fill spout closure means;

FIG. 4 is a side elevational view taken along line 4—4 of FIG. 2;

FIG. 5 is an end view taken along line 5—5 of FIG. 4;

FIG. 6 is a top plan view of the novel funnel member;

FIG. 7 is a side elevational view of the funnel member taken along line 7—7 of FIG. 6;

FIG. 8 is a side elevational view, like that of FIG. 4, which shows the funnel member engaging the fill spout of the container body;

FIG. 9 is a side elevational view of the novel funnel downspout extension member with the rigid sleeve in its locked position;

FIG. 10 is a side elevational view of the funnel downspout extension member with the rigid sleeve in its unlocked position;

FIG. 11 is a side elevational view showing the extension member operatively coupled to the funnel member with the sleeve in its locked position; and

FIG. 12 is a side elevational view showing the extension member operatively coupled to the funnel member with the sleeve in its unlocked position.

Similar reference numerals refer to similar parts throughout the several views of the drawings.

DETAILED DESCRIPTION OF THE
PREFERRED EMBODIMENT

Referring now to FIG. 1, it will there be seen that an illustrative embodiment of the invention is designated by the reference numeral 10 as a whole. The container body 12 has a parallelepiped construction when seen in

4

perspective. Visible in FIG. 1 are the container's top wall 14, bottom wall 16, its left and right end walls 18, 20, a side wall 22, and support members collectively designated 26.

The novel funnel is indicated generally by the numeral 28. Funnel 28 includes downspout 30, sloping or converging walls 32, and an annular rim 34.

A vent closure member 36 is integrally formed with the rim 34 and extends therefrom as shown. The closure member 36 overlies a vent shroud 38 which is shown in phantom lines in FIG. 1.

A latch 40 is also integrally formed with the funnel rim 34 and is on the opposite side thereof relative to the vent closure member 36. The latch 40 has an "L" shape as shown. The horizontal leg of the latch abuts the top wall 14 of the container 12 and extends radially with respect to the axis of symmetry S of the funnel 28. It terminates in an upstanding leg (shown in phantom lines in FIG. 1) that extends into a cavity 42, which cavity 42 is an annular recess as shown in FIG. 2.

Referring again to FIG. 1, fill spout cap 44 is internally threaded to mate with the external threads of the fill spout 46. The annular latch-receiving recess 42 is formed by the provision of annular wall 48 that surrounds the spout 46, said annular wall depending on the periphery of the top wall of cap 44. The diameter of the top wall of cap 44 is greater than the diameter of the fill spout 46 by an amount substantially equal to the width of the latch-receiving recess 42.

The placement of the upstanding portion of latch 40 in the annular cavity 42 maintains the funnel 28 in its inverted, stored position until the cap 44 is removed.

The space designated 54 in FIG. 1 is a display space and accommodates a label which may have imprinted thereon the trademark of the device and other information.

Returning now to FIG. 2, it will there be seen that the longitudinal axis of symmetry of the device 10 is indicated by the centerline C. It bisects the vent 58 which is formed in the top wall 14 of the container 10 and which is surrounded by vent shroud 38, the fill spout 46, and the longitudinally spaced handles 60, 62 of the invention. The width of the handles 60, 62 is sufficient to accommodate four fingers of a human hand. Both of the label-accommodating recesses 54, 54 mentioned in connection with the description of FIG. 1 are shown in FIG. 2 as well.

The vent closure member 36 slideably and snugly engages the outer walls of the shroud 38, thereby closing the vent opening 58, when funnel 28 is in the inverted storaage position, as aforesaid.

FIG. 3 shows the internal threads 64 on the cap 44 and the annular wall 48 that depends to the periphery of the cap top wall to define the annular cavity 42 into which the upstanding portion of latch 40 extends.

The externally threaded fill spout 46 is shown in FIG. 4, which FIG. shows the container 12 with funnel 28 and cap 44 separated therefrom.

The handles 60, 62 include concave surfaces 61, 63, respectively, and convex surfaces 65, 67, the former of which are abutted by fingertips when the container is carried and the latter of which provide a comfortable rounded weight bearing surface.

An end view of the container 12 is provided in FIG. 5.

A top view of the novel funnel 28 appears in FIG. 6. A strainer 66 formed by a pair of cross bars is formed where the downwardly sloping walls 32 of the funnel 28

4,557,395

5

merge with the funnel's downspout. The generally rectangular planform of the funnel 28 conforms to the planform of the container body 12 as shown in FIG. 2, but the corresponding dimensions of the funnel are smaller.

The downspout 30 of funnel 28 is internally threaded as indicated by the reference numeral 68 appearing in FIG. 7, and is thus adapted for screw threaded engagement with the externally threaded fill spout 46. Accordingly, the downspout 30 of the funnel 28 is coupled to fill spout 46 when it is desired to charge the container with dirty oil. This operative positioning of the funnel 28 and fill spout 46 is depicted in FIG. 8. A comparison of FIGS. 1 and 8 indicates that the removal of cap 44 from spout 46 releases latch 40 so that funnel 28 can be separated from its engagement with top wall 14 of container 12, restored to its upright configuration, and coupled with the spot 46. The separation of the funnel 28 and the container body top wall 14 also separates the vent closure member 36 from vent shroud 38, which separation exposes vent 58 (FIG. 2) to ambient. The internal threads 68 of downspout 30 are formed in outer wall 31 thereof. An inner wall 29 is spaced radially inwardly of outer wall 31, and is concentric therewith. Accordingly, dirty oil contacts inner wall 29 only.

The truncate downspout 30 of funnel 28 is provided because some vehicle are built close to the ground. However, other vehicles are built higher from the ground and the use of a downspout extension member becomes advisable.

An improved downspout extension member is shown in FIGS. 9-12, and is designated 70 as a whole. It includes an externally threaded adapter 72 which is coupled to the internally threaded downspout 30 of funnel 28 when in use, as shown in FIGS. 11 and 12. Another adapter 74 at the lower end of the extension member 70 is internally threaded as at 75 (FIG. 10) to mate with the external threads of the fill spout 46. An elongate medial portion 76 interconnects the upper and lower adapters 72 and 74.

A slideably mounted rigid sleeve member 78 is shown mid-length of the medial portion 76 in FIG. 9. When the sleeve member 78 is locked into this position by means disclosed hereinafter, the novel extension member 70 can be used in the same manner as conventional downspout extension members, which use is depicted in FIG. 11.

However, when the sleeve 78 is unlocked and slideably displaced to its lowermost position, which position is depicted in FIG. 10, such displacement frees a flexible member 80 from confinement so that it is free to bend. More specifically, upper portion 82 of the downspout extension member medial portion 76 and lower portion 84 thereof may be displaced from their axial alignment with each other, i.e., their respective axes of longitudinal symmetry may be made oblique to one another. As shown in FIG. 12, when the flexible member 80 is free, funnel 28 can be moved in any direction relative to lower coupling 74, or vice versa.

FIGS. 10 and 12 both show the means employed to lock and unlock sleeve 78 as desired. A pair of vertically spaced beads, collectively designated 86, are formed on upper and lower portions 82, 84 of the extension member medial portion 76. A pair of vertically spaced bead-receiving cavities, collectively designated 88, are formed internally of sleeve member 78, so that the sleeve 78 is locked into overlying relation to the flexible member 80 when beads 86 are disposed therein.

6

To unlock the sleeve 78, the user of the inventive apparatus grasps sleeve 78 and slides it upwardly by a distance equal to the depth of the bead-receiving cavities 88. Each bead 86 will then be positioned in channels 90 which are also formed internally of sleeve 78. The user of the device then rotates the sleeve 78 until the beads 86 have traveled the length of the arcuate channels 90, which length could be a quarter of an inch, for example. This rotation of sleeve 78 will bring the beads 86 into registration with a vertically extending channel 92 so that the sleeve 78 can be moved to the position shown in FIGS. 10 and 12.

It will thus be seen that the objects set forth above, and those made apparent from the foregoing description, are effectively attained and since certain changes may be made in the above construction without departing from the scope of the invention, it is intended that all matters contained in the foregoing description or shown in the accompanying drawings shall be interpreted as illustrative and not in a limiting sense.

It is also to be understood that the following claims are intended to cover all of the generic and specific features of the invention herein described, and all statements of the scope of the invention which, as a matter of language, might be said to fall therebetween.

Now that the invention has been described,

What is claimed is:

1. A container of the type having a small fill spout and having increased utility when used in conjunction with a funnel, comprising:
   a container body member of generally parallelepiped configuration,
   a fill spout formed in a top wall of said container body member and projecting upwardly therefrom,
   a vent means in the form of an aperture formed in said top wall,
   a funnel member having a rim, converging sidewalls, and a downspout,
   said fill spout and funnel downspout adapted for releasable engagement with one another,
   a vent closure member secured to said funnel rim and projecting outwardly therefrom,
   said vent closure member closing said vent when brought into registration therewith.

2. The container of claim 1, further comprising,
   a fill spout closure means in the form of a cap member,
   a latch member secured to and projecting outwardly from said funnel rim,
   said cap member adapted to releasably engage said latch member when said funnel member is inverted and disposed atop said container top wall and when said cap member is releasably engaged to said fill spout.

3. The container of claim 2, wherein said vent closure member and said latch member are secured to said rim in opposed relation to each other.

4. The container of claim 3, further comprising,
   a sleeve-shaped shroud member disposed in surrounding relation to said aperture and projecting upwardly from said container top wall,
   said vent closure member adapted to engage said shroud member when said funnel is inverted and said vent closure member is brought into releasable engagement with said shroud member.

5. The container of claim 4, further comprising,
   a first handle means formed in said container body member at a first end thereof,

4,557,395

7

a second handle means formed in said container body member at a second end thereof which is longitudinally spaced from said first end,

each of said first and second handle means defined by a concavity formed in the bottom wall of said container body member and by a convexity contiguous thereto and continuous therewith, said convexity merging with an end wall of said container body member.

6. The container of claim 5, wherein the depth of the concavity forming a handle means is greater than the height of the convexity contiguous thereto.

7. The container of claim 5, wherein said first and second handle means are disposed transverse to and are bisected by the longitudinal axis of symmetry of said container body member.

8. The container of claim 3, wherein said cap member has a top wall having a diameter greater than the outer diameter of said fill spout, wherein an annular wall depends to the periphery of said cap top wall, wherein an annular cavity is defined between said fill spout and said depending wall, and wherein said latch member is specifically configured to enter into said annular cavity when brought into registration therewith.

9. The container of claim 8, wherein said latch member has a generally L-shaped configuration.

10. The container of claim 3, wherein said fill spout, said vent and said funnel member, latch member and vent closer member are collectively aligned with the longitudinal axis of symmetry of said container body member when said funnel member is inverted, when said vent closure member is disposed in engaging relation to said vent, and when said latch member is disposed in engaging relation to said fill spout cap.

11. The container of claim 1, wherein a strainer means is positioned within said funnel member at the juncture of said converging sidewalls and said downspout.

12. The container of claim 1, wherein said funnel member has a generally rectangular configuration when seen in plan view, and wherein said latch member and vent closure member are disposed mid-length of the opposite truncate sidewalls of said funnel member.

13. The container of claim 1, wherein said fill spout is externally threaded and wherein said funnel member downspout is internally threaded.

8

14. The container of claim 1, further comprising,

an elongate funnel downspout extension member having a first end adapted to releasably engage said funnel downspout and a second end adapted to releasably engage said fill spout,

and said downspout extension member having a flexible medial portion.

15. The container of claim 14, further comprising,

a rigid sleeve-shaped locking member, having a length greater than the length of said flexible medial portion and having an inside diameter slightly greater than the outside diameter of said downspout extension member, disposed in ensleeving relation to said flexible medial portion and restricting said downspout extension member from flexing at said medial portion.

16. The container of claim 15, further comprising,

means for selectively locking and unlocking said sleeve member into and out of its restricting engagement with said medial portion, respectively.

17. The container of claim 16, wherein said means for selectively locking and unlocking said sleeve member includes a pair of vertically spaced bead members formed on said downspout extension member, one of which is positioned above said flexible medial portion and one of which is positioned below said flexible medial portion, and wherein said sleeve member has a pair of cooperatively spaced bead-receiving cavities formed therein, which cavities are interconnected by a vertical slot and which cavities are formed at the end of associated channels orthogonal to said vertical slot.

18. The container of claim 13, wherein said funnel member downspout further comprises a cylindrical outer wall within which said internal threads are formed, and a cylindrical inner wall spaced radially inwardly of said outer wall so that dirty oil contacts only said inner wall when the container is used.

19. The container of claim 18, wherein said downspout inner wall is concentric with said downspout outer wall.

20. The container of claim 19, wherein the spacing between said downspout outer and inner walls is sufficient to receive therebetween said externally threaded fill spout.

* * * * *

United States Patent

Des. 237,427
Patented Nov. 4, 1975

237,427

SHOE

William H. Thornberry, Newtown, Conn., assignor to
Uniroyal, Inc.

Filed July 26, 1974, Ser. No. 492,307

Term of patent 14 years

Int. Cl. D2—*04*

U.S. Cl. D2—310

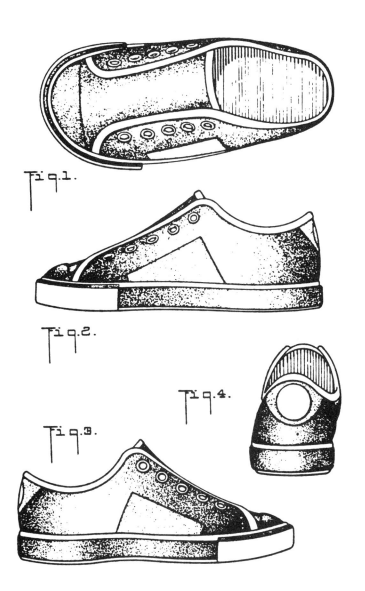

Fig.1.

Fig.2.

Fig.4.

Fig.3.

FIG. 1 is a plan view of a shoe embodying my new design;

FIG. 2 is a side elevational view of the FIG. 1 article;

FIG. 3 is a side elevational view of the FIG. 1 article; and

FIG. 4 is an end elevational view of the FIG. 1 article.

I claim:

The ornamental design for a shoe, substantially as shown and described.

References Cited

UNITED STATES PATENTS

D. 118,131	12/1939	Pick	D2—313
D. 173,699	12/1954	Hosker	D2—310
D. 226,461	3/1973	Nelson	D2—309

LOIS S. LANIER, Primary Examiner

95

Examples of Complete Disclosure

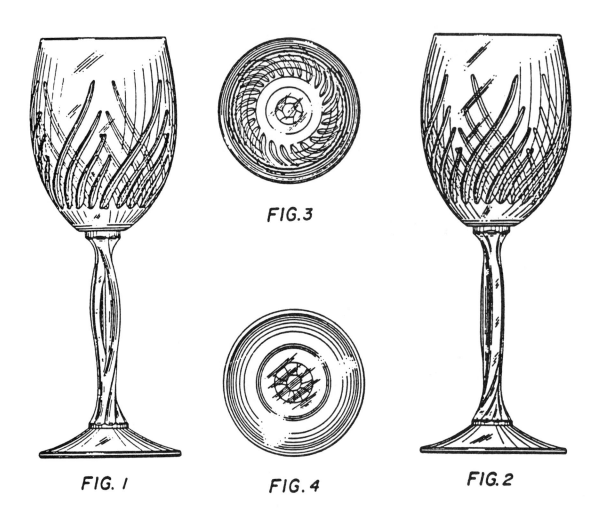

FIG. 3

FIG. 1

FIG. 4

FIG. 2

This is an example only. Formal drawings may not include figure descriptions. Descriptions must be typed on a separate page.

Des. 286,841

FIG. 1 is a front elevational view of a wine goblet showing my new design, the rear elevational view being identical;
FIG. 2 is a right side elevational view thereof, the left side elevational view being identical;
FIG. 3 is a top plan view thereof; and
FIG. 4 is a bottom plan view thereof.

Examples of Shading

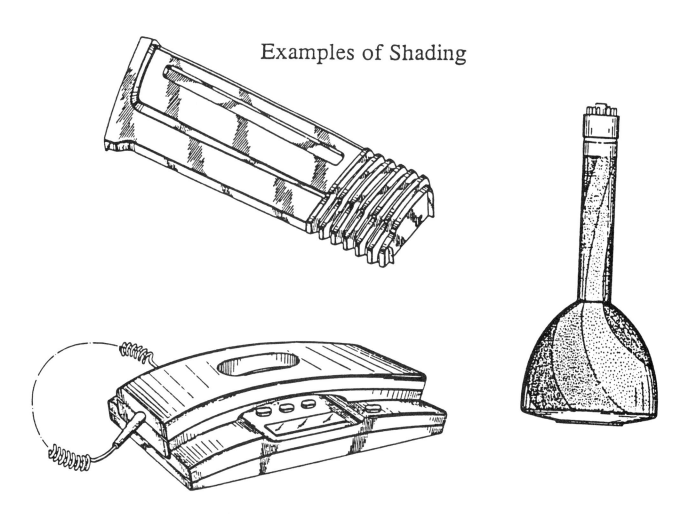

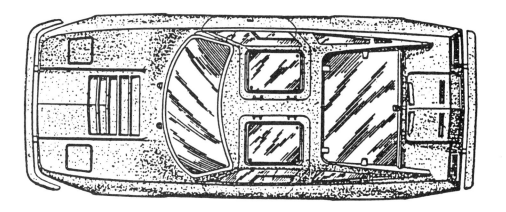

Design Patent Drawing Figures:

Figures in a design drawing must be shaded to show the character or contour of the surfaces represented as required by 37CFR 1.152. Shading is also necessary for compliance with the disclosure requirement of 35USC 112, paragraphs 1 and 2. As illustrated by the above examples, linear or stippled shading may be used.

Examples of Color Illustration

Fig.2.

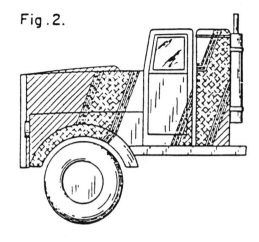

Fig.1.

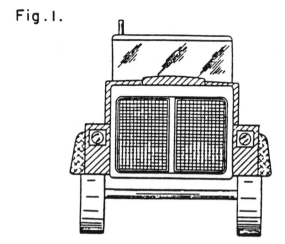

Fig.3.

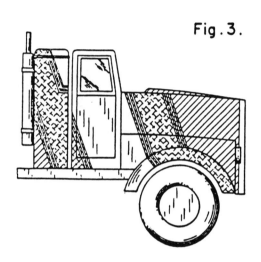

Fig.4.

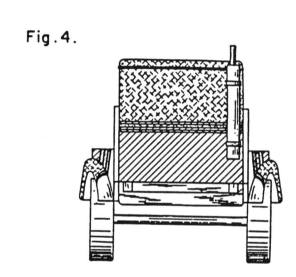

Fig.5.

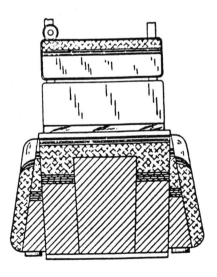

Des. 275,272

FIG. 1 is a front elevational view of a truck tractor showing my new design;
FIG. 2 is a right side elevational view thereof.
FIG. 3 is a left side elevational view thereof.
FIG. 4 is a rear elevational view thereof; and
FIG. 5 is a top plan view thereof.

The characteristic feature of my design rides in the particular pattern of the color scheme on the surface of the truck tractor.

This is an example only. Formal drawings may not include figure descriptions. Descriptions must be typed on a separate page.

UNITED STATES PATENT OFFICE

HENRY F. BOSENBERG, OF NEW BRUNSWICK, NEW JERSEY, ASSIGNOR TO LOUIS C. SCHUBERT, OF NEW BRUNSWICK, NEW JERSEY

CLIMBING OR TRAILING ROSE

Application filed August 6, 1930. Serial No. 473,410.

My invention relates to improvements in roses of the type known as climbing or trailing roses in which the central or main stalks acquire considerable length and when given 5 moderate support "climb" and branch out in various directions.

In roses it is very desirable to have a long period of blooming. This has been acquired in non-climbing roses of the type ordinarily 10 called monthly roses or everblooming roses. My invention now gives the true everblooming character to climbing roses.

The following description and accompanying illustrations apply to my improvements 15 upon the well known variety Dr. Van Fleet, with which my new plant is identical as respects color and form of flower, general climbing qualities, foliage and hardiness, but from which it differs radically in flowering habits 20 —but the same everblooming habits may be attained by breeding this new quality into other varieties of climbing roses.

Figure I shows (1) a flower that is just dropping its petals, (2) a bud about to open, 25 (3) a terminal bud just forming on a large side shoot, and (4) a new shoot which has not yet finished its growth and formed buds at its terminus. This shoot would not appear on the branch illustrated until several weeks 30 later than the stage of development shown, when it would grow out ordinarily from the axil of the first or second leaf below the bloomed-off flower. (5) shows a second way in which new flowering shoots form, by 35 branching off on a short stem immediately or closely adjacent to the blossom that has just finished blooming. Figure II shows a further method of branching and bud formation in cases where the bloom has been 40 cut off, but the formation of new flowering shoots is not dependent upon pruning off the old blossoms. It is evident that this succession of blooms continuously or intermittently supplied by new shoots branching 45 out throughout the summer and fall gives the true everblooming character. When grown in the latitude of New Brunswick, New Jersey, my new climbing rose named "The New Dawn" and illustrated herewith in 50 exact drawings from photographs, provides a

succession of blossoms on a single plant from about the end of May to the middle of November, or until stopped by frost.

No claim is made as to novelty in color or other physical characteristics of the individ- 55 ual blossoms, nor as to the foliage or growing habits of this rose other than as described above.

I claim:

A climbing rose as herein shown and de- 60 scribed, characterized by its everblooming habit.

In testimony whereof I affix my signature hereunto.

HENRY F. BOSENBERG. 65

70

75

80

85

90

95

100

Aug. 18, 1931.   H. F. BOSENBERG   Plant Pat.   1

CLIMBING OR TRAILING ROSE

Filed Aug. 6, 1930

Fig. 1

Fig 2

INVENTOR,

Henry F. Bosenberg.

Per

agent.

Orville M. Kile

PATENT APPLICATION TRANSMITTAL LETTER	Docket Number (Optional)

To the Commissioner of Patents and Trademarks:

Transmitted herewith for filing under 35 U.S.C. 111 and 37 CFR 1.53 is the patent application of

_____

entitled _____

_____

Enclosed are:

☐ _____ pages of written description, claims and abstract.

☐ _____ sheets of drawings.

☐ an assignment of the invention to _____

☐ executed declaration of the inventors.

☐ a certified copy of a _____ application.

☐ associate power of attorney.

☐ a verified statement to establish small entity status under 37 CFR 1.9 and 1.27.

☐ information disclosure statement

☐ preliminary amendment

☐ other: _____ .

CLAIMS AS FILED

	NUMBER FILED	NUMBER EXTRA	RATE	FEE
BASIC FEE			$710	$710
TOTAL CLAIMS	- 20 =	*	x $22	
INDEPENDENT CLAIMS	- 3 =	*	x $74	
MULTIPLE DEPENDENT CLAIM PRESENT			$230	

* NUMBER EXTRA MUST BE ZERO OR LARGER	TOTAL	$
If applicant has small entity status under 37 CFR 1.9 and 1.27, then divide total fee by 2, and enter amount here.	SMALL ENTITY TOTAL	$

☐ A check in the amount of $ _____ to cover the filing fee is enclosed.

☐ The Commissioner is hereby authorized to charge and credit Deposit Account No. _____ as described below. I have enclosed a duplicate copy of this sheet.

　　☐ Charge the amount of $ _____ as filing fee.

　　☐ Credit any overpayment.

　　☐ Charge any additional filing fees required under 37 CFR 1.16 and 1.17.

　　☐ Charge the issue fee set in 37 CFR 1.18 at the mailing of the Notice of Allowance, pursuant to 37 CFR 1.311(b).

_____　　　　_____
　　　Date　　　　　　　　　　　　　Signature

　　　　　　　　　　　　　　　　_____
　　　　　　　　　　　　　　　　Typed or printed name

　　　　　　　　　　　　　　　　_____
　　　　　　　　　　　　　　　　Address

RECORDATION FORM COVER SHEET
PATENTS ONLY

Tab settings ⇨ ⇨ ⇨ ▼   ▼   ▼   ▼   ▼   ▼   ▼

To the Honorable Commissioner of Patents and Trademarks: Please record the attached original documents or copy thereof.

1. Name of conveying party(ies):

Additional name(s) of conveying party(ies) attached? ❑ Yes ❑ No

2. Name and address of receiving party(ies)

Name:_____

Internal Address:_____

_____

_____

Street Address:_____

_____

City:_____ State: _____ ZIP:_____

3. Nature of conveyance:

❑ Assignment   ❑ Merger

❑ Security Agreement   ❑ Change of Name

❑ Other _____

Execution Date: _____

Additional name(s) & address(es) attached? ❑ Yes ❑ No

4. Application number(s) or patent number(s):

If this document is being filed together with a new application, the execution date of the application is: _____

A. Patent Application No.(s)

B. Patent No.(s)

Additional numbers attached? ❑ Yes ❑ No

5. Name and address of party to whom correspondence concerning document should be mailed:

Name:_____

Internal Address:_____

_____

_____

Street Address:_____

_____

City:_____ State: _____ ZIP:_____

6. Total number of applications and patents involved: ▭

7. Total fee (37 CFR 3.41)............$_____

❑ Enclosed

❑ Authorized to be charged to deposit account

8. Deposit account number:

_____

(Attach duplicate copy of this page if paying by deposit account)

DO NOT USE THIS SPACE

9. Statement and signature.
To the best of my knowledge and belief, the foregoing information is true and correct and any attached copy is a true copy of the original document.

_____   _____   _____
Name of Person Signing   Signature   Date

Total number of pages including cover sheet, attachments, and document: ▭

Mail documents to be recorded with required cover sheet information to:
Commissioner of Patents & Trademarks, Box Assignments
Washington, D.C. 20231

Guidelines for Completing Patents Cover Sheets

Cover Sheet information must be submitted with each document to be recorded. If the document to be recorded concerns both patents and trademarks, separate patent and trademark cover sheets, including any attached pages for continuing information, must accompany the document. All pages of the cover sheet should be numbered consecutively, for example, if both a patent and trademark cover sheet is used, and information is continued on one additional page for both patents and trademarks, the pages of the cover sheet would be numbered form 1 to 4.

Item1. Name of Conveying Party(ies).

Enter the full name of the party(ies) conveying the interest. If there is insufficient space, enter a check mark in the "Yes" box to indicate that additional information is attached. The name of the additional conveying party(ies) should be placed on an attached page clearly identified as a continuation of the information Item1. Enter a check mark in the "No" box, if no information is contained on an attached page.

Item 2. Name and Address of Receiving Party(ies).

Enter the name and full address of the first party receiving the interest. If there is more than one party receiving the interest, enter a check mark in the "Yes" box to indicate that additional information is attached. Enter a check mark in the "No" box, if no information is contained on an attached page.

Item 3. Nature of Conveyance.

Place a check mark in the appropriate box describing the nature of the conveying document. If the "Other" box is checked, specify the nature of the conveyance. Enter the execution date of the document. It is preferable to use the name of the month, or an abbreviation of that name, in order that confusion over dates is minimized.

Item 4. Application Number(s) or Patent Number(s).

Indicate the application number(s), and/or patent number(s) against which the document is to be recorded. National application numbers must include both the series code and the serial number; and international application numbers must be complete, e.g., 07/123,456 for national application numbers, and PCTUS91/12345 for international application numbers. Enter a check mark in the appropriate box: "Yes" or "No" if additional numbers appear on attached pages. Be sure to identify numbers included on attached pages as the continuation of Item 4.

Item 5. Name and Address of Party to whom correspondence concerning the document should be mailed.

Enter the name and full address of the party to whom correspondence is to be mailed.

Item 6. Total Applications and Patents Involved.

Enter the total number of applications and patents identified for recordation. Be sure to include all applications and patents identified on the cover sheet and on additional pages.

Block 7. Total Fee Enclosed.

Enter the total fee enclosed or authorized to be charged. A fee is required for each application and patent against which the document is recorded.

Item 8. Deposit Account Number.

Enter the deposit account number to authorize charges. Attach a duplicate copy of cover sheet to be used for the deposit charge account transaction.

Item 9. Statement and Signature.

Enter the name of the person submitting the document. The submitter must sign and date the cover sheet, confirming that to the best of the persons knowledge and belief, the information contained on the cover sheet is correct and that any copy of the document is a true copy of the original document. Enter the total number of pages including the cover sheet, attachments, and document.

Public burden reporting for this sample cover sheet is estimated to average approximately 30 minutes per document to be recorded, including time for reviewing the document and gathering the data needed to complete the sample cover sheet. Send your comments regarding this burden estimate to the U.S. Patent and Trademark Office, Office of Information Systems, PK2-1000C, Washington, D.C. 20231 and to the Office of Management and Budget, Paperwork Reduction Project (065-0011), Washington, D.C. 20503 (DO NOT SEND COMPLETED COVER SHEETS WITH ASSIGNMENTS TO THIS ADDRESS).

DECLARATION FOR PATENT APPLICATION

Docket Number (Optional)

As a below named inventor, I hereby declare that.

My residence, post office address and citizenship are as stated below next to my name.

I believe I am the original, first and sole inventor (if only one name is listed below) or an original, first and joint inventor (if plural names are listed below) of the subject matter which is claimed and for which a patent is sought on the invention entitled _____ , the specification of which

is attached hereto unless the following box is checked:

☐ was filed on _____ as United States Application Number or PCT International Application
Number _____ and was amended on _____ (if applicable).

I hereby state that I have reviewed and understand the contents of the above identified specification, including the claims, as amended by any amendment referred to above.

I acknowledge the duty to disclose information which is material to patentability as defined in Title 37, Code of Federal Regulations, § 1.56.

I hereby claim foreign priority benefits under Title 35, United States Code, § 119 of any foreign application(s) for patent or inventor's certificate listed below and have also identified below any foreign application for patent or inventor's certificate having a filing date before that of the application on which priority is claimed.

Prior Foreign Application(s)

Priority Claimed

(Number)	(Country)	(Day/Month/Year Filed)	☐ Yes ☐ No
(Number)	(Country)	(Day/Month/Year Filed)	☐ Yes ☐ No
(Number)	(Country)	(Day/Month/Year Filed)	☐ Yes ☐ No

I hereby claim the benefit under Title 35, United States Code, § 120 of any United States application(s) listed below and, insofar as the subject matter of each of the claims of this application is not disclosed in the prior United States application in the manner provided by the first paragraph of Title 35, United States Code, § 112, I acknowledge the duty to disclose information which is material to patentability as defined in Title 37, Code of Federal Regulations, § 1.56 which became available between the filing date of the prior application and the national or PCT international filing date of this application.

| (Application Number) | (Filing Date) | (Status – patented, pending, abandoned) |
| (Application Number) | (Filing Date) | (Status – patented, pending, abandoned) |

I hereby appoint the following attorney(s) and/or agent(s) to prosecute this application and to transact all business in the Patent and Trademark Office connected therewith:

Address all telephone calls to _____ at telephone number _____
Address all correspondence to _____

I hereby declare that all statements made herein of my own knowledge are true and that all statements made on information and belief are believed to be true; and further that these statements were made with the knowledge that willful false statements and the like so made are punishable by fine or imprisonment, or both, under Section 1001 of Title 18 of the United States Code and that such willful false statements may jeopardize the validity of the application or any patent issued thereon.

Full name of sole or first inventor (given name, family name) _____
Inventor's signature _____ Date _____
Residence _____ Citizenship _____
Post Office Address _____

Full name of second joint inventor, if any (given name, family name) _____
Second Inventor's signature _____ Date _____
Residence _____ Citizenship _____
Post Office Address _____

☐ Additional inventors are being named on separately numbered sheets attached hereto.

SUPPLEMENTAL DECLARATION FOR PATENT APPLICATION

Docket Number (Optional)

As a below named inventor, I hereby declare that:

My residence, post office address and citizenship are as stated below next to my name.

I believe I am the original, first and sole inventor (if only one name is listed below) or an original, first and joint inventor (if plural names are listed below) of the subject matter which is claimed and for which a patent is sought on the invention entitled

_____ , the specification of which was filed

on _____ as United States Application Number or PCT International Application Number _____ .

I hereby declare that the subject matter of the ☐ attached amendment ☐ amendment filed on _____ , was part of my or our invention and was invented before the filing date of the original application, above identified, for such invention.

I hereby state that I have reviewed and understand the contents of the above identified specification, including the claims, as amended by any amendment referred to above.

I acknowledge the duty to disclose information which is material to patentability as defined in Title 37, Code of Federal Regulations, § 1.56.

I hereby claim foreign priority benefits under Title 35, United States Code, § 119 of any foreign application(s) for patent or inventor's certificate listed below and have also identified below any foreign application for patent or inventor's certificate having a filing date before that of the application on which priority is claimed.

Prior Foreign Application(s)

Priority Claimed

_____ _____ _____ ☐ Yes ☐ No
(Number)   (Country)   (Day/Month/Year Filed)

_____ _____ _____ ☐ Yes ☐ No
(Number)   (Country)   (Day/Month/Year Filed)

I hereby claim the benefit under Title 35, United States Code, § 120 of any United States application(s) listed below and, insofar as the subject matter of each of the claims of this application is not disclosed in the prior United States application in the manner provided by the first paragraph of Title 35, United States Code, § 112, I acknowledge the duty to disclose information which is material to patentability as defined in Title 37, Code of Federal Regulations, § 1.56 which became available between the filing date of the prior application and the national or PCT international filing date of this application.

_____ _____ _____
(Application Number)   (Filing Date)   (Status – patented, pending, abandoned)

_____ _____ _____
(Application Number)   (Filing Date)   (Status – patented, pending, abandoned)

I hereby appoint the following attorney(s) and/or agent(s) to prosecute this application and to transact all business in the Patent and Trademark Office connected therewith:

_____

Address all telephone calls to _____ at telephone number _____

Address all correspondence to _____

_____

_____

I hereby declare that all statements made herein of my own knowledge are true and that all statements made on information and belief are believed to be true; and further that these statements were made with the knowledge that willful false statements and the like so made are punishable by fine or imprisonment, or both, under Section 1001 of Title 18 of the United States Code and that such willful false statements may jeopardize the validity of the application or any patent issued thereon.

Full name of sole or first inventor (given name, family name) _____

Inventor's signature _____ Date _____

Residence _____ Citizenship _____

Post Office Address _____

Full name of second joint inventor, if any (given name, family name) _____

Second Inventor's signature _____ Date _____

Residence _____ Citizenship _____

Post Office Address _____

☐ Additional inventors are being named on separately numbered sheets attached hereto.

(2-92)   Patent and Trademark Office: U.S. DEPARTMENT OF COMMERCE

VERIFIED STATEMENT CLAIMING SMALL ENTITY STATUS (37 CFR 1.9(f) & 1.27(b))--INDEPENDENT INVENTOR

Docket Number (Optional)

Applicant or Patentee: _____

Serial or Patent No.: _____

Filed or Issued: _____

Title: _____

As a below named inventor, I hereby declare that I qualify as an independent inventor as defined in 37 CFR 1.9(c) for purposes of paying reduced fees to the Patent and Trademark Office described in:

☐ the specification filed herewith with title as listed above.

☐ the application identified above.

☐ the patent identified above.

I have not assigned, granted, conveyed or licensed and am under no obligation under contract or law to assign, grant, convey or license, any rights in the invention to any person who would not qualify as an independent inventor under 37 CFR 1.9(c) if that person had made the invention, or to any concern which would not qualify as a small business concern under 37 CFR 1.9(d) or a nonprofit organization under 37 CFR 1.9(e).

Each person, concern or organization to which I have assigned, granted, conveyed, or licensed or am under an obligation under contract or law to assign, grant, convey, or license any rights in the invention is listed below:

☐ No such person, concern, or organization exists.

☐ Each such person, concern or organization is listed below.

Separate verified statements are required from each named person, concern or organization having rights to the invention averring to their status as small entities. (37 CFR 1.27)

I acknowledge the duty to file, in this application or patent, notification of any change in status resulting in loss of entitlement to small entity status prior to paying, or at the time of paying, the earliest of the issue fee or any maintenance fee due after the date on which status as a small entity is no longer appropriate. (37 CFR 1.28(b))

I hereby declare that all statements made herein of my own knowledge are true and that all statements made on information and belief are believed to be true; and further that these statements were made with the knowledge that willful false statements and the like so made are punishable by fine or imprisonment, or both, under section 1001 of Title 18 of the United States Code, and that such willful false statements may jeopardize the validity of the application, any patent issuing thereon, or any patent to which this verified statement is directed.

NAME OF INVENTOR	NAME OF INVENTOR	NAME OF INVENTOR
Signature of inventor	Signature of inventor	Signature of inventor
Date	Date	Date

VERIFIED STATEMENT CLAIMING SMALL ENTITY STATUS (37 CFR 1.9(f) & 1.27(c))--SMALL BUSINESS CONCERN

Docket Number (Optional)

Applicant or Patentee: _____
Serial or Patent No.: _____
Filed or Issued: _____
Title: _____

I hereby declare that I am

☐ the owner of the small business concern identified below:

☐ an official of the small business concern empowered to act on behalf of the concern identified below:

NAME OF SMALL BUSINESS CONCERN _____

ADDRESS OF SMALL BUSINESS CONCERN _____

_____

I hereby declare that the above identified small business concern qualifies as a small business concern as defined in 13 CFR 121.12, and reproduced in 37 CFR 1.9(d), for purposes of paying reduced fees to the United States Patent and Trademark Office, in that the number of employees of the concern, including those of its affiliates, does not exceed 500 persons. For purposes of this statement, (1) the number of employees of the business concern is the average over the previous fiscal year of the concern of the persons employed on a full-time, part-time or temporary basis during each of the pay periods of the fiscal year, and (2) concerns are affiliates of each other when either, directly or indirectly, one concern controls or has the power to control the other, or a third party or parties controls or has the power to control both.

I hereby declare that rights under contract or law have been conveyed to and remain with the small business concern identified above with regard to the invention described in:

☐ the specification filed herewith with title as listed above.

☐ the application identified above.

☐ the patent identified above.

If the rights held by the above identified small business concern are not exclusive, each individual, concern or organization having rights in the invention must file separate verified statements averring to their status as small entities, and no rights to the invention are held by any person, other than the inventor, who would not qualify as an independent inventor under 37 CFR 1.9(c) if that person made the invention, or by any concern which would not qualify as a small business concern under 37 CFR 1.9(d), or a nonprofit organization under 37 CFR 1.9(e).

Each person, concern or organization having any rights in the invention is listed below

☐ no such person, concern, or organization exists.

☐ each such person, concern or organization is listed below.

Separate verified statements are required from each named person, concern or organization having rights to the invention averring to their status as small entities. (37 CFR 1.27)

I acknowledge the duty to file, in this application or patent, notification of any change in status resulting in loss of entitlement to small entity status prior to paying, or at the time of paying, the earliest of the issue fee or any maintenance fee due after the date on which status as a small entity is no longer appropriate. (37 CFR 1.28(b))

I hereby declare that all statements made herein of my own knowledge are true and that all statements made on information and belief are believed to be true; and further that these statements were made with the knowledge that willful false statements and the like so made are punishable by fine or imprisonment, or both, under section 1001 of Title 18 of the United States Code, and that such willful false statements may jeopardize the validity of the application, any patent issuing thereon, or any patent to which this verified statement is directed

NAME OF PERSON SIGNING _____

TITLE OF PERSON IF OTHER THAN OWNER _____

ADDRESS OF PERSON SIGNING _____

SIGNATURE _____ DATE _____

Patent and Trademark Office; U.S. DEPARTMENT OF COMMERC

VERIFIED STATEMENT CLAIMING SMALL ENTITY STATUS
(37 CFR 1.9(f) & 1.27(d))--NONPROFIT ORGANIZATION

Docket Number (Optional)

Applicant or Patentee: _____

Serial or Patent No. : _____

Filed or Issued: _____

Title: _____

I hereby declare that I am an official empowered to act on behalf of the nonprofit organization identified below:

NAME OF NONPROFIT ORGANIZATION _____

ADDRESS OF NONPROFIT ORGANIZATION _____
_____

TYPE OF NONPROFIT ORGANIZATION:

☐ UNIVERSITY OR OTHER INSTITUTION OF HIGHER EDUCATION

☐ TAX EXEMPT UNDER INTERNAL REVENUE SERVICE CODE (26 U.S.C. 501(a) and 501(c)(3))

☐ NONPROFIT SCIENTIFIC OR EDUCATIONAL UNDER STATUTE OF STATE OF THE UNITED STATES OF AMERICA
   (NAME OF STATE _____)
   (CITATION OF STATUTE _____)

☐ WOULD QUALIFY AS TAX EXEMPT UNDER INTERNAL REVENUE SERVICE CODE (26 U.S.C. 501(a) and 501(c)(3)) IF LOCATED IN THE UNITED STATES OF AMERICA

☐ WOULD QUALIFY AS NONPROFIT SCIENTIFIC OR EDUCATIONAL UNDER STATUTE OF STATE OF THE UNITED STATES OF AMERICA IF LOCATED IN THE UNITED STATES OF AMERICA
   (NAME OF STATE _____)
   (CITATION OF STATUTE _____)

I hereby declare that the nonprofit organization identified above qualifies as a nonprofit organization as defined in 37 CFR 1.9(e) for purposes of paying reduced fees to the United States Patent and Trademark Office regarding the invention described in:

☐ the specification filed herewith with title as listed above.
☐ the application identified above.
☐ the patent identified above.

I hereby declare that rights under contract or law have been conveyed to and remain with the nonprofit organization regarding the above identified invention. If the rights held by the nonprofit organization are not exclusive, each individual, concern or organization having rights in the invention must file separate verified statements averring to their status as small entities and that no rights to the invention are held by any person, other than the inventor, who would not qualify as an independent inventor under 37 CFR 1.9(c) if that person made the invention, or by any concern which would not qualify as a small business concern under 37 CFR 1.9(d) or a nonprofit organization under 37 CFR 1.9(e).

Each person, concern or organization having any rights in the invention is listed below:

☐ no such person, concern, or organization exists.
☐ each such person, concern or organization is listed below.

I acknowledge the duty to file, in this application or patent, notification of any change in status resulting in loss of entitlement to small entity status prior to paying, or at the time of paying, the earliest of the issue fee or any maintenance fee due after the date on which status as a small entity is no longer appropriate. (37 CFR 1.28(b))

I hereby declare that all statements made herein of my own knowledge are true and that all statements made on information and belief are believed to be true; and further that these statements were made with the knowledge that willful false statements and the like so made are punishable by fine or imprisonment, or both, under section 1001 of Title 18 of the United States Code, and that such willful false statements may jeopardize the validity of the application, any patent issuing thereon, or any patent to which this verified statement is directed.

NAME OF PERSON SIGNING _____

TITLE IN ORGANIZATION OF PERSON SIGNING _____

ADDRESS OF PERSON SIGNING _____

SIGNATURE _____ DATE _____

Patent and Trademark Office: U.S. DEPARTMENT OF COMMER

Request for Statutory Invention Registration

Application Number _____ , or ☐ attached hereto

Filed: _____

Titled:
_____

Applicant(s): _____

A. In the above identified patent application, I hereby:

1. Request and authorize the Commissioner of Patents and Trademarks to publish the above identified regularly filed patent application as a Statutory Invention Registration. (35 U.S.C. 157)

2. Waive the right to receive a United States patent on the same invention claimed in the above identified patent application. These rights, which are waived, include those specified in 35 U.S.C. 183 and 271 through 289 as well as all attributes specified for patents in any other provisions of law other than title 35, United States Code. The waiver includes, but is not limited to, the remedies under 19 U.S.C. 1337 and 1337a, 22 U.S.C. 2356 and 28 U.S.C. 1498. (35 U.S.C. 157(c))

3. Understand that the above waiver will be effective pursuant to 37 CFR 1.293 upon publication of the Statutory Invention Registration to waive the inventor's right to receive a United States patent on the invention claimed in the Statutory Invention Registration. (37 CFR 1.293(b)(1))

4. State that, in my opinion, the disclosure and claims of the above identified patent application meet the requirements of 35 U.S.C. 112. (37 CFR 1.293(b)(3))

5. State that, in my opinion, the above identified patent application complies with the requirements for printing as set forth in the Rules of Practice for Patent Cases, 37 CFR Part 1. (37 CFR 1.293(b)(4))

6. Enclose the fee set forth in 37 CFR 1.17(n) or (o) for requesting publication of a Statutory Invention Registration:

   ☐ A first Office Action <u>has not been mailed</u> in the above application, 37 CFR 1.17(n) $820.00*

   ☐ A first Office Action <u>has been mailed</u> in the above application, 37 CFR 1.17(o) $1640.00

   Request fee $ _____

   <u>MINUS BASIC FILING FEE, IF PREVIOUSLY PAID</u>
   Basic filing fee for utility patent application set forth in 37 CFR 1.16(a)
   or
   Basic filing fee for design patent application set forth in 37 CFR 1.16(f)
   or
   Basic filing fee for plant patent application set forth in 37 CFR 1.16(g)   Minus basic filing fee $ _____

   Amount due $ _____

   ☐ Amount enclosed by check or money order _____ .
   ☐ Please charge Deposit Account No. _____ the amount of $ _____ .
   ☐ If payment of any additional fee is required for publication of the Statutory Invention Registration, charge such amount to Deposit Account No. _____ .

B. For printing of the Statutory Invention Registration front page, if desired, list below the name(s) of not more than 3 registered patent attorneys and agents OR alternatively, the name of a firm having as a member a registered patent attorney or agent. If no name is listed below, no name will be printed on the Statutory Invention Registration.
_____
_____ .

C. Name of assignee, if any, for printing on the Statutory Invention Registration _____ .
   Address (City and State or Country) _____ .
   State of incorporation, if assignee is a corporation _____ .

* Where this request is submitted at the time the application   _____
  is filed, the filing fee is included in the $720.00 fee.   Signature(s) (37 CFR 1.293(a))
                                                                      ☐ attorney or agent of record   ☐ applicant(s) and any assignee

(2-92)   Patent and Trademark Office; U.S. DEPARTMENT OF COMMERCE

REQUEST FOR FILING A PATENT APPLICATION UNDER 37 CFR 1.60

(2-92)

DOCKET NUMBER	ANTICIPATED CLASSIFICATION OF THIS APPLICATION		PRIOR APPLICATION EXAMINER	ART UNIT
	CLASS	SUBCLASS		

Address to:
Commissioner of Patents and Trademarks
Washington, D.C. 20231

This is a request for filing a ☐ continuation ☐ divisional application under 37 CFR 1.60, of pending prior application Number ___ / _____ , filed on _____ entitled _____

1. Enclosed is a copy of the latest inventor-signed prior application, including a copy of the oath or declaration showing the original signature or an indication it was signed. I hereby verify that the papers are a true copy of the latest signed prior application number ___ / _____ , and further that all statements made herein of my own knowledge are true; and further that these statements were made with the knowledge that willful false statements and the like are made punishable by fine or imprisonment, or both, under section 1001 of Title 18 of the United States Code and that such willful statements may jeopardize the validity of the application or any patent issuing thereon.

CLAIMS	(1) FOR	(2) NUMBER FILED	(3) NUMBER EXTRA	(4) RATE	(5) CALCULATIONS
	TOTAL CLAIMS	- 20 =		x $22.00 =	S
	INDEPENDENT CLAIMS	- 3 =		x $74.00 =	
	MULTIPLE DEPENDENT CLAIMS (if applicable)			+ $230.00 =	
			BASIC FEE		+ 710.00
			Total of above Calculations =		
	Reduction by 50% for filing by small entity (Note 37 CFR 1.9, 1.27, 1.28).				
			TOTAL =		

2. ☐ A verified statement to establish small entity status under 37 CFR 1.9 and 1.27

   ☐ is enclosed.
   ☐ was filed in prior application number ___ / _____ and such status is still proper and desired (37 CFR 1.28(a)).

3. ☐ The Commissioner is hereby authorized to charge any fees which may be required under 37 CFR 1.16 and 1.17, or credit any overpayment to Deposit Account No. _____ . A duplicate copy of this sheet is enclosed.

4. ☐ A check in the amount of $ _____ is enclosed.

5. ☐ Cancel in this application original claims _____ of the prior application before calculating the filing fee. (At least one original independent claim must be retained for filing purposes.)

6. ☐ Amend the specification by inserting before the first line the sentence: "This application is a ☐ continuation ☐ division of application number ___ / _____ , filed _____ , (status, abandoned, pending, etc.)."

7. ☐ Transfer the drawings from the pending prior application to this application and abandon said prior application as of the filing date accorded this application. A duplicate copy of this sheet is enclosed for filing in the prior application. (May only be used if signed by person authorized by 37 CFR 1.138 and before payment of issue fee.)

(REQUEST FOR FILING A PATENT APPLICATION UNDER 37 CFR 1.60, PAGE 2)

8. ☐ New formal drawings are enclosed.

9. ☐ Priority of foreign application number _____, filed on _____ in _____
is claimed under 35 U.S.C. 119.

☐ The certified copy has been filed in prior application number ___ / _____ , filed _____ .

10. ☐ A preliminary amendment is enclosed.

11. ☐ The prior application is assigned of record to _____
_____ .

12. ☐ Also enclosed:

13. ☐ The power of attorney in the prior application is to: _____
_____ .

   a. ☐ The power of attorney appears in the original papers in the prior application.

   b. ☐ Since the power does not appear in the original papers, a copy of the power in the prior
   application is enclosed.

   c. ☐ Address all future correspondence to: (May only be completed by applicant, or attorney
   or agent of record.)

   _____

   _____

   _____

_____   _____
Date   Signature

                          _____
                                          Typed or printed name

☐ Inventor(s)
☐ Assignee of complete interest
☐ Attorney or agent of record
☐ Filed under 37 CFR 1.34(a)
   Registration number if acting under 37 CFR 1.34(a)._____ .

REQUEST FORM FOR FILING A PATENT APPLICATION UNDER 37 CFR 1.62

DOCKET NUMBER	ANTICIPATED CLASSIFICATION OF THIS APPLICATION		PRIOR APPLICATION EXAMINER	ART UNIT
	CLASS	SUBCLASS		

Address to:
Commissioner of Patents and Trademarks
Box FWC
Washington, D.C. 20231

This is a Request for filing a ☐ continuation-in-part, ☐ continuation, ☐ divisional application under 37 CFR 1.62 of prior application Number __ / _____ , filed on _____ entitled _____

_____ by the following named inventor(s):

FULL NAME OF INVENTOR	FAMILY NAME	FIRST GIVEN NAME	SECOND GIVEN NAME
RESIDENCE & CITIZENSHIP	CITY	STATE OR FOREIGN COUNTRY	COUNTRY OF CITIZENSHIP
POST OFFICE ADDRESS	POST OFFICE ADDRESS	CITY	STATE & ZIP CODE / COUNTRY
FULL NAME OF INVENTOR	FAMILY NAME	FIRST GIVEN NAME	SECOND GIVEN NAME
RESIDENCE & CITIZENSHIP	CITY	STATE OR FOREIGN COUNTRY	COUNTRY OF CITIZENSHIP
POST OFFICE ADDRESS	POST OFFICE ADDRESS	CITY	STATE & ZIP CODE / COUNTRY
FULL NAME OF INVENTOR	FAMILY NAME	FIRST GIVEN NAME	SECOND GIVEN NAME
RESIDENCE & CITIZENSHIP	CITY	STATE OR FOREIGN COUNTRY	COUNTRY OF CITIZENSHIP
POST OFFICE ADDRESS	POST OFFICE ADDRESS	CITY	STATE & ZIP CODE / COUNTRY

☐ Additional inventors are being named on separately numbered sheet(s) attached hereto.

The above identified prior application in which no payment of the issue fee, abandonment of, or termination of proceedings has occurred, is hereby expressly abandoned under 37 CFR 1.62(g) as of the filing date of this new application. Please use all the contents of the prior application file wrapper, including the drawings, as the basic papers for the new application. (No new specification is required, 37 CFR 1.62(e).) (note: 37 CFR 1.60 may only be used for continuation or divisional applications where the prior application is not to be abandoned.)

1. ☐ Enter the unentered amendment previously filed on _____ under 37 CFR 1.116 in the prior application.

2. ☐ A preliminary amendment is enclosed.

The filing fee is calculated on the basis of the claims existing in the prior application as amended at 1 and 2 above.

CLAIMS	(1) FOR	(2) NUMBER FILED	(3) NUMBER EXTRA	(4) RATE	(5) CALCULATIONS
	TOTAL CLAIMS	- 20 =		x $22.00 =	$
	INDEPENDENT CLAIMS	- 3 =		x $74.00 =	
	MULTIPLE DEPENDENT CLAIMS (if applicable)			+ $230.00 =	
			BASIC FEE		+ 710.00
			Total of above Calculations =		
	Reduction by 50% for filing by small entity (Note 37 CFR 1.9, 1.27, 1.28).				
			TOTAL =		

(2-92) [Page 1 of 2] Patent and Trademark Office, U.S. DEPARTMENT OF COMMERCE

(REQUEST FORM FOR FILING A PATENT APPLICATION UNDER 37 CFR 1.62, Page 2)

3. ☐ A verified statement to establish small entity status under 37 CFR 1.9 and 1.27
    ☐ is enclosed.
    ☐ was filed in the prior application and such status is still proper and desired (37 CFR 1.28(a)).

4. ☐ The Commissioner is hereby authorized to charge fees under 37 CFR 1.16 and 1.17 which may be required, or credit any overpayment to Deposit Account No. _____ . (A duplicate copy of this form is enclosed.)

5. ☐ A check in the amount of $ _____ is enclosed.

6. ☐ A new oath or declaration in compliance with 37 CFR 1.63 is included since this application is a continuation-in-part which discloses and claims additional matter.

7. ☐ Amend the specification by inserting before the first line the sentence:

This application is a ☐ continuation-in-part, ☐ continuation, ☐ division, of application number ___ / _____ , filed _____ , now abandoned.

8. ☐ Priority of foreign application number _____ , filed on _____ in (country) _____ is claimed under 35 U.S.C. 119.

9. ☐ The prior application is assigned of record to _____
_____ .

10. ☐ The power of attorney in the prior application is to: (name & address) _____
_____ .

11. ☐ Also enclosed:

Address all future correspondence to: (May only be completed by applicant, or attorney or agent of record)

_____

_____

_____

It is understood that secrecy under 35 U.S.C. 122 is hereby waived to the extent that if information or access is available to any one of the applications in the file wrapper of a 37 CFR 1.62 application, be it either this application or a prior application in the same file wrapper, the Patent and Trademark Office may provide similar information or access to all the other applications in the same file wrapper.

Date	Signature
☐ Inventor(s)	Typed or printed name
☐ Assignee of complete interest	
☐ Attorney or agent of record	
☐ Filed under 37 CFR 1.34(a)	

Registration number if acting under 37 CFR 1.34(a). _____ .

(2-92)   [Page 2 of 2]   Patent and Trademark Office, U.S. DEPARTMENT OF COMMERCE

POWER OF ATTORNEY OR AUTHORIZATION OF AGENT, NOT ACCOMPANYING APPLICATION	Docket Number (Optional)

In re Application of	
Application Number	Filed
For	
Group Art Unit	Examiner

To: Commissioner of Patents and Trademarks
   Washington, D.C. 20231

I, the undersigned, having applied for a patent in the application identified above, hereby appoint

_____

_____

_____

of   _____

Registration Number _____ as my/our attorney(s) or agent(s) to prosecute said

application, and to transact all business in the Patent and Trademark Office connected therewith.

I am:

☐   the applicant.

☐   the assignee of record of the entire interest.

              _____
              Signature(s) of all inventor(s) or assignee

              _____
              Typed or printed name(s) of person(s) signing

              _____
              Date

REVOCATION OF POWER OF ATTORNEY OR AUTHORIZATION OF AGENT	Docket Number (Optional)

In re Application of

Application Number	Filed

Group Art Unit	Examiner

To: Commissioner of Patents and Trademarks
Washington, DC 20231

I hereby revoke the power of attorney or authorization of agent given to

_____ in

the above identified application.

I appointed said attorney or agent on the following date: _____ .

The residence of said attorney or agent is _____

_____ in the State of _____ .

I am the:

☐ Applicant.

☐ Assignee of record of the entire interest.

_____   _____
Signature   Date

_____
Typed or printed name

Patent and Trademark Office; U.S. DEPARTMENT OF COMMERCE

DECLARATION FOR PLANT PATENT APPLICATION (35 U.S.C. 161)

Docket Number (Optional)

As a below named inventor, I hereby declare that:

My residence, post office address and citizenship are as stated below next to my name.

I believe I am the original, first and sole inventor (if only one name is listed below) or an original, first and joint inventor (if plural names are listed below) of the new and distinct variety of _____ plant named _____ which is claimed and for which a plant patent is sought, the specification of which is attached hereto unless the following box is checked:

☐ was filed on _____ as United States Application Number _____ and was amended on _____ (if applicable).

I hereby state that I have reviewed and understand the contents of the above identified specification, including the claim, as amended by any amendment referred to above.

I have asexually reproduced the plant to which this application applies.
☐ Said plant was found in a cultivated area. *(Check this box for newly found plant only)*

I acknowledge the duty to disclose information which is material to patentability as defined in Title 37, Code of Federal Regulations, § 1.56.

I hereby claim foreign priority benefits under Title 35, United States Code, § 119 of any foreign application(s) for patent or inventor's certificate listed below and have also identified below any foreign application for patent or inventor's certificate having a filing date before that of the application on which priority is claimed.

Prior Foreign Application(s)

Priority Claimed

☐ Yes ☐ No

_____   _____   _____
(Number)   (Country)   (Day/Month/Year Filed)

☐ Yes ☐ No

_____   _____   _____
(Number)   (Country)   (Day/Month/Year Filed)

I hereby claim the benefit under Title 35, United States Code, § 120 of any United States application(s) listed below and, insofar as the subject matter of the claim of this application is not disclosed in the prior United States application in the manner provided by the first paragraph of Title 35, United States Code, § 112, I acknowledge the duty to disclose information which is material to patentability as defined in Title 37, Code of Federal Regulations, § 1.56 which became available between the filing date of the prior application and the filing date of this application.

_____   _____   _____
(Application Number)   (Filing Date)   (Status – patented, pending, abandoned)

_____   _____   _____
(Application Number)   (Filing Date)   (Status – patented, pending, abandoned)

I hereby appoint the following attorney(s) and/or agent(s) to prosecute this application and to transact all business in the Patent and Trademark Office connected therewith:

_____

Address all telephone calls to _____ at telephone number _____
Address all correspondence to _____
_____
_____
_____

I hereby declare that all statements made herein of my own knowledge are true and that all statements made on information and belief are believed to be true; and further that these statements were made with the knowledge that willful false statements and the like so made are punishable by fine or imprisonment, or both, under Section 1001 of Title 18 of the United States Code and that such willful false statements may jeopardize the validity of the application or any patent issued thereon.

Full name of sole or first inventor (given name, family name) _____
Inventor's signature _____ Date _____
Residence _____ Citizenship _____
Post Office Address _____
_____

Full name of second joint inventor, if any (given name, family name) _____
Second Inventor's signature _____ Date _____
Residence _____ Citizenship _____
Post Office Address _____
_____

☐ Additional inventors are being named on separately numbered sheets attached hereto.

(2-92)   Patent and Trademark Office; U.S. DEPARTMENT OF COMMERCE

Form PTO-1449	Docket Number (Optional)		Application Number
INFORMATION DISCLOSURE CITATION IN AN APPLICATION *(Use several sheets if necessary)*	Applicant		
	Filing Date		Group Art Unit

U. S. PATENT DOCUMENTS

EXAMINER INITIAL	DOCUMENT NUMBER	DATE	NAME	CLASS	SUBCLASS	FILING DATE IF APPROPRIATE

FOREIGN PATENT DOCUMENTS

DOCUMENT NUMBER	DATE	COUNTRY	CLASS	SUBCLASS	Translation YES	Translation NO

OTHER DOCUMENTS   *(Including Author, Title, Date, Pertinent Pages, Etc.)*

EXAMINER	DATE CONSIDERED

EXAMINER: Initial if citation considered, whether or not citation is in conformance with MPEP § 609; Draw line through citation if not in conformance and not considered. Include copy of this form with next communication to the applicant.

Patent and Trademark Office; U.S.DEPARTMENT OF COMMERCE

REQUEST FOR FILING A CONTINUATION OR DIVISION OF AN INTERNATIONAL APPLICATION

DOCKET NUMBER	ANTICIPATED CLASSIFICATION OF THIS APPLICATION		PRIOR APPLICATION EXAMINER	ART UNIT
	CLASS	SUBCLASS		

Address to:
Commissioner of Patents and Trademarks
Washington, D.C. 20231

This is a request for filing a ☐ continuation ☐ divisional application under 37 CFR 1.53, of pending prior international application Number PCT__ / _____ , filed on _____ entitled _____

_____ ,

which designated the United States.

Note: 37 CFR 1.60 and 1.62 cannot be used to file a continuation or divisional application of an international application which has not entered the national stage.

CLAIMS	(1) FOR	(2) NUMBER FILED	(3) NUMBER EXTRA	(4) RATE	(5) CALCULATIONS
	TOTAL CLAIMS	- 20 =		x $22.00 =	$
	INDEPENDENT CLAIMS	- 3 =		x $74.00 =	
	MULTIPLE DEPENDENT CLAIMS (if applicable)			+ $230.00 =	
			BASIC FEE		+ 710.00
			Total of above Calculations =		
	Reduction by 50% for filing by small entity (Note 37 CFR 1.9, 1.27, 1.28).				
			TOTAL =		

1. Enclosed are the specification, claims and drawing(s).

2. ☐ A verified statement to establish small entity status under 37 CFR 1.9 and 1.27 is enclosed.

3. ☐ The Commissioner is hereby authorized to charge any fees which may be required under 37 CFR 1.16 and 1.17, or credit any overpayment to Deposit Account No. _____ . A duplicate copy of this sheet is enclosed.

4. ☐ A check in the amount of $ _____ is enclosed.

5. ☐ Amend the specification by inserting before the first line the sentence: "This application is a ☐ continuation ☐ division of international application number PCT__ / _____ , filed _____ , (status, abandoned, pending, etc.)."

(REQUEST FOR FILING A CONTINUATION OR DIVISION OF AN INTERNATIONAL APPLICATION, PAGE 2)

6. ☐ A declaration under 37 CFR 1.63 is enclosed.

7. ☐ Priority of foreign application number _____, filed on _____ in _____
   is claimed under 35 U.S.C. 119.

   ☐ The certified copy is enclosed.

8. ☐ A preliminary amendment is enclosed.

9. ☐ Also enclosed:

Address all future correspondence to: (May only be completed by applicant, or attorney or agent of record.)

_____

_____

_____

_____   _____
       Date   Signature

                                 _____
                                      Typed or printed name

☐ Inventor(s)
☐ Assignee of complete interest
☐ Attorney or agent of record
☐ Filed under 37 CFR 1.34(a)
   Registration number if acting under 37 CFR 1.34(a)._____.

PETITION FOR EXTENSION OF TIME UNDER 37 CFR 1.136(a)	Docket Number (Optional)

In re Application of	
Application Number	Filed
For	
Group Art Unit	Examiner

This is a request under the provisions of 37 CFR 1.136(a) to extend the period for filing a response in the above identified application.

The requested extension and appropriate non-small-entity fee are as follows (check time period desired):

☐ One month $110.00

☐ Two months $350.00

☐ Three months $810.00

☐ Four months $1,280.00

☐ Applicant is a small entity under 37 CFR 1.9 and 1.27, therefore the fee amount shown above is reduced by one-half, and the resulting fee is: $ _____ .

A verified statement of small entity status as a small entity under 37 CFR 1.27:

☐   is enclosed.

☐   has already been filed in this application.

☐ A check in the amount of the fee is enclosed.

☐ The Commissioner has already been authorized to charge fees in this application to a Deposit Account.

☐ The Commissioner is hereby authorized to charge any fees which may be required, or credit any overpayment, to Deposit Account Number _____ . I have enclosed a duplicate copy of this sheet.

I am the ☐ assignee of record of the entire interest.

☐ applicant

☐ attorney or agent of record.

☐ attorney or agent under 37 CFR 1.34(a).
      Registration number if acting under 37 CFR 1.34(a). _____ .

_____
Date

_____
Signature

_____
Typed or printed name

    Patent and Trademark Office; U.S. DEPARTMENT OF COMMERCE

Patent Worksheet

TRADE MARK (WORKING): _____

TAG LINE:_____

DESCRIPTION:_____

_____

_____

_____

PATENT NOTES:_____

_____

_____LODE #_____

WHO/DATE CONCEIVED:_____

WITNESSED:_____

SKETCH/PHOTO:

NOTES:_____

_____

_____

LICENSES/TIE-INS/SPIN-OFFS/ACCESSORIES/LINE CONCEPTS:___

_____

_____

_____

POTENTIAL MANUFACTURERS:_____

_____

_____

SEEN BY/DATE: _____ _____

_____ _____ _____

_____ _____ _____

U. S. PATENT AND TRADEMARK OFFICE
Effective October 1, 1994

The U. S. Patent and Trademark Office has amended its rules of practice in patent cases, Part 1 of Title 37, Code of Federal Regulations to adjust certain patent fee amounts to reflect fluctuations in the Consumer Price Index (CPI).

Any fee payment due and paid on or after October 1, 1994, must be paid in the revised amount. The date of mailing indicated on a proper Certificate of Mailing or Transmission under 37 CFR 1.8 will be considered to be the date of receipt and payment in the Office.

As this fee sheet is a summary and the content of rules also may be changing, you should refer to the notice published in the *Federal Register* on August 25, 1994 in Volume 59, Number 164, pages 43736 through 43745. See also the *Official Gazette of the United States Patent and Trademark Office* of August 30, 1994.

The fees which are subject to reduction for small entities who have established status (37 CFR 1.27) are shown in a separate column.

For additional information, please contact the PTO Public Service Center at (703) 308-HELP.

Fee Code	37 CFR	Description	Fee	Small Entity Fee if applicable

PATENT FEES

Filing Fees

101 / 201	1.16(a)	Basic filing fee - utility	730.00	365.00
102 / 202	1.16(b)	Independent claims in excess of three	76.00	38.00
103 / 203	1.16(c)	Claims in excess of twenty	22.00	11.00
104 / 204	1.16(d)	Multiple dependent claim	240.00	120.00
105 / 205	1.16(e)	Surcharge- Late filing fee or oath or declaration	130.00	65.00
106 / 206	1.16(f)	Design filing fee	300.00	150.00
107 / 207	1.16(g)	Plant filing fee	490.00	245.00
108 / 208	1.16(h)	Reissue filing fee	730.00	365.00
109 / 209	1.16(i)	Reissue independent claims over original patent	76.00	38.00
110 / 210	1.16(j)	Reissue claims in excess of 20 and over original patent	22.00	11.00
139	1.17(k)	Non-English specification	130.00	

Issue Fees

142 / 242	1.18(a)	Utility issue fee	1,210.00	605.00
143 / 243	1.18(b)	Design issue fee	420.00	210.00
144 / 244	1.18(c)	Plant issue fee	610.00	305.00

Maintenance Fees : Applications Filed on or after December 12, 1980

183 / 283	1.20(e)	Due at 3.5 years	960.00	480.00
184 / 284	1.20(f)	Due at 7.5 years	1,930.00	965.00
185 / 285	1.20(g)	Due at 11.5 years	2,900.00	1,450.00
186 / 286	1.20(h)	Surcharge - Late payment within 6 months	130.00	65.00
187	1.20(i)(1)	Surcharge after expiration - Late payment is unavoidable	640.00	
188	1.20(i)(2)	Surcharge after expiration - Late payment is unintentional	1,500.00	

Miscellaneous Fees

111	1.20(j)	Extension of term of patent	1,030.00	
112	1.17(n)	Requesting publication of SIR - Prior to examiner's action	840.00*	
113	1.17(o)	Requesting publication of SIR - After examiner's action	1,690.00*	
145	1.20(a)	Certificate of correction	100.00	
147	1.20(c)	For filing a request for reexamination	2,320.00	
148 / 248	1.20(d)	Statutory Disclaimer	110.00	55.00

*Reduced by Basic Filing Fee Paid

REMITTANCES FROM FOREIGN COUNTRIES MUST BE PAYABLE AND IMMEDIATELY NEGOTIABLE IN THE UNITED STATES FOR THE FULL AMOUNT OF THE FEE REQUIRED

Fee Code	37 CFR	Description	Fee	Small EntityFee if applicable

Extension Fees

Fee Code	37 CFR	Description	Fee	Small Entity Fee
115 / 215	1.17(a)	Extension for response within first month	110.00	55.00
116 / 216	1.17(b)	Extension for response within second month	370.00	185.00
117 / 217	1.17(c)	Extension for response within third month	870.00	435.00
118 / 218	1.17(d)	Extension for response within fourth month	1,360.00	680.00

Appeals/Interference Fees

119 / 219	1.17(e)	Notice of appeal	280.00	140.00
120 / 220	1.17(f)	Filing a brief in support of an appeal	280.00	140.00
121 / 221	1.17(g)	Request for oral hearing	240.00	120.00

Patent Petition Fees

122		Petitions to the Commissioner, unless otherwise specified	130.00	
126	1.17(p)	Submission of an Information Disclosure Statement (§1.97(c))	210.00	
138	1.17(j)	Petition to institute a public use proceeding	1,390.00	
140 / 240	1.17(l)	Petition to revive unavoidably abandoned application	110.00	55.00
141 / 241	1.17(m)	Petition to revive unintentionally abandoned application	1,210.00	605.00

PCT Fees - National Stage

154 / 254	1.492(e)	Surcharge - Late filing fee or oath or declaration	130.00	65.00
156	1.492(f)	English translation - after twenty months	130.00	
956 / 957	1.492(a)(1)	IPEA - U.S.	660.00	330.00
958 / 959	1.492(a)(2)	ISA - U.S.	730.00	365.00
960 / 961	1.492(a)(3)	PTO not ISA or IPEA	980.00	490.00
962 / 963	1.492(a)(4)	Claims meet PCT Article 33(1)-(4) - IPEA - U.S.	92.00	46.00
964 / 965	1.492(b)	Claims - extra independent (over three)	76.00	38.00
966 / 967	1.492(c)	Claims - extra total (over twenty)	22.00	11.00
968 / 969	1.492(d)	Claims - multiple dependent	240.00	120.00
970 / 971	1.492(a)(5)	For filing with EPO or JPO search report	850.00	425.00

PCT Fees - International Stage

150	1.445(a)(1)	Transmittal fee	210.00	
151	1.445(a)(2)	PCT search fee - no U.S. application	640.00	
152	1.445(a)(3)	Supplemental search per additional invention	180.00	
153	1.445(a)(2)	PCT search- prior U.S. application	420.00	
190	1.482(a)(1)	Preliminary examination fee - ISA was the U.S.	460.00	
191	1.482(a)(1)	Preliminary examination fee - ISA not the U.S.	690.00	
192	1.482(a)(2)	Additional invention - ISA was the U.S.	140.00	
193	1.482(a)(2)	Additional invention - ISA not the U.S.	240.00	

PCT Fees to WIPO

800		Basic fee (first thirty pages)	530.00*	
801		Basic supplemental fee (for each page over thirty)	10.00*	
803		Handling fee	162.00*	
805 - 898		Designation fee per country	128.00*	

PCT Fee to EPO

802		International search	1,537.00*	

*WIPO and EPO fees subject to periodic change due to fluctuations in exchange rate. Refer to Patent Official Gazette for current amounts.

REMITTANCES FROM FOREIGN COUNTRIES MUST BE PAYABLE AND IMMEDIATELY NEGOTIABLE IN THE UNITED STATES FOR THE FULL AMOUNT OF THE FEE REQUIRED

Fee Code	37 CFR	Description	Fee

Patent Service Fees

Fee Code	37 CFR	Description	Fee
561	1.19(a)(1)(i)	Printed copy of patent w/o color, regular service	3.00
562	1.19(a)(1)(ii)	Printed copy of patent w/o color, expedited local service	6.00
563	1.19(a)(1)(iii)	Printed copy of patent w/o color, ordered via EOS, expedited service	25.00
564	1.19(a)(2)	Printed copy of plant patent, in color	12.00
565	1.19(a)(3)	Copy of utility patent or SIR, with color drawings	24.00
566	1.19(b)(1)(i)	Certified or uncertified copy of patent application as filed, regular service	12.00
567	1.19(b)(1)(ii)	Certified or uncertified copy of patent application, expedited local service	24.00
568	1.19(b)(2)	Certified or uncertified copy of patent-related file wrapper and contents	150.00
569	1.19(b)(3)	Certified or uncertified copy of document, unless otherwise provided	25.00
570	1.19(b)(4)	For assignment records, abstract of title and certification, per patent	25.00
571	1.19(c)	Library Service	50.00
572	1.19(d)	List of U.S. patents and SIRs in subclass	3.00
573	1.19(e)	Uncertified statement re status of maintenance fee payments	10.00
574	1.19(f)	Copy of non-U.S. document	25.00
575	1.19(g)	Comparing and Certifying Copies, Per Document, Per Copy	25.00
576	1.19(h)	Additional filing receipt, duplicate or corrected due to applicant error	25.00
577	1.21(c)	Filing a Disclosure Document	10.00
578	1.21(d)	Local delivery box rental, per annum	50.00
579	1.21(e)	International type search report	40.00
580	1.21(g)	Self-service copy charge, per page	0.25
581	1.21(h)	Recording each patent assignment, agreement or other paper, per property	40.00
583	1.21(i)	Publication in Official Gazette	25.00
584	1.21(j)	Labor charges for services, per hour or fraction thereof	30.00
585	1.21(k)	Unspecified other services	AT COST
586	1.21(l)	Retaining abandoned application	130.00
587	1.21(n)	Handling fee for incomplete or improper application	130.00
588	1.21(o)	APS-Text terminal session time, per hour	40.00
591	1.21(k)	APS-Text terminal session time at the PTDLs	ANNUAL SUBSCRIPTION
592	1.21(k)	APS-CSIR terminal session time, per hour	50.00
590	1.24	Patent coupons	3.00
589	1.296	Handling fee for withdrawal of SIR	130.00

Patent Enrollment Fees

Fee Code	37 CFR	Description	Fee
609	1.21(a)(1)	Admission to examination	300.00
610	1.21(a)(2)	Registration to practice	100.00
611	1.21(a)(3)	Reinstatement to practice	15.00
612	1.21(a)(4)	Copy of certificate of good standing	10.00
613	1.21(a)(4)	Certificate of good standing - suitable for framing	20.00
615	1.21(a)(5)	Review of decision of Director, Office of Enrollment and Discipline	130.00
616	1.21(a)(6)	Regrading of Examination	130.00

GENERAL FEES

Finance Service Fees

Fee Code	37 CFR	Description	Fee
607	1.21(b-1)	Establish deposit account	10.00
608	1.21(b)(2)	Service charge for below minimum balance	25.00
608	1.21(b)(3)	Service charge for below minimum balance restricted subscription deposit account	25.00
617	1.21(m)	Processing returned checks	50.00

Computer Service Fees

Fee Code	37 CFR	Description	Fee
618		Computer records	AT COST

REMITTANCES FROM FOREIGN COUNTRIES MUST BE PAYABLE AND IMMEDIATELY NEGOTIABLE IN THE UNITED STATES FOR THE FULL AMOUNT OF THE FEE REQUIRED

Fee Code	37 CFR	Description	Fee

TRADEMARK FEES

Trademark Processing Fees

Fee Code	37 CFR	Description	Fee
361	2.6(a)(1)	Application for registration, per class	245.00
362	2.6(a)(2)	Filing an Amendment to Allege Use under § 1(c), per class	100.00
363	2.6(a)(3)	Filing a Statement of Use under § 1(d)(1), per class	100.00
364	2.6(a)(4)	Filing a Request for a Six-month Extension of Time for Filing a Statement of Use under § 1(d)(1), per class	100.00
365	2.6(a)(5)	Application for renewal, per class	300.00
366	2.6(a)(6)	Additional fee for late renewal, per class	100.00
367	2.6(a)(7)	Publication of mark under §12(c), per class	100.00
368	2.6(a)(8)	Issuing new certificate of registration	100.00
369	2.6(a)(9)	Certificate of Correction, registrant's error	100.00
370	2.6(a)(10)	Filing disclaimer to registration	100.00
371	2.6(a)(11)	Filing amendment to registration	100.00
372	2.6(a)(12)	Filing § 8 affidavit, per class	100.00
373	2.6(a)(13)	Filing § 15 affidavit, per class	100.00
374	2.6(a)(14)	Filing combined §§ 8 & 15 affidavit, per class	200.00
375	2.6(a)(15)	Petition to the Commissioner	100.00
376	2.6(a)(16)	Petition for cancellation, per class	200.00
377	2.6(a)(17)	Notice of opposition, per class	200.00
378	2.6(a)(18)	Ex parte appeal, per class	100.00
379	2.6(a)(19)	Dividing an application, per new application (file wrapper) created	100.00

Trademark Service Fees

Fee Code	37 CFR	Description	Fee
461	2.6(b)(1)(i)	Printed copy of each registered mark, regular service	3.00
462	2.6(b)(1)(ii)	Printed copy of each registered mark, expedited local service	6.00
463	2.6(b)(1)(iii)	Printed copy of each registered mark ordered via EOS, expedited service	25.00
464	2.6(b)(4)(i)	Certified copy of registered mark, with title and/or status, regular service	10.00
465	2.6(b)(4)(ii)	Certified copy of registered mark, with title and/or status, expedited local service	20.00
466	2.6(b)(2)(i)	Certified or uncertified copy of trademark application as filed, regular service	12.00
467	2.6(b)(2)(ii)	Certified or uncertified copy of trademark application as filed, expedited local service	24.00
468	2.6(b)(3)	Certified or uncertified copy of trademark-related file wrapper and contents	50.00
469	2.6(b)(5)	Certified or uncertified copy of trademark document, unless otherwise provided	25.00
470	2.6(b)(7)	For assignment records, abstracts of title and certification per registration	25.00
475	1.19(g)	Comparing and certifying copies, per document, per copy	25.00
480	2.6(b)(9)	Self-service copy charge, per page	0.25
481	2.6(b)(6)	Recording trademark assignment, agreement or other paper, first mark per document	40.00
482	2.6(b)(6)	For second and subsequent marks in the same document	25.00
484	2.6(b)(10)	Labor charges for services, per hour or fraction thereof	30.00
485	2.6(b)(11)	Unspecified other services	AT COST
488	2.6(b)(8)	Each hour of T-SEARCH terminal session time	40.00
490	1.24	Trademark coupons	3.00

REMITTANCES FROM FOREIGN COUNTRIES MUST BE PAYABLE AND IMMEDIATELY NEGOTIABLE IN THE UNITED STATES FOR THE FULL AMOUNT OF THE FEE REQUIRED

PATENT APPLICATION FEE DETERMINATION RECORD

Application or Docket Number

CLAIMS AS FILED - PART I

FOR	NUMBER FILED (Column 1)	NUMBER EXTRA (Column 2)	SMALL ENTITY RATE	SMALL ENTITY FEE	OR	OTHER THAN SMALL ENTITY RATE	OTHER THAN SMALL ENTITY FEE
BASIC FEE					OR		
TOTAL CLAIMS	minus 20 =	*	x =		OR x =		
INDEPENDENT CLAIMS	minus 3 =	*	x =		OR x =		
MULTIPLE DEPENDENT CLAIM PRESENT			x		OR x		
			TOTAL		OR TOTAL		

* If the difference in column 1 is less than zero, enter "0" in column 2

CLAIMS AS AMENDED - PART II

AMENDMENT A

	CLAIMS REMAINING AFTER AMENDMENT (Column 1)	HIGHEST NUMBER PREVIOUSLY PAID FOR (Column 2)	PRESENT EXTRA (Column 3)	SMALL ENTITY RATE	SMALL ENTITY ADDITIONAL FEE	OR	OTHER THAN SMALL ENTITY RATE	OTHER THAN SMALL ENTITY ADDITIONAL FEE
Total	*	Minus **	=	x =		OR x =		
Independent	*	Minus ***	=	x =		OR x =		
FIRST PRESENTATION OF MULTIPLE DEPENDENT CLAIM				x		OR x		
				TOTAL ADDIT. FEE		OR	TOTAL ADDIT. FEE	

AMENDMENT B

	CLAIMS REMAINING AFTER AMENDMENT (Column 1)	HIGHEST NUMBER PREVIOUSLY PAID FOR (Column 2)	PRESENT EXTRA (Column 3)	RATE	ADDITIONAL FEE	OR	RATE	ADDITIONAL FEE
Total	*	Minus **	=	x =		OR x =		
Independent	*	Minus ***	=	x =		OR x =		
FIRST PRESENTATION OF MULTIPLE DEPENDENT CLAIM				x		OR x		
				TOTAL ADDIT. FEE		OR	TOTAL ADDIT. FEE	

AMENDMENT C

	CLAIMS REMAINING AFTER AMENDMENT (Column 1)	HIGHEST NUMBER PREVIOUSLY PAID FOR (Column 2)	PRESENT EXTRA (Column 3)	RATE	ADDITIONAL FEE	OR	RATE	ADDITIONAL FEE
Total	*	Minus **	=	x =		OR x =		
Independent	*	Minus ***	=	x =		OR x =		
FIRST PRESENTATION OF MULTIPLE DEPENDENT CLAIM				x		OR x		
				TOTAL ADDIT. FEE		OR	TOTAL ADDIT. FEE	

* If the entry in column 1 is less than the entry in column 2, write "0" in column 3.
** If the "Highest Number Previously Paid For" IN THIS SPACE is less than 20, enter "20".
*** If the "Highest Number Previously Paid For" IN THIS SPACE is less than 3, enter "3".
The "Highest Number Previously Paid For" (Total or Independent) is the highest number found in the appropriate box in column 1.

Patent and Trademark Office: U.S. DEPARTMENT OF COMMERCE

MAINTENANCE FEE TRANSMITTAL FORM

Address to:
Commissioner of Patents and Trademarks
Box M Fee
Washington, D.C. 20231

I hereby certify that this correspondence is being deposited with the United States Postal Service as first class mail in an envelope address to "Commissioner of Patents and Trademarks, Box M Fee, Washington, D.C. 20231" on _____ .

Signature _____ .

Typed or printed name _____ .

Enclosed herewith is the payment of the maintenance fee(s) for the listed patent(s).

1. ☐ A check for the amount of $ _____ for the full payment of the maintenance fee(s) and any necessary surcharge on the following patents is enclosed.
2. ☐ The Commissioner is hereby authorized to charge $ _____ to cover the payment of the fee(s) indicated below to Deposit Account No. _____ .
3. ☐ The Commissioner is hereby authorized to charge any deficiency in the payment of the required fee(s) or credit any overpayment to Deposit Account No. _____ .

* Information required by 37 CFR 1.366(c)(columns 1 & 5). Information requested under 37 CFR 1.366(d) (columns 2-4 & 6-9)

Item	Patent Number* 1	Fee Code (see below) 2	Maintenance Fee Amount (37 CFR 1.20) 3	Surcharge Amount 4	U.S. Application Number* [06/555,555] 5	Patent Date mm/dd/yy 6	Application Filing Date mm/dd/yy 7	Payment Year 8	Small Entity? 9
1									
2									
3									
4									
5									
6									
7									
8									

Sub-totals __ Columns 3 & 4

Total Payment   Use additional sheets for listing additional patents.

Maintenance Fee Codes:
183 (283 for small entity)....................Due at 3.5 years
184 (284 for small entity)....................Due at 7.5 years
185 (285 for small entity)....................Due at 11.5 years
186 (286 for small entity)....................Surcharge - Late payment within 6 months
187...Surcharge after expiration

[For Office Accounting Use Only]

Respectfully submitted**:   PAYOR'S NUMBER (if assigned) _____

(Payor's name):   FEE ADDRESS _____

_____

(Payor's Signature): _____

Note: All correspondence will be forwarded to the "Fee Address" or to the "Correspondence Address" if no "Fee Address" has been provided. 37 CFR 1.363.

**WHERE MAINTENANCE FEE PAYMENTS ARE TO BE MADE BY AUTHORIZATION TO CHARGE A DEPOSIT ACCOUNT, PAYOR'S NAME AND SIGNATURE SHOULD BOTH APPEAR IN THE BOTTOM LEFT CORNER OF THIS FORM.

(2-92)   Patent and Trademark Office; U.S. DEPARTMENT OF COMMERCE

127

"FEE ADDRESS" INDICATION FORM

Address to:
Commissioner of Patents and Trademarks
Box M. Fee
Washington, D.C. 20231

Please recognize as the "Fee Address" under the provisions of 37 CFR 1.363 the following address:

Payor Number if assigned _____   Payor's Telephone Number _____

in the following listed application(s) or patent(s) for which the Issue Fee has been paid.

PATENT NUMBER (if known)	APPLICATION NUMBER	PATENT DATE (if known)	U.S. FILING DATE

(check one)
☐ Patentee
☐ Owner of record
☐ Owner's attorney or agent of record _____
                                    (Reg. No.)
☐ Assignment recorded at Reel _____ Frame _____

_____
Signature

_____
Typed or printed name

_____
Date

Address of signer:
_____
_____
_____

Patent and Trademark Office, U.S.DEPARTMENT OF COMMERCE

Request for Payor Number

Address to:
Commissioner of Patents and Trademarks
Box M. Fee
Washington, DC 20231

To the Commissioner of Patents and Trademarks:

Please assign a Payor Number to the Fee Address indicated below:
(May be completed by the individual or organization which will be responsible for paying maintenance fees.)

Fee Address:

Payor's name: _____

Payor's Telephone number: _____

Name of person signing request: _____

Respectfully submitted _____
                          (Signature)   (Date)

(Note: Any Patent and Trademark Office notices relating to maintenance fees in a particular patent will be mailed to the "Fee Address" set forth in 37 CFR 1.363 of record in that particular patent. The entry of a "Fee Address" in a particular patent, or application in which the issue fee has been paid must be requested by filing a paper identifying the patent or application and signed by the owner of record or attorney or agent of record.)

(Also form PTO-1550)

Patent and Trademark Office; U.S. DEPARTMENT OF COMMERCE

8

What Happens to Your Application at the PTO?

Reports Don Coster of the Nevada Inventor Association after a visit to the PTO: "Our patent applications go through a process that is so thorough and so efficient that it is hard to believe unless you see it in action. The application does not go directly to an examiner. It must first be examined for content and completeness. The drawings are checked and screened for things like military sensitivity or unlawful usage. Once accepted as legal and complete, [the application] is classified for the proper art group. This is very critical. If the wrong examiner ends up with it on his or her table, it might be months before he or she even gets a first look at it, because applications are taken in the order that they're received.... Those people are so conscientious that it rarely ever happens."

Many of life's failures are people who did not realize how close they were to success when they gave up.

—THOMAS EDISON

If your application passes initial muster, it will be assigned to the appropriate examining group and then to an examiner. Applications are handled in the order received.

The application examination inspects for compliance with the legal requirements and includes a search through U.S. patents, prior foreign patent documents which are available in the PTO, and available literature to ensure that the invention is new. A decision is reached by the examiner in light of the study and the result of the search.

First Office Action

You or your attorney will be notified of the examiner's decision by what the PTO refers to as an "action." An action is actually a letter that gives the reasons for any adverse action or any objection or requirement. Noted will be any appropriate references or information that you'll find useful in making the decision to continue the prosecution of the application or to drop it.

If the invention is not considered patentable subject matter, the claims will be rejected. If the examiner finds that the invention is not new, the claims will be rejected; but the claims may also be rejected if they depict an object that is found to be obvious. It is not uncommon for some or all of the claims to be rejected on the examiner's first action; very few applications sail through as first submitted.

Your First Response

Let's say the examiner gives you the thumbs down on all or some of your claims. Your next move, if you wish to continue seeking the patent, is to respond, specifically pointing out the supposed errors in the examiner's action. Patent examiners have a lot on their plates and their units are typically understaffed for the amount of work they handle.

Examiners must process a specific number of patents to be considered productive by their superiors for periodic job performance ratings. The bottom line is that as careful as they try to be, they make mistakes that can be reversed with careful argument.

Your response should address every ground of the objection and/or rejection. Show where the examiner is wrong. The mere allegation that the examiner has erred is not enough.

Your response will cause the examiner to reconsider, and you'll be notified if the claims are rejected, or objections or requirements made, in the same manner as after the first examination. This second action usually will be the final one.

Feel free to call your examiner up on the telephone to discuss your case. I have always found them to be most hospitable and helpful. His or her telephone number will apear at the end of the PTO action or you can look it up in the PTO Information Contacts telephone directory (see Appendix 2).

Depending upon how serious the matter, you or your attorney might wish to make an appointment to personally visit the examiner. Don't just drop in unannounced. It is to your benefit that the examiner has the time to prepare for your visit and get up to speed on the case. Remember that personal interviews do not remove the necessity for response to PTO actions within the required time, and the action of the PTO is based solely on the written record.

Final Rejection

On the second or later consideration, the rejection of claims may be made final. Your response is then limited to appeal and further amendment is restricted. You may petition the commissioner in the case of objections or requirements not involved in the rejection of any claim. Response to a final rejection must include cancellation of, or appeal from, the rejection of each claim so rejected and, if any claim stands allowed, compliance with any requirement or objection as to its form.

In determining such final rejection, your examiner will repeat or state all grounds of rejection then considered applicable to your claims as stated in the application.

The odds? As in the case of the examination by the PTO, patents are granted in about two of every three applications filed.

Making Amendments to Your Application

The preceding section referred to amendments to an application. Following are some details concerning amendments:

1. The applicant may amend before or after the first examination and action as specified in the rules, or when and as specifically required by the examiner.

2. After final rejection or action, amendments may be made canceling claims or complying with any requirement of form which has been made, but the admission of any such amendment or its refusal, and any proceedings relative thereto, shall not operate to relieve the application from its condition as subject to appeal or to save it from abandonment.

3. If amendments touching the merits of the application are presented after final rejection, or after appeal has been taken, or when such amendment might not otherwise be proper, they may be admitted upon a showing of good and sufficient reasons why they are necessary and were not earlier presented.

4. No amendment can be made as a matter of right in appealed cases. After decision on appeal, amendments can only be made as provided in the rules.

Erasures, additions, insertions, or alterations of the papers and records in a patent application must not be made by the applicant. Amendments are made by filing a paper directing or requesting that specified changes or additions be made. The exact word or words to be stricken out or inserted in the application must be specified and the precise point indicated where the deletion or insertion is to be made.

Amendments are "entered" by the PTO through the making of proposed deletions by drawing a line in red ink through the word or words canceled and by making the proposed substitutions or insertions in red ink, small insertions being written in at the designated place and larger insertions being indicated by reference.

5. The specifications, claims, and drawing must be amended and revised when required to correct inaccuracies of description and definition of unnecessary words and to secure correspondence between the claims, the description, and the drawing.

All amendments of the drawings or specifications, and all additions thereto, must conform to at least one of them as it was at the time of the filing of the application. Matter not found in either, involving a departure from or an addition to the original disclosure, cannot be added to the application even though supported by a supplemental oath or declaration, and can be shown or claimed only in a separate application.

The claims may be amended by canceling particular claims, by presenting new claims, or by amending the language of particular claims (such amended claims being in effect new claims). In presenting new or amended claims, applicants must point out how they avoid any reference or ground rejection of record which may be pertinent.

No change in the drawing may be made except by permission of the PTO. Permissible changes in the construction shown in any drawing may be made only by the bonded draftsmen. A sketch in permanent ink showing proposed changes, to become part of the record, must be filed for approval by the PTO before the corrections are made. The paper requesting amendments to the drawing should be separate from other papers.

The original numbering of the claims must be preserved throughout the prosecution. When claims are canceled, the remaining claims must not be renumbered. When claims are added by amendment or substituted for canceled claims, they must be numbered by the applicant consecutively beginning with the number next following the highest numbered claim previously presented. When the application is ready for allowance, the examiner, if necessary, will renumber the claims consecutively in the order in which they appear or in such order as may have been requested by applicant.

Time for Response and Abandonment

The maximum period given for response is six months, but the commissioner has the right to shorten the period to no less than thirty days. The typical response time allowed to a PTO action is three months. If you want a longer time, you usually have to pay some extra money for an extension. The amount of the fee depends upon the response time desired. If you miss any target date, your application will be abandoned by the PTO and made no longer pending. However, if you can show whereby your failure to prosecute was unavoidable or unintentional, the application can be revived by the commissioner.

The revival requires a petition to the commissioner and a fee for petition, which should be filed without delay. The proper response must also accompany the petition if it has not yet been filed.

How to Make Appeals

If the examiner circles his or her wagons and begins to stonewall, there is a higher court. Rejections that have been made final may be appealed to the Board of Patent Appeals and Interferences. This august body consists of the commissioner of patents and trademarks, the deputy commissioner, the assistant commissioners, and the examiners-in-chief, but typically each appeal is heard by only three members. An appeal fee is required and you must file a brief in support of your position. You can even get an oral hearing if you pay enough.

If the board goes against you, there is yet a higher court, the Court of Appeals for the Federal Circuit. Or you might file a civil action against the commissioner in the U.S. District Court for the District of Columbia. He won't take it personally; it goes with the territory. The Court of Appeals for the Federal Circuit will review the record made in the PTO and may affirm or reverse the PTO's action. In a civil action, you may present testimony in the court, and the court will make a decision.

What Are Interference Proceedings?

Parallel development is a phenomenon that should not be discounted. On numerous occasions a company executive has said to me, "I've seen that concept twice in the last month," or something to this effect. At times, two or more applications may be filed by different inventors claiming substantially the same patentable invention. A patent can only be granted to one of them, and a proceeding known as an "interference" is instituted by the PTO to determine who the original inventor is, and who is entitled to the patent. About one percent of all applications filed become engaged in an interference proceeding.

Interference proceedings may also be instituted between an application and a patent already issued, if the patent has not been issued for more than one year prior to the filing of the conflicting application, and if the conflicting application is not barred from being patentable for some other reason.

The priority question is determined by a board of three examiners-in-chief on the evidence submitted. From the decision of the Board of Patent Appeals and Interferences, the losing party may appeal to the Court of Appeals for the Federal Circuit or file a civil action against the winning party in the appropriate U.S. district court.

The terms "conception of the invention" and "reduction to practice" are encountered in connection with priority questions. "Conception of the invention" refers to the completion of the devising of the means for accomplishing the result. "Reduction to practice" refers to the actual construction of the invention in physical form. In the case of a machine it includes the actual building of the machine. In the case of an article or composition it includes the actual carrying out of the steps in the process; actual operation, demonstration, or testing for the intended use is usually required. The filing of a regular application for patent completely disclosing the invention is treated as equivalent to reduction to practice. The inventor who proves to be the first to conceive the invention and the first to reduce it to practice will be held to be the prior inventor, but more complicated situations cannot be stated this simply.

This is why it is important to have evidence that proves when you first had an idea and when the prototype was made. It is critical that you keep careful and accurate records throughout the development of an idea. The Disclosure Document Program was established by the PTO for this purpose.

Allowance and Award of Patents

If your utility patent is found to be allowable, a notice of allowance will be sent to you or your attorney. Within three months from the date of the notice you must pay an issue fee.

All Good Things Come to an End

After a patent has expired, anyone may make, use, or sell the invention without permission of the person who first obtained the patent, provided the matter covered by other unexpired patents is not used. Patent terms may not be extended except by a special act of Congress.

What Rights Does a Patent Give You?

It's a pretty exciting moment when you get your first patent. It comes bound inside a beautiful oyster-white folder that has the U.S. Constitution screened in blue as its background. The large official gold seal of the Patent and Trademark Office is embossed thereon.

Between the covers of that folder is your patent, a grant that gives you—the inventor(s)—"the right to exclude others from making, using or selling the invention throughout the United States" and its territories and possessions for a designated period of time (17 or 20 years, depending on the actions of Congress), subject to the payment of maintenance fees as provided by law. Having a patent does not guarantee your ability—nor does it explicitly give you the right—to make, use, or sell the invention. Any person is ordinarily free to make, use, or sell anything he or she pleases, and a grant from Uncle Sam is not required. Having a patent **does** give you the right to exclude others from commercial exploitation of the invention; therefore, the patentee is the only one who may make, use, or sell the invention. Others may not do so without authorization. You may assign your rights in the invention to another person or company.

If you receive a patent for a new soda pop and the marketing of said beverage is prohibited by law, the patent will not help you. Nor may you market said soda pop if by doing so you infringe on the prior rights of others.

Maintenance Fees

All utility patents which issued from applications filed on or after December 12, 1980, are subject to the payment of maintenance fees which must be paid to keep the patent in force. These fees are due at 3½, 7½, and 11½ years from the date the patent is granted and can be paid without a surcharge during the six-month period preceding each due date. The amounts of the maintenance fees are subject to change every three years.

Be advised that the PTO does not mail notices to patent owners advising them that a maintenance fee is due. If you have a patent attorney tracking your business, he or she will let you know when the money is due. An attorney gets paid every time your business moves across his or her desk. But if you are doing it by yourself and you miss a payment, it may result in the expiration of the patent. A six-month grace period is provided, during which the maintenance fee may be paid with a surcharge.

Can Two People Own A Patent?

Yes. Two or more people may jointly own patents as either inventors, investors, or licensees. Most of my patents are joint ownerships. Anyone who shares in the ownership of a patent, no matter how small a part they might own, has the right to make, use, or sell it for his or her own profit, unless prohibited from doing so by prior agreement. It is accordingly dangerous to assign part interest in a patent of yours without having a definite agreement hammered out vis-a-vis respective rights and obligations to each other.

Can A Patent Be Sold?

Yes. The patent law provides for the transfer or sale of a patent, or of an application for patent, by a contract. When assigned the patent, the assignee becomes the owner of the patent and has rights identical to those of the original patentee.

Should you wish to assign your patent or patent application to a third party (manufacturer, investor, university, employer, etc.), this is possible by filing the appropriate form, Assignment of Patent, or Assignment of Patent Application (see forms at the end of this chapter).

You can sell all or part of the interest in a patent. If you prefer, you could even sell it by geographic region. I consider patents valuable properties—personal assets. Never assume that because you have been unsuccessful in selling a patent it has no value. You might sell it eventually or find someone infringing it, thus turning it to positive account.

Infringement of Patents

Infringement of a patent consists in the unauthorized making, using, or selling of the patented invention within the territory of the United States during the term of the patent. If your patent is infringed, that is, if someone uses it without your permission, it is your right to seek relief in the appropriate federal court.

When I see an apparent infringement of a patent of ours, as has occurred occasionally over the years, the first thing I do is call the company and set up a meeting. I am not litigious. Things can often be worked out between parties. Thus far, I have always been able to do this. Court battles over patents can be long and expensive affairs. And, if you want to continue working in your particular field, it is wise to avoid making too many corporate enemies.

Several years ago I saw an infringement of a patent we hold. One call to the company's president and a quick fax of our patent brought immediate relief in the form of a royalty on all items made to date and in the future. Not only that, but I was invited to submit ideas for licensing consideration.

If your friendly approach is turned away, and you are sure of your position, then the next step is to get a lawyer and decide if a Temporary Restraining Order (TRO) is appropriate. A TRO is an injunction to prevent the continuation of the infringement. You may also ask the court for an award of damages because of the infringement. In such an infringement suit, the defendant may raise the question of the validity of the patent, which is then decided by the court. The defendant may also aver that what is being done does not constitute infringement.

Infringement is determined primarily by the language of the claims of the patent, and if what the defendant is making does not fall within the language of any of the claims of the patent, there is no infringement.

The PTO has no jurisdiction over questions relating to infringement of patents. In examining applications for patent, no determination is made as to whether the patent-seeking invention infringes any prior patent.

To Sue or Not to Sue

If you do catch someone infringing your patent, you may decide to sue for damages. This can be a costly exercise. According to Stephen R. May, manager of Intellectual Property Services Department at Pacific Northwest Laboratory in Richland, Washington, "a full-blown patent lawsuit that actually goes to trial will probably cost a minimum of $75,000 to $100,000, although a very simple case could cost less." In most instances, May reports, the costs can be $250,000 and up.

His advice to inventors: "If you believe someone is infringing your patent, an attorney can draft a 'cease and desist' letter, possible for as little as a few hundred dollars. This might resolve the matter if the infringer ceases, but in many cases it does not."

The expensive part of any lawsuit is "discovery," in which the parties exchange documents and take depositions of potential witnesses. The photocopying bill alone could run into the thousands of dollars and the process could last anywhere from six months to several years. Trials tend to run from one to six weeks with decisions rendered in a matter of days in the case of a jury, or as long as several months if the verdict is by a judge. If you lose, appeals take more time and money.

Patent Enforcement Insurance

The only right a patent gives the inventor is the right to defend it in a court of law. And, as mentioned above, patent infringement litigation can be costly. How much? Ask Diane B. Loisel, a nurse from Bowie, Maryland. After obtaining a patent for a cap she had invented for use in neonatal respiratory therapy, she claimed that a company to whom she had presented the concept had begun to manufacture and market it without her permission.

Loisel was told by her patent attorney that to litigate would cost her $250,000 in legal fees. "If you're going to get a patent, you're going to have to fight," her lawyer had told her previously. "But he never told me it would cost so much money," she said.

To help inventors shoulder the risk and responsibility for enforcing their patents against infringers, there are insurance companies that market policies designed to reimburse the litigation expenses incurred by a patent owner in enforcing his or her U.S. patents. For information on such insurance, you may try calling your state insurance commissioner.

Patent attorney Robert W. Faris, a partner in Nixon & Venderhye of Arlington, Virginia, says of this kind of policy, "One of the downsides of this type of insurance is that the insurance company is reimbursed for its expenses out of the settlement or judgment. This means that if the recovery is on the order of the legal expenses incurred for the litigation, the patent owner could come away with practically no financial recovery, although his patent rights will have been vindicated."

Faris adds that the program seems pretty risky for the insurance company. "I don't know how they are able to predict with any certainty what the risks would be beforehand. They would have to be only taking on patents whose chances for infringement are very remote."

One such company, Intellectual Property Insurance Services, permits its insured to choose patent counsel. However, before the company will open the tap and start paying bills, the policy holder must provide a written opinion from his or her attorney attesting to the fact that the matter is one that can be litigated and the policy holder must show proof that the alleged infringement will cause economic damage.

Would Faris recommend patent infringement insurance? "I might well recommend that certain clients look into it because it is the only way some small businesses might be able to enforce their patent rights," he concludes.

He points out that patent infringement insurance does not cover the inventor should his or her U.S. patent infringe an existing patent. In other words, it could protect you (the inventor) from someone infringing your patent, but is of no help to you if your patent infringes the patent of another inventor.

Abandonment of Patents

There are two kinds of abandonment: **intentional** (it's your fault), and **unintentional** (due to circumstances beyond your control). If the reason for abandonment is your fault, for example, you simply lost track of dates and missed a deadline, then you must pay the due fee ($185 for a small entity design patent; $525 for a small entity utility patent) plus a penalty for your mistake (an additional $525). A costly error!

If your reason is unintentional, for example, you claim to never have received the notice from the PTO, then you must pay $50 to have your petition considered plus you still have to pay the required fee ($185 for a small entity design patent; $525 for a small entity utility patent). You may wish to add a notarized letter to form PTO/SB/64 (5-91) (see forms at the end of chapter) explaining your story in detail.

I was involved once in a case that involved the PTO using the wrong zip code on our paperwork, an error that caused a one-month delay in the delivery of our paperwork, and ultimately resulted in our getting slapped with abandonment papers.

I was able to cure this with a phone call to a senior PTO official. It was an open-and-shut case, as far as he was concerned. Upon seeing proof of the typo, he personally ordered the abandonment to be withdrawn. We never even had to pay the petition fees; there was just no question about who was wrong.

On pages 138-39 you will find the official PTO form used to petition for revival of an application for a patent abandoned unintentionally.

If you have any questions about how to handle a petition, do not hesitate to call the PTO at (703) 557-4282 for the latest information. I have had occasion to revive patents and the folks who answer this line are extremely helpful.

Mail petitions for revival to: Commissioner, Patent and Trademark Office, Box DAC, Washington, D.C. 20231.

PTO Log

It is a good idea to log all of your PTO correspondence in and out of your office. This will help you keep track of deadlines as well as give you a record of paper flow. Losing paperwork or missing a deadline can be both costly and time consuming.

PETITION FOR REVIVAL OF AN APPLICATION FOR PATENT ABANDONED UNINTENTIONALLY UNDER 37 CFR 1.137(b), 37 CFR 1.155(c) OR 37 CFR 1.316(c)	Docket Number (Optional)

First named inventor:

Serial No.:   Group Art Unit:

Filed: Examiner:

Title:

Attention: Assistant Commissioner for Patents
Commissioner of Patents and Trademarks
Washington, D.C. 20231

> NOTE: If information or assistance is needed in completing this form, please contact Petitions Information at (703)557-4282.

The above-identified application became abandoned for failure to file a timely and proper response to the Office action mailed on _____ , which set a _____ month/day period for response. The abandonment date of this application is _____ (i.e., the day after the expiration date of the period set for response plus any extensions of time obtained therefore).

APPLICANT HEREBY PETITIONS FOR REVIVAL OF THIS APPLICATION

> NOTE: A grantable petition requires the following items:
> (1) Petition fee
> (2) Proposed response and/or issue fee
> (3) Verified statement that the abandonment was unintentional

1. Petition fee
   ☐ Small entity - fee $525.00
        ☐ Small entity statement enclosed herewith.
        ☐ Small entity statement previously filed.
   ☐ Other than small entity - fee $1,050.00

2. Proposed response and/or fee

   A. The proposed response and/or fee to the above-noted Office action in
        the form of _____ (identify type of response):
        ☐ has been filed previously on _____ .
        ☐ is enclosed herewith.

   B. The issue fee of $ _____
        ☐ has been paid previously on _____ .
        ☐ is enclosed herewith.

3. Verified statement that abandonment was unintentional

This application became abandoned unintentionally.

I hereby declare that all statements made herein of my own knowledge are true and that all statements made on information and belief are believed to be true; and further that these statements were made with the knowledge that willful false statements and the like so made are punishable by fine or imprisonment, or both, under Section 1001 of Title 18 of the United States Code, and that such willful false statements may jeopardize the validity of the application, any patent issuing thereon, or any patent to which this verified statement is directed.

_____
Date

_____
Signature

Telephone
Number: (____) _____

_____
Typed or printed name

_____
Address

_____

Enclosures: ☐ Response

☐ Fee Payment

☐ Small Entity Status Form

☐ _____

By completing the Certificate of Mailing, below, the date mailed will be considered the date this paper is filed.

CERTIFICATE OF MAILING [37 CFR 1.8(a)]

I hereby certify that this paper is being deposited with the United States Postal Service on the date shown below with sufficient postage as first class mail in an envelope addressed to the: Commissioner of Patents and Trademarks, Washington, DC 20231.

_____
Date

_____
Signature

_____
Typed or printed name

[Page 2 of 2]

Patent and Trademark Office: U.S. DEPARTMENT OF COMMERCE

AMENDMENT TRANSMITTAL LETTER	Docket Number (Optional)

Application Number	Filing Date	Examiner	Group Art Unit

Invention Title

TO THE COMMISSIONER OF PATENTS AND TRADEMARKS

Transmitted herewith is an amendment in the above - identified application.

☐ Small Entity status of this application has been established under 37 CFR 1.27 by a verified statement previously submitted.

☐ A verified statement to establish Small Entity status under 37 CFR 1.27 is enclosed.

☐ No additional fee is required.

☐ The fee has been calculated as shown below:

CLAIMS AS AMENDED

	(1) CLAIMS REMAINING AFTER AMENDMENT		(2) HIGHEST NUMBER PREVIOUSLY PAID FOR	(3) PRESENT NUMBER EXTRA	RATE	FEE
TOTAL CLAIMS	*	minus	**		x $22	
INDEPENDENT CLAIMS	*	minus	***		x $74	
MULTIPLE DEPENDENT CLAIM ADDED					$230	
					TOTAL	$
If applicant has small entity status under 37 CFR 1.9 and 1.27, then divide total fee by 2, and enter amount here.					SMALL ENTITY TOTAL	$

* If the entry in column 1 is less than the entry in column 2, write "0" in column 3

** If the highest number previously paid for IN THIS SPACE is less than 20, enter "20".

*** If the highest number previously paid for IN THIS SPACE is less than 3, enter "3".

The "highest number previously paid for" (total or independent) is the highest number found in the appropriate box in column 1.

☐ Please charge Deposit Account Number _____ in the amount of $ _____ . A duplicate copy of this sheet is enclosed.

☐ A check in the amount of $ _____ to cover the filing fee is enclosed.

☐ The Commissioner is hereby authorized to charge payment of the following fees associated with this communication or credit any overpayment to Deposit Account Number _____ . A duplicate copy of this sheet is enclosed.

☐ Any additional filing fees required under 37 CFR 1.16.

☐ Any patent application processing fees under 37 CFR 1.17.

_____   _____
Date   Signature

Patent and Trademark Office; U.S. DEPARTMENT OF COMMERCE

How to Apply for U.S. Trademarks

What's in a name? That which we call a rose by any other name would smell as sweet.

—SHAKESPEARE

Brand Recognition

What do Kleenex, Ford, Trivial Pursuit, IBM, Barbie, the characters Mr. Clean, Poppin' Fresh, Indian Maid, and the Gerber Baby have in common with the distinctive buildings that house McDonald's, Fotomat, and Wendy's? They are all registered trademarks, and are very important to their owners as tools to sell their products and services.

Mark My Words

A trademark can be a coined word (Trix, Acura), an everyday word with no connection to the product it promotes (Ivory, Apple), a phrase (All the News That's Fit to Print, Catch the Wave), a word that describes a quality or product function (Blockbuster, Marks-a-Lot), or a coined word that suggests product performance (Timex, Jell-O). It can be a foreign word with or without significance to the item (Volkswagen, Dos Equis). Many trademarks are the names of a product inventor or company founder (Ford, Piaget), the name of a celebrated person selected for a positive image (Lincoln, Raleigh), or a name from literature (Peter Pan, Atlas). Initials are an option (CVS, MCA), as are numbers (4711, 66), or a combination of letters and numbers (WD-40, S-500). A pictoral mark (Gerber Baby, Playboy Bunny) can also be registered. A distinct color (John Deere green) can be registered. A sound (NBC chimes) can be registered. And since 1990 scents can be registered.

Clark's Oh-Sew-Easy Needles of Goleta, California (invented by Clark Osewez), was awarded the first fragrance trademark registration for its line of sewing threads perfumed with "a high impact fresh floral fragrance reminiscent of plumeria blossoms."

Explaining the decision, J. David Sams, then-chairman of the PTO's Trademark Trial and Appeal Board, offered, "If it looks like a trademark and quacks like a trademark, it is a trademark."

In reviewing the Oh-Sew-Easy application, the board found that no one else was using "high impact . . . plumeria blossoms" fragrance—or any fragrance whatsoever—in the marketing of sewing thread. That made the use of a fragrance distinctive.

Supreme Court Makes A Colorful Decision

Since the 1950s, Qualitex Company has used a distinctive green-gold color on the pads it makes and sells to dry-cleaning firms for use on dry-cleaning presses. In 1989, Jacobson Products, a Qualitex rival, began to sell its own press pads to dry-cleaning firms; its pads were a similar green-gold color.

In 1991, Qualitex registered the special green-gold on press pads with the PTO as a trademark (Registration No. 1,633,711). Further, it brought a trademark infringement suit challenging Jacobson's use of the green-gold color.

The case made it all the way to the U.S. Supreme Court which, in a unanimous opinion, delivered by Justice Breyer on March 28, 1995, determined that a color may sometimes meet the basic legal requirements for use as a trademark.

John Deere green. Toro red. Corning Fiberglass pink. They're golden.

The Name Is The Game

Do not underestimate the contribution a good trademark can make to your licensing effort. It always creates fast product identification and can help communicate the item's story. I, therefore, spend time coming up with the most appropriate trademark for every product we develop.

For example, my pitch to Proctor & Gamble that it license our unique big wheel trike design would not have had the same impact without our trademark, The Fluorider. Proctor & Gamble placed our unique ride-on in 33,000 supermarkets under our trademark.

When we invented and patented a toy truck that transformed from a conventional wheel configuration to an in-line configuration via a variable geometry chassis, the trademark Switchblade was licensed by Remco as part of the deal.

And our line of Micro Machines that was built around authentic state police cruisers, motorcycles, and aircraft was contingent upon our licensing to the manufacturer our trademark Troopers.

Trademarks can also be very important to inventors as a source of additional income should the trademark on the principal invention become valuable. For example, if your invention becomes a hit under a trademark you control, you may have opportunities to license the trademark on collateral merchandise, an arena that your licensee(s) may or may not participate in with you. For example, Caterpillar is more than earth-moving equipment today: it is also a line of rugged footwear. Mighty Morphin' Power Rangers, the hit toy line, has its trademark licensed for much more than toys. It is on everything from pencils to popcorn.

Lost in the Translation

Since your item may be marketed overseas one day, when selecting a trademark do your best to see what it means in other languages, and/or how it will translate. Why is this important? Ask General Motors what happened when it found out its Chevy Nova was the laughing stock of Latin America. In Spanish the words "no va" mean "does not run." Ask Coca-Cola: after entering the Chinese market, they discovered that Coca-Cola —translated literally—means "bite the wax tadpole."

Trademark Office

Trademarks fall under the jurisdiction of the Trademark Office, a division of the Patent and Trademark Office. In fiscal year 1994, the PTO received 155,376 applications from which it issued a record 86,122 trademark registrations. This is a 24 percent increase in registrations from fiscal year 1992 and is primarily due to the large number of applications filed following enactment of the intent-to-use legislation. The average time between filing an application and its disposition (in other words, registration, abandonment, or issuance of notice of allowance) was 14.4 months at the end of fiscal year 1993 and 16.3 months at the end of fiscal year 1994.

PTO commissioner Bruce A. Lehman says that the current team is one of the most experienced ever retained in the Trademark Office. I can personally attest to this. In fact, I call the office often for information and have never once been disappointed or frustrated. You would never know it is a bureaucracy.

Correspondence and Information

All correspondence about trademark matters should be addressed to Commissioner of Patents and Trademarks, Washington, D.C. 20231, unless you have the name of a particular examiner or other official.

For the phone numbers and specific mail stops of various departments at the Trademark Office, please refer to Appendix 2 for the PTO Information Contacts directory.

Information Numbers

General Trademark or Patent Information	(703) 308-HELP
Automated (Recorded) General Trademark or Patent Information	(703) 557-INFO
Automated Line for Status Information on Trademark Applications	(703) 305-8747
Additional Status Information	(703) 308-9400
Assignment Branch	(703) 308-9711
Certified Copies of Registrations	(703) 308-9726
Information Regarding Renewals [Sec. 9], Affidavits of Use [Sec. 8], Incontestability [Sec. 15], or Correcting a Mistake on a Registration	(703) 308-9500
Information Regarding Applications Based on International Agreements or for Certification, Collective, or Collective Membership Marks	(703) 308-9400
Trademark Trial and Appeal Board	(703) 308-9300

What Kinds of Trademarks Are Available?

Trademark. A "trademark," as defined in section 45 of the 1946 Trademark Act (Lanham Act) "includes any word, name, symbol, or device, or any combination thereof adopted and used by a manufacturer or merchant to identify his goods and distinguish them from those manufactured or sold by others." Examples of trademarks include Coca-Cola, Barbie, Ford, and Adver*teasing*.

Service Mark. A mark used in the sale or advertising of services to identify the services of one person and distinguish them from the services of others. Titles, character names, and other distinctive features of radio or television programs may be registered as service marks notwithstanding that they, or the programs, may advertise the goods of the sponsor. Examples of service marks include American Express, Mr. Goodwrench, and The Cosby Show.

Certification Mark. A mark used upon or in connection with the products or services of one or more persons other than the owner of the mark to certify regional or other origin, material, mode of manufacture, quality, accuracy, or other characteristics of such goods or services, or that the work or labor on the goods or services was performed by members of a union or other organization. Examples of certification marks of quality include the UL symbol of Underwriters Laboratories and 100% Pure Florida's Seal of Approval. Examples of certification marks of service include the Automobile Association of America's Approved Auto Repair and the Motion Picture Association of America's movie ratings.

A Skirmish in the Cola Wars

"The Uncola" was registered as a trademark by the Seven-Up Company over the objections of the Coca-Cola Company, which had made the case that no entity should be permitted exclusive rights to a term that was the equivalent of *noncola*.

Collective Mark. A trademark or service mark used by the members of a cooperative, an association, or other collective group or organization. Marks used to indicate membership in a union, an association, or other organization may be registered as collective membership marks. Examples of collective marks include the National Collegiate Athletic Association, the National Rifle Association of American Member, the Automobile Association of America, and Sigma Delta Chi.

Trade and Commercial Name. Marks differ from trade and commercial names that are used by manufacturers, industrialists, merchants, agriculturists, and others to identify their businesses, vocations, or occupations, or other names or titles lawfully adopted by persons, firms, associations, companies, unions, and other organizations. The latter are not subject to registration unless actually used as trademarks. Examples of trade and commercial names include Coca-Cola Company, Gale Research Incorporated, and Sony Corporation of America.

Function of Trademarks

The primary function of a trademark is to indicate the origin of a product or service; however, trademarks also serve to guarantee the quality of the goods bearing the mark and, through advertising, create and maintain a demand for the product. Trademark rights are acquired only through use of the trademark; this use must continue if the rights you can acquire are to be preserved. Registration of a trademark in the PTO does not in itself create or establish any exclusive rights, but it is recognition by the government of your rights to use the mark in commerce to distinguish your goods from those of others.

Do You Need a Federal Trademark Registration?

While federal registration is not necessary for trademark protection, registration on the Principal Register does provide certain advantages:

1. A constructive date of first use of the mark in commerce. This gives registrant nationwide priority as of that date, except as to certain prior users or prior applicants

2. The right to sue in federal court for trademark infringement

3. Recovery of profits, damages, and costs in a federal court infringement action and the possibility of treble damages and attorneys' fees

4. Constructive notice of a claim of ownership. This eliminates a good faith defense for a party adopting the trademark subsequent to the registrant's date of registration

5. The right to deposit the registration with Customs in order to stop the importation of goods bearing an infringing mark

6. *Prima facie* evidence of the validity of the registration, registrant's ownership of the mark, and of registrant's exclusive right to use the mark in commerce in connection with the use of goods or services specified in the certificate

7. The possibility of incontestability, in which case the registration constitutes conclusive evidence of the registrant's exclusive right, with certain limited exceptions, to use the registered mark in commerce

8. Limited grounds for attacking a registration once it is five years old

9. Availability of criminal penalties and treble damages in an action for counterfeiting a registered trademark

10. A basis for filing trademark applications in foreign countries

Marks Not Subject to Registration

A trademark cannot be registered if it:

1. Consists of or comprises immoral, deceptive, or scandalous matter or matter that may disparage or falsely suggest a connection with persons, living or dead, institutions, beliefs, or national symbols, or bring them into contempt or disrepute

2. Consists of or comprises the flag or coat of arms or other insignia of the United States, or of any state or municipality, or of any foreign nation, or any simulation thereof

3. Consists of or comprises a name, portrait, or signature identifying a particular living individual except by his written consent, or the name, signature, or portrait of a deceased president of the United States during the life of his widow, if any, except by the written consent of the widow

4. Consists of or comprises a mark which so resembles a mark registered in the Patent and Trademark Office or a mark or trade name previously used in the United States by another and not abandoned, as to be likely when applied to the goods of another person, to cause confusion, or to cause mistake, or to deceive

Registrable Marks

Principal Register

The trademark, if otherwise eligible, may be registered on the *Principal Register* unless it consists of a mark which, 1) when applied to the goods/services of the applicant is merely descriptive or deceptively misdescriptive of them, except as indications of regional origin, or 2) is primarily merely a surname.

Such marks, however, may be registered on the *Principal Register*, provided they have become distinctive as applied to the applicant's goods in commerce. The Commissioner may accept as *prima facie* evidence that the mark has become distinctive as applied to applicant's goods/services in commerce, proof of substantially exclusive and continuous use thereof as a mark by the applicant in commerce for the five years preceding the date of filing the application for registration.

Supplemental Register

All marks capable of distinguishing your goods and not registrable on the *Principal Register*, which have been in lawful use in commerce for the year preceding your filing for registration, may be registered on the *Supplemental Register*. A mark on this register may consist of any trademark, symbol, label, package, configuration of goods, name, word, slogan, phrase, surname, geographical name, numeral, device, or any combination of the foregoing.

Lawyers and professional searchers often charge more for logo searches. In my opinion, trademark searches and logo searches take an equal amount of effort. Yet you'll find trademark searches costing $75 and logo searches priced at $150 from the same firm.

Searches for Conflicting Marks

An applicant is not required to conduct a search for conflicting marks prior to applying with the PTO. However, some people find it useful. In evaluating an application, an examining attorney conducts a search and notifies the applicant if a conflicting mark is found. The application fee, which covers processing and search costs, will not be refunded even if a conflict is found and the mark cannot be registered.

To determine whether there is a conflict between two marks, the PTO determines whether there would be likelihood of confusion, that is, whether relevant consumers would be likely to associate the goods or services of one party with those of the other party as a result of the use of the marks at issue by both parties. The principal factors to be considered in reaching this decision are the similarity of the marks and the commercial relationship between the goods and services identified by the marks. To find a conflict, the marks need not be identical, and the goods and services do not have to be the same.

Three Ways to Conduct a Search

There are several ways to do a search. You could have your patent attorney do it, engage the services of a professional trademark search firm, or do it yourself.

Law Firm Trademark Search

Your patent attorney will gladly handle a trademark search. But understand, he or she will not do it personally. Lawyers rely on the services of a professional trademark search firm and then add a premium of 40 percent to 100 percent or more, depending on what the market will bear and the firm's overhead.

If you decide to hire a lawyer for this task, get an estimate first. Also find out how much the law firm charges for copies. Avoid surprises.

Professional Trademark Search

If you want to save the lawyer's mark-up, you may consider hiring a trademark searcher yourself. They are listed in the yellow pages under "Trademark Searchers." The largest national firm of this kind is Thomson & Thomson (T & T). For current fees, telephone 1-800-356-8630.

I employ Greentree Information Service (GIS) for our federal trademark searches. GIS may be reached by writing: 7342 Greentree Road, Bethesda, Maryland 20817, or calling: (301) 469-0902. The president is George Harvill. Ask for a current rate sheet and compare the GIS prices with what you are quoted elsewhere.

Do-It-Yourself Trademark Search

There are a number of ways you could approach your search if you decide to go it alone:

1. You may visit the Trademark Office's Public Search Room, located on the second floor of the South Tower Building, 2900 Crystal Drive, Arlington, Virginia 22202. The staff up there is very helpful, and once you learn the layout, you can breeze through the search process by hand or computer terminal.

2. Trademark information on CD-ROM has been distributed to the Patent and Trademark Depository Libraries (PTDLs). These products include T®ADE-MARKS Registrations, which contains all currently registered U.S. trade-

marks, T®ADEMARKS Pending Applications, and T®ADEMARKS Assignment File, which contains ownership information.

By the way, these products are available for sale to the public and may be ordered from the PTO's Office of Information Products Development, Crystal Plaza 2, Room 9D30, Washington, D.C. 20231. The telephone number is (703) 308-0322.

The rest of this chapter is designed to be a guide to the trademark registration process, including a step-by-step guide to filling out the applications. This information comes to you courtesy of the Trademark Office, from the booklet entitled *Basic Facts about Registering a Trademark*.

Basic Facts

Establishing Trademark Rights

Trademark rights arise from either (1) actual use of the mark, or (2) the filing of a proper application to register a mark in the PTO stating that the applicant has a bona fide intention to use the mark in commerce regulated by the U.S. Congress. (See below, under Types of Applications, for a discussion of what is meant by the terms "commerce" and "use in commerce.") Federal registration is not required to establish rights in a mark, nor is it required to begin use of a mark. However, federal registration can secure benefits beyond the rights acquired by merely using a mark. For example, the owner of a federal registration is presumed to be the owner of the mark for the goods and services specified in the registration, and to be entitled to use the mark nationwlde.

There are two related but distinct types of rights in a mark: the right to register and the right to use. Generally, the first party who either uses a mark in commerce or files an application in the PTO has the ultimate right to register that mark. The PTO's authority is limited to determining the right to register. The right to use a mark can be more complicated to determine. This is particularly true when two parties have begun use of the same or similar marks without knowledge of one another and neither has a federal registration. Only a court can render a decision about the right to use, such as issuing an injunction or awarding damages for infringement. It should be noted that a federal registration can provide significant advantages to a party involved in a court proceeding. The PTO cannot provide advice concerning rights in a mark. Only a private attorney can provide such advice.

Terms of a Trademark

Unlike copyrights or patents, trademark rights can last indefinitely if the owner continues to use the mark to identify its goods or services. The term of a federal trademark registration is 10 years, with 10-year renewal terms. However, between the fifth and sixth year after the date of initial registration, the registrant must file an affidavit setting forth certain information to keep the registration alive. If no affidavit is filed, the registration is cancelled.

Types of Applications for Federal Registration

You may apply for federal registration in three principal ways.

1. An applicant who *has already commenced using a mark in commerce* may file based on that use (a "use" application)

2.   An applicant who has not yet used the mark may apply based on *a bona fide intention to use the mark in commerce* (an "intent-to- use" application). For the purpose of obtaining federal registration, commerce means all commerce which may lawfully be regulated by the U.S. Congress, for example, interstate commerce or commerce between the U.S. and another country. The use in commerce must be a bona fide use in the ordinary course of trade, and not made merely to reserve a right in a mark. Use of a mark in promotion or advertising before the product or service is actually provided under the mark on a normal commercial scale does not qualify as use in commerce. If an applicant files based on a bona fide intention to use in commerce, the applicant will have to use the mark in commerce and submit an allegation of use to the PTO before the PTO will register the mark.

3.   Additionally, under certain international agreements, an applicant may file in the United States based on an application or registration in another country. For information regarding applications based on international agreements please call the information number listed on page 143 (under the heading Information Numbers).

Who May File an Application?

The application must be filed in the name of the owner of the mark; usually an individual, corporation, or partnership. The owner of a mark controls the nature and quality of the goods or services identified by the mark.

The owner may submit and prosecute its own application for registration, or may be represented by an attorney. The PTO cannot help select an attorney.

Foreign Applicants

Applicants not living in the United States must designate in writing the name and address of a domestic representative—a person residing in the United States "upon whom notices of process may be served for proceedings affecting the mark." This person will receive all communications from the PTO unless the applicant is represented by an attorney in the United States.

Where to Send the Application and Correspondence

The application and all other correspondence should be addressed to The Commissioner of Patents and Trademarks, Washington, D.C., 20231.

- The initial application should be directed to Box Trademark.
- An Amendment to Allege Use should be directed to the attention of AAU.
- A Statement of Use or Request for an Extension of Time to File a Statement of Use should be directed to Box ITU.

Indicate your telephone number on the application form. Once a serial number is assigned to the application, you should refer to the serial number in all written and telephone communications concerning the application.

It is advisable to submit a stamped, self-addressed postcard with your application specifically listing each item in the mailing, that is, the written application, the drawing, the fee, and the specimens (if appropriate). The PTO will stamp the filing date and serial number of the application on the postcard to acknowledge receipt. This will help you if any item is later lost or if you wish to inquire about the application. The PTO will send a separate official notification of the filing date and serial number for every application about two months after receipt.

Anyone who claims rights in a mark may use the TM (trademark) or SM (service mark) designation with the mark to alert the public to the claim. *It is not necessary to have a registration, or even a pending application, to use these designations.* The claim may or may not be valid. The registration symbol, ®, may only be used when the mark is registered in the PTO. It is improper to use this symbol at any point before the registration issues. Please omit all symbols from the mark in the drawing you submit with your application; the symbols are not considered part of the mark.

The Registration Process

Filing Date-Filing Receipt

The PTO is responsible for the federal registration of trademarks. When an application is received, the PTO reviews it to determine if it meets the minimum requirements for receiving a filing date. If the application meets the filing requirements, the PTO assigns it a serial number and sends the applicant a receipt about two months after filing. If the minimum requirements are *not* met, the entire mailing, including the filing fee, is returned to the applicant.

Examination

About three months after filing, an examining attorney at the PTO reviews the application and determines whether the mark may be registered. If the examining attorney determines that the mark cannot be registered, the examining attorney will issue a letter listing any grounds for refusal and any corrections required in the application. The examining attorney may also contact you by telephone if only minor corrections are required. If a correction or modification can be done by phone, that's how the examiners prefer to handle it. I have always found trademark examiners to be most helpful and professional. Unlike many bureaucrats, they tend to favor less paper.

You must respond to any objections within six months of the mailing date of the letter, or the application will be abandoned. If your response does not overcome all objections, the examining attorney will issue a final refusal. The applicant may then appeal to the Trademark Trial and Appeal Board, an administrative tribunal within the PTO.

A common ground for refusal is likelihood of confusion between your mark and a registered mark. Marks which are merely descriptive in relation to the applicant's goods or services, or a feature of the goods or services, may also be refused. Marks consisting of geographic terms or surnames may also be refused. Marks may be refused for other reasons as well.

Publication for Opposition

If there are no objections, or if you overcome all objections, the examining attorney will approve the mark for publication in the *Official Gazette,* a weekly publication of the PTO. The PTO will send a Notice of Publication to the applicant indicating the date of publication. In the case of two or more applications for similar marks, the PTO will publish the application with the earliest effective filing date first. Any party who believes it may be damaged by the registration of the mark has 30 days from the date of publication to file an opposition to registration. An opposition is similar to a formal proceeding in the federal courts, but is held before the Trademark Trial and Appeal Board. If no opposition is filed, the application enters the next stage of the registration process.

Issuance of Certificate of Registration or Notice of Allowance

If your application was based upon *the actual use of the mark in commerce prior to approval for publication,* the PTO will register the mark and issue a registration certificate about 12 weeks after the date the mark was published, if no opposition was filed.

If, instead, the mark was published based upon your statement of having a *bonafide intention to use the mark in commerce,* the PTO will issue a Notice of Allowance about 12 weeks after the date the mark was published, again provided no opposition was filed. The applicant then has six months from the date of the Notice of Allowance to either (1) use the mark in commerce and submit a Statement of Use, or (2) request a six-month Extension of Time to File a Statement of Use (see forms and instructions at the end of this chapter). You may request additional extensions of time only as noted in the instructions. If the Statement of Use is filed and approved, the PTO will then issue the registration certificate.

Filing Requirements

Warning: Before completing an application, read the instructions carefully and study the examples provided. Errors or omissions may result in the denial of a filing date and the return of application papers, or the denial of registration and forfeiture of the filing fee.

To receive a filing date, the applicant must provide all of the following:

1. A written application form

2. A drawing of the mark on a separate piece of paper

3. The required filing fee

4. If the application is filed based upon prior use of the mark in commerce, three specimens for each class of goods or services. The specimens must show actual use of the mark with the goods or services.

Written Application Form [PTO Form 1478]

The application must be in English. A separate application must be filed for each mark you wish to register. Likewise, if you wish to register more than one version of the same mark, a separate application must be filed for each version. PTO Form 1478, included on pages 160–61 may be used for either a trademark or service mark application. It may be photocopied for your convenience. See the examples of completed applications on pages 170 and 171 with references to the following line-by-line instructions. These instructions are for filling out PTO Form 1478, entitled "Trademark/Service Mark Application, Principal Register, with Declaration."

Space 1: The Mark. Indicate the mark (for example, "Theorytec" or "Pinstripes and Design"). This should agree with the mark shown on the drawing page.

Space 2: Classification. It is not necessary to fill in this box. The PTO will determine the proper International Classification based upon the identification of the goods and services in the application. However, if you know the International Class number(s) for the goods and services, you may place the number(s) in this box. The International Classes are listed on page 157. If the PTO determines that the

Copies of the *Trademark Official Gazette* may be obtained from the Superintendent of Documents, Government Printing Office, Washington, D.C. 20402. For more information, call (202) 783-3238.

goods and services listed are in more than one class, the PTO will notify you during examination of the application, and you will have the opportunity to pay the fees for any additional classes or to limit the goods and services to one or more classes.

Space 3: The Owner of the Mark. The name of the owner of the mark must be entered in this box. The application must be filed in the name of the owner of the mark or the application will be void, and the applicant will forfeit the filing fee. The owner of the mark is the party who controls the nature and quality of the goods sold, or services rendered, under the mark. The owner may be an individual, a partnership, a corporation, or an association or similar firm. If the applicant is a corporation, the applicant's name is the name under which it is incorporated. If the applicant is a partnership, the applicant's name is the name under which it is organized.

Space 4: The Owner's Address. Enter your business address. If you are applying as an individual, enter either your business or home address.

Space 5: Entity Type and Citizenship/Domicile. The applicant must check the box which indicates the type of entity applying. In addition, in the blank following the box, the applicant must specify the following information:

Space 5(a): for an individual, the applicant's national citizenship

Space 5(b): for a partnership, the names and national citizenships of the general partners and the state where the partnership is organized (if a U.S. partnership) or country (if a foreign partnership)

Space 5(c): for a corporation, the state of incorporation (if a U.S. corporation), or country (if a foreign corporation)

Space 5(d): for another type of entity, specify the nature of the entity and the state where it is organized (if in the U.S.) or country where it is organized (if a foreign entity).

Space 6: Identification of the Goods and/or Services. In this blank you must state the specific goods and services for which registration is sought and with which you have actually used the mark in commerce, or in the case of an "intent-to-use" application, have a bona fide intention to use the mark in commerce. Use clear and concise terms specifying the actual goods and services by their common commercial names. A mark can only be registered for specific goods and services. The goods and services listed will establish the scope of your rights in the relevant mark.

The goods and services listed must be your actual "goods in trade" or the actual services you render for the benefit of others. Use language that would be readily understandable to the general public. For example, if you use or intend to use the mark to identify "toy," "word processors," or "baseballs and baseball bats," the identification should clearly and concisely list each such item. If you use indefinite terms, such as "accessories," "components," "devices," "equipment," "food," "materials," "parts," "systems," "products," or the like, then those words must be followed by the word "namely" and the goods or services listed by their common commercial name(s).

You must be very careful when identifying the goods and services. Because the filing of an application establishes certain presumptions of rights as of the filing date, the application may not be amended later to add any products or services not within the scope of the identiflcation. For example, the identification of "clothing" could be amended to "shirts and jackets," which narrows the scope, but could not be amended to "retail clothing store services," which would change the scope.

Characteristics of an Effective Logo

- Reflective of brand identification

- Reflective of brand improvement

- Novel

- Simple

- Telegenic

- Charismatic

- Promotable on multiple levels

Similarly, "physical therapy services" could not be changed to "medical services" because this would broaden the scope of the identification.

The identification of goods and services must not describe the mode of use of the mark, such as on labels, stationery, menus, signs, containers, or in advertising. There is another place on the application, called the "method-of-use clause," for this kind of information. (See information under Space 7a, fourth blank.) For example, in the identification of goods and services, the term "advertising" usually is intended to identify a service rendered by advertising agencies. Moreover, "labels," "menus," "signs," and "containers" are specific goods. If you identify these goods or services by mistake, you may not amend your identification to the actual goods or services. Thus, if the identification indicates "menus," it could not be amended to "restaurant services." Similarly, if the goods are identified as "containers or labels for jam," the identification could not be amended to "jam."

Note: If nothing appears in this blank, or if the identification does not identify any recognizable goods or services, the application will be denied a filing date and returned to the applicant. For example, if you specify the mark itself or wording such as "company name," "corporate name," or "company logo," and nothing else, the application will be denied a filing date and returned to the applicant. If you identify the goods and services too broadly as, for example, "advertising and business," "miscellaneous goods and services," or just "products," or "services," the application will also be denied a filing date and returned to the applicant.

Space 7: Basis for Filing. You must check at least one of the four boxes to specify a basis for filing the application. You should also fill in all blanks which follow the checked box(es). Usually an application is based upon either (1) use of the mark in commerce (the first box), or (2) a bona fide intention to use the mark in commerce (the second box). You may not check both the first and second box. If both the first and second boxes are checked, the PTO will not accept the application and will return it. If you wish to apply to register a mark for certain goods and services for which it is already using the mark in commerce, and also for other goods and services based on future use, separate applications must be filed to separate the relevant goods and services from each other.

Space 7(a): if you are using the mark in commerce in relation to all of the goods and services listed in the application, check this first box and fill in the blanks.

In the **first blank** specify the date the trademark was first used to identify the goods and services in a type of commerce which may be regulated by Congress.

In the **second blank** specify the type of commerce, specifically a type of commerce which may be regulated by Congress, in which the goods were sold or shipped, or the services were rendered. For example, indicate "interstate commerce" (commerce between two or more states) or commerce between the United States and a specific foreign country, for example, "commerce between the U.S. and Canada."

In the **third blank** specify the date that the mark was first used anywhere to identify the goods or services specified in the application. This date will be the same as the date of first use in commerce unless the applicant made some use, for example, within a single state, before the first use in commerce.

In the **fourth blank** specify how the mark is placed on the goods or used with the services. This is referred to as the "method-of-use clause," and should not be confused with the identification of the goods and services described under Space 6. For example, in relation to goods, state "the mark is used on labels affixed to the goods," or "the mark is used on containers for the goods," whichever is accurate. In relation to services, state "the mark is used in advertisements for the services."

Space 7(b): if you have a bona fide intention to use the mark in commerce in relation to the goods or services specified in the application, check this second box and fill in the blank. You should check this box if the mark has not been used at all or if the mark has been used on the specified goods or services only within a single state.

In the blank, state how the mark is intended to be placed on the goods or used with the services. For example, for goods, state "the mark will be used on labels affixed to the goods," or "the mark will be used on containers for the goods," whichever is accurate. For services, state "the mark will be used in advertisements for the services."

Spaces 7(c) and (d): these spaces are usually used only by applicants from foreign countries who are filing in the United States under international agreements. These applications are less common. For further information about treaty-based applications, call (703) 308-9400, or contact a private trademark attorney.

Space 8: Verification and Signature. You must verify the truth and accuracy of the information in the application and must sign the application. If the application is not signed, the application will not be granted a filing date and will be returned to you. If the application is not signed by an appropriate person, the application will be found void and the filing fee will be forfeited. Therefore, it is important that the proper person sign the application. Who should sign?

- If the applicant is an individual, that individual must sign.
- If the applicant is a partnership, a general partner must sign.
- If the applicant is a corporation, association, or similar organization, an officer of the corporation, association, or organization must sign. An officer is a person who holds an office established in the articles of incorporation or the bylaws. Officers may not delegate this authority to nonofficers.
- If the applicants are joint applicants, all joint applicants must sign.

The person who signs the application must indicate the date signed, provide a telephone number to be used if it is necessary to contact the applicant, and clearly print or type his or her name and position.

The Drawing Page

Every application must include a single drawing page. If there is no drawing page, the application will be denied a filing date and returned to the applicant. The PTO uses the drawing to file the mark in the PTO search records and to print the mark in the *Official Gazette* ad on the registration.

The drawing must be on pure white, durable, non-shiny paper that is 8½ inches (21.59 cm) wide by 11 inches (27.94 cm) long. There must be at least a one-inch (2.54 cm) margin on the sides, top, and bottom of the page, and at least one inch between the heading and the display of the mark.

At the top of the drawing there must be a heading, listing on separate lines the applicant's complete name, address, the goods and services specified in the application, and in applications based on use in commerce, the date of first use of the mark and the date of first use of the mark in commerce. This heading should be typewritten. If the drawing is in special form, the heading should include a description of the essential elements of the mark.

The drawing of the mark should appear at the center of the page. The drawing of the mark may be typewritten, as shown on page 173 or it may be in special form, as shown on page 172.

If the mark includes words, numbers, or letters, the applicant can usually elect to submit either a typewritten or a special-form drawing. To register a mark consisting of only words, letters, or numbers, without indicating any particular style or design, provide a typewritten drawing. In a typewritten drawing the mark must be typed entirely in capital letters, even if the mark, as used, includes lowercase letters. Use a standard typewriter or type of the same size and style as that on a standard typewriter.

To register a word mark in the form in which it is actually used or intended to be used in commerce, or any mark including a design, submit a special-form drawing. In a special-form drawing, the mark must not be larger than 4 inches by 4 inches (10.16 cm by 10.16 cm). If the drawing of the mark is larger than 4 inches by 4 inches, the application will be denied a filing date and returned to the applicant. In addition, the drawing must appear only in black and white, with every line and letter black and clear. No color or gray is allowed. Do not combine typed matter and special form in the same drawing.

The drawing in special form must be a substantially exact representation of the mark as it appears on the specimens. The applicant may apply to register any portion of a mark consisting of more than one element, provided the mark displayed in the drawing creates a separate impression apart from other elements it appears with on the specimens. For example, generally it is possible to register a word mark by itself even though the specimen shows the word mark used in combination with a design or a part of a logo. Do not include nontrademark matter in the drawing, such as informational matter which may appear on a label. In the end, the applicant must decide exactly what to register and in what form. The PTO considers the drawing central in determining exactly what mark the application covers.

To indicate color, use the color linings shown in the "Symbols for Draftsmen" in Chapter 7. The appropriate lining should appear in the area where the relevant color would appear. If the drawing is lined for color, insert a statement in the written application to indicate so, for example, "The mark is lined for the colors red and green." A plain black-and-white drawing is acceptable even if the mark is used in color. Most drawings do not indicate specific colors.

While it may be possible to make some minor changes, the rules prohibit any material change to the drawing of the mark after filing.

Fees

Filing Fee

As of this writing, the application filing fee is $245 for each class of goods or services listed. At least $245 must accompany the application, or the application will be denied a filing date and all the papers returned to the applicant. Fee increases, when necessary, usually take effect on October 1 of any given year. Please call (703) 557-INFO for up-to-date fee information. The PTO receives no taxpayer funds. The PTO's operations are supported entirely from fees paid by applicants and registrants.

Additional Fees Related to Intent-To-Use Applications

In addition to the application filing fee, applicants filing based on having a bona fide intention to use a mark in commerce must submit a fee of $100 for each class of goods or services in the application when filing any of the following:

- an Amendment to Allege Use
- a Statement of Use
- a Request for an Extension of Time to File a Statement of Use

Form of Payment

All payments must be made in United States currency, by check, post office money order, or certified check. Personal or business checks may be submitted. Make checks and money orders payable to: The Commissioner of Patents and Trademarks.

Note: fees are not refundable.

Specimens

The following information is designed to provide guidance regarding the specimens required to show use of the mark in commerce.

When to File the Specimens

If you have already used the mark in commerce and files based on this use in commerce, then you must submit three specimens per class showing use of the mark in commerce with the application. If, instead, the application is based on having a bona fide intention to use mark in commerce, you must submit three specimens per class at the time the applicant files either an Amendment to Allege Use or a Statement of Use.

What to File as a Specimen

The specimens must be actual samples of how the mark is actually being used in commerce.

If the mark is used on goods, examples of acceptable specimens are tags or labels which are attached to the goods, containers for the goods, displays associated with the goods, or photographs of the goods showing use of the mark on the goods themselves. If it is impractical to send an actual specimen because of its size, photographs or other acceptable reproductions that show the mark on the goods, or packaging for the goods, must be furnished. Invoices, announcements, order forms, bills of lading, leaflets, brochures, catalogs, publicity releases, letterhead, and business cards generally are not acceptable specimens for goods.

If the mark is used for services, examples of acceptable specimens are signs, brochures about the services, advertisements for the services, business cards or stationery showing the mark in connection with the services, or photographs which show the mark as it is used either in the rendering or advertising of the services. In the case of a service mark, the specimens must either show the mark and include some clear reference to the type of services rendered under the mark in some form of advertising, or show the mark as it is used in the rendering of the service, for example on a store front or the side of a delivery or service truck. The three specimens may be identical or they may be examples of three different uses showing the same mark.

Specimens may not be larger than 8½ inches by 11 inches (21.59 cm by 27.94 cm) and must be flat. Smaller specimens, such as labels, may be stapled to a sheet of paper and labeled "specimens." A separate sheet can be used for each class.

Additional Requirements for Intent-to-Use Applications

An applicant who files his or her application based on having a bona fide intention to use a mark in commerce must make use of the mark in commerce before the mark can register. After use in commerce begins, you must submit:

1. Three specimens evidencing use as discussed above

2. A fee of $100 per class of goods or services in the application

3. Either (1) an Amendment to Allege Use if the application has not yet been approved for publication (use PTO Form 1579), or (2) a Statement of Use if the mark has been published and the PTO has issued a Notice of Allowance (use PTO Form 1580).

If you will not make use of the mark in commerce within six months of the Notice of Allowance, you must file a Request for an Extension of Time to File a Statement of Use, or the application is abandoned. (Use PTO Form 1581, which is intended only for this purpose.)

See the instructions and information following the forms. The previous information about specimens, identifications of goods and services, and dates of use is also relevant to filing an Amendment to Allege Use or Statement of Use. Follow the instructions on these forms carefully. Failure to file the necessary papers in proper form within the time provided may result in abandonment of the application.

International schedule of classes of goods and services

Goods

1 Chemicals used in industry, science and photography, as well as in agriculture, horticulture and forestry; unprocessed artificial resins, unprocessed plastics; manures; fire extinguishing compositions; tempering and soldering preparations; chemical substances for preserving foodstuffs; tanning substances; adhesives used in industry.

2 Paints, varnishes, lacquers; preservatives against rust and against deterioration of wood; colorants; mordants; raw natural resins; metals in foil and powder form for painters, decorators, printers and artists.

3 Bleaching preparations and other substances for laundry use; cleaning polishing, scouring and abrasive preparations; soaps; perfumery, essential oils, cosmetics, hair lotions; dentifrices.

4 Industrial oils and greases; lubricants; dust absorbing, wetting and binding compositions; fuels (including motor spirit) and illuminants; candles, wicks.

5 Pharmaceutical, veterinary and sanitary preparations; dietetic substances adapted for medical use, food for babies; plasters, materials for dressings; material for stopping teeth, dental wax; disinfectants; preparations for destroying vermin; fungicides, herbicides.

6 Common metals and their alloys; metal building materials; transportable buildings of metal; materials of metal for railway tracks; non-electric cables and wires of common metal; ironmongery, small items of metal hardware; pipes and tubes of metal; safes; goods of common metal not included in other classes; ores.

7 Machines and machine tools; motors and engines (except for land vehicles); machine coupling and transmission components (except for land vehicles); agricultural implements; incubators for eggs.

8 Hand tools and implements (hand operated); cutlery; side arms; razors.

9 Scientific, nautical, surveying, electric, photographic, cinematographic, optical, weighing, measuring, signalling, checking (supervision), life-saving and teaching apparatus and instruments; apparatus for recording, transmission or reproduction of sound or images; magnetic data carriers, recording discs; automatic vending machines and mechanisms for coin operated apparatus; cash registers, calculating machines, data processing equipment and computers; fire-extinguishing apparatus.

10 Surgical, medical, dental and veterinary apparatus and instruments, artificial limbs, eyes and teeth; orthopedic articles; suture materials.

11 Apparatus for lighting, heating, steam generating, cooking, refrigerating, drying, ventilating, water supply and sanitary purposes.

12 Vehicles; apparatus for locomotion by land, air or water.

13 Firearms; ammunition and projectiles; explosives; fireworks.

14 Precious metals and their alloys and goods in precious metals or coated therewith, not included in other classes; jewellery, precious stones; horological and chronometric instruments.

15 Musical instruments.

16 Paper, cardboard and goods made from these materials, not included in other classes; printed matter; bookbinding material; photographs; stationery; adhesives for stationery or household purposes; artists' materials; paint brushes; typewriters and office requisites (except furniture); instructional and teaching material (except apparatus); playing cards; printers' type; printing blocks.

17 Rubber, gutta-percha, gum asbestos, mica and goods made from these materials and not included in other classes; plastics in extruded form for use in manufacture; packing, stopping and insulating materials; flexible pipes, not of metal.

18 Leather and imitations of leather, and goods made of these materials and not included in other classes; animal skins, hides; trunks and travelling bags; umbrellas, parasols and walking sticks; whips, harness and saddlery.

19 Building materials (non-metallic); non-metallic rigid pipes for building; asphalt, pitch and bitumen; non-metallic transportable buildings; monuments, not of metal.

20 Furniture, mirrors, picture frames; goods (not included in other classes) of wood, cork, reed, cane, wicker, horn, bone, ivory, whalebone, shell, amber, mother-of-pearl, meerschaum and substitutes for all these materials, or of plastics.

21 Household or kitchen utensils and containers (not of precious metal or coated therewith); combs and sponges; brushes (except paint brushes); brush-making materials; articles for cleaning purposes; steelwool; unworked or semi-worked glass (except glass used in building); glassware, porcelain and earthenware not included in other classes.

22 Ropes, string, nets, tents, awnings, tarpaulins, sails, sacks and bags (not included in other classes); padding and stuffing materials (except of rubber or plastics); raw fibrous textile materials.

23 Yarns and threads, for textile use.

24 Textiles and textile goods, not included in other classes; bed and table covers.

25 Clothing, footwear, headgear.

26 Lace and embroidery, ribbons and braid; buttons, hooks and eyes, pins and needles; artificial flowers.

27 Carpets, rugs, mates and matting, linoleum and other materials for covering existing floors; wall hangings (non-textile).

28 Games and playthings; gymnastic and sporting articles not included in other classes; decorations for Christmas trees.

29 Meat, fish, poultry and game; meat extracts; preserved, dried and cooked fruits and vegetables; jellies, james, fruit sauces; eggs, milk and milk products; edible oils and fats.

30 Coffee, tea, cocoa, sugar, rice, tapioca, sago, artificial coffee; flour and preparations made from cereals, bread, pastry and confectionery, honey, treacle; yeast, baking-powder; salt, mustard; vinegar, sauces (condiments); spices; ice.

31 Agricultural, horticultural and forestry products and grains not included in other classes; live animals; fresh fruits and vegetables; seeds, natural plants and flowers; foodstuffs for animals, malt.

32 Beers; mineral and aerated waters and other non-alcoholic drinks; fruit drinks and fruit juices; syrups and other preparations for making beverages.

33 Alcoholic beverages (except beers).

34 Tobacco; smokers' articles; matches.

Services

35 Advertising; business management; business administration; office functions.

36 Insurance; financial affairs; monetary affairs; real estate affairs.

37 Building construction; repair; installation services.

38 Telecommunications.

39 Transport; packaging and storage of goods; travel arrangement.

40 Treatment of materials.

41 Education; providing of training; entertainment; sporting and cultural activities.

42 Providing of food and drink; temporary accommodation; medical, hygienic and beauty care; veterinary and agricultural services; legal services; scientific and industrial research; computer programming; services that cannot be placed in other classes.

FORM **PTO-1594**
(Rev. 6-93)
OMB No. 0651-0011 (exp. 4/94)

RECORDATION FORM COVER SHEET
TRADEMARKS ONLY

U.S. DEPARTMENT OF COMMERCE
Patent and Trademark Office

Tab settings ⇨ ⇨ ⇨ ▼   ▼   ▼   ▼   ▼   ▼   ▼

To the Honorable Commissioner of Patents and Trademarks: Please record the attached original documents or copy thereof.

1. Name of conveying party(ies):	2. Name and address of receiving party(ies)

1. Name of conveying party(ies):

❑ Individual(s)   ❑ Association
❑ General Partnership   ❑ Limited Partnership
❑ Corporation-State
❑ Other_____

Additional name(s) of conveying party(ies) attached? ❑ Yes ❑ No

2. Name and address of receiving party(ies)

Name:_____

Internal Address:_____

Street Address:_____

City:_____ State:_____ ZIP:_____

❑ Individual(s) citizenship_____
❑ Association _____
❑ General Partnership_____
❑ Limited Partnership_____
❑ Corporation-State_____
❑ Other_____

If assignee is not domiciled in the United States, a domestic represetative designation is attached: ❑ Yes ❑ No
(Designations must be a separate document from assignment)
Additional name(s) & address(es) attached? ❑ Yes ❑ No

3. Nature of conveyance:

❑ Assignment   ❑ Merger
❑ Security Agreement   ❑ Change of Name
❑ Other _____

Execution Date: _____

4. Application number(s) or patent number(s):

A. Trademark Application No.(s)

B. Trademark Registration No.(s)

Additional numbers attached? ❑ Yes ❑ No

5. Name and address of party to whom correspondence concerning document should be mailed:

Name:_____

Internal Address:_____

_____

_____

Street Address:_____

_____

City:_____ State:_____ ZIP:_____

6. Total number of applications and registrations involved: ....

7. Total fee (37 CFR 3.41)............$_____

❑ Enclosed

❑ Authorized to be charged to deposit account

8. Deposit account number:

_____

(Attach duplicate copy of this page if paying by deposit account)

DO NOT USE THIS SPACE

9. Statement and signature.
To the best of my knowledge and belief, the foregoing information is true and correct and any attached copy is a true copy of the original document.

| _____ | _____ | _____ |
| Name of Person Signing | Signature | Date |

Total number of pages including cover sheet, attachments, and document:

Mail documents to be recorded with required cover sheet information to:
Commissioner of Patents & Trademarks, Box Assignments
Washington, D.C. 20231

Guidelines for Completing Trademarks Cover Sheets

Cover Sheet information must be submitted with each document to be recorded. If the document to be recorded concerns both patents and trademarks, separate patent and trademark cover sheets, including any attached pages for continuing information, must accompany the document. All pages of the cover sheet should be numbered consecutively, for example, if both a patent and trademark cover sheet is used, and information is continued on one additional page for both patents and trademarks, the pages of the cover sheet would be numbered form 1 to 4.

Item1. Name of Conveying Party(ies).

Enter the full name of the party(ies) conveying the interest. If there is more than one conveying party, enter a check mark in the "Yes" box to indicate that addtional information is attached. The name of the second and any subsequent conveying party(ies) should be placed on an attached page clearly identified as a continuation of the information in Item 1. Enter a check mark in the "No" box, if no informtion is contained on an attached page.

Item 2. Name and Address of Receiving Party(ies).

Enter the name and full address of the first party receiving the interest. If there is more than one party receiving the interest, enter a check mark in the "Yes" box to indicate that additional information is attached. If the receiving party is an assignee not domiciled in the United States, a designation of domestic representative is required. Place a check mark in appropriate box to indicate whether or not a designation of docmestic representative is attached. Enter a check mark in the "No" box if no information is contained on an attached page.

Item 3. Nature of Conveyance.

Place a check mark in the appropriate box describing the nature of the conveying document. If the "Other" box is checked, specify the nature of the conveyance. Enter the execution date of the document. It is preferable to use the name of the month, or an abbreviation of that name, in order that confusion over dates is minimized.

Item 4. Application Number(s) or Registration Number(s).

Indicate the application number(s) including series code and serial number, and/or registration number(s) against which the document is to be recorded. Enter a check mark in the appropriate box: "Yes" or "No" if addtional numbers appear on attached pages. Be sure to identify numbers included on attached pages as the continuation of Item 4.

Item 5. Name and Address of Party to whom correspondence concerning the document should be mailed.

Enter the name and full address of the party to whom correspondence is to be mailed.

Item 6. Total Applications and Patents Involved.

Enter the total number of applications and trademarks identified for recordation. Be sure to include all applications and registrations identified on the cover sheet and on additional pages.

Block 7. Total Fee Enclosed.

Enter the total fee enclosed or authorized to be charged. A fee is required for each application and patent against which the document is recorded.

Item 8. Deposit Account Number.

Enter the deposit account number to authorize charges. Attach a duplicate copy of cover sheet to be used for the deposit charge account transaction.

Item 9. Statement and Signature.

Enter the name of the person submitting the document. The submitter must sign and date the cover sheet, confirming that to the best of the persons knowledge and belief, the information contained on the cover sheet is correct and that any copy of the document is a true copy of the original document. Enter the total number of pages including the cover sheet, attachments, and document.

TRADEMARK/SERVICE MARK APPLICATION, PRINCIPAL REGISTER, WITH DECLARATION	MARK (Word(s) and/or Design)	CLASS NO. (If known)

TO THE ASSISTANT SECRETARY AND COMMISSIONER OF PATENTS AND TRADEMARKS:

APPLICANT'S NAME:

APPLICANT'S BUSINESS ADDRESS: _____
(Display address exactly as
it should appear on registration) _____
_____
_____

APPLICANT'S ENTITY TYPE: (**Check one** and supply requested information)

	Individual - Citizen of (Country):
	Partnership - State where organized (Country, if appropriate): _____ Names and Citizenship (Country) of General Partners: _____ _____ _____
	Corporation - State (Country, if appropriate) of Incorporation:
	Other (Specify Nature of Entity and Domicile):

GOODS AND/OR SERVICES:

Applicant requests registration of the trademark/service mark shown in the accompanying drawing in the United States Patent and Trademark Office on the Principal Register established by the Act of July 5, 1946 (15 U.S.C. 1051 et. seq., as amended) for the following goods/services (**SPECIFIC GOODS AND/OR SERVICES MUST BE INSERTED HERE**):
_____
_____
_____
_____
_____

BASIS FOR APPLICATION: (Check boxes which apply, **but never both the first AND second boxes**, and supply requested information related to each box checked.)

[ ]	Applicant is using the mark in commerce on or in connection with the above identified goods/services. (15 U.S.C. 1051(a), as amended.) Three specimens showing the mark as used in commerce are submitted with this application. •Date of first use of the mark in commerce which the U.S. Congress may regulate (for example, interstate or between the U.S. and a foreign country): _____ •Specify the type of commerce: _____ (for example, interstate or between the U.S. and a specified foreign country) •Date of first use anywhere (the same as or before use in commerce date): _____ •Specify manner or mode of use of mark on or in connection with the goods/services: _____ _____ (for example, trademark is applied to labels, service mark is used in advertisements)
[ ]	Applicant has a bona fide intention to use the mark in commerce on or in connection with the above identified goods/services. (15 U.S.C. 1051(b), as amended.) •Specify intended manner or mode of use of mark on or in connection with the goods/services: _____ _____ (for example, trademark will be applied to labels, service mark will be used in advertisements)
[ ]	Applicant has a bona fide intention to use the mark in commerce on or in connection with the above identified goods/services, and asserts a claim of priority based upon a foreign application in accordance with 15 U.S.C. 1126(d), as amended. • Country of foreign filing: _____ • Date of foreign filing: _____
[ ]	Applicant has a bona fide intention to use the mark in commerce on or in connection with the above identified goods/services and, accompanying this application, submits a certification or certified copy of a foreign registration in accordance with 15 U.S.C. 1126(e), as amended. • Country of registration: _____ • Registration number: _____

NOTE: Declaration, on Reverse Side, MUST be Signed

If submitted on one page, side two of the form should be "Upside Down" in relation to page 1.

DECLARATION

The undersigned being hereby warned that willful false statements and the like so made are punishable by fine or imprisonment, or both, under 18 U.S.C. 1001, and that such willful false statements may jeopardize the validity of the application or any resulting registration, declares that he/she is properly authorized to execute this application on behalf of the applicant; he/she believes the applicant to be the owner of the trademark/service mark sought to be registered, or, if the application is being filed under 15 U.S.C. 1051(b), he/she believes applicant to be entitled to use such mark in commerce; to the best of his/her knowledge and belief no other person, firm, corporation, or association has the right to use the above identified mark in commerce, either in the identical form thereof or in such near resemblance thereto as to be likely, when used on or in connection with the goods/services of such other person, to cause confusion, or to cause mistake, or to deceive; and that all statements made of his/her own knowledge are true and that all statements made on information and belief are believed to be true.

_____ _____
DATE   SIGNATURE

_____ _____
TELEPHONE NUMBER   PRINT OR TYPE NAME AND POSITION

INSTRUCTIONS AND INFORMATION FOR APPLICANT

TO RECEIVE A FILING DATE, THE APPLICATION MUST BE COMPLETED AND SIGNED BY THE APPLICANT AND SUBMITTED ALONG WITH:

1. The prescribed **FEE ($245.00 effective 10/1/93)*** for each class of goods/services listed in the application;
2. A **DRAWING PAGE** displaying the mark in conformance with 37 CFR 2.52;
3. If the application is based on use of the mark in commerce, **THREE (3) SPECIMENS** (evidence) of the mark as used in commerce for each class of goods/services listed in the application. All three specimens may be in the nature of: (a) labels showing the mark which are placed on the goods; (b) photographs of the mark as it appears on the goods, (c) brochures or advertisements showing the mark as used in connection with the services.
4. An **APPLICATION WITH DECLARATION** (this form) - The application must be signed in order for the application to receive a filing date. Only the following person may sign the declaration, depending on the applicant's legal entity: (a) the individual applicant; (b) an officer of the corporate applicant; (c) one general partner of a partnership applicant; (d) all joint applicants.

SEND APPLICATION FORM, DRAWING PAGE, FEE, AND SPECIMENS (IF APPROPRIATE) TO:
Commissioner of Patents and Trademarks
Box TRADEMARK
Washington, D.C. 20231

Additional information concerning the requirements for filing an application is available in a booklet entitled **Basic Facts About Registering a Trademark**, which may be obtained by writing to the above address or by calling: (703) 308-HELP.

*Fees are subject to change; changes usually take effect on October 1. If filing on or after October 1, 1994, please call the PTO to confirm the correct fee.

This form is estimated to take an average of 1 hour to complete, including time required for reading and understanding instructions, gathering necessary information, recordkeeping, and acutally providing the information. Any comments on this form, including the amount of time required to complete this form, should be sent to the Office of Management and Organization, U.S. Patent and Trademark Office, U.S. Department of Commerce, Washington, D.C. 20231, and to Paperwork Reduction Project 0651-0009, Office of Information and Regulatory Affairs, Office of Management and Budget, Washington, D.C. 20503. Do NOT send completed forms to either of these addresses.

<table>
<tr><td>STATEMENT OF USE
UNDER 37 CFR 2.88, WITH
DECLARATION</td><td>MARK (Identify the mark)</td></tr>
<tr><td></td><td>SERIAL NO.</td></tr>
</table>

TO THE ASSISTANT SECRETARY AND COMMISSIONER OF PATENTS AND TRADEMARKS:

APPLICANT NAME:

NOTICE OF ALLOWANCE ISSUE DATE:

Applicant requests registration of the above-identified trademarks/service mark in the United States Patent and Trademark Office on the Principal Register established by the Act of July 5, 1946 (15 U.S.C. 1051 et. seq., as amended). Three specimens per class showing the mark as used in commerce are submitted with this statement.

☐ Check here if a Request to Divide under 37 C.F.R. 2.87 is being submitted with this statement.

Applicant is using the mark in commerce on or in connection with the following goods/services: (Check One)

☐ Those goods/services identified in the Notice of Allowance in this application.

☐ Those goods/services identified in the Notice of Allowance in this application except: (Identify goods/services to be deleted from application) _____

_____

_____

Date of first use of mark in commerce
which the U.S. Congress may regulate: _____

Specify type of commerce: (e.g., interstate, between the U.S. and a specified foreign country) _____

_____

Date of first use anywhere: _____
(the same as or before use-in-commerce date)

Specify manner or mode of use of mark on or in connection with the goods/services: (e.g., trademark is applied to labels, service mark is used in advertisements) _____

_____

The undersigned being hereby warned that willful false statements and the like so made are punishable by fine or imprisonment, or both, under 18 U.S.C. 1001, and that such willful false statements may jeopardize the validity of the application or any resulting registration, declares that he/she is properly authorized to execute this Statement of Use on behalf of the applicant; he/she believes the applicant to be the owner of the trademark/service mark sought to be registered; the trademark/service mark is now in use in commerce; and all statements made of his/her own knowledge are true and all statements made on information and belief are believed to be true.

_____
Date

_____
Signature

_____
Telephone Number

_____
Print or Type Name and Position

PTO Form 1580 (REV. 8-93)
OMB No. 0651-0009
Exp. 6-30-95

U.S. Department of Commerce/Patent and Trademark Office

INSTRUCTIONS AND INFORMATION FOR APPLICANT

In an application based upon a bona fide intention to use a mark in commerce, applicant must use its mark in commerce before a registration will be issued. After use begins, the applicant must submit, along with evidence of use (specimens) and the prescribed fee(s), either:

(1) an Amendment to Allege Use under 37 CFR 2.76, or
(2) a Statement of Use under 37 CFR 2.88.

The difference between these two filings is the timing of the filing. Applicant may file an Amendment to Allege Use before approval of the mark for publication for opposition in the **Official Gazette,** or, if a final refusal has been issued, prior to the expiration of the six-month response period. Otherwise, applicant must file a Statement of Use after the Office issues a Notice of Allowance. The Notice of Allowance will issue after the opposition period is completed if no successful opposition is filed. Neither Amendment to Allege Use or Statement of Use papers will be accepted by the Office during the period of time between approval of the mark for publication for opposition in the **Official Gazette** and the issuance of the Notice of Allowance. Applicant may call (703) 305-8747 to determine whether the mark has been approved for publication for opposition in the **Official Gazette.**

Before filing an Amendment to Allege Use or a Statement of Use, applicant must use the mark in commerce on or in connection with **all** of the goods/services for which applicant will seek registration, **unless** applicant submits with the papers, a request to divide out from the application the goods or services to which the Amendment to Allege Use or Statement of Use pertains. (See: 37 CFR 2.87, Dividing an application)

Applicant **must** submit with an Amendment to Allege Use or a Statement of Use:

(1) the appropriate fee of $100.00* per class of goods/services listed in the Amendment to Allege Use or the Statement of Use, and
(2) three (3) specimens or facsimiles of the mark as used in commerce for each class of goods/services asserted (e.g., photograph of mark as it appears on goods, label containing mark which is placed on goods, or brochure or advertisement showing mark as used in connection with services).

Cautions/Notes concerning completion of this Statement of Use form:

(1) The statement of Use must be received in the PTO within six months of the mailing of the Notice of Allowance or within a granted extension period.
(2) The goods/services identified in the Statement of Use must be identical to the goods/services identified in the Notice of Allowance. Applicant may delete goods/services. Deleted goods/services may not be reinstated in the application at a later time.
(3) Applicant may list dates of use for only one item in each class of goods/services identified in the Statement of Use. However, applicant must have used the mark in commerce on all the goods/services in the class. Applicant must identify the particular item to which the dates apply.
(4) Only the following person may sign the verification of the Statement of Use, depending on the applicant's legal entity: (a) the individual applicant; (b) an officer of corporate applicant; (c) one general partner of partnership applicant; (d) all joint applicants.

MAIL COMPLETED FORM TO:

**COMMISSIONER OF PATENTS AND TRADEMARKS
BOX ITU
WASHINGTON, D.C. 20231**

*Fees are effective through 9/30/94 and subject to change, usually on October 1.

<table>
<tr><td>

REQUEST FOR EXTENSION OF TIME UNDER 37 CFR 2.89, TO FILE A STATEMENT OF USE, WITH DECLARATION

</td><td>

MARK (Identify the mark)

SERIAL NO.

</td></tr>
</table>

TO THE ASSISTANT SECRETARY AND COMMISSIONER OF PATENTS AND TRADEMARKS:

APPLICANT NAME:

NOTICE OF ALLOWANCE MAILING DATE:

Applicant requests a six-month extension of time to file the Statement of Use under 37 CFR 2.89 in this application.

☐   Check here if a Request to Divide under 37 C.F.R. 2.87 is being submitted with this request.

Applicant has a continued bona fide intention to use the mark in commerce on in connection with the following goods/services: (Check One below)

    ☐ Those goods/services identified in the Notice of Allowance in this application.

    ☐ Those goods/services identified in the Notice of Allowance in this application except: (Identify goods/services to be **deleted** from application) _____

_____

_____

This is the _____ request for an Extension of Time following mailing of the Notice of Allowance.
      (Specify: First - Fifth)

If this is not the first request for an Extension of Time, check one box below. If the first box is checked explain the circumstance(s) of the non-use in the space provided:

☐   Applicant has not used the mark in commerce yet on all goods/services specified in the Notice of Allowance; however, applicant has made the following ongoing efforts to use the mark in commerce on or in connection with each of the goods/services specified above:

_____

_____

      If additional space is needed, please attach a separate sheet to this form

☐   Applicant believes that it has made valid use of the mark in commerce, as evidenced by the Statement of Use submitted with this request; however, if the Statement of Use does not meet minimum requirements under 37 CFR 2.88(e), applicant will need additional time in which to file a new statement.

The undersigned being hereby warned that willful false statements and the like so made are punishable by fine or imprisonment, or both, under 18 U.S.C. 1001, and that such willful false statements may jeopardize the validity of the application or any resulting registration, declares that he/she is properly authorized to execute this Request for an Extension of Time to File a Statement of Use on behalf of the applicant; and that all statements made of his/her own knowledge are true and all statements made on information and belief are believed to be true.

<table>
<tr><td>Date</td><td>Signature</td></tr>
<tr><td>Telephone Number</td><td>Print or Type Name and Position</td></tr>
</table>

PTO Form 1580 (REV. 8-93)   U.S. Department of Commerce/Patent and Trademark Office
OMB No. 0651-0009
Exp. 6-30-95

INSTRUCTIONS AND INFORMATION FOR APPLICANT

Applicant must file a Statement of Use within six months after the mailing of the Notice of Allowance in an application based upon a bona fide intention to use a mark in commerce, UNLESS, within that same period, applicant submits a request for a six-month extension of time to file the Statement of Use. The written request **must**:

    (1) be received in the PTO within six months after the mailing of the Notice of Allowance,

    (2) include applicant's verified statement of continued bona fide intention to use the mark in commerce,

    (3) specify the goods/services to which the request pertains as they are identified in the Notice of Allowance, and

    (4) include a fee of $100* for each class of goods/services.

Applicant may request four further six-month extensions of time. No extensions may extend beyond 36 months from the issue date of the Notice of Allowance. Each further request must be received in the PTO within the previously granted six-month extension period and must include, in addition to the above requirements, a showing of **GOOD CAUSE**. This good cause showing must include:

    (1)   applicant's statement that the mark has not been used in commerce yet on all the goods or services specified in the Notice of Allowance with which applicant has a continued bona fide intention to use the mark in commerce, **and**

    (2)   applicant's statement of ongoing efforts to make such use, which may include the following: (a) product or service research or development, (b) market research, (c) promotional activities, (d) steps to acquire distributors, (e) steps to obtain required governmental approval, or (f) similar specified activity.

Applicant may submit one additional six-month extension request during the existing period in which applicant files the Statement of Use, unless the granting of this request would extend the period beyond 36 months from the issue date of the Notice of Allowance. As a showing of good cause for such a request, applicant should state its belief that applicant has made valid use of the mark in commerce, as evidenced by the submitted Statement of Use, but that if the Statement is found by the PTO to be defective, applicant will need additional time in which to file a new statement of use.

Only the following person may sign the verification of the Request for Extension of Time, depending on the applicant's legal entity: (a) the individual applicant; (b) an officer of corporate applicant; (c) one general partner of partnership applicant; (d) all joint applicants.

*Fees are effective through 9/30/94 and subject to change, usually on October 1.

MAILING INSTRUCTIONS

MAIL COMPLETED FORM TO:

**COMMISSIONER OF PATENTS AND TRADEMARKS
BOX ITU
WASHINGTON, D.C. 20231**

You can ensure timely filing of this form by following the procedure described in 37 CFR 1.10 as follows: (1) on or before the due date for filing this form, deposit the completed form with the U.S. Post Office using the "Express Mail Post Office to Addressee" Service; (2) include a certificate of "Express Mail" under 37 CFR 1.10. Papers properly mailed under 37 CFR 1.10 are considered received by the PTO on the date that they are deposited with the Post Office.

When placing the certificate directly on the correspondence, use the following language:

Certificate of Express Mail Under 37 CFR 1.10

"Express Mail" mailing label number: _____

Date of Deposit: _____

I hereby certify that this paper and fee is being deposited with the United States Postal Service "Express Mail Post Office to Addressee" service under 37 CFR 1.10 on the date indicated above and is addressed to the Commissioner of Patents and Trademarks, Washington, D.C. 20231.

_____

(Typed or printed name of person mailing paper & fee)

_____

(Signature of person mailing paper & fee)

This form is estimated to take 15 minutes to complete including time required for reading and understanding instructions, gathering necessary information, record keeping and actually providing the information. Any comments on the amount of time you require to complete this form should be sent to the Office of Management and Organization, U.S. Patent and Trademark Office, U.S. Department of Commerce, Washington D.C. 20231, and to the Office of Information and Regulatory Affairs, Office of Management and Budget, Washington, D.C. 20503. Do not send forms to either of these addresses.

<table>
<tr><td>**AMENDMENT TO ALLEGE USE UNDER 37 CFR 2.76, WITH DECLARATION**</td><td colspan="2">MARK (Identify the mark)</td></tr>
<tr><td></td><td colspan="2">SERIAL NO.</td></tr>
</table>

TO THE ASSISTANT SECRETARY AND COMMISSIONER OF PATENTS AND TRADEMARKS:

APPLICANT NAME:

Applicant requests registration of the above-identified trademarks/service mark in the United States Patent and Trademark Office on the Principal Register established by the Act of July 5, 1946 (15 U.S.C. 1051 et. seq., as amended). Three specimens per class showing the mark as used in commerce are submitted with this amendment.

☐ Check here if Request to Divide under 37 C.F.R. 2.87 is being submitted with this amendment.

Applicant is using the mark in commerce on or in connection with the following goods/services:

_____

_____

(NOTE: Goods/Services listed above may not be broader than the goods/services identified in this application currently)

Date of first use of mark in commerce
which the U.S. Congress may regulate: _____

Specify type of commerce: (e.g., interstate, between the U.S. and a specified foreign country) _____

_____

Date of first use anywhere: _____
(the same as or before use-in-commerce date)

Specify manner or mode of use of mark on or in connection with the goods/services: (e.g., trademark is applied to labels, service mark is used in advertisements) _____

_____

The undersigned being hereby warned that willful false statements and the like so made are punishable by fine or imprisonment, or both, under 18 U.S.C. 1001, and that such willful false statements may jeopardize the validity of the application or any resulting registration, declares that he/she is properly authorized to execute this Amendment to Allege Use on behalf of the applicant; he/she believes the applicant to be the owner of the trademark/service mark sought to be registered; the trademark/service mark is now in use in commerce; and that all statements made of his/her own knowledge are true and all statements made on information and belief are believed to be true.

_____   _____
Date   Signature

_____   _____
Telephone Number   Print or Type Name and Position

PTO Form 1579 (REV. 8-93)   U.S. Department of Commerce/Patent and Trademark Office
OMB No. 0651-0009
Exp. 6-30-95

INSTRUCTIONS AND INFORMATION FOR APPLICANT

In an application based upon a bona fide intention to use a mark in commerce, applicant must use its mark in commerce before a registration will be issued. After use begins, the applicant must submit, along with evidence of use (specimens) and the prescribed fee(s), **either**:

    (1) an Amendment to Allege Use under 37 CFR 2.76, or
    (2) a Statement of Use under 37 CFR 2.88.

The difference between these two filings is the timing of the filing. Applicant may file an Amendment to Allege Use before approval of the mark for publication for opposition in the **Official Gazette,** or, if a final refusal has been issued, prior to the expiration of the six-month response period. Otherwise, applicant must file a Statement of Use after the Office issues a Notice of Allowance. The Notice of Allowance will issue after the opposition period is completed if no successful opposition is filed. Neither Amendment to Allege Use or Statement of Use papers will be accepted by the Office during the period of time between approval of the mark for publication for opposition in the **Official Gazette** and the issuance of the Notice of Allowance.

Applicant may call (703) 305-8747 to determine whether the mark has been approved for publication for opposition in the **Official Gazette**.

Before filing an Amendment to Allege Use or a Statement of Use, applicant must use the mark in commerce on or in connection with **all** of the goods/services for which applicant will seek registration, **unless** applicant submits with the papers, a request to divide out from the application the goods or services to which the Amendment to Allege Use or Statement of Use pertains. (See: 37 CFR 2.87, Dividing an application)

Applicant **must** submit with an Amendment to Allege Use or a Statement of Use:

    (1) the appropriate fee of $100.00* per class of goods/services listed in the Amendment to Allege Use or the Statement of Use, and
    (2) three (3) specimens or facsimiles of the mark as used in commerce for each class of goods/services asserted (e.g., photograph of mark as it appears on goods, label containing mark which is placed on goods, or brochure or advertisement showing mark as used in connection with services).

Cautions/Notes concerning completion of this Amendment to Allege Use form:

    (1) The goods/services identified in the Amendment to Allege Use must be identical to the goods/services identified in the application currently. Applicant may delete goods/services. Deleted goods/services may not be reinstated in the application at a later time.
    (2) Applicant may list dates of use for only one item in each class of goods/services identified in the Statement of Use. However, applicant must have used the mark in commerce on all the goods/services in the class. Applicant must identify the particular item to which the dates apply.
    (3) Only the following person may sign the verification of the Amendment to Allege Use, depending on the applicant's legal entity: (a) the individual applicant; (b) an officer of corporate applicant; (c) one general partner of partnership applicant; (d) all joint applicants.

> MAIL COMPLETED FORM TO:
>
> **COMMISSIONER OF PATENTS AND TRADEMARKS**
> **ATTN: AAU**
> **WASHINGTON, D.C. 20231**

*Fees are effective through 9/30/94 and subject to change, usually on October 1.

This form is estimated to take 15 minutes to complete including time required for reading and understanding instructions, gathering necessary information, record keeping and actually providing the information. Any comments on the amount of time you require to complete this form should be sent to the Office of Management and Organization, U.S. Patent and Trademark Office, U.S. Department of Commerce, Washington D.C. 20231, and to the Office of Information and Regulatory Affairs, Office of Management and Budget, Washington, D.C. 20503. Do not send forms to either of these addresses.

<table>
<tr>
<td rowspan="2">

DECLARATION OF USE OF A MARK UNDER SECTION 8 OF THE TRADEMARK ACT OF 1946, AS AMENDED

</td>
<td colspan="2">MARK (Identify the mark)</td>
</tr>
<tr>
<td>REGISTRATION NO.</td>
<td>DATE OF REGISTRATION:</td>
</tr>
</table>

TO THE ASSISTANT SECRETARY AND COMMISSIONER OF PATENTS AND TRADEMARKS:

REGISTRANT'S NAME:[1]

REGISTRANT'S CURRENT MAILING ADDRESS: _____
_____
_____

GOODS AND/OR SERVICES AND USE IN COMMERCE STATEMENT:

The mark shown in Registration No. _____ owned by the above-identified registrant is in use in

_____ commerce on or in connection with all of the goods and/or services identified in the
 (type of)[2]

registration, (*except* for the following)[3] _____

_____

as evidenced by the attached specimen(s)[4] showing the mark as currently used.

DECLARATION

The undersigned being hereby warned that willful false statements and the like so made are punishable by fine or imprisonment, or both, under 18 U.S.C. 1001, and that such willful false statements may jeopardize the validity of this document, declares that he/she is properly authorized to execute this document on behalf of the registrant; he/she believes the registrant to be the owner of the above identified registration; the trademark/service mark is in use in commerce; and all statements made of his/her own knowledge are true and all statements made on information and belief are believed to be true.

_____   _____
Date   Signature

_____   _____
Telephone Number   Print or Type Name and Position
                                                      [if applicable][5]

PTO Form 1583 (Rev. 1/93)   U.S. DEPARTMENT OF COMMERCE/Patent and Trademark Office
OMB No. 0651-0009 (Exp. 6/30/95)

168

FOOTNOTES

1. The present owner of the registration must file this form between the 5th and 6th year after the date of registration. If ownership of the registration has changed since the registration date, provide supporting documentation if available or a verified explanation. The present owner should refer to itself as the registrant.

2. "Type of Commerce" must be specified as "interstate," "territorial," "foreign," or such other commerce as may lawfully be regulated by Congress. Foreign registrants must specify commerce which Congress may regulate, using wording such as "foreign commerce between the U.S. and a foreign country."

3. List only those goods and/or services for which registrant is no longer using the mark. You should fill in this blank only if you are no longer using the mark on all the goods or services in the registration.

4. A specimen showing current use of the registered mark for at least one product or service in each class of the registration must be submitted with this form. Examples of specimens are tags or labels for goods, and advertisements for services.

5. If the present owner is an individual, the individual should sign the declaration.

   If the present owner is a partnership, the declaration should be signed by a General Partner.

   If the present owner is a corporation or similar juristic entity, the declaration should be signed by an officer of the corporation/entity. Please print or type the officer title of the person signing the declaration.

NOTE: If the registration is owned by more than one party, as joint owners, each owner must sign this declaration.

FEES

For each declaration under Section 8, the required fee is $100.00 per international class. Please be aware that our fees may change. Changes, if any, are normally effective October 1 of each year. If this declaration is intended to cover less than the total number of classes in the registration, please specify the classes for which the declaration is submitted. The declaration, with appropriate fee(s), should be sent to:

```
Assistant Commissioner for Trademarks
2900 Crystal Drive
Arlington, VA  22202-3513
```

MAILING INSTRUCTION BOX

You can ensure timely filing of this form by following the procedure described in 37 CFR 1.10 as follows: (1) on or before the due date for filing this form, deposit the completed form with the U.S. Post Office using the "Express Mail Post Office to Addressee" Service; (2) include a certificate of "Express Mail" under 37 CFR 1.10. Papers properly mailed under 37 CFR 1.10 are considered received by the PTO on the date that they are deposited with the Post Office.

When placing the certificate directly on the correspondence, use the following language:

Certificate of Express Mail Under 37 CFR 1.10

"Express Mail" mailing label number: _____
Date of Deposit: _____
I hereby certify that this paper and fee is being deposited with the United States Postal Service "Express Mail Post Office to Addressee" service under 37 CFR 1.10 on the date indicated above and is addressed to the Commissioner of Patents and Trademarks, Washington, D.C. 20231.

_____   _____
(Typed or printed name of person mailing   (Signature of person mailing paper & fee)
paper & fee)

This form is estimated to take 15 minutes to complete. Time will vary depending upon the needs of the individual case. Any comments on the amount of time you require to complete this form should be sent to the Office of Management and Organization, U.S. Patent and Trademark Office, U.S. Department of Commerce, Washington, D.C. 20231, and to the Office of Information and Regulatory Affairs, Office of Management and Budget, Washington, D.C. 20503. **DO NOT SEND FORMS TO EITHER OF THESE ADDRESSES.**

SAMPLE WRITTEN APPLICATION BASED ON USE IN COMMERCE
(Two classes)

TRADEMARK/SERVICE MARK APPLICATION, PRINCIPAL REGISTER, WITH DECLARATION	MARK (Word(s) and/or Design) 1 PINSTRIPES AND DESIGN	CLASS NO. 2 (If known) 16 & 35

TO THE ASSISTANT SECRETARY AND COMMISSIONER OF PATENTS AND TRADEMARKS:

APPLICANT'S NAME: Pinstripes Inc.   3

APPLICANT'S BUSINESS ADDRESS: 100 Main Street   4
(Display address exactly as it should appear on registration)   Anytown, Missouri 12345

APPLICANT'S ENTITY TYPE: (Check one and supply requested information)

	Individual - Citizen of (Country):	5a
	Partnership - State where organized (Country, if appropriate): _____ Names and Citizenship (Country) of General Partners: _____	5b
X	Corporation - State (Country, if appropriate) of Incorporation: Missouri	5c
	Other (Specify Nature of Entity and Domicile):	5d

GOODS AND/OR SERVICES:

Applicant requests registration of the trademark/service mark shown in the accompanying drawing in the United States Patent and Trademark Office on the Principal Register established by the Act of July 5, 1946 (15 U.S.C. 1051 et. seq., as amended) for the following goods/services (SPECIFIC GOODS AND/OR SERVICES MUST BE INSERTED HERE): 6
Magazines in the field of buisness management (Class 16); business management consulting services (Class 35)

BASIS FOR APPLICATION: (Check boxes which apply, but never both the first AND second boxes. and supply requested information related to each box checked.)

XX 7(a)	Applicant is using the mark in commerce on or in connection with the above identified goods/services. (15 U.S.C. 1051(a), as amended.) Three specimens showing the mark as used in commerce are submitted with this application. •Date of first use of the mark in commerce which the U.S. Congress may regulate (for example, interstate or between the U.S. and a foreign country): (Class 16) 1/15/92; (Class 35) 8/27/90 •Specify the type of commerce: Interstate (for example. interstate or between the U.S. and a specified foreign country) •Date of first use anywhere (the same as or before use in commerce date): Cl 16-1/15/92; Cl 35-8/27/90 •Specify manner or mode of use of mark on or in connection with the goods/services: On the magazines and in advertisements for the services (for example. trademark is applied to labels. service mark is used in advertisements)
[ ] 7(b)	Applicant has a bona fide intention to use the mark in commerce on or in connection with the above identified goods/services. (15 U.S.C. 1051(b), as amended.) •Specify intended manner or mode of use of mark on or in connection with the goods/services: _____ (for example. trademark will be applied to labels. service mark will be used in advertisements)
[ ] 7(c)	Applicant has a bona fide intention to use the mark in commerce on or in connection with the above identified goods/services, and asserts a claim of priority based upon a foreign application in accordance with 15 U.S.C. 1126(d), as amended. • Country of foreign filing: _____ • Date of foreign filing: _____
[ ] 7(d)	Applicant has a bona fide intention to use the mark in commerce on or in connection with the above identified goods/services and, accompanying this application, submits a certification or certified copy of a foreign registration in accordance with 15 U.S.C. 1126(e), as amended. • Country of registration: _____ • Registration number: _____

NOTE: Declaration, on Reverse Side, MUST be Signed

and the like so made are , and that such willful false ulting registration, declares that of the applicant; he/she believes t to be registered, or, if the applicant to be entitled to use such other person, firm, corporation, or erce, either in the identical form thereof or in such near resemblance thereto as to be likely, when used on or in connection with the goods/services of such other person, to cause confusion, or to cause mistake, or to deceive; and that all statements made of his/her own knowledge are true and that all statements made on information and belief are believed to be true.

January 16, 1992
DATE

SIGNATURE

(123) 456-7890
TELEPHONE NUMBER

John Doe, Jr., President
PRINT OR TYPE NAME AND POSITION

SAMPLE WRITTEN APPLICATION
BASED ON INTENT TO USE IN COMMERCE
(One class)

TRADEMARK/SERVICE MARK APPLICATION, PRINCIPAL REGISTER, WITH DECLARATION	MARK (Word(s) and/or Design) THEORYTEC	1	CLASS NO. (If known) 9	2

TO THE ASSISTANT SECRETARY AND COMMISSIONER OF PATENTS AND TRADEMARKS:

APPLICANT'S NAME: A-OK Software Development Group   3

APPLICANT'S BUSINESS ADDRESS: 100 Main Street   4
(Display address exactly as   Anytown, Missouri 12345
it should appear on registration)

APPLICANT'S ENTITY TYPE: (**Check one** and supply requested information)

	Individual - Citizen of (Country):	5a
X	Partnership - State where organized (Country, if appropriate): _____ Names and Citizenship (Country) of General Partners: Mary Baker, citizen of the USA; Harry Parker, citizen of the USA; and Jane Witlow, citizen of the USA	5b
	Corporation - State (Country, if appropriate) of Incorporation:	5c
	Other (Specify Nature of Entity and Domicile):	5d

GOODS AND/OR SERVICES:

Applicant requests registration of the trademark/service mark shown in the accompanying drawing in the United States Patent and Trademark Office on the Principal Register established by the Act of July 5, 1946 (15 U.S.C. 1051 et. seq., as amended) for the following goods/services (**SPECIFIC GOODS AND/OR SERVICES MUST BE INSERTED HERE**):
Computer software for analyzing statistics   6

BASIS FOR APPLICATION: (Check boxes which apply, but **never both the first AND second boxes**, and supply requested information related to each box checked.)

[ ] 7(a)	Applicant is using the mark in commerce on or in connection with the above identified goods/services. (15 U.S.C. 1051(a), as amended.) Three specimens showing the mark as used in commerce are submitted with this application. • Date of first use of the mark in commerce which the U.S. Congress may regulate (for example, interstate or between the U.S. and a foreign country): _____ • Specify the type of commerce: _____ (for example, interstate or between the U.S. and a specified foreign country) • Date of first use anywhere (the same as or before use in commerce date): _____ • Specify manner or mode of use of mark on or in connection with the goods/services: _____ (for example, trademark is applied to labels, service mark is used in advertisements)
[X] 7(b)	Applicant has a bona fide intention to use the mark in commerce on or in connection with the above identified goods/services. (15 U.S.C. 1051(b), as amended.) • Specify intended manner or mode of use of mark on or in connection with the goods/services: On labels affixed to the software (for example, trademark will be applied to labels, service mark will be used in advertisements)
[ ] 7(c)	Applicant has a bona fide intention to use the mark in commerce on or in connection with the above identified goods/services, and asserts a claim of priority based upon a foreign application in accordance with 15 U.S.C. 1126(d), as amended. • Country of foreign filing: _____ • Date of foreign filing: _____
[ ] 7(d)	Applicant has a bona fide intention to use the mark in commerce on or in connection with the above identified goods/services and, accompanying this application, submits a certification or certified copy of a foreign registration in accordance with 15 U.S.C. 1126(e), as amended. • Country of registration: _____ • Registration number: _____

NOTE: Declaration, on Reverse Side, MUST be Signed

and the like so made are 1, and that such willful false sulting registration, declares that of the applicant; he/she believes it to be registered, or, if the applicant to be entitled to use such mark in commerce; to the best of his/her knowledge and belief no other person, firm, corporation, or association has the right to use the above identified mark in commerce, either in the identical form thereof or in such near resemblance thereto as to be likely, when used on or in connection with the goods/services of such other person, to cause confusion, or to cause mistake, or to deceive; and that all statements made of his/her own knowledge are true and that all statements made on information and belief are believed to be true.

February 2, 1992
DATE

(123) 456-7890
TELEPHONE NUMBER

Mary Baker
SIGNATURE

Mary Baker, General Partner
PRINT OR TYPE NAME AND POSITION

SAMPLE DRAWING - SPECIAL FORM

8½" (21.6 cm)

APPLICANT'S NAME: Pinstripes Inc.

APPLICANT'S ADDRESS: 100 Main Street, Anytown, MO 12345

GOODS AND SERVICES: Magazines in the field of business
management; business management
consulting services

FIRST USE: Magazines (Class 16) January 15, 1992
Consulting (Class 35) August 27, 1990

FIRST USE IN COMMERCE: Magazines (Class 16) January 15, 1992
Consulting (Class 35) August 27, 1990

DESIGN: A zebra

11"
(27.9
cm)

Pinstripes

SAMPLE DRAWING - TYPEWRITTEN

8½" (21.6 cm)

11"
(27.9
cm)

APPLICANT'S NAME: A-OK Software Development Group

APPLICANT'S ADDRESS: 100 Main Street, Anytown, MO 12345

GOODS: Computer software for analyzing statistics.

DATE OF FIRST USE: Intent-to-Use Application

DATE OF FIRST USE IN COMMERCE: Intent-to-Use Application

THEORYTEC

How to Apply for Copyrights

The Copyright Office

The Copyright Office is one of the major service units of the Library of Congress. It employs more than 500 people. Its archives contain over 41 million individual cards. In fiscal year 1994, the Copyright Office registered more than 520,000 claims to copyright.

Execution becomes content in a work of genius.

What do Hal David's song "Do You Know the Way to San Jose?," Cadaco's hit game "Adver*teasing,*" *Baby Talk* magazine, the film *The Graduate,* and this book have in common? If you said they are all protected by copyright, you are correct.

Copyrights are very different from patents and trademarks. A patent primarily prevents inventions, discoveries, or advancements of useful processes from being manufactured, used, or marketed by anyone other than the patentee. A trademark is a word, name, or symbol to indicate origin, and in so doing distinguish the products and services of one company from those of another. Copyrights protect the form of expression rather than the subject matter of the writing. They protect the original works of authors and other creative people against copying and unauthorized public performance.

The Copyright Office

Copyrights are not handled by the Patent and Trademark Office (PTO). For this discussion we move across the Potomac River from the PTO's Crystal City, Virginia, headquarters, up Independence Avenue, and onto Capitol Hill to the Library of Congress, which is primarily responsible for administering copyright law. When sending mail to the Copyright Office, use the following address: Register of Copyrights, Copyright Office, Library of Congress, Washington, D.C. 20559-6000.

Copyright Office Shoots. It Scores. All 'Net.

As this book goes to press, the Copyright Office is developing the Electronic Copyright Management System (ECMS), a new mechanism that would allow creators to register works electronically, transmitting both electronic registration information as well as the work itself via the Internet right into the Library of Congress Copyright Office.

According to Mary Berghaus Levering, associate register for national copyright programs, ECMS would add value by improving current services. It is designed to:

1. Expand the national centralized copyright registration database, identifying copyrighted works and their owners

2. Streamline and eventually eliminate the current cumbersome process for handling physical objects and documents throughout the copyright process

3. Gradually reduce the need for vast amounts of physical storage space for copyright deposits

New services available through ECMS will include:

- A new mechanism for creators to register their works electronically, transmitting both electronic registration information as well as digital work

- A national centralized online database for recording transfers and assignments of copyright ownership and provide Internet access to this database

- A centralized recording of licenses and licensing information of digital copyrighted works, including basic terms and conditions for access and royalty payments for use

- A platform for the Copyright Registration System to connect electronically with other national Rights Management Systems to simplify licensing of digital works in the future

- Authorized access through Rights Management Systems to digital works in the LC/Copyright Office repository

- Authorized access over the *Nil and Gil* to digital copyrighted works in other repositories

Of particular interest and benefit to inventors and/or creators of copyrightable material will be the ability of the system to receive digital submissions via the Internet, verify the authenticity and completeness of each, initiate debiting of the applicant's Copyright Office deposit account, create a tracking record, acknowledge receipt of the claim, make the claim and material available for online processing by examiners and catalogers, and return an electronic acknowledgment to the claimant.

Testbed System

The complete testbed system will be available in mid to late 1995 for trial use for copyright registration and deposit of unpublished journal articles on computer science from the following educational institutions: Carnegie-Mellon University, Cornell University, Massachusetts Institute of Technology, Stanford University, and the University of California at Berkeley.

Future Plans

After completion of the testbed project, the Copyright Office plans to build on the basic system developed during the test, incorporate changes based on analysis of the results, and start to expand the program systematically to other formats of copyrighted materials including books and published journal articles, music, sound recordings, and other visual materials submitted by representative groups of copyright remitters.

How to Check the Status of ECMS

For copyright questions, contact Mary Berghaus Levering, Associate Register for National Copyright Programs, U.S. Copyright Office, Library of Congress, Department 17, Washington, D.C. 20540-6001. Her telephone number is (202) 707-8350; the fax number is (202) 707-8366. She can also be reached by e-mail at: levering@mail.loc.gov.

Reach Out and Touch Someone

The public may visit the Copyright Public Information Office at 101 Independence Avenue SE, Washington, D.C. or call (202) 707-3000. Recorded information on copyright is available 24 hours a day, 7 days a week. Information specialists are on duty to answer queries by phone or in person from 8:30 a.m. to 5:00 p.m., Monday through Friday, except holidays. Mail should be sent to Register of Copyrights, Copyright Office, Library of Congress, Washington, D.C. 20559-6000.

For technical questions about the system, contact the Corporation for National Research Initiatives, 1895 Preston White Drive, Reston, Virginia 22091-5434.

What is a Copyright?

Copyright is a form of protection by the laws of the United States (Title 17, U.S. Code) for the creators of "original works of authorship" including literary, dramatic, musical, artistic, and certain other intellectual works.

What Can Be Copyrighted?

Copyright protection is available for both published and unpublished works. I slap copyright notices on everything I create. Copyrights can be as important to an inventor as to an author, which is why I have included it in this book. Instructions and other written instruments such as background papers, concept papers, drawings, photographs, and the like that relate to inventions are all protected under U.S. copyright laws.

How Do You Secure a Copyright?

The way in which copyright protection is secured under the present law is frequently misunderstood. In years past it was required that you fill out special forms and send them to the Library of Congress together with a check and a number of copies of the original work. Today, no publication or registration or other action in the Copyright Office is required to secure copyright under the new law.

Under present law, copyright is secured "automatically" when the work is created, and the work is "created" when it is fixed in a copy or phonographically recorded for the first time. In general, "copies" are material objects from which a work can be read or visually perceived either directly or with the aid of a machine or device, such as books, manuscripts, sheet music, film, videotape, or microfilm.

However, it is still prudent to make a formal application with the Library of Congress. This is to establish a "public record" of your claim and to receive a certificate of registration, which is required if you ever have to go into a court of law over infringement. If you receive a copyright within five years of publication, it will be considered to be *prima facie* evidence in a court of law.

Notice of Copyright

Before you publicly show or distribute your work, notice of copyright is required. The use of the copyright notice is your responsibility and does not need any special advance permission from, or registration with, the Copyright Office.

The notice for visually perceptible copies should contain the following three elements:

1. The symbol © or the word "Copyright," or the abbreviation "Copr."

2. The year of first publication of said work. In the case of complications or derivative works incorporating previously published material, the year of first publication of the compilation or derivative work is enough. The year may be omitted where a pictorial, graphic, or sculptural work, with accompanying text (if any), is reproduced in or on greeting cards, postcards, stationary, jewelry, dolls, toys, or any useful article.

Notable Dates

• August 18, 1787: James Madison submitted to the framers of the Constitution a provision "to secure to literary authors their copyrights for a limited time."

• May 31, 1790: First copyright law enacted under the new U.S. Constitution. Term of 14 years with privilege of renewal for term of 14 additional years.

• June 9, 1790: First copyright entry, *The Philadelphia Spelling Book* by John Barry, registered in the U.S. District Court of Pennsylvania.

3. The name of the owner of copyright in the work, or an abbreviation by which the name can be recognized, or a generally known alternative of the owner.

Example: © 1995 Richard C. Levy (You should affix the notice in such a way as to give it reasonable notice of the claim of copyright.)

How Long Does Copyright Protection Endure?

A work that was created on or after January 1, 1978, is automatically protected from the moment of its creation, and is usually given a term enduring for the author's life, plus an extra 50 years after the author's death. In the case of a "joint work prepared by two or more authors who did not work for hire," the term lasts for fifty years after the last surviving author's death. For works made for hire, and for anonymous and pseudonymous works (unless the author's identity is revealed in Copyright Office records), the duration of copyright will be seventy-five years from the publication or one hundred years from creation, whichever is shorter.

Works that were created before the present law came into effect, but had neither been published nor registered for copyright before January 1, 1978, have been automatically brought under the statute and are now given federal copyright protection. The duration of copyright in these works will generally be computed in the same way as for works created on or after January 1, 1978, (the life-plus-50 or 75/100-year terms apply to them as well). However, all works in this category are guaranteed at least 25 years of statutory protection.

Under the law in effect before 1978, copyright was secured either on the date a work was published, or on the date of registration if the work was registered in unpublished form. In either case, the copyright endured for a first term of 28 years from the date secured. During the last (28th) year of the first term, the copyright was eligible for renewal. The new copyright law has extended the renewal term from 28 to 47 years for copyrights that were subsisting on January 1, 1978, making these works eligible for a total term of protection of 75 years. However, the copyright must be renewed on time to receive the 47-year period of added protection.

What Copyrights Do Not Protect

Ideas cannot be copyrighted. The same is true of the name or title given to a product (these are protected by trademarks) and the method or methods for doing something.

Copyright protects only the particular manner in which you express yourself in a literary, artistic, or musical form. Copyright protection does not extend to ideas, systems, devices, or trademark material involved in the development, merchandising, or usage of a product.

Application Forms

The forms you will likely require include Form TX, Form VA, Form CA, and Form RE, all of which can be found at the end of this chapter. The cost is $20 per registration of copyright.

Form TX: For published and unpublished non-dramatic literary works. This comprises the broadest category, covering everything from novels to computer programs, game instructions, and invention proposals.

GATT on Copyrights

The General Agreement on Tariffs and Trade (GATT), which led to sweeping changes in U.S. patent law (effective June 8, 1995) is expected to impact copyright law beginning in January of 1996, according to attorney David Levy of the Library of Congress. For information on GATT, contact an attorney specializing in intellectual property, or ask your U.S. Representative in Washington, D.C. for an update.

Form VA: For published and unpublished works of the visual arts. This would be for artwork you may have developed as an adjunct to your invention, charts, technical drawings, diagrams, models, and works of artistic craftsmanship.

Form CA: For application for supplementary copyright registration. Use when an earlier registration has been made in the Copyright Office and some of the facts given in that registration are incorrect or incomplete. Form CA allows you to place the correct or complete fact on record.

Form RE: Renewal registration. Use when you wish to renew a copyright.

How to Investigate the Copyright Status of a Work

There are several ways to investigate whether a work is under copyright protection and, if so, the facts of the copyright.

1. Examine a copy of the work (or, if the work is a sound recording, examine the disk, tape cartridge, or cassette in which the recorded sound is fixed, or the album cover, sleeve, or container in which the recording is sold) for such elements as a copyright notice, place and date of publication, author and publisher;

2. Make a personal search of the Copyright Office catalogs and other record; or

3. Have the Copyright Office make a search for you.

Individual Searches of Copyright Records

The Copyright Office is located in the Library of Congress, James Madison Memorial Building, 101 Independence Avenue SE, Washington, D.C.

Most records in the Copyright Office are open to public inspection and searching from 8:30 a.m. to 5 p.m., Monday through Friday (except legal holidays). The various records freely available to the public include an extensive card catalog, an automated catalog containing records from 1978 on, record books, and microfilm records of assignments and related documents. Other records, including correspondence files and deposit copies, are not open to the public for searching. However, they may be inspected upon request and payment of a $20 per hour search fee.

If you wish to do your own searching in the Copyright Office's public files, you will be given assistance in locating the records you need and in learning searching procedures. If the Copyright Office staff actually makes the search for you, a search fee of $20 per hour will be charged. The search will not be done while you wait.

To save you time, I have included a copy of the Copyright Office's Search Request Form on page 179. Make as many copies of this form as you need and fill them out before you go to the Library of Congress.

search request form

Copyright Office
Library of Congress
Washington, D.C.
20559-6000

Reference & Bibliograpy
Section
(202) 707-6850
8:30 a.m.-5 p.m. Monday-Friday
Eastern Time

Type of work:

☐ Book   ☐ Music   ☐ Motion Picture   ☐ Drama   ☐ Sound Recording
☐ Photograph/Artwork   ☐ Map   ☐ Periodical   ☐ Contribution   ☐ Architectural Work

Search information you require:

☐ Registration   ☐ Renewal   ☐ Assignment   ☐ Address

Specifics of work to be searched:

TITLE: _____

AUTHOR: _____

COPYRIGHT CLAIMANT: _____
(name in © notice)

APPROXIMATE YEAR DATE OF PUBLICATION/CREATION: _____

REGISTRATION NUMBER (if known): _____

OTHER IDENTIFYING INFORMATION: _____

If you need more space please attach additional pages.

*Estimates are based on the Copyright Office fee of $20.00 an **hour or fraction of an hour** consumed.* The more information you furnish as a basis for the search the better service we can provide.*

Names, titles, and short phrases are not copyrightable.

Please read Circular 22 for more information on copyright searches.

YOUR NAME: _____ DATE: _____

ADDRESS: _____

DAYTIME TELEPHONE NO. (_____) _____

Convey results of estimate/search by telephone
☐ yes   ☐ no

Fee enclosed? ☐ yes   Amount $ _____
         ☐ no

*The Copyright Office has the authority to adjust fees at 5-year intervals, based on changes in the Consumer Price Index. The next adjustment is due in 1996. Please contact the Copyright Office after July 1995 to determine the actual fee schedule.

January 1994

☒ Filling Out Application Form TX

Detach and read these instructions before completing this form.
Make sure all applicable spaces have been filled in before you return this form.

BASIC INFORMATION

When to Use This Form: Use Form TX for registration of published or unpublished non-dramatic literary works, excluding periodicals or serial issues. This class includes a wide variety of works: fiction, nonfiction, poetry, textbooks, reference works, directories, catalogs, advertising copy, compilations of information, and computer programs. For periodicals and serials, use Form SE.

Deposit to Accompany Application: An application for copyright registration must be accompanied by a deposit consisting of copies or phonorecords representing the entire work for which registration is to be made. The following are the general deposit requirements as set forth in the statute:

Unpublished Work: Deposit one complete copy (or phonorecord).

Published Work: Deposit two complete copies (or one phonorecord) of the best edition.

Work First Published Outside the United States: Deposit one complete copy (or phonorecord) of the first foreign edition.

Contribution to a Collective Work: Deposit one complete copy (or phonorecord) of the best edition of the collective work.

The Copyright Notice: For works first published on or after March 1, 1989, the law provides that a copyright notice in a specified form "may be placed on all publicly distributed copies from which the work can be visually per-ceived." Use of the copyright notice is the responsibility of the copyright owner and does not require advance permission from the Copyright Office. The required form of the notice for copies generally consists of three elements: (1) the symbol "©," or the word "Copyright," or the abbreviation "Copr."; (2) the year of first publication; and (3) the name of the owner of copyright. For example: "© 1993 Jane Cole." The notice is to be affixed to the copies "in such manner and location as to give reasonable notice of the claim of copyright." Works first published prior to March 1, 1989, **must** carry the notice or risk loss of copyright protection.

For information about notice requirements for works published before March 1, 1989, or other copyright information, write: Information Section, LM-401, Copyright Office, Library of Congress, Washington, D.C. 20559.

LINE-BY-LINE INSTRUCTIONS

Please type or print using black ink.

1 SPACE 1: Title

Title of This Work: Every work submitted for copyright registration must be given a title to identify that particular work. If the copies or phonorecords of the work bear a title or an identifying phrase that could serve as a title, transcribe that wording *completely* and *exactly* on the application. Indexing of the registration and future identification of the work will depend on the information you give here.

Previous or Alternative Titles: Complete this space if there are any additional titles for the work under which someone searching for the registration might be likely to look or under which a document pertaining to the work might be recorded.

Publication as a Contribution: If the work being registered is a contribution to a periodical, serial, or collection, give the title of the contribution in the "Title of this Work" space. Then, in the line headed "Publication as a Contribution," give information about the collective work in which the contribution appeared.

2 SPACE 2: Author(s)

General Instructions: After reading these instructions, decide who are the "authors" of this work for copyright purposes. Then, unless the work is a "collective work," give the requested information about every "author" who contributed any appreciable amount of copyrightable matter to this version of the work. If you need further space, request Continuation sheets. In the case of a collective work such as an anthology, collection of essays, or encyclopedia, give information about the author of the collective work as a whole.

Name of Author: The fullest form of the author's name should be given. Unless the work was "made for hire," the individual who actually created the work is its "author." In the case of a work made for hire, the statute provides that "the employer or other person for whom the work was prepared is considered the author."

What is a "Work Made for Hire"? A "work made for hire" is defined as (1) "a work prepared by an employee within the scope of his or her employment"; or (2) "a work specially ordered or commissioned for use as a contribution to a collective work, as a part of a motion picture or other audiovisual work, as a translation, as a supplementary work, as a compilation, as an instructional text, as a test, as answer material for a test, or as an atlas, if the parties expressly agree in a written instrument signed by them that the works shall be considered a work made for hire." If you have checked "Yes" to indicate that the work was "made for hire," you must give the full legal name of the employer (or other person for whom the work was prepared). You may also include the name of the employee along with the name of the employer (for example: "Elster Publishing Co., employer for hire of John Ferguson").

"Anonymous" or "Pseudonymous" Work: An author's contribution to a work is "anonymous" if that author is not identified on the copies or phonorecords of the work. An author's contribution to a work is "pseudonymous" if that author is identified on the copies or phonorecords under a fictitious name. If the work is "anonymous" you may: (1) leave the line blank; or (2) state "anonymous" on the line; or (3) reveal the author's identity. If the work is "pseudonymous" you may: (1) leave the line blank; or (2) give the pseudonym and identify it as such (for example: "Huntley Haverstock, pseudonym"); or (3) reveal the author's name, making clear which is the real name and which is the pseudonym (for example, "Judith Barton, whose pseudonym is Madeline Elster"). However, the citizenship or domicile of the author **must** be given in all cases.

Dates of Birth and Death: If the author is dead, the statute requires that the year of death be included in the application unless the work is anonymous or pseudonymous. The author's birth date is optional but is useful as a form of identification. Leave this space blank if the author's contribution was a "work made for hire."

Author's Nationality or Domicile: Give the country of which the author is a citizen or the country in which the author is domiciled. Nationality or domicile **must** be given in all cases.

Nature of Authorship: After the words "Nature of Authorship," give a brief general statement of the nature of this particular author's contribution to the work. Examples: "Entire text"; "Coauthor of entire text"; "Chapters 11-14"; "Editorial revisions"; "Compilation and English translation"; "New text."

3 SPACE 3: Creation and Publication

General Instructions: Do not confuse "creation" with "publication." Every application for copyright registration must state "the year in which creation of the work was completed." Give the date and nation of first publication only if the work has been published.

Creation: Under the statute, a work is "created" when it is fixed in a copy or phonorecord for the first time. Where a work has been prepared over a period of time, the part of the work existing in fixed form on a particular date constitutes the created work on that date. The date you give here should be the year in which the author completed the particular version for which registration is now being sought, even if other versions exist or if further changes or additions are planned.

Publication: The statute defines "publication" as "the distribution of copies or phonorecords of a work to the public by sale or other transfer of ownership, or by rental, lease, or lending"; a work is also "published" if there has been an "offering to distribute copies or phonorecords to a group of persons for purposes of further distribution, public performance, or public display." Give the full date (month, day, year) when, and the country where, publication first occurred. If first publication took place simultaneously in the United States and other countries, it is sufficient to state "U.S.A."

4 SPACE 4: Claimant(s)

Name(s) and Address(es) of Copyright Claimant(s): Give the name(s) and address(es) of the copyright claimant(s) in this work even if the claimant is the same as the author. Copyright in a work belongs initially to the author of the work (including, in the case of a work made for hire, the employer or other person for whom the work was prepared). The copyright claimant is either the author of the work or a person or organization to whom the copyright initially belonging to the author has been transferred.

Transfer: The statute provides that, if the copyright claimant is not the author, the application for registration must contain "a brief statement of how the claimant obtained ownership of the copyright." If any copyright claimant named in space 4 is not an author named in space 2, give a brief statement explaining how the claimant(s) obtained ownership of the copyright. Examples: "By written contract"; "Transfer of all rights by author"; "Assignment"; "By will." Do not attach transfer documents or other attachments or riders.

5 SPACE 5: Previous Registration

General Instructions: The questions in space 5 are intended to show whether an earlier registration has been made for this work and, if so, whether there is any basis for a new registration. As a general rule, only one basic copyright registration can be made for the same version of a particular work.

Same Version: If this version is substantially the same as the work covered by a previous registration, a second registration is not generally possible unless: (1) the work has been registered in unpublished form and a second registration is now being sought to cover this first published edition; or (2) someone other than the author is identified as copyright claimant in the earlier registration, and the author is now seeking registration in his or her own name. If either of these two exceptions apply, check the appropriate box and give the earlier registration number and date. Otherwise, do not submit Form TX; instead, write the Copyright Office for information about supplementary registration or recordation of transfers of copyright ownership.

Changed Version: If the work has been changed and you are now seeking registration to cover the additions or revisions, check the last box in space 5, give the earlier registration number and date, and complete both parts of space 6 in accordance with the instructions below.

Previous Registration Number and Date: If more than one previous registration has been made for the work, give the number and date of the latest registration.

6 SPACE 6: Derivative Work or Compilation

General Instructions: Complete space 6 if this work is a "changed version," "compilation," or "derivative work" and if it incorporates one or more earlier works that have already been published or registered for copyright or that have fallen into the public domain. A "compilation" is defined as "a work formed by the collection and assembling of preexisting materials or of data that are selected, coordinated, or arranged in such a way that the resulting work as a whole constitutes an original work of authorship." A "derivative work" is "a work based on one or more preexisting works." Examples of derivative works include translations, fictionalizations, abridgments, condensations, or "any other form in which a work may be recast, transformed, or adapted." Derivative works also include works "consisting of editorial revisions, annotations, or other modifications" if these changes, as a whole, represent an original work of authorship.

Preexisting Material (space 6a): For derivative works, complete this space **and** space 6b. In space 6a identify the preexisting work that has been recast, transformed, or adapted. An example of preexisting material might be: "Russian version of Goncharov's 'Oblomov'." Do not complete space 6a for compilations.

Material Added to This Work (space 6b): Give a brief, general statement of the new material covered by the copyright claim for which registration is sought. **Derivative work** examples include: "Foreword, editing, critical annotations"; "Translation"; "Chapters 11-17." If the work is a **compilation**, describe both the compilation itself and the material that has been compiled. Example: "Compilation of certain 1917 Speeches by Woodrow Wilson." A work may be both a derivative work and compilation, in which case a sample statement might be: "Compilation and additional new material."

7 SPACE 7: Manufacturing Provisions

Due to the expiration of the Manufacturing Clause of the copyright law on June 30, 1986, this space has been deleted.

8 SPACE 8: Reproduction for Use of Blind or Physically Handicapped Individuals

General Instructions: One of the major programs of the Library of Congress is to provide Braille editions and special recordings of works for the exclusive use of the blind and physically handicapped. In an effort to simplify and speed up the copyright licensing procedures that are a necessary part of this program, section 710 of the copyright statute provides for the establishment of a voluntary licensing system to be tied in with copyright registration. Copyright Office regulations provide that you may grant a license for such reproduction and distribution solely for the use of persons who are certified by competent authority as unable to read normal printed material as a result of physical limitations. The license is entirely voluntary, nonexclusive, and may be terminated upon 90 days notice.

How to Grant the License: If you wish to grant it, check one of the three boxes in space 8. Your check in one of these boxes together with your signature in space 10 will mean that the Library of Congress can proceed to reproduce and distribute under the license without further paperwork. For further information, write for Circular 63.

9,10,11 SPACE 9,10,11: Fee, Correspondence, Certification, Return Address

Fee: The Copyright Office has the authority to adjust fees at 5-year intervals, based on changes in the Consumer Price Index. The next adjustment is due in 1996. Please contact the Copyright Office after July 1995 to determine the actual fee schedule.

Deposit Account: If you maintain a Deposit Account in the Copyright Office, identify it in space 9. Otherwise leave the space blank and send the fee of $20 with your application and deposit.

Correspondence (space 9) This space should contain the name, address, area code, and telephone number of the person to be consulted if correspondence about this application becomes necessary.

Certification (space 10): The application can not be accepted unless it bears the date and the **handwritten signature** of the author or other copyright claimant, or of the owner of exclusive right(s), or of the duly authorized agent of author, claimant, or owner of exclusive right(s).

Address for Return of Certificate (space 11): The address box must be completed legibly since the certificate will be returned in a window envelope.

FORM TX

For a Literary Work
UNITED STATES COPYRIGHT OFFICE

REGISTRATION NUMBER

_____

TX _____ TXU _____

EFFECTIVE DATE OF REGISTRATION

_____

Month _____ Day _____ Year _____

DO NOT WRITE ABOVE THIS LINE. IF YOU NEED MORE SPACE, USE A SEPARATE CONTINUATION SHEET.

1

TITLE OF THIS WORK ▼

PREVIOUS OR ALTERNATIVE TITLES ▼

PUBLICATION AS A CONTRIBUTION If this work was published as a contribution to a periodical, serial, or collection, give information about the collective work in which the contribution appeared. **Title of Collective Work ▼**

If published in a periodical or serial give: **Volume ▼**   **Number ▼**   **Issue Date ▼**   **On Pages ▼**

2

a

NAME OF AUTHOR ▼

DATES OF BIRTH AND DEATH
Year Born ▼   Year Died ▼

Was this contribution to the work a "work made for hire"?
☐ Yes
☐ No

AUTHOR'S NATIONALITY OR DOMICILE
Name of Country
OR { Citizen of ▶ _____
Domiciled in ▶ _____

WAS THIS AUTHOR'S CONTRIBUTION TO THE WORK
Anonymous? ☐ Yes ☐ No
Pseudonymous? ☐ Yes ☐ No
If the answer to either of these questions is "Yes," see detailed instructions.

NATURE OF AUTHORSHIP Briefly describe nature of material created by this author in which copyright is claimed. ▼

NOTE

Under the law, the "author" of a "work made for hire" is generally the employer, not the employee (see instructions). For any part of this work that was "made for hire" check "Yes" in the space provided, give the employer (or other person for whom the work was prepared) as "Author" of that part, and leave the space for dates of birth and death blank.

b

NAME OF AUTHOR ▼

DATES OF BIRTH AND DEATH
Year Born ▼   Year Died ▼

Was this contribution to the work a "work made for hire"?
☐ Yes
☐ No

AUTHOR'S NATIONALITY OR DOMICILE
Name of Country
OR { Citizen of ▶ _____
Domiciled in ▶ _____

WAS THIS AUTHOR'S CONTRIBUTION TO THE WORK
Anonymous? ☐ Yes ☐ No
Pseudonymous? ☐ Yes ☐ No
If the answer to either of these questions is "Yes," see detailed instructions.

NATURE OF AUTHORSHIP Briefly describe nature of material created by this author in which copyright is claimed. ▼

c

NAME OF AUTHOR ▼

DATES OF BIRTH AND DEATH
Year Born ▼   Year Died ▼

Was this contribution to the work a "work made for hire"?
☐ Yes
☐ No

AUTHOR'S NATIONALITY OR DOMICILE
Name of Country
OR { Citizen of ▶ _____
Domiciled in ▶ _____

WAS THIS AUTHOR'S CONTRIBUTION TO THE WORK
Anonymous? ☐ Yes ☐ No
Pseudonymous? ☐ Yes ☐ No
If the answer to either of these questions is "Yes," see detailed instructions.

NATURE OF AUTHORSHIP Briefly describe nature of material created by this author in which copyright is claimed. ▼

3

a **YEAR IN WHICH CREATION OF THIS WORK WAS COMPLETED** This information must be given _____ ◀ Year in all cases.

b **DATE AND NATION OF FIRST PUBLICATION OF THIS PARTICULAR WORK**
Complete this information ONLY if this work has been published.
Month ▶ _____ Day ▶ _____ Year ▶ _____
◀ Nation

4

See instructions before completing this space.

COPYRIGHT CLAIMANT(S) Name and address must be given even if the claimant is the same as the author given in space 2. ▼

TRANSFER If the claimant(s) named here in space 4 is (are) different from the author(s) named in space 2, give a brief statement of how the claimant(s) obtained ownership of the copyright. ▼

DO NOT WRITE HERE OFFICE USE ONLY

APPLICATION RECEIVED

ONE DEPOSIT RECEIVED

TWO DEPOSITS RECEIVED

FUNDS RECEIVED

MORE ON BACK ▶ • Complete all applicable spaces (numbers 5-11) on the reverse side of this page.
• See detailed instructions.   • Sign the form at line 10.

DO NOT WRITE HERE
Page 1 of _____ pages

DO NOT WRITE ABOVE THIS LINE. IF YOU NEED MORE SPACE, USE A SEPARATE CONTINUATION SHEET.

PREVIOUS REGISTRATION Has registration for this work, or for an earlier version of this work, already been made in the Copyright Office?

☐ **Yes** ☐ **No** If your answer is "Yes," why is another registration being sought? (Check appropriate box) ▼

a. ☐ This is the first published edition of a work previously registered in unpublished form.

b. ☐ This is the first application submitted by this author as copyright claimant.

c. ☐ This is a changed version of the work, as shown by space 6 on this application.

If your answer is "Yes," give: **Previous Registration Number** ▼   **Year of Registration** ▼

5

DERIVATIVE WORK OR COMPILATION Complete both space 6a and 6b for a derivative work; complete only 6b for a compilation.

a. Preexisting Material Identify any preexisting work or works that this work is based on or incorporates. ▼

b. Material Added to This Work Give a brief, general statement of the material that has been added to this work and in which copyright is claimed. ▼

6

See instructions before completing this space.

—space deleted—

7

REPRODUCTION FOR USE OF BLIND OR PHYSICALLY HANDICAPPED INDIVIDUALS A signature on this form at space 10 and a check in one of the boxes here in space 8 constitutes a non-exclusive grant of permission to the Library of Congress to reproduce and distribute solely for the blind and physically handicapped and under the conditions and limitations prescribed by the regulations of the Copyright Office: (1) copies of the work identified in space 1 of this application in Braille (or similar tactile symbols); or (2) phonorecords embodying a fixation of a reading of that work; or (3) both.

a ☐ Copies and Phonorecords   **b** ☐ Copies Only   **c** ☐ Phonorecords Only

8

See instructions.

DEPOSIT ACCOUNT If the registration fee is to be charged to a Deposit Account established in the Copyright Office, give name and number of Account.

Name ▼   **Account Number** ▼

9

CORRESPONDENCE Give name and address to which correspondence about this application should be sent.   Name/Address/Apt/City/State/ZIP ▼

Area Code and Telephone Number ▶

Be sure to give your daytime phone number ◀

CERTIFICATION* I, the undersigned, hereby certify that I am the

Check only one ▶ {
☐ author
☐ other copyright claimant
☐ owner of exclusive right(s)
☐ authorized agent of _____

Name of author or other copyright claimant, or owner of exclusive right(s) ▲

of the work identified in this application and that the statements made by me in this application are correct to the best of my knowledge.

10

Typed or printed name and date ▼ If this application gives a date of publication in space 3, do not sign and submit it before that date.

_____ date ▶ _____

☞ Handwritten signature (X) ▼

MAIL CERTIFI-CATE TO

Certificate will be mailed in window envelope

Name ▼

Number/Street/Apartment Number ▼

City/State/ZIP ▼

YOU MUST:
• Complete all necessary spaces
• Sign your application in space 10
SEND ALL 3 ELEMENTS IN THE SAME PACKAGE:
1. Application form
2. Nonrefundable $20 filing fee in check or money order payable to *Register of Copyrights*
3. Deposit material
MAIL TO:
Register of Copyrights
Library of Congress
Washington, D.C. 20559-6000

The Copyright Office has the authority to adjust fees at 5-year intervals, based on changes in the Consumer Price Index. The next adjustment is due in 1996. Please contact the Copyright Office after July 1995 to determine the actual fee schedule.

11

*17 U.S.C. § 506(e): Any person who knowingly makes a false representation of a material fact in the application for copyright registration provided for by section 409, or in any written statement filed in connection with the application, shall be fined not more than $2,500.

July 1993—300,000   ⊕ PRINTED ON RECYCLED PAPER   ☆U.S. GOVERNMENT PRINTING OFFICE: 1993-342-582/80,019

⬛Filling Out Application Form VA

Detach and read these instructions before completing this form.
Make sure all applicable spaces have been filled in before you return this form.

BASIC INFORMATION

When to Use This Form: Use Form VA for copyright registration of published or unpublished works of the visual arts. This category consists of "pictorial, graphic, or sculptural works," including two-dimensional and three-dimensional works of fine, graphic, and applied art, photographs, prints and art reproductions, maps, globes, charts, technical drawings, diagrams, and models.

What Does Copyright Protect? Copyright in a work of the visual arts protects those pictorial, graphic, or sculptural elements that, either alone or in combination, represent an "original work of authorship." The statute declares: "In no case does copyright protection for an original work of authorship extend to any idea, procedure, process, system, method of operation, concept, principle, or discovery, regardless of the form in which it is described, explained, illustrated, or embodied in such work."

Works of Artistic Craftsmanship and Designs: "Works of artistic craftsmanship" are registrable on Form VA, but the statute makes clear that protection extends to "their form" and not to "their mechanical or utilitarian aspects." The "design of a useful article" is considered copyrightable "only if, and only to the extent that, such design incorporates pictorial, graphic, or sculptural features that can be identified separately from, and are capable of existing independently of, the utilitarian aspects of the article."

Labels and Advertisements: Works prepared for use in connection with the sale or advertisement of goods and services are registrable if they contain "original work of authorship." Use Form VA if the copyrightable material in the work you are registering is mainly pictorial or graphic; use Form TX if it consists mainly of text. **NOTE:** Words and short phrases such as names, titles, and slogans cannot be protected by copyright, and the same is true of standard symbols, emblems, and other commonly used graphic designs that are in the public domain. When used commercially, material of that sort can sometimes be protected under state laws of unfair competition or under the Federal trademark laws. For information about trademark registration, write to the Commissioner of Patents and Trademarks, Washington, D.C. 20231.

Architectural Works: Copyright protection extends to the design of buildings created for the use of human beings. Architectural works created on or after December 1, 1990, or that on December 1, 1990, were unconstructed and embodied only in unpublished plans or drawings are eligible. Request Circular 41 for more information.

Deposit to Accompany Application: An application for copyright registration must be accompanied by a deposit consisting of copies representing the entire work for which registration is to be made.

Unpublished Work: Deposit one complete copy.

Published Work: Deposit two complete copies of the best edition.

Work First Published Outside the United States: Deposit one complete copy of the first foreign edition.

Contribution to a Collective Work: Deposit one complete copy of the best edition of the collective work.

The Copyright Notice: For works first published on or after March 1, 1989, the law provides that a copyright notice in a specified form "may be placed on all publicly distributed copies from which the work can be visually perceived." Use of the copyright notice is the responsibility of the copyright owner and does not require advance permission from the Copyright Office. The required form of the notice for copies generally consists of three elements: (1) the symbol "©", or the word "Copyright," or the abbreviation "Copr."; (2) the year of first publication; and (3) the name of the owner of copyright. For example: "© 1991 Jane Cole." The notice is to be affixed to the copies "in such manner and location as to give reasonable notice of the claim of copyright." Works first published prior to March 1, 1989, **must** carry the notice or risk loss of copyright protection.

For information about notice requirements for works published before March 1, 1989, or other copyright information, write: Information Section, LM-401, Copyright Office, Library of Congress, Washington, D.C. 20559-6000.

LINE-BY-LINE INSTRUCTIONS
Please type or print using black ink.

1 SPACE 1: Title

Title of This Work: Every work submitted for copyright registration must be given a title to identify that particular work. If the copies of the work bear a title (or an identifying phrase that could serve as a title), transcribe that wording *completely* and *exactly* on the application. Indexing of the registration and future identification of the work will depend on the information you give here. For an architectural work that has been constructed, add the date of construction after the title; if unconstructed at this time, add "not yet constructed."

Previous or Alternative Titles: Complete this space if there are any additional titles for the work under which someone searching for the registration might be likely to look, or under which a document pertaining to the work might be recorded.

Publication as a Contribution: If the work being registered is a contribution to a periodical, serial, or collection, give the title of the contribution in the "Title of This Work" space. Then, in the line headed "Publication as a Contribution," give information about the collective work in which the contribution appeared.

Nature of This Work: Briefly describe the general nature or character of the pictorial, graphic, or sculptural work being registered for copyright. Examples: "Oil Painting"; "Charcoal Drawing"; "Etching"; "Sculpture"; "Map"; "Photograph"; "Scale Model"; "Lithographic Print"; "Jewelry Design"; "Fabric Design."

2 SPACE 2: Author(s)

General Instruction: After reading these instructions, decide who are the "authors" of this work for copyright purposes. Then, unless the work is a "collective work," give the requested information about every "author" who contributed any appreciable amount of copyrightable matter to this version of the work. If you need further space, request Continuation Sheets. In the case of a collective work, such as a catalog of paintings or collection of cartoons by various authors, give information about the author of the collective work as a whole.

Name of Author: The fullest form of the author's name should be given. Unless the work was "made for hire," the individual who actually created the work is its "author." In the case of a work made for hire, the statute provides that "the employer or other person for whom the work was prepared is considered the author."

What is a "Work Made for Hire"? A "work made for hire" is defined as: (1) "a work prepared by an employee within the scope of his or her employment"; or (2) "a work specially ordered or commissioned for use as a contribution to a collective work, as a part of a motion picture or other audiovisual work, as a translation, as a supplementary work, as a compilation, as an instructional text, as a test, as answer material for a test, or as an atlas, if the parties expressly agree in a written instrument signed by them that the work shall be considered a work made for hire." If you have checked "Yes" to indicate that the work was "made for hire," you must give the full legal name of the employer (or other person for whom the work was prepared). You may also include the name of the employee along with the name of the employer (for example: "Elster Publishing Co., employer for hire of John Ferguson").

"Anonymous" or "Pseudonymous" Work: An author's contribution to a work is "anonymous" if that author is not identified on the copies or phonorecords of the work. An author's contribution to a work is "pseudonymous" if that author is identified on the copies or phonorecords under a fictitious name. If the work is "anonymous" you may: (1) leave the line blank; or (2) state "anonymous" on the line; or (3) reveal the author's identity. If the work is "pseudonymous" you may: (1) leave the line blank; or (2) give the pseudonym and identify it as such (for example: "Huntley Haverstock, pseudonym"); or (3) reveal the author's name, making clear which is the real name and which is the pseudonym (for example: "Henry Leek, whose pseudonym is Priam Farrel"). However, the citizenship or domicile of the author **must** be given in all cases.

Dates of Birth and Death: If the author is dead, the statute requires that the year of death be included in the application unless the work is anonymous or pseudonymous. The author's birth date is optional but is useful as a form of identification. Leave this space blank if the author's contribution was a "work made for hire."

Author's Nationality or Domicile: Give the country of which the author is a citizen or the country in which the author is domiciled. Nationality or domicile **must** be given in all cases.

Nature of Authorship: Catagories of pictorial, graphic, and sculptural authorship are listed below. Check the box(es) that best describe(s) each author's contribution to the work.

3-Dimensional sculptures: fine art sculptures, toys, dolls, scale models, and sculptural designs applied to useful articles.

2-Dimensional artwork: watercolor and oil paintings; pen and ink drawings; logo illustrations; greeting cards; collages; stencils; patterns; computer graphics; graphics appearing in screen displays; artwork appearing on posters, calendars, games, commercial prints and labels, and packaging, as well as 2-dimensional artwork applied to useful articles.

Reproductions of works of art: reproductions of preexisting artwork made by, for example, lithography, photoengraving, or etching.

Maps: cartographic representations of an area such as state and county maps, atlases, marine charts, relief maps, and globes.

Photographs: pictorial photographic prints and slides and holograms.

Jewelry designs: 3-dimensional designs applied to rings, pendants, earrings, necklaces, and the like.

Designs on sheetlike materials: designs reproduced on textiles, lace, and other fabrics; wallpaper; carpeting; floor tile; wrapping paper; and clothing.

Technical drawings: diagrams illustrating scientific or technical information in linear form such as architectural blueprints or mechanical drawings.

Text: textual material that accompanies pictorial, graphic, or sculptural works such as comic strips, greeting cards, games rules, commercial prints or labels, and maps.

Architectural works: designs of buildings, including the overall form as well as the arrangement and composition of spaces and elements of the design. NOTE: Any registration for the underlying architectural plans must be applied for on a separate Form VA, checking the box "Technical drawing."

3 SPACE 3: Creation and Publication

General Instructions: Do not confuse "creation" with "publication." Every application for copyright registration must state "the year in which creation of the work was completed." Give the date and nation of first publication only if the work has been published.

Creation: Under the statute, a work is "created" when it is fixed in a copy or phonorecord for the first time. Where a work has been prepared over a period of time, the part of the work existing in fixed form on a particular date constitutes the created work on that date. The date you give here should be the year in which the author completed the particular version for which registration is now being sought, even if other versions exist or if further changes or additions are planned.

Publication: The statute defines "publication" as "the distribution of copies or phonorecords of a work to the public by sale or other transfer of ownership, or by rental, lease, or lending"; a work is also "published" if there has been an "offering to distribute copies or phonorecords to a group of persons for purposes of further distribution, public performance, or public display." Give the full date (month, day, year) when, and the country where, publication first occurred. If first publication took place simultaneously in the United States and other countries, it is sufficient to state "U.S.A."

4 SPACE 4: Claimant(s)

Name(s) and Address(es) of Copyright Claimant(s): Give the name(s) and address(es) of the copyright claimant(s) in this work even if the claimant is the same as the author. Copyright in a work belongs initially to the author of the work (including, in the case of a work make for hire, the employer or other person for whom the work was prepared). The copyright claimant is either the author of the work or a person or organization to whom the copyright initially belonging to the author has been transferred.

Transfer: The statute provides that, if the copyright claimant is not the author, the application for registration must contain "a brief statement of how the claimant obtained ownership of the copyright." If any copyright claimant named in space 4 is not an author named in space 2, give a brief statement explaining how the claimant(s) obtained ownership of the copyright. Examples: "By written contract"; "Transfer of all rights by author"; "Assignment"; "By will." Do not attach transfer documents or other attachments or riders.

5 SPACE 5: Previous Registration

General Instructions: The questions in space 5 are intended to find out whether an earlier registration has been made for this work and, if so, whether

there is any basis for a new registration. As a rule, only one basic copyright registration can be made for the same version of a particular work.

Same Version: If this version is substantially the same as the work covered by a previous registration, a second registration is not generally possible unless: (1) the work has been registered in unpublished form and a second registration is now being sought to cover this first published edition; or (2) someone other than the author is identified as a copyright claimant in the earlier registration, and the author is now seeking registration in his or her own name. If either of these two exceptions apply, check the appropriate box and give the earlier registration number and date. Otherwise, do not submit Form VA; instead, write the Copyright Office for information about supplementary registration or recordation of transfers of copyright ownership.

Changed Version: If the work has been changed and you are now seeking registration to cover the additions or revisions, check the last box in space 5, give the earlier registration number and date, and complete both parts of space 6 in accordance with the instruction below.

Previous Registration Number and Date: If more than one previous registration has been made for the work, give the number and date of the latest registration.

6 SPACE 6: Derivative Work or Compilation

General Instructions: Complete space 6 if this work is a "changed version," "compilation," or "derivative work," and if it incorporates one or more earlier works that have already been published or registered for copyright, or that have fallen into the public domain. A "compilation" is defined as "a work formed by the collection and assembling of preexisting materials or of data that are selected, coordinated, or arranged in such a way that the resulting work as a whole constitutes an original work of authorship." A "derivative work" is "a work based on one or more preexisting works." Examples of derivative works include reproductions of works of art, sculptures based on drawings, lithographs based on paintings, maps based on previously published sources, or "any other form in which a work may be recast, transformed, or adapted." Derivative works also include works "consisting of editorial revisions, annotations, or other modifications" if these changes, as a whole, represent an original work of authorship.

Preexisting Material (space 6a): Complete this space **and** space 6b for derivative works. In this space identify the preexisting work that has been recast, transformed, or adapted. Examples of preexisting material might be "Grunewald Altarpiece" or "19th century quilt design." Do not complete this space for compilations.

Material Added to This Work (space 6b): Give a brief, general statement of the **additional** new material covered by the copyright claim for which registration is sought. In the case of a derivative work, identify this new material. Examples: "Adaptation of design and additional artistic work"; "Reproduction of painting by photolithography"; "Additional cartographic material"; "Compilation of photographs." If the work is a compilation, give a brief, general statement describing both the material that has been compiled **and** the compilation itself. Example: "Compilation of 19th century political cartoons."

7,8,9 SPACE 7,8,9: Fee, Correspondence, Certification, Return Address

Fee: The Copyright Office has the authority to adjust fees at 5-year intervals, based on changes in the Consumer Price Index. The next adjustment is due in 1996. Please contact the Copyright Office after July 1995 to determine the actual fee schedule.

Deposit Account: If you maintain a Deposit Account in the Copyright Office, identify it in space 7. Otherwise leave the space blank and send the fee of $20 with your application and deposit.

Correspondence (space 7): This space should contain the name, address, area code, and telephone number of the person to be consulted if correspondence about this application becomes necessary.

Certification (space 8): The application cannot be accepted unless it bears the date and the **handwritten signature** of the author or other copyright claimant, or of the owner of exclusive right(s), or of the duly authorized agent of the author, claimant, or owner of exclusive right(s).

Address for Return of Certificate (space 9): The address box must be completed legibly since the certificate will be returned in a window envelope.

FORM VA
For a Work of the Visual Arts
UNITED STATES COPYRIGHT OFFICE

REGISTRATION NUMBER

_____

VA   VAU

EFFECTIVE DATE OF REGISTRATION

_____

Month   Day   Year

DO NOT WRITE ABOVE THIS LINE. IF YOU NEED MORE SPACE, USE A SEPARATE CONTINUATION SHEET.

1

TITLE OF THIS WORK ▼

NATURE OF THIS WORK ▼ See instructions

PREVIOUS OR ALTERNATIVE TITLES ▼

PUBLICATION AS A CONTRIBUTION If this work was published as a contribution to a periodical, serial, or collection, give information about the collective work in which the contribution appeared.   **Title of Collective Work ▼**

If published in a periodical or serial give: **Volume ▼**   **Number ▼**   **Issue Date ▼**   **On Pages ▼**

2

a

NAME OF AUTHOR ▼

DATES OF BIRTH AND DEATH
Year Born ▼   Year Died ▼

Was this contribution to the work a "work made for hire"?
☐ Yes
☐ No

AUTHOR'S NATIONALITY OR DOMICILE
Name of Country
OR { Citizen of ▶ _____
Domiciled in ▶ _____

WAS THIS AUTHOR'S CONTRIBUTION TO THE WORK
Anonymous? ☐ Yes ☐ No
Pseudonymous? ☐ Yes ☐ No
If the answer to either of these questions is "Yes," see detailed instructions.

NATURE OF AUTHORSHIP Check appropriate box(es). **See instructions**
☐ 3-Dimensional sculpture   ☐ Map   ☐ Technical drawing
☐ 2-Dimensional artwork   ☐ Photograph   ☐ Text
☐ Reproduction of work of art   ☐ Jewelry design   ☐ Architectural work
☐ Design on sheetlike material

NOTE

Under the law, the "author" of a "work made for hire" is generally the employer, not the employee (see instructions). For any part of this work that was "made for hire" check "Yes" in the space provided, give the employer (or other person for whom the work was prepared) as "Author" of that part, and leave the space for dates of birth and death blank.

b

NAME OF AUTHOR ▼

DATES OF BIRTH AND DEATH
Year Born ▼   Year Died ▼

Was this contribution to the work a "work made for hire"?
☐ Yes
☐ No

AUTHOR'S NATIONALITY OR DOMICILE
Name of Country
OR { Citizen of ▶ _____
Domiciled in ▶ _____

WAS THIS AUTHOR'S CONTRIBUTION TO THE WORK
Anonymous? ☐ Yes ☐ No
Pseudonymous? ☐ Yes ☐ No
If the answer to either of these questions is "Yes," see detailed instructions.

NATURE OF AUTHORSHIP Check appropriate box(es). **See instructions**
☐ 3-Dimensional sculpture   ☐ Map   ☐ Technical drawing
☐ 2-Dimensional artwork   ☐ Photograph   ☐ Text
☐ Reproduction of work of art   ☐ Jewelry design   ☐ Architectural work
☐ Design on sheetlike material

3

a

YEAR IN WHICH CREATION OF THIS WORK WAS COMPLETED   This information must be given
◀ Year in all cases.

b

DATE AND NATION OF FIRST PUBLICATION OF THIS PARTICULAR WORK
Complete this information ONLY if this work has been published.
Month ▶ _____ Day ▶ _____ Year ▶ _____
◀ Nation

4

See instructions before completing this space.

COPYRIGHT CLAIMANT(S) Name and address must be given even if the claimant is the same as the author given in space 2. ▼

TRANSFER If the claimant(s) named here in space 4 is (are) different from the author(s) named in space 2, give a brief statement of how the claimant(s) obtained ownership of the copyright. ▼

DO NOT WRITE HERE OFFICE USE ONLY

APPLICATION RECEIVED

ONE DEPOSIT RECEIVED

TWO DEPOSITS RECEIVED

FUNDS RECEIVED

MORE ON BACK ▶
- Complete all applicable spaces (numbers 5-9) on the reverse side of this page.
- See detailed instructions.
- Sign the form at line 8.

DO NOT WRITE HERE

Page 1 of _____ pages

186

DO NOT WRITE ABOVE THIS LINE. IF YOU NEED MORE SPACE, USE A SEPARATE CONTINUATION SHEET.

PREVIOUS REGISTRATION Has registration for this work, or for an earlier version of this work, already been made in the Copyright Office?

☐ **Yes** ☐ **No** If your answer is "Yes," why is another registration being sought? (Check appropriate box) ▼

a. ☐ This is the first published edition of a work previously registered in unpublished form.

b. ☐ This is the first application submitted by this author as copyright claimant.

c. ☐ This is a changed version of the work, as shown by space 6 on this application.

If your answer is "Yes," give: **Previous Registration Number** ▼   **Year of Registration** ▼

5

DERIVATIVE WORK OR COMPILATION   Complete both space 6a and 6b for a derivative work; complete only 6b for a compilation.
a. Preexisting Material Identify any preexisting work or works that this work is based on or incorporates. ▼

b. Material Added to This Work Give a brief, general statement of the material that has been added to this work and in which copyright is claimed. ▼

6

See instructions before completing this space.

DEPOSIT ACCOUNT If the registration fee is to be charged to a Deposit Account established in the Copyright Office, give name and number of Account.
Name ▼   **Account Number** ▼

7

CORRESPONDENCE Give name and address to which correspondence about this application should be sent.   Name/Address/Apt/City/State/ZIP ▼

Area Code and Telephone Number ▶

Be sure to give your daytime phone ◀ number

CERTIFICATION* I, the undersigned, hereby certify that I am the

check only one ▼

☐ author

☐ other copyright claimant

☐ owner of exclusive right(s)

☐ authorized agent of _____
              Name of author or other copyright claimant, or owner of exclusive right(s) ▲

8

of the work identified in this application and that the statements made
by me in this application are correct to the best of my knowledge.

Typed or printed name and date ▼ If this application gives a date of publication in space 3, do not sign and submit it before that date.

                                                                                    Date ▶ _____

✍ **Handwritten signature** (X) ▼

MAIL CERTIFI- CATE TO

Name ▼

Number/Street/Apt ▼

Certificate will be mailed in window envelope

City/State/ZIP ▼

9

*17 U.S.C. § 506(e): Any person who knowingly makes a false representation of a material fact in the application for copyright registration provided for by section 409, or in any written statement filed in connection with the application, shall be fined not more than $2,500.

July 1993—300,000   ♲ PRINTED ON RECYCLED PAPER   ☆U.S. GOVERNMENT PRINTING OFFICE: 1993-342-582/80,021

187

Filling Out Application Form CA

Detach and read these instructions before completing this form.
Make sure all applicable spaces have been filled in before you return this form.

BASIC INFORMATION

Use Form CA When:

An earlier registration has been completed in the Copyright Office; and

Some of the facts given in that registration are incorrect or incomplete; and

You want to place the correct or complete facts on record.

Purpose of Supplementary Copyright Registration:
As a rule, only one basic copyright registration can be made for the same work. To take care of cases where information in the basic registration turns out to be incorrect or incomplete, section 408(d) of the copyright law provides for "the filing of an application for supplementary registration, to correct an error in a copyright registration or to amplify the information given in a registration."

Who May File:
Once basic registration has been made for a work, any author or other copyright claimant or owner of any exclusive right in the work or the duly authorized agent of any such author, other claimant, or owner who wishes to correct or amplify the information given in the basic registration may submit Form CA.

Please Note:

Do not use Form CA to correct errors in statements on the copies or phonorecords of the work in question or to reflect changes in the content of the work. If the work has been changed substantially, you should consider making an entirely new registration for the revised version to cover the additions or revisions.

Do not use Form CA as a substitute for renewal registration. Renewal of copyright cannot be accomplished by using Form CA. For information on renewal of copyright, write the Copyright Office for Circular 15.

Do not use Form CA as a substitute for recording a transfer of copyright or other document pertaining to rights under a copyright. Recording a document under section 205 of the statute gives all persons constructive notice of the facts stated in the document and may have other important consequences in cases of infringement or conflicting transfers. Supplementary registration does not have that legal effect.

For information on recording a document, request Circular 12 from the Copyright Office. To record a document in the Copyright Office, request the Document Cover Sheet.

How to Apply for Supplementary Registration:

First: Study the information on this page to make sure that filing an application on Form CA is the best procedure to follow in your case.

Second: Read the back of this page for the specific instructions on filling out Form CA. Before starting to complete the form, make sure that you have all of the necessary detailed information from the certificate of the basic registration.

Third: Complete all applicable spaces on this form following the line-by-line instructions on the back of this page. Use a typewriter or print the information in black ink.

Fourth: Detach this sheet and send your completed Form CA to: Register of Copyrights, Library of Congress, Washington, D.C. 20559-6000. Unless you have a Deposit Account in the Copyright Office, your application must be accompanied by a nonrefundable filing fee in the form of a check or money order for $20 payable to: *Register of Copyrights.* Do not send copies, phonorecords, or supporting documents with your application. They cannot be made part of the record of a supplementary registration.

What Happens When a Supplementary Registration is Made?
When a supplementary registration is completed, the Copyright Office will assign it a new registration number in the appropriate registration category and will issue a certificate of supplementary registration under that number. The basic registration will not be cancelled. The two registrations will stand in the Copyright Office records. The supplementary registration will have the effect of calling the public's attention to a possible error or omission in the basic registration and of placing the correct facts or the additional information on official record.

LINE-BY-LINE INSTRUCTIONS
Please type or print using black ink.

A PART A: Identification of Basic Registration

General Instructions: The information in this part identifies the basic registration that will be corrected or amplified. Even if the purpose of filing Form CA is to change one of these items, each item must agree exactly with the information as it already appears in the basic registration, that is, as it appears in the registration you wish to correct. Do not give any new information in this part.

Title of Work: Give the title as it appears in the basic registration.

Registration Number: Give the registration number (the series of numbers preceded by one or more letters) that appears in the upper right-hand corner of the certificate of registration.

Registration Date: Give the year when the basic registration was completed.

Name(s) of Author(s) and Name(s) of Copyright Claimant(s): Give all of the names as they appear in the basic registration.

B PART B: Correction

General Instructions: Complete this part **only** if information in the basic registration **was incorrect at the time that basic registration was made.** Leave this part blank and complete Part C, instead, if your purpose is to add, update, or clarify information rather than to rectify an actual error.

Location and Nature of Incorrect Information: Give the line number and the heading or description of the space in the basic registration where the error occurs. Example: "Line number 3 . . . Citizenship of author."

Incorrect Information as it Appears in Basic Registration: Transcribe the incorrect statement exactly as it appears in the basic registration, even if you have already given this information in Part A.

Corrected Information: Give the statement as it should have appeared in the application of the basic registration.

Explanation of Correction: You may need to add an explanation to clarify this correction.

C PART C: Amplification

General Instructions: Complete this part if you want to provide any of the following: (1) information that was omitted at the time of basic registration; (2) changes in facts other than ownership but including changes such as title or address of claimant, that have occurred since the basic registration; or (3) explanations clarifying information in the basic registration.

Location and Nature of Information to be Amplified: Give the line number and the heading or description of the space in the basic registration where the information to be amplified appears.

Amplified Information: Give a statement of the additional, updated, or explanatory information as clearly and succinctly as possible.

Explanation of Amplification: You should add an explanation of the amplification if it is necessary to clarify the amplification.

D,E,F,G PARTS D,E,F,G: Continuation, Fee, Certification, Return Address

Continuation (Part D): Use this space if you do not have enough room in Parts B or C.

Fee: The Copyright Office has the authority to adjust fees at 5-year intervals, based on changes in the Consumer Price Index. The next adjustment is due in 1996. Please contact the Copyright Office after July 1995 to determine the actual fee schedule.

Deposit Account and Mailing Instructions (Part E): If you maintain a Deposit Account in the Copyright Office, identify it in Part E. Otherwise, you will need to send the nonrefundable filing fee of $20 with your form. The space headed "Correspondence" should contain the name, address, and telephone number with area code of the person to be consulted if correspondence about the form becomes necessary.

Certification (Part F): The application is not acceptable unless it bears the handwritten signature of the author, or other copyright claimant, or of the owner of exclusive right(s), or of the duly authorized agent of such author, claimant, or owner.

Address for Return of Certificate (Part G): The address box must be completed legibly, since the certificate will be returned in a window envelope.

PRIVACY ACT ADVISORY STATEMENT
Required by the Privacy Act of 1974 (Public Law 93-579)

AUTHORITY FOR REQUESTING THIS INFORMATION:
● Title 17, U.S.C., Sec. 408(d)

FURNISHING THE REQUESTED INFORMATION IS:
● Voluntary

BUT IF THE INFORMATION IS NOT FURNISHED:
● It may be necessary to delay or refuse supplementary registration

PRINCIPAL USES OF REQUESTED INFORMATION:
● Establishment and maintenance of a public record
● Examination for compliance with legal requirements

OTHER ROUTINE USES:
● Public inspection and copying
● Preparation of public indexes
● Preparation of public catalogs of copyright registrations
● Preparation of search reports upon request

NOTE:
● No other advisory statement will be given you in connection with this application
● Please keep this statement and refer to it if we communicate with you regarding this application

FORM CA

For Supplementary Registration
UNITED STATES COPYRIGHT OFFICE

REGISTRATION NUMBER

TX	TXU	PA	PAU	VA	VAU	SR	SRU	RE

EFFECTIVE DATE OF SUPPLEMENTARY REGISTRATION

_____ _____ _____
Month   Day   Year

DO NOT WRITE ABOVE THIS LINE. IF YOU NEED MORE SPACE, USE A SEPARATE CONTINUATION SHEET.

A

TITLE OF WORK ▼

REGISTRATION NUMBER OF THE BASIC REGISTRATION ▼   YEAR OF BASIC REGISTRATION ▼

NAME(S) OF AUTHOR(S) ▼   NAME(S) OF COPYRIGHT CLAIMANT(S) ▼

B

LOCATION AND NATURE OF INCORRECT INFORMATION IN BASIC REGISTRATION ▼

Line Number . . . . Line Heading or Description . . . .

INCORRECT INFORMATION AS IT APPEARS IN BASIC REGISTRATION ▼

CORRECTED INFORMATION ▼

EXPLANATION OF CORRECTION ▼

C

LOCATION AND NATURE OF INFORMATION IN BASIC REGISTRATION TO BE AMPLIFIED ▼

Line Number . . . . Line Heading or Description . . . .

AMPLIFIED INFORMATION ▼

EXPLANATION OF AMPLIFIED INFORMATION ▼

MORE ON BACK ▶   • Complete all applicable spaces (D -G) on the reverse side of this page.   **DO NOT WRITE HERE**
                 • See detailed instructions.   • Sign the form at space F.   Page 1 of _____ pages

FORM CA RECEIVED	FORM CA

FUNDS RECEIVED DATE

EXAMINED BY _____

CHECKED BY _____

CORRESPONDENCE ☐

REFERENCE TO THIS REGISTRATION ADDED TO
BASIC REGISTRATION   ☐ YES   ☐ NO

FOR
COPYRIGHT
OFFICE
USE
ONLY

DO NOT WRITE ABOVE THIS LINE. IF YOU NEED MORE SPACE, USE A SEPARATE CONTINUATION SHEET.

CONTINUATION OF: (Check which)   ☐ PART B OR   ☐ PART C

D

DEPOSIT ACCOUNT: If the registration fee is to be charged to a Deposit Account established in the Copyright Office, give name and number of Account.

Name _____

_____

Account Number _____

CORRESPONDENCE: Give name and address to which correspondence about this application should be sent.

Name _____

Address _____
(Apt)

(City)   (State)   (ZIP)

Area Code and Telephone Number ▶ _____

Be sure to give your daytime phone number

E

CERTIFICATION* I, the undersigned, hereby certify that I am the: (Check one)

☐ author ☐ other copyright claimant ☐ owner of exclusive right(s) ☐ duly authorized agent of _____
(Name of author or other copyright claimant, or owner of exclusive right(s) ▲

of the work identified in this application and that the statements made by me in this application are correct to the best of my knowledge.

Typed or printed name ▼   Date ▼

_____   _____

☞ Handwritten signature (X) ▼

_____

F

MAIL CERTIFI-CATE TO

Certificate will be mailed in window envelope

Name ▼

Number/Street/Apt ▼

City/State/ZIP ▼

YOU MUST:
• Complete all necessary spaces
• Sign your application in space F

SEND ALL ELEMENTS IN THE SAME PACKAGE:
1. Application form
2. Nonrefundable $20 filing fee in check or money order payable to *Register of Copyrights*

MAIL TO:
Register of Copyrights
Library of Congress
Washington, D.C. 20559-6000

The Copyright Office has the authority to adjust fees at 5-year intervals, based on changes in the Consumer Price Index. The next adjustment is due in 1996. Please contact the Copyright Office after July 1995 to determine the actual fee schedule.

G

December 1993—25,000   ☆U.S. GOVERNMENT PRINTING OFFICE: 1993-301-241/80,049

◉ Filling Out Application Form RE

Detach and read these instructions before completing this form.
Make sure all applicable spaces have been filled in before you return this form.

—BASIC INFORMATION—

How to Register a Renewal Claim:

First: Study the information on this page and make sure you know the answers to two questions:
(1) What is the renewal filing period in your case?
(2) Who can claim the renewal?

Second: Read through the specific instructions for filling out Form RE. Before starting to complete the form, make sure that the copyright is now eligible for renewal, that you are authorized to file a renewal claim, and that you have all of the information about the copyright you will need.

Third: Complete all applicable spaces on Form RE, following the line-by-line instructions. Use typewriter or print the information in black ink.

Fourth: Detach this sheet and send your completed Form RE to: Register of Copyrights, Library of Congress, Washington, D.C. 20559. Unless you have a Deposit Account in the Copyright Office, your application must be accompanied by a check or money order for $20, payable to: *Register of Copyrights*. Do not send copies, phonorecords, or supporting documents with your renewal application unless specifically requested to do so by the Copyright Office.

What Is Renewal of Copyright?
For works copyrighted before January 1, 1978, the copyright law provides a first term of copyright protection lasting 28 years. These works were required to be renewed within strict time limits in order to obtain a second term of copyright protection lasting 47 years. If copyright originally secured before January 1, 1964, was not renewed at the proper time, copyright protection expired permanently at the end of the 28th year and could not be renewed.

Public Law 102-307, enacted June 26, 1992, amended the copyright law to extend automatically the term of copyrights secured between January 1, 1964, and December 31, 1977, to a further term of 47 years. This recent legislation makes renewal registration optional. The first term of copyright protection expires on December 31st of the 28th year of the original term of the copyright and the 47-year renewal term automatically vests in the party entitled to claim renewal as of that date.

Some Basic Points About Renewal:
(1) A work is eligible for renewal registration at the beginning of the 28th year of the first term of copyright.
(2) There is no requirement to make a renewal filing in order to extend the original 28-year copyright term to the full term of 75 years; however, there are some benefits from making a renewal registration during the 28th year of the original term. (For more information, write to the Copyright Office for Circular 15.)
(3) Only certain persons who fall into specific categories named in the law can claim renewal.
(4) For works originally copyrighted on or after January 1, 1978, the copyright law has eliminated all renewal requirements and established a single copyright term and different methods for computing the duration of a copyright. (For further information, write the Copyright Office for Circular 15a.)

Renewal Filing Period:
The amended copyright statute provides that, in order to register a renewal copyright, the renewal application and fee must be received in the Copyright Office
—within the last (28th) calendar year before the expiration of the original term of copyright or
—at any time during the renewed and extended term of 47 years.

To determine the filing period for renewal in your case:
(1) First, find out the date of original copyright for the work. (In the case of works originally registered in unpublished form, the date of copyright is the date of registration; for published works, copyright begins on the date of first publication.)
(2) Then add 28 years to the year the work was originally copyrighted.
Your answer will be the calendar year during which the copyright will become eligible for renewal. Example: A work originally copyrighted on April 19, 1966, will be eligible for renewal in the calendar year 1994.

To renew a copyright during the original copyright term, the renewal application and fee **must** be received in the Copyright Office within 1 year prior to the expiration of the original copyright. All terms of the original copyright run through the end of the 28th calendar year making the period for renewal registration during the original term from December 31st of the 27th year of the copyright through December 31st of the following year.

Who May Claim Renewal:
Renewal copyright may be claimed only by those persons specified in the law. Except in the case of four specific types of works, the law gives the right to claim renewal to the individual author of the work, regardless of who owned the copyright during the original term. If the author is dead, the statute gives the right to claim renewal to certain of the author's beneficiaries (widow and children, executors, or next of kin, depending on the circumstances). The present owner (proprietor) of the copyright is entitled to claim renewal only in four specified cases as explained in more detail on the reverse of this page.

—LINE-BY-LINE INSTRUCTIONS—

Please type or print using black ink.

1 SPACE 1: Renewal Claimant(s)

General Instructions: In order for this application to result in a valid renewal, space 1 must identify one or more of the persons who are entitled to renew the copyright under the statute. Give the full name and address of each claimant, with a statement of the basis of each claim, using the wording given in these instructions.

For registration in the 28th year of the original copyright term, the renewal claimant is the individual(s) or entity who is entitled to claim renewal copyright on the date filed.

For registration after the 28th year of the original copyright term, the renewal claimant is the individual(s) or entity who is entitled to claim renewal copyright on December 31st of the 28th year.

Persons Entitled to Renew:

A. The following persons may claim renewal in all types of works except those enumerated in Paragraph B below:

1. The author, if living. State the claim as: *the author*

2. The widow, widower, and/or children of the author, if the author is not living. State the claim as:
the widow (widower) of the author . . . .
(Name of author)
and/or the child (children) of the deceased author . . . .
(Name of author)

3. The author's executor(s), if the author left a will and if there is no surviving widow, widower, or child. State the claim as:
the executor(s) of the author . . . .
(Name of author)

4. The next of kin of the author, if the author left no will and if there is no surviving widow, widower, or child. State the claim as:
the next of kin of the deceased author . . . . *there being no will.*
(Name of author)

B. In the case of the following four types of works, the proprietor (owner of the copyright at the time of renewal registration) may claim renewal:

1. Posthumous work (a work published after the author's death as to which no copyright assignment or other contract for exploitation has occurred during the author's lifetime). State the claim as: *proprietor of copyright in a posthumous work.*

2. Periodical, cyclopedic, or other composite work. State the claim as: *proprietor of copyright in a composite work.*

3. "Work copyrighted by a corporate body otherwise than as assignee or licensee of the individual author." State the claim as: *proprietor of copyright in a work copyrighted by a corporate body otherwise than as assignee or licensee of the individual author.* (This type of claim is considered appropriate in relatively few cases.)

4. Work copyrighted by an employer for whom such work was made for hire. State the claim as: *proprietor of copyright in a work made for hire.*

2 SPACE 2: Work Renewed

General Instructions: This space is to identify the particular work being renewed. The information given here should agree with that appearing in the certificate of original registration.

Title: Give the full title of the work, together with any subtitles or descriptive wording included with the title in the original registration. In the case of a musical composition, give the specific instrumentation of the work.

Renewable Matter: Copyright in a new version of a previously published or copyrighted work (such as an arrangement, translation, dramatization, compilation, or work republished with new matter) covers only the additions, changes, or other new material appearing for the first time in that version. If this work was a new version, state in general the new matter upon which copyright was claimed.

Contribution to Periodical, Serial, or other Composite Work: Separate renewal registration is possible for a work published as a contribution to a periodical, serial, or other composite work, whether the contribution was copyrighted independently or as part of the larger work in which it appeared. Each contribution published in a separate issue ordinarily requires a separate renewal registration. However, the law provides an alternative, permitting groups of periodical contributions by the same individual author to be combined under a single renewal application and fee in certain cases.

If this renewal application covers a single contribution, give all of the requested information in space 2. If you are seeking to renew a group of contributions, include a reference such as "See space 5" in space 2 and give the requested information about all of the contributions in space 5.

3 SPACE 3: Author(s)

General Instructions: The copyright secured in a new version of a work is independent of any copyright protection in material published earlier. The only "authors" of a new version are those who contributed copyrightable matter to it. Thus, for renewal purposes, the person who wrote the original version on which the new work is based cannot be regarded as an "author" of the new version, unless that person also contributed to the new matter.

Authors of Renewable Matter: Give the full names of all authors who contributed copyrightable matter to this particular version of the work.

4 SPACE 4: Facts of Original Registration

General Instructions: Each item in space 4 should agree with the information appearing in the original registration for the work. If the work being renewed is a single contribution to a periodical or composite work that was not separately registered, give information about the particular issue in which the contribution appeared. You may leave this space blank if you are completing space 5.

Original Registration Number: Give the full registration number, which is a series of numerical digits, preceded by one or more letters. The registration number appears in the upper right hand corner of the front of the certificate of registration.

Original Copyright Claimant: Give the name in which ownership of the copyright was claimed in the original registration.

Date of Publication or Registration: Give only one date. If the original registration gave a publication date, it should be transcribed here; otherwise the registration was for an unpublished work, and the date of registration should be given.

NOTE: An original registration is not required but there are supplemental deposit requirements. You may call or write the Renewals Section for details. Phone 202-707-8180, or FAX 202-707-3849.
Renewals Section, LM 449
Copyright Office
Library of Congress
Washington, D.C. 20559

5 SPACE 5: Group Renewals

General Instructions: A renewal registration using a single application and $20 fee can be made for a group of works if **all** of the following statutory conditions are met: (1) all of the works were written by the same author, who is named in space 3 and who is or was an individual (not an employer for hire); (2) all of the works were first published as contributions to periodicals (including newspapers) and were copyrighted on their first publication; (3) the renewal claimant or claimants and the basis of claim or claims, as stated in space 1, are the same for all of the works; (4) the renewal application and fee are received not less than 27 years after the 31st day of December of the calendar year in which all of the works were first published; and (5) the renewal application identifies each work separately, including the periodical containing it and the date of first publication.

Time Limits for Group Renewals: To be renewed as a group, all of the contributions must have been first published during the same calendar year. For example, suppose six contributions by the same author were published on April 1, 1965, July 1, 1965, November 1, 1965, February 1, 1966, July 1, 1966, and March 1, 1967. The three 1965 copyrights can be combined and renewed at any time during 1993, and the two 1966 copyrights can be renewed as a group during 1994, but the 1967 copyright must be renewed by itself, in 1995.

Identification of Each Work: Give all of the requested information for each contribution. The registration number should be that for the contribution itself if it was separately registered, and the registration number for the periodical issue if it was not.

6,7,8 SPACE 6,7,8: Fee, Correspondence, Certification, Return Address

Fee: The Copyright Office has the authority to adjust fees at 5-year intervals, based on changes in the Consumer Price Index. The next adjustment is due in 1996. Please contact the Copyright Office after July 1995 to determine the actual fee schedule.

Deposit Account and Correspondence (Space 6): If you maintain a Deposit Account in the Copyright Office, identify it in space 6. Otherwise, you will need to send the renewal registration fee of $20 with your form. The space headed "Correspondence" should contain the name and address of the person to be consulted if correspondence about the form becomes necessary.

Certification (Space 7): The renewal application is not acceptable unless it bears the handwritten signature of the renewal claimant or the duly authorized agent of the renewal claimant.

Address for Return of Certificate (Space 8): The address box must be completed legibly, since the certificate will be returned in a window envelope.

FORM RE
For Renewal of a Work
UNITED STATES COPYRIGHT OFFICE

REGISTRATION NUMBER

EFFECTIVE DATE OF RENEWAL REGISTRATION

Month	Day	Year

DO NOT WRITE ABOVE THIS LINE. IF YOU NEED MORE SPACE, USE A SEPARATE CONTINUATION SHEET(RE/CON).

1

RENEWAL CLAIMANT(S), ADDRESS(ES), AND STATEMENT OF CLAIM ▼ (See Instructions)

1
Name ....
Address ....
Claiming as ....
(Use appropriate statement from instructions)

2
Name ....
Address ....
Claiming as ....

3
Name ....
Address ....
Claiming as ....

2

TITLE OF WORK IN WHICH RENEWAL IS CLAIMED ▼

RENEWABLE MATTER ▼

PUBLICATION AS A CONTRIBUTION If this work was published as a contribution to a periodical, serial, or other composite work, give information about the collective work in which the contribution appeared. **Title of Collective Work ▼**

If published in a periodical or serial give: **Volume ▼**   **Number ▼**   **Issue Date ▼**

3

AUTHOR(S) OF RENEWABLE MATTER ▼

4

ORIGINAL REGISTRATION NUMBER ▼ ORIGINAL COPYRIGHT CLAIMANT ▼

ORIGINAL DATE OF COPYRIGHT

If the original registration for this work was made in published form, give:
DATE OF PUBLICATION: _____
(Month)   (Day)   (Year)

OR

If the original registration for this work was made in unpublished form, give:
DATE OF REGISTRATION: _____
(Month)   (Day)   (Year)

MORE ON BACK ▶ • Complete all applicable spaces (numbers 5-8) on the reverse side of this page.
• See detailed instructions.   • Sign the form at space 7.

DO NOT WRITE HERE

Page 1 of _____ pages

194

CORRESPONDENCE ☐ YES

EXAMINED BY _____

CHECKED BY _____

FOR
COPYRIGHT
OFFICE
USE
ONLY

DO NOT WRITE ABOVE THIS LINE. IF YOU NEED MORE SPACE, USE A SEPARATE CONTINUATION SHEET (RE/CON).

RENEWAL FOR GROUP OF WORKS BY SAME AUTHOR: To make a single registration for a group of works by the same individual author published as contributions to periodicals (see instructions), give full information about each contribution. If more space is needed, request continuation sheet (Form RE/CON).

5

1
Title of Contribution: ....

Title of Periodical: .... Vol: .... No: .... Issue Date: ....

Date of Publication: .... Registration Number: ....
(Month) (Day) (Year)

2
Title of Contribution: ....

Title of Periodical: .... Vol: .... No: .... Issue Date: ....

Date of Publication: .... Registration Number: ....
(Month) (Day) (Year)

3
Title of Contribution: ....

Title of Periodical: .... Vol: .... No: .... Issue Date: ....

Date of Publication: .... Registration Number: ....
(Month) (Day) (Year)

4
Title of Contribution: ....

Title of Periodical: .... Vol: .... No: .... Issue Date: ....

Date of Publication: .... Registration Number: ....
(Month) (Day) (Year)

DEPOSIT ACCOUNT: If the registration fee is to be charged to a Deposit Account established in the Copyright Office, give name and number of Account.

Name _____

_____

Account Number _____

CORRESPONDENCE: Give name and address to which correspondence about this application should be sent.

Name _____

Address _____
(Apt)

(City) _____ (State) _____ (ZIP)

Area Code and Telephone Number ▶ _____

6

Be sure to
give your
daytime phone
◀ number

CERTIFICATION* I, the undersigned, hereby certify that I am the: (Check one)
☐ renewal claimant ☐ duly authorized agent of _____
(Name of renewal claimant) ▲
of the work identified in this application and that the statements made by me in this application are correct to the best of my knowledge.

Typed or printed name ▼ _____ Date ▼ _____

👉 Handwritten signature (X) ▼ _____

7

**MAIL
CERTIFI-
CATE TO**

**Certificate
will be
mailed in
window
envelope**

Name ▼ _____

Number/Street/Apt ▼ _____

City/State/ZIP ▼ _____

YOU MUST:
• Complete all necessary spaces
• Sign your application in space 7
**SEND ALL ELEMENTS
IN THE SAME PACKAGE:**
1. Application form
2. Nonrefundable $20 filing fee
   in check or money order
   payable to
   Register of Copyrights
MAIL TO:
Register of Copyrights
Library of Congress
Washington, D.C. 20559

8

The Copyright Office
has the authority to ad-
just fees at 5-year intervals, based on changes
in the Consumer Price
Index. The next adjust-
ment is due in 1996.
Please contact the
Copyright Office after
July 1995 to determine
the actual fee schedule.

April 1993—40,000

☆U.S. GOVERNMENT PRINTING OFFICE: 1993-342-581/60,513

Commercialization Strategies

HOW TO LICENSE YOUR INVENTION

Just Say Yes

Before you consider licensing your invention, you must be able to answer yes to all of these questions:

1. Do you have a patent, copyright, or other legal protection?

2. Do you have a working model, or better yet, an engineering prototype?

3. Do you have credible data about the size of the market, including probable impact of selling price on quantity demanded?

4. Do you know what it will cost to produce your invented product at various levels of output?

Mere shape determines whether iron shall float or sink.

To get your invention to market, somebody has to produce it, and somebody has to sell it. In fact, as your invention moves toward the marketplace, business skills become more critical than technical skills. You will require increasing quantities of time from people who have these skills, and of course, your product will demand more and more money as it moves along the road to commercialization.

If you opt to go the licensing route—always my choice—your invention becomes more important than you in the corporate decision-making process. In other words, the manufacturer typically will not feel you are necessary once it has an understanding of your invention. On the other hand, the inventor who seeks venture capital faces a maxim that says: *Better to take a chance on a first-rate manager with a second-rate product, than on a first-rate product in the hands of a second-rate manager*. First-rate managers are, by definition, first-rate planners. Investors are comforted by knowing that their investment is in the hands of a well-grounded, seasoned manager.

There are two ways to commercialize an invention: you can license someone else to manufacture and market it, or you can do it yourself. Most other options are variations of these two possibilities.

Licensing makes sense in my case because it does not require that I raise venture capital and dedicate myself to a single enterprise. It keeps my exposure to lawyers and bankers very limited. And it allows me to do what I do best, dream up new concepts. However, this does not mean that licensing is best for you.

Some people thrive on building businesses, crunching numbers, and all that this entails. Others are control freaks, people who do not like giving up dominion over their inventions. The do-it-yourself option allows them to be the boss.

I have managed small and large enterprises, and I want no part of it anymore. In my last "real job" many moons ago, I was a principal architect in the establishment of a 24-hour, global, interactive, satellite TV broadcast operation and responsible for a staff of 230 people and a multi-million-dollar budget. I made a good wage and traveled around the world on jet planes. People met me when I landed, carried my bags, and whisked me away in cars. I had all kinds of prestige, power, and perks. But I was also up to my keister in meetings, personnel issues, and administrivia.

I spent each day going from meeting to meeting, barely having a chance to return phone calls or take a breath. When it got to the point that I felt the oxygen flow to my brain was beginning to shut down, I quit and returned to the unpredictable world of entrepreneurism.

To help you make a decision, here are some positives and negatives to consider about the licensing option versus the do-it-yourself option. This information was prepared by the Argonne National Laboratory for the Department of Energy.

The Licensing Option

Licensing tempts many inventors because the amount of money, as well as the catalog of tasks, skills, and people required, may seem considerably less than in running your own business. That doesn't necessarily mean it's the right alternative for you. In the first place, you may not find a licensee, and you can bet none will find you. In the second place, even when it's possible, licensing has its pros and cons. Here are some considerations:

First, the Negative Side

- You lose control of the technology. Usually total control, for a long time, and often forever.

- Your own involvement is reduced. In most cases, you'll have no further direct involvement at all. You may stay around as a consultant to the licensee, but usually for a limited time only.

- Finding the right licensee is tough. The right one may make you rich. The wrong one may bury your technology, or butcher it. Even if you can eventually get it back, it may be too late.

- Protecting your interests is crucial. But it's also extremely difficult to do. Negotiating with licensees means playing with the big boys. They confront you with the immense staff resources of the corporation—lawyers, market analysts, production engineers—a tough team for you to take on by yourself. Licensing agreements, when properly done, result from tough negotiations between two parties. The other side has professionals to represent it, so you better have one of your own. If you're an amateur at the game—and you almost certainly are— you need the help of a lawyer with experience in such negotiations.

Now, the Positive

Licensing multiplies the resources to develop your invention. The licensee, if it's a dynamic firm—and you don't want to license any other kind—can immediately put whole teams of professionals to work developing, producing, and marketing the technology. Insurmountable financial mountains to you may be petty cash molehills to them.

They see things you don't. Licensees often perceive uses—and therefore markets—for your invention that you didn't see. One licensee turned a salt-water taffy machine into a new and highly efficient type of concrete mixer. The more markets, the more potential income.

You may make some money and you may make it soon. The licensee may pay you money up front, although probably not as much as you hope. In addition, they may agree to a minimum amount of royalties for some period.

Licensing frees you to do something else. If what you want to do is retire, or go back to inventing, then giving up control of the technology may serve your interests rather than defeat them.

If you have a technology with a demonstrably strong potential market, thriving businesses out there may want your invention. Some large corporations regularly acquire new products that way, but you should also keep in mind that smaller firms, though they may be less well known, offer possibilities as well. Many of them can't afford expensive research and development programs, but nonetheless need new products. Furthermore, smaller firms often operate much more dynamically than big ones, so don't write them off.

Before considering licensing, however, you should be able to answer yes to all these questions:

❏ **Do you have a patent, copyright, or other legal protection?**

If not, you won't get far, because no company will risk investing in an unprotected innovation. Why should they pay you for something you don't own?

❏ **Do you have a working model, or better yet, an engineering prototype?**

If not, you can't prove the thing will work with competitive efficiency (unless it's self-evident that it will, which doesn't happen often). If you haven't made it work, your licensee will have to, which will cost them money, which will weaken your bargaining position. Indeed, licensing may succeed or fail on the basis of your technical development prior to licensing, for your licensee may have neither the skill nor the commitment that you bring to the task.

❏ **Do you have credible data about the size of the market, including probable impact of selling price on quantity demanded?**

❏ **Do you know what it will cost to produce at various levels of output?**

You may have thought licensing would enable you to avoid the last two of these questions. On the contrary, if you don't know the answers, then you don't know what your invention will be worth to your licensee; therefore, you don't know what payments you can reasonably demand. Your licensee will work up his version of all these figures. If he's reputable, he won't cheat you, but his estimates of sales and profits will be on the low end, and costs on the high side. You can count on it.

In short, you not only have to demonstrate technical feasibility, you also have to prepare a package of information about production and marketing so close to that required for a business plan that you might consider writing one. Such a document will help you decide whether you want to venture or license in the first place, and then help you carry out that decision by supplying you with the data you need to raise money for your own business, or to persuade a prospective licensee to talk you out of it.

At the very least, if you decide to license your invention, you'll have to build a working model; reaching the engineering prototype stage would greatly increase both your chances of finding a licensee and the amount of money you may convince him or her to pay. By contrast, if you want to start your own business, or develop the technology within a business you already operate, you'll have to do even more.

To get a better idea of the various types of prototypes—working model, engineering prototype, and so on—see the following table, which explains the purpose and characteristics of each type of model.

FAMILY OF SCALE MODELS/PROTOTYPES			
Mockup	Working Model	Engineering Prototype	Production Prototype
Purpose			
Built to demonstrate relations of adjacent parts	Built to illustrate the concept functioning-- especially the relative motions of connected moving parts	Built to demonstrate important design parameters; used to test reliability, speed, accuracy, etc.	Built to illustrate performance requirements met, production problems resolved, quality control achievable, etc.
Characteristics			
Built to scale; need not work	Not always to full scale; probably does not operate optimally	Usually hand built; often full scale; durable to withstand testing	Built as like the mass produced item as possible, differing only in the volume level of production

Doing It Yourself: The Venturing Strategy

Starting your own business, or "venturing," as it's often called, will require more from you, but has its own advantages and disadvantages to consider:

First the Disadvantages

It's risky. Many new businesses fail. A new business built around a new product runs a double risk, especially since the list of reasons for new business failures reads like a catalog of many inventors' weaknesses. These include (among many, many others):

- inadequate financing
- lack of management skills, such as personnel, accounting
- overestimating the market
- poor choice of location
- inability to delegate responsibility

- Resources remain limited. You'll have whatever money you yourself can raise, and raising the kind of money required to set up production and marketing usually takes a professional. If you aren't one, you'll have to find one.
- You'll be spread increasingly thin. As the number of tasks and skills required multiplies—and it does, with a vengeance—you'll spend more and more time either doing them, or finding someone who can—and will.
- You probably won't make much money for quite a while. Building a business gobbles cash, and a lot of it will continue to be yours. If you can found a company and finance it adequately, you may be able to pay yourself a salary, but it'll probably be modest—your backers will expect you to be frugal with their money.

Now for the Advantages

- Running a company can be exciting. If you have the will and skill, you may enjoy it more than inventing. Some inventors are entrepreneurs by experience, and some by instinct. The inventor/entrepreneur can sometimes achieve powerful

Patience is a Virtue

Consider the Mighty Morphin' Power Rangers. These superhero action figures are one of the biggest hits in the history of the United States toy industry. According to Makoto Yamashina, president of Bandai, Japan's largest toy company, and marketer of the toys, the Rangers took in $330 million in 1994. Sales are expected to increase to $400 million in 1995 with the Twentieth Century Fox release of the Power Ranger movie.

But if you think it was an overnight success, think again. Haim Saban, the man who sold the series to American television, said that the show almost did not make it. "I had been schlepping the Ranger series for nine years until I got one buyer," he told the *New York Times*.

Most American TV executives rejected the live action story line as nonsensical. Then the Fox Children's Network, in need of product to fill its schedule, decided to air the series in the fall of 1993. The rest, as they say, is history.

things, as Edwin Land at Polaroid and Steven Jobs at Apple have shown. The combination, however, occurs rarely.

- In the long run, you may make a lot more money. If your invention turns out to be a big success, your rewards could vastly exceed the royalties you could expect from any licensing agreement.

- Even if it's your company, you may not have to run it. Building a successful business involves hiring all kinds of people. This could include a chief operating officer. There are plenty of examples of inventors who retained a large or controlling interest in their companies, but turned the management of it over to someone else. Edwin Land did it several years ago, and Steven Jobs did it recently.

Obviously, being in business for yourself can mean a lot of different things. You may decide you want a company that engages in the whole range of activities involved in designing, manufacturing, and selling your product. More likely, you will focus on some parts of the process while making arrangements with other firms to do the rest of it. (After all, even General Motors buys a lot of its parts from independent suppliers, and lets franchised dealers do the retailing.)

As the sponsor of an invention, you may already be in business formally. Even if you think that you don't have a company in the legal sense, the day you commit yourself to making a financial success of your invention you embark on a business enterprise in the eyes of the Internal Revenue Service—however small and informal that enterprise may seem to you. Therefore, if you haven't yet thought of the time and money you've invested getting this far in terms of a business proposition, start now, whether you think your business will stay small or grow. If you haven't created a structure that provides you with limited liability (that is, a structure that legally insulates your personal assets against losses you may incur in your business) you should see a lawyer soon. Prospective investors will concern themselves with this issue, even if you haven't.

If you intend to develop your business around your technology, experience suggests that your company will have to grow, even if it's sometimes possible to get an invention into the marketplace without involving yourself in the complexities of building a large company. If, for example, you've invented a specialized tool with a large profit per sale, you may be able to "bootstrap" your business by selling one, taking the proceeds and making two more, selling them and making four, and so on. Even in such rare cases, however, you will ultimately have to decide to stay small (running the risk that some larger firm, seeing your success, may invade the market with a competitive product), or to expand.

If you run a growing business you'll eventually need capital from outside sources, which means you'll need a formal business structure providing limited liability to investors—one in which tasks are subdivided functionally (manufacturing, marketing, etc.) and assigned to professionals hired to carry them out. The two things intertwine, because no rational investor will put up the kind of money you'll need for a company of even modest size unless you have at least a plan for such a formal structure. Investors know, even if you won't admit it, that inventors generally prefer doing everything themselves; moreover, they know that building a successful enterprise absolutely requires genuine delegation of authority, something most inventors find extremely difficult to do. If you hope to grow a business, therefore, you must accept the ironic proposition that to keep overall control yourself, you'll have to delegate a lot of specific authority to other people.

Successful management of a business requires launching, mastering, and controlling a dynamic process, as well as dealing with continuous change caused by such things as the business's growth, new technology in the industry, revisions in tax laws, and behavior of competitors. A successful, growing, and dynamic business rests on a foundation of continuous planning, involving constant updating to reflect such things as changing circumstances, goals, and organization. The plan will help keep you on track, and it's an invaluable tool with which to sell yourself and your business to prospective investors, customers, and suppliers—as well as to the people you want to recruit for your company. This last has crucial importance, because you can't grow much without first-class help, and people worth hiring want to know what they're getting into, especially in terms of future prospects.

Prerequisites Common to Licensing and Venturing

Despite the apparently great differences between licensing and venturing as commercialization strategies, they prove to have a lot in common, including certain prerequisites. Some things you simply have to do whether you hope to persuade someone else to buy the rights to produce and distribute your invention, or decide to do it yourself. Remember that either way somebody will have to spend money, a lot of money. Whatever you may have spent so far will shrink in comparison with what's required henceforth. So whether you want to go on and market it yourself, or convince someone else to buy the rights to do it, you have to put together a convincing package. This includes:

- Proof that it works. This means a working model, or better yet, an engineering prototype. There's no substitute for showing investors or would-be licensees something they can see, touch, and watch do its stuff. Without at least a working model, you haven't much chance of interesting people beyond your family and friends who put their trust in you personally. Strangers (and friends who are experienced investors) demand:

- A market analysis. This means a serious breakdown of who the potential customers are, how many of them there are, how much they will pay, what the competition is, and how you will beat it. In addition, you need to know exactly what the market channels are through which products like yours reach the market. You should be able to show three significant points of difference between your product and the competition. If you can't, you've got a problem. You had better be sure your invention has no fatal flaws. For example, one inventor had a device that depended on a manufacturer converting an experimental glass product into a mass production item. When the manufacturer quit making the glass, he effectively killed the invention at the same time. Above all, you have to be able to show why people will buy your product, and show this through statements from prospective customers, backed up with believable figures in dollars and cents. The surest way to turn off any prospective investor who asks about the market is to say, "When they see it, they'll jump for it." It ain't necessarily so. Your market analysis determines whether it's worth going on with your invention, regardless of its technical elegance, and that analysis forms the basis for the next thing you need, which is:

- A commercialization plan. This is a detailed analysis showing what you intend to do to develop, market, and sell your technology, how much all this will cost, and who will do the work required—with all this information translated into a year-by-year, dollars and cents projection five years into the future. Investors (other than friends and family) will absolutely demand such a plan; prospective licensees may insist on one. And even if they don't, you should have one.

Without it you have little ammunition with which to combat their campaign to beat down your price.

Other Factors in Choosing a Commercialization Strategy

In deciding to license or venture, you should accept that, either way, you will have to give up some measure of ownership and/or control. In a sense, therefore, you're not deciding whether to get out, but when, how completely, under what circumstances, and by what method. In other words, you're looking for an exit strategy at the same time you're looking for a commercialization strategy.

In addition, no matter which commercialization strategy you follow, you will increasingly have to involve yourself with people from the business world. These folks have different imperatives, different expectations, and speak a different language. Many of them care nothing about technology except as a possible money spinner. Like it or not, you will increasingly need these people, so you have to learn to deal with them pretty much on their terms. They're no more inclined to translate their professional language for you than Parisians are to speak English to American tourists. Understanding these realities of the business world is just one of the skills of the entrepreneur, a role you'll have to understand and that someone—you, a partner, a licensee—will have to play. Building a business absolutely requires the skills of the entrepreneur; that is, the know-how to assemble all the components, make them function harmoniously, and sustain growth. If you yourself have run a business, you have a first-hand idea of what it takes. If you haven't, then you have a lot to learn. Whether you have the aptitude for it is something you have to ask yourself, and answer honestly.

If you decide that you aren't cut out to be an entrepreneur, or that you don't want to be one, that doesn't mean you can't create a business around your invention. It **does** mean you'll have to get an entrepreneur on your team, and soon. They don't come easy; you'll have to do sufficient spade-work to turn up enough evidence to persuade one to cast his or her lot with you and your technology. And they don't come cheap; he or she will want a piece of the action, probably a big piece. But it may be worth it: Chester Carlson was an inventor who couldn't balance his checkbook, much less run a company, but an entrepreneur named Joe Wilson made him a multi-millionaire by building a company called Xerox.

Every library has do-it-yourself handbooks containing self-administered tests that will help you decide, but you can begin dealing with the question of whether you want to be an entrepreneur by looking at the Innovation Process table on page 203 and answering these questions:

1. Which four tasks do you do best?

2. Which four do you do worst, or think you would do worst?

3. Which four tasks do you enjoy most, or think you would enjoy most?

4. Which four do you enjoy least, or think you would enjoy least?

If at least half your answers to Question 1 don't come from columns other than the "Technical" column, or if more than half of your answers to Question 2 don't come from the "Business" or "Market" columns, you probably **aren't** much of an entrepreneur.

If at least half your answers to Question 3 don't come from columns other than the "Technical" column, or at least half the answers to Question 4 don't come

THE INNOVATION PROCESS

Pre-Product Stage: Concept to Engineering Prototype

Technical Steps	Market Steps	Business Steps	Skills Required	People Involved
Concept	Does Market Exist	Decide to Develop	Intuition to Technical	Inventor
Concept Analysis	Define Market	Find Money	Technical to Engineering	Inventor
Working Model	Define Three Points of Difference	Find More Money	Engineering	Inventor Local Technicians Friends as Investors
Engineering Prototype Test Refine	Identify Market Barriers Decide to License or Venture	Find Even More Money Protect: Patent/ Trade Secret Start Business Plan	Engineering Legal Market Analysis Capital Acquisition	Inventor Engineer Patent Attorney More Investors Market Analysis Business Planner

Entrepreneurial Stage: Prototype to Production

Technical Steps	Market Steps	Business Steps	Skills Required	People Involved
Production Prototype Scale Up Test Refine Production Engineering Product Safety Engineering	Full Market Analysis and Plan Niches Barriers Pricing Competition Cost Data Distribution Method Alternative Product Applications Risk Analysis Sales Projections	Find Big Money Complete Business Plan Form Business Meet State and Federal Regulations Arrange Insurance Price Production Facility	Engineering Production Product Safety Entrepreneurial Financing Marketing Cost Analysis Legal Management	Inventor (?) Entrepreneur Investors Engineers Production Safety Attorneys Patent Corporate Accountants Consultants Marketing Business Management Financial Insurance Brokers Trade Union Officers
Limited Production Qualification testing Running changes	Contact Customers Commence Distribution Seek Product Endorsements Follow-up Sales Advertise Publish in Technical Journals	Find Big, Big Money Start-up Business Build Plant Buy Equipment Hire Foreman and Labor Arrange Product Service Purchasing Transportation Record Keeping	All of the above -PLUS- Speciality Engineering Systems Engineering Sales Analysis Supervisory	All of the above -PLUS Foreman Labor Sales People Speciality Engineers Systems Engineers
Full Production Start-up	All of the above -PLUS- Expand Distribution Analyze Competitor Response	All of the Above -PLUS- Monitor Costs Finance Cash-Flow Deficit Refine Production System	All of the Above -PLUS- Delegation Market Forecasting Strategic Planning Long-Term Financial Projections	All of the Above -PLUS- Expanding Management Sales Labor Force
Initial Growth	Increasingly Complex		Increasingly Complex	

Managerial Stage: Production for Major Market Penetration

Technical Steps	Market Steps	Business Steps	Skills Required	People Involved
Product Improvement New Products Sustained Growth	Complexities Intensify		Complex Management	Entrepreneur (?) Fully Bureaucratized management R&D Staff National Investment Firm

from the "Business" or "Market" columns, then you probably **don't want to be** an entrepreneur. (Of course the reverse applies as well.)

These questions about what you do best and enjoy most aren't just a gimmick to help you decide if you're an entrepreneur, or whether you want to license your invention or run your own business. They also serve to introduce another dimension you should consider carefully when deciding how to commercialize your technology—the dimension of costs.

Think about Costs at All Costs

As you know, there are three kinds of costs: money, time, and personal. You also realize that the three of them are intertwined and to some extent interchangeable. If you think you can't afford to hire a model maker, for example, you may decide to save money by building it yourself at a cost of your time, which in turn often involves a personal cost to your health, your marriage, and so on, not to mention the fact that you may produce a poor model.

To measure these costs accurately in relationship to one another, you must understand and apply the principle of "opportunity cost." In terms of money, it's the interest lost by putting money somewhere other than in the safest investment you can find, such as in U.S. government securities. That's exactly what you've done when you've put your money into your invention. It's also what you'll be asking investors to do, and you can bet your last dime that professional investors never lose sight of opportunity costs. Since the current rate of return on sure-fire investments runs between 7 and 10 percent, they'll demand a steep price for putting money into your high-risk venture.

Opportunity costs, however, also apply to time and personal costs. While you're doing one thing, you can't be doing something else, and if you spend a lot of time doing things you don't do well, you may be wasting something more precious than money. In the long run, money costs may be the least expensive of all because, if you run out of money, there's always bankruptcy. If you run out of time, there's only the grave. Financial bankruptcy is as American as apple pie, and plenty of people have survived it to go on to later success. Bankruptcy in time or spirit, on the other hand, is a disaster from which there often is no recovery.

All of this argues for riding the expert express instead of the do-it-yourself local. The Innovation Process table should convince you that eventually you'll have to get expert help. (If you have a patent attorney, in fact, you already have.) Look at the table and at your answers to the previous set of questions. Keeping in mind the interplay of the three kinds of costs, including the opportunity cost factor, ask yourself again: "What's the best way to commercialize my invention, and what help do I need first to get the show on the road?" Many innovators will of course respond, "Whatever strategy I choose, whatever step I decide to take next, whatever role I see for myself, the help I need is money."

Venture Capital

If you opt to raise venture capital and go the entrepreneurial route, a subject I am not qualified to address in anything other than a superficial way, I recommend you seek specialized assistance. There are many books on the ins and outs of venture capital. Any major library will provide you with a plethora of up-to-date works.

To find out about an ongoing university venture capital or entrepreneurism course, contact your nearest university or adult education center.

To learn about state and federal programs, call the Small Business Administration information line at: 1-800-827-5722.

Finding the Right Licensee

Because my expertise is in licensing, here are some tips on how to find the right company for your product.

Put a lot of time into your decision on which manufacturer to approach. Don't make the error of insufficient options. Some rejection is to be expected, so you'll want as many targets as possible.

Study corporate product lines. Do store checks. Get new product catalogues. Companies do what they do, and bringing them something out of their discipline is usually a lost effort. You would not, for example, go to Black & Decker with a new type of record player. Black & Decker manufactures innovative tools and labor–saving devices. Round pegs don't fit square holes.

Corporations exist to make money. Executives, especially those in lucrative profit-sharing plans and incentive programs, want their corporations to be successful, but at what risk? I have found that most senior executives will listen to any scheme that rings of potential profit. That profit can come in the form of a new product or a labor-saving device. But it is rare that an executive will rock the proverbial boat for untested, unfamiliar products, especially those that fall outside the company's expertise or channels of distribution. Unless you are dealing with an executive in charge of new business opportunities, trying to get a company to purchase a product that is inappropriate to its line-up or retail outlet channels is not only a waste of everyone's time, but it does nothing for your reputation. And that reputation is more precious than your invention.

After a while in this business you learn that many large firms are guided by numbers more than by products. Lawyers and accountants tend to become CEOs before research and development executives do. These bean counters see products in terms of SKUs (stock keeping units) only. They like to keep the pipeline filled with line extensions of already proven and profitable products. Their attitude is that a major breakthrough in medicine, for example, would be nice, but let's keep the mouthwashes and toothpastes coming. These types would rather take a popular pudding and put it on a stick than gamble on creating a new novelty food. They are so busy listening to statistics that they forget companies can create them.

One company licensed a product of ours and was going to produce it until a consumer study showed that its popularity would upset existing business. The company opted for the status quo and dropped the item. "Why should we erode our market share in an industry we almost totally control?" reasoned the company president. "If we bring your product out, the consumers will obviously love it, but we'll just point to opportunities our competition does not realize exists."

Proctor & Gamble spends about $1 billion per year on research and development, yet it is still willing to entertain outside submissions from independent inventors. Numerous Proctor & Gamble products were based upon outside submissions.

The original Crest toothpaste was based upon an invention made at Indiana University by a team of professors. Proctor & Gamble paid significant royalties to the university which then shared them with the professors. The invention was the use of anticaries stannous fluoride with a special compatible abrasive.

My success is reflected by the amount of research I have done," says Howard Jay Fleisher, New York City inventor of the Polygonzo puzzle. "I basically get every one of the trade magazines in the gift and toy industries. I go to every trade show from the gift show to the toy show, walk around, meet the people and find out who's who and what's what.

"I know what's out there and know the right company to approach. That's the biggest mistake most inventors and designers make. They don't really research who is the best company for their product . . . instead they just do a shotgun approach."

A senior research and development executive from a billion-dollar company told me once, "It's my responsibility to waste $2 million per year on long shots. I would not be doing my job if I didn't." Unfortunately, there are too few corporate executives with this entrepreneurial attitude.

The first move is yours. It is not an easy one. Going to the wrong company with the wrong product can cost valuable time, do nothing to enhance your contacts, and even bring grief. You want your idea in the right hands because, as advertising legend Bill Burnbach said, "An idea can turn to dust or magic depending on the talent that rubs against it."

NIH Syndrome

NIH, as used here, is not the acronym for the government's National Institute of Health. NIH is corporate jargon for "Not Invented Here," a syndrome from which many companies suffer. It means that such companies have in-house research and development staffs and do not entertain outside submissions at all; or it means that they do accept submissions, but only to see what is being done independent of themselves, not to license.

It is hard to tell from the outside which companies fall into the NIH category. Those you think would, do not, and those you would bet don't, do. This learning curve is part of every selection exercise.

The NIH issue has two sides, and corporate policy can change with different administrations. Many executives feel that no insulated group of salaried product development people, no matter how brilliant, can come up with winning products day in and day out.

Many companies cannot afford to pay engineers and designers to sit around and come up with ideas all day long. These kinds of companies are always worth approaching.

Other firms are against outside licensing of patented ideas because they would rather see the millions of dollars paid in royalties kept inside for its own research and development activities. They are not typically structured to interact with independent inventors.

I like to tell executives who believe they can do it all in-house that it is the spirit of the independent inventor that built America. To completely shut the independent inventor out is to severely limit one's opportunities and horizons. Then I remind them of these stories:

Kodak, America's largest manufacturer of photographic products, should have developed the instant camera. It didn't. The 60-second camera was invented and produced by Edwin Land, a maverick inventor. And when Kodak ultimately decided to imitate Land's invention, it was stopped in its tracks by the courts.

IBM, a name synonymous with computer innovation, completely missed the handheld calculator market. It should not have. The Japanese captured the lucrative market and never gave an inch.

And the U.S. television networks, with worldwide news-gathering operations, let CNN get started because they thought no one would pay to see news 24 hours a day. Nevertheless, CNN has become a tremendous success by any standards.

The Deep Pocket Problem

The larger the manufacturer, the deeper its pockets. The deeper its pockets, the more fearful it is of being sued. And law suits are another reason outside inventors have a problem approaching potential licensees. Major corporations are taken for all kinds of money by independent inventors who claim to have given ideas to the companies. And even when a company wins in court against the inventor, the cost to defend itself can run into the hundreds of thousands of dollars.

My company is quite small, yet we do not permit people to present ideas to us for the same reason. The last thing we want is to invent something and have a person come out of the woodwork claiming to have given us the idea.

About a year after our game Adver*teasing* had been on the market, Cadaco's then president, Waymon Wittman, called to say that a lawyer from Iowa had written the company claiming that his clients had invented our game. The alleged inventors had never submitted the concept to Cadaco. And we had never heard of the people. The lawyer said that they had told the idea to a toy salesman, who may have passed it on to us or Cadaco. My notebook held the date and proof of invention, and that was that. Nonetheless, it was a lesson to us in what can happen.

We do not allow the most innocent communication of ideas. At a high school invention competition that I was judging, a father came up to me with his son; he asked if they could have my business card so the son could call me with his ideas. I politely responded no, and then went on to explain why we cannot entertain outside concepts.

Attack of the Killer Lawyers

Another factor that is contributing more and more to the problem of approaching large manufacturers with outside submissions is the business some quick-buck lawyers are engaged in: the business of litigating patents.

What these unscrupulous attorneys do is buy from small companies and independent inventors the rights to portfolios of their existing but unlicensed patents. They tend to look for patents covering things like early bar code readers, liquid crystal display technologies, and electronic musical instruments, especially synthesizers. In other words, they look for innovations that are heavily used in one form or another today by industry.

The sole and unfortunate purpose of these predators is to find large manufacturers of products which might be perceived to violate their newly acquired patents, and quickly initiate law suits for patent infringement. They make their money not through royalties their patents generate, but through out-of-court settlements paid by large manufacturers to get rid of them.

"I would call it a scam," says Robert W. Faris, a partner in the patent law firm of Nixon & Vanderhye in Arlington, Virginia. "They'll (attorneys) sue at the drop of a hat. Their only purpose is to exploit patents through litigation."

Faris explains that such attorneys know that the courts today are pro-patents. They also know that all anyone needs to get a patent is a well-written description. No inventor has to prove his or her idea with a prototype anymore. When a patent issues, it is assumed valid.

Did You Know ...

You can learn a lot about publicly held companies just by perusing a copy of their annual reports. Form 10-K, the most useful of all reports filed with the Securities and Exchange Commission, reveals the following important tidbits:

- When the company was organized and incorporated.

- What the company produces and percentages of sales any one item may be.

- How the company markets: via independent sales representatives, or its own regional staff offices?

- Whether or not it pays royalties, and how much per year.

- What amount of money the company spends to advertise and promote its products.

- Details on design and development.

- Significant background on production capabilities.

- Terms of long-term leases.

- If the company is involved in any legal proceedings.

- The security ownership of certain owners and management.

- An accurate picture of the competition.

Breaking the Code

Many companies that do work with outside developers do not encourage "unknowns" and return inquiries with a letter containing a paragraph like this:

"Our advertising, research, marketing, and new product planning staffs are primarily responsible for creativity and development.

Corporate policy precludes us from either encouraging or accepting unsolicited ideas from persons outside the Company. While an idea may seem feasible to the submitter, there are usually a number of factors that would make it impractical for us to implement it. Moreover, many of the unsolicited ideas that we receive from both nonprofessional and professional sources have previously been submitted in one form or another."

Reading between the lines, this letter is not as negative as it initially appears to be. The first clue that the company does not do everything internally is that its internal staffs are not "exclusively responsible," but rather "primarily responsible" for creativity and development. This means that outside people back up their company's research and development.

The next good news comes when the letter states that the company cannot encourage or accept "unsolicited ideas from persons outside the Company." I read this to signify that the company probably solicits ideas from a trusted base of outside creative sources. Companies that send back letters similar to this are worth a second look.

Public Companies vs. Private Companies

I have no preference for either one. My decision is based upon what I am able to find out about a company and whether it is best for my product. I maintain, at all costs, the frame of mind that I am evaluating the company rather than viewing myself in the inferior position of being considered by the company.

Finding Information on Public Companies

The best place I know to obtain deep and detailed information on a publicly traded company is at the Securities and Exchange Commission (SEC) in Washington, D.C. This independent, bipartisan, quasi-judicial federal agency was created on July 2, 1934, by act of Congress. It requires a public disclosure of financial and other data about companies whose securities are offered for public sale. Some 11,000 companies are registered.

All companies whose securities are registered on a national securities exchange and, in general, companies whose assets exceed $3 million with a class of equity securities held by 500 or more investors must register their securities under the 1934 act.

In the national capital area, the SEC operates a public reference room (public reference rooms are also located in Chicago and New York City). This specially staffed and equipped facility provides, for your inspection, all of the publicly available records of the SEC.

These include corporate registration statements, periodic company reports, annual reports to shareholders, tender offers and acquisition reports, and much more.

Requests by mail. If you find it inconvenient to visit one of the public reference rooms, the SEC will, upon written request, send you copies of any document or information. In your request, state the documents or information needed, and indicate a willingness to pay the copying and shipping charges. Also include a daytime telephone number. Address all correspondence to: Securities and Exchange Commission, Public Reference Branch, Stop 1-2, 450 Fifth Street NW, Washington, D.C. 20549.

Bechtel Information Services also provides prompt and low-cost research and copying services. It is located at 15740 Shady Grove Road, Gaithersburg, Maryland 20877-1454. In Maryland, the telephone number is (301) 258-4300. Outside Maryland, phone toll free: 1-800-231-DATA.

Annual reports: research pay dirt. Corporate annual reports are on file with the SEC, or can be obtained directly from the company. There is no charge for annual reports that are ordered from the company; if you get annual reports from the SEC, it will require copying, which costs money. Contact the executive in charge of Investor Relations or the Senior Vice President and Chief Financial Officer at the particular company that you are researching.

Somewhere within an annual report it will say something like this: A copy of the company's annual report on Form 10-K, as filed with the SEC, will be furnished without charge upon written request to the Office of the Corporate Secretary. The 10-K is research pay dirt!

I find the 10-K to be the most useful of all SEC filings. In summary, it will tell you the registrant's state of business. This form is filed within 90 days after the end of the company's fiscal year. The SEC retains 10-Ks for ten years.

Part one of the 10-K reveals, among other things:

1. When the company was organized and incorporated. You will want to know how long the company has been in business to gauge its experience. What you would expect from an established company may vary from what you would tolerate at a start-up firm.

2. What the company produces; percentages of sales any one item could constitute; seasonal/nonseasonal, etc. It is critical to have a complete picture of the company's product lines, their strengths and markets, any seasonality or other restrictions to the appropriateness of your item, and if and how your product could be positioned.

3. How the company markets, for instance via independent sales representatives or its own regional staff offices. It is important to know how a company gets something onto the market and where sales staff loyalty is. For example, company employees typically have more loyalty than independent sales reps who handle more than one company's line.

4. Whether or not it pays royalties and how much per year. You can often see how much work the company does with outside developers and whether it licenses anything at all. An example of such wording is this from one corporate 10-K: "We review several thousand ideas from professionals outside the Company each year." I recall another 10-K that read, "The Company is actively planning to expand its business base as a licenser of its products." Statements such as these show that doors are open!

5. What amount of money the company spends to advertise and promote its products. If your product will require heavy promotion, and the company does not promote its lines, you may be at the wrong place. It is counterproductive to take promotional products to companies that don't advertise.

Ask the Right Questions

You probably won't have details at your fingertips about private companies like you do with publicly traded companies. But, by doing some digging, you can come up with some useful information. Before approaching a private company about licensing your invention, there are certain questions you should ask:

1. Is the company a corporation, partnership, or sole propietorship?

2. When was the company organized or incorporated?

3. Who are the company's owners, partners, or officers?

4. What are the company's bank and credit references?

5. Is the manufacturer the source for raw material?

6. How many plants does the company own/lease, and what is the total square footage?

7. What products are currently being manufactured or distributed?

8. How does the company distribute?

6. Details on design and development. You should know before approaching a company whether an internal design and development group exists and how strong it is. I found one 10-K in which a company stated, "Management believes that expansion of its R&D department will reduce expenses associated with the use of independent designers and engineers and enable the Company to exert greater control over the design and quality of its products." It could not be more obvious that outside inventors were not wanted.

7. Significant background on production capabilities. Often it is valuable to know in advance what the company's in-house production capabilities are, and what its outsourcing experiences are in your field of invention. It's no use taking a technology to a company that does not have the experience to produce it.

8. Terms of long-term leases. It can be important to know whether a company owns or rents its facilities as a measurement of its strength and capabilities. An inventory of real estate can also give you an excellent overview of warehouses, plants, offices, and so forth.

9. If the company is involved in any legal proceedings, such as law suits. You may not want to go with a manufacturer that is being sued right and left. Maybe it has just risen from a bankruptcy and is still not strong financially. All of this kind of information is an excellent indicator of corporate health.

10. The security ownership of certain beneficial owners and management. This is vital to understanding the pecking order and power structure. Here is where you'll see who owns how much stock (including family members), and what percentage of the company this represents. Age and years with the company are also shown.

11. Competition. This section will give you a frank assessment of the company's competition and its ability to compete. One 10-K I read once admitted, "The Company competes with many larger, better capitalized companies in design and development . . ." It is unlawful to paint a rosy picture when it doesn't exist. The 10-K is one of the few places you can get an accurate picture of the company's competitiveness. Would you want to license a product to a company that states, for example, "Most of the Company's competitors have financial resources, manufacturing capability, volume, and marketing expertise which the Company does not have." This tells me to check out the competition!

Another helpful document to consult about publicly held companies is Form 10-Q. This report is filed quarterly by most registered companies. It includes unaudited financial statements and provides a continuing view of the company's financial position during the year.

Private Companies: Questions You Need to Ask

Detailed information on privately held companies is harder to come by. There are no regulations requiring that they fill out the kinds of revealing reports public companies must. Nevertheless, it is important to gather as much background information as possible.

Here are some questions I get answers to **before** approaching a private company. The answers come to me from a combination of sources ranging from state incorporation records to interviews with competition, suppliers, retailers (as appropriate), and the owners themselves. Finding the answers requires some digging, but acquiring this background information on a company you hope to deal with is critical to your long-term success.

1. Is the company a corporation, partnership, or sole proprietorship? This can have legal ramifications from the standpoint of liabilities the licensee assumes. A lawyer can advise you on the pluses and minuses of each situation.

2. When was the company organized or incorporated? If a corporation, in which state is it registered? When a company was organized will give you some idea as to its experience. The more years in business, the more tracks in the sand are left. The state in which it is registered to do business will tell you where you may have to go to sue it.

3. Who are the company's owners, partners, or officers? Always know with whom you are going into business. In the end, companies are people, not just faceless institutions.

4. What are the company's bank and credit references? How a company pays its bills is important for obvious reasons, and its capital base is worth assessing.

5. Is the manufacturer the source for raw material? Does it do the fabrication? Such information will help estimate a company's capabilities for bringing your invention to the marketplace.

6. How many plants does the company own (lease), and what is the total square footage? Does it warehouse? This kind of information will help complete the corporate picture.

7. What products are currently being manufactured or distributed? You don't want to waste time pitching to companies that do not manufacture your type of invention. Maybe a company you thought to be a manufacturer is really only a distributor.

8. How does the company distribute? Find out about the direct sales force. Make inquiries concerning outside sales representatives and number of jobbers. Does the company use mail order, house-to-house, mass marketing, or some other form of distribution? This information will quickly reveal how a company delivers its product and whether its system is appropriate for your product. With a mass-market item, it would be foolish to approach a firm that markets door-to-door, regardless of its success.

Company Product and Corporate Profiles: Where to Find Them

The Thomas Register

One of the best sources for product and corporate profiles is the *Thomas Register.* Available in most public library reference rooms, "Thomcat," as it is known, contains information on more than 145,000 U.S. companies in alphabetical order, including addresses and phone numbers, asset ratings, company executives, the location of sales offices, distributors, plants, and service and engineering offices. If you know a brand name, you can locate it in the *Thomcat Brand Names Index.*

The complete *Thomas Register* is also available on compact disc. Should you wish to purchase a *Thomas Register,* for the most current price information call: 1-800-222-7900.

"I'm not too happy with inventor expos," says E. D. Young, a seasoned veteran of inventor organization management and founder of The Inventors Network. "I would rather see the inventor hire a professional to market his product . . . and I have yet to see excellent brokers—I mean good, solid, reliable brokers—at expos. . . . For some reason they do not attend these things." Young adds that these shows primarily consist of inventors looking at each other's products. "And," he adds, "that's the blind leading the blind."

Another excellent source is *Standard & Poor's Register of Corporations, Directors and Executives,* available at public libraries. Consisting of three volumes, it carries data on more than 45,000 corporations, including zip codes, telephone numbers, and names, titles, and functions of approximately 400,000 officers, directors, and executives. A separate volume selects 70,000 key executives for special biographical sketches. The last volume contains a classified industrial index.

Should you wish to purchase a set of S&P reference books, for the most current price information call: 1-800-221-5277.

Does the Shoe Fit?

After you have read all the literature and investigated the company inside and out, you must ask yourself: can the company deliver? And will you be comfortable working with its people? The abbreviation "Inc." after a company's name is not significant. Nice offices, a few secretaries, a fax machine, and a copying machine do not a successful licensee make. The way you are treated should be in good taste. Make sure the shoe fits!

Marketing Your Inventions via Inventor Expos

Some inventor expositions are worthwhile and well-meaning, such as the annual National Inventors Day Exhibit sponsored by the Patent and Trademark Office or regional shows sponsored by inventor organizations. The most reputable and best organized invention expo is an event co-sponsored by the Patent and Trademark Office. The National Inventors Expo is held annually at Crystal Plaza Building 3, 2021 Jefferson Davis Highway in Arlington, Virginia. It features some 50 exhibits by independent inventors who display their patented inventions, and an array of large and small businesses. Admission is free, and the Expo is open to the public. For current information, contact the Office of Public Affairs, PTO, Washington, D.C. 20231, or call (703) 305-8341.

But be aware that many slick operators disguise their motives by organizing inventor expositions and fairs. Know thy promoters and their motives. Some wolves in sheep's clothing invite inventors suffering from "sell-itus" to display prototypes, even drawings or photographs of their inventions. They charge for booth exhibition space and advertising in publications that are released in conjunction with the event. The general public is charged an admission fee to see the inventions. Then to top it all off, some promoters hit exhibitors with broker or commission fees should an invention be licensed through its exposure at the show. Another scheme is to offer the inventor a large cash buyout should a product sell, and the promoter walks off with the royalty points.

When all is said and done, my experience confirms that few if any meaningful contacts ever come out of these shows from the inventor's standpoint, while the promoters make lots of money and get publicity to boot.

Respected inventor Calvin D. MacCracken, president of Calmac Manufacturing, feels that inventor expos are generally not worth it. "My reasoning is simple," he explains. "Product should be marketed within the field for which it is intended to be used. For example, if you have something for the plumbing industry, you show it at a plumbing industry—not an inventor's—expo, no matter how good the product is."

Marketing Your Inventions via Trade Shows

Do not confuse inventor expos with trade shows. Every industry takes part in trade fairs, including the butchers, the bakers, and the candlestick makers. According to the Trade Show Bureau, more than 9,000 trade shows took place in the United States during 1993.

National, regional, and local events promote the sale of almost anything you can imagine. There are trade shows for everything from hardware, consumer electronics, apparel, and aircraft to nuclear medicine, dental equipment, toys, comic books, and musical instruments. If it has been manufactured and sold, you can be sure that it has been marketed at a trade show somewhere, sometime.

One-Stop Research Shopping

While I do not recommend trade shows as the best places to present or license patented inventions, they are a must for getting the beat on any particular market and its dynamics. It's all there for you to peruse at your leisure, with the convenience of one-stop shopping. Competitors line up side-by-side for the important buyers to compare products and pricing and to look for industry innovations and trends.

If you go to the shows, remember that companies have paid many thousands of dollars to exhibit. Their primary reason for being there is to ring up sales or create leads. They are not there to license concepts, by and large.

Anyway, the sales force does not review new concepts. It is responsible for selling, not developing, product. It is both fruitless and dangerous to impose on and expose inventions to salespeople. They are, however, excellent sources of information on companies and the industry and normally delighted to chat about their products, the state of the market, and so on.

Exceptions exist, of course. In some industries, research and development executives attend trade shows to get a feel for the competition as well as host "invited" outside inventors with product to show. Presentations are typically conducted in hotel suites away from the exhibition site. It is best to call the corporate headquarters in advance of the trade show and check policy.

I attend the shows to scout for new product introductions, to pick up information handouts and samples, and to make personal contacts. There is no better or more cost-effective way to acquire product literature than at a trade show. Manufacturers publish flyers and information kits just for trade show distribution. And most come with a price list!

To transport home the booty that I collect, I take empty flight bags with me. Sturdy bags. I never rely on the paper or plastic bags some of the companies supply. The material is too valuable (and heavy). Thanks to trade shows over the past decade, I now have a comprehensive reference library.

Trade shows are also an excellent place to meet and network with executives to whom you otherwise would not have access. They rarely take their "bodyguards" to trade shows; it is too expensive and, after all, they are also there to meet new people. They even make it easier by wearing name tags. I have made super contacts in convention hotel elevators, lobby queues, and shared taxi rides to and from the exhibition centers.

The best kinds of shows at which to meet senior executives are the national or international events. The smaller regional or local trade shows are typically

Important!

Should you be contemplating showing your inventions at inventor expos or trade shows, keep one very important thing in mind. If you intend to apply for patent protection on an invention, its display may foul your chances. Exhibiting a new invention in public for sale or otherwise will make it ineligible for patent protection if a patent application is not filed within one year from the time of the invention's first public exposure.

At most conferences, you will pay an admission fee to attend. For most trade shows, however, admission is free "to the trade." All you usually need is a business card to enter the exhibition area.

staffed by salespeople alone. Nevertheless, such shows provide a less hectic atmosphere and many of the same resource materials.

How to Locate the Right Trade Show

There are several methods for discovering where and when trade shows for any particular industry will be taking place.

1. Ask a manufacturer or distributor in your field of invention. The sales and marketing people will have such information at their fingertips.

2. Contact the trade association that covers your field of invention. More than 3,600 trade associations operate on the national level in the United States. A great way to start is to peruse Gale Research's annual *Encyclopedia of Associations,* available at most libraries. More than 22,000 active associations, organizations, and other nonprofit membership groups are described, in virtually every field of human endeavor.

3. Another source to check is Gale Research's *Trade Shows Worldwide.* The directory provides a full description of each show, including number of exhibitors and significant events taking place during the show.

Conferences and Meetings: Networking Meccas

Perhaps there are even more conferences and meetings going on than there are trade shows. It doesn't take much to have either. All you technically require is so many experts sitting around a table discussing a field of interest.

The biggest difference between trade shows and conferences is that you almost always pay to attend conferences. This is because the primary reasons to attend conferences include hearing experts speak, picking their brains, sharing your ideas, and networking.

Conferences are excellent places to get to know the people behind the products. Socializing is encouraged and the atmosphere is calmer than that at trade shows. There is no pressure to sell. There is no pressure to buy. The object is to brainstorm and exchange ideas. Participants can increase their "know how" and "know who" at the same time.

Many trade fairs have conferences or seminars scheduled. And many conferences offer simultaneous resource fairs.

How Can You Find Out about Conferences in Your Area?

Several methods are available for finding out where and when conferences for a particular industry will take place.

1. Ask a manufacturer or distributor in your field of invention. Many larger manufacturers have training departments that can provide helpful information.

2. As with trade shows, contact the trade association that covers your field of invention. Once again, check Gale's *Encyclopedia of Associations.*

3. Ask department heads and professors at nearby universities where your field of interest is taught. Universities, particularly those teaching engineering and kindred technical fields, will have a current schedule of conferences on hand.

12

Prototyping and Presenting Product
MANUFACTURERS DON'T LICENSE IDEAS

It takes courage to be creative; just as soon as you have a new idea, you're in the minority of one.

—E. PAUL TORRANCE

No one licenses an idea. It is, therefore, required that you present a prototype to potential licensees. A prototype is defined as an original model on which something is patterned. If you do not have the time, money, skills, or commitment to build a prototype of your idea, the odds of your ever licensing it are reduced to practically zero.

The most effective kind of prototype is what is called a "looks like–works like" model. There are no short cuts. Manufacturers react to physical matter, not theories. Don't count on people being able to "imagine" what your product will look like or how it will operate. Even if they could, busy executives do not usually have the time or interest to engage in such typically futile exercises.

Executives love to touch and feel prototypes. Kick the tires, so to speak. Knowing this, do your best to have prototypes that resemble and operate like a production model. Go that extra mile to ensure the prototypes are solid and have perceived value.

Prototypes must be well made because often they take quite a beating at the hands of executives. Don't be surprised when prototypes come back broken because they were mishandled or poorly packed for shipment. It happens at the best of companies. It comes with the territory.

There can be a positive aspect to these unfortunate incidents, however. I have received from $750 to $15,000 as payments from companies that either broke or lost our submissions. This does not happen often, but it does occur more than you would expect.

Don't take the construction and presentation of your prototype lightly. You must be as sophisticated and slick in your presentation to potential licensees as they will have to be in their pitch to the trade and/or the consumer. While getting a product known is relatively easy, marketing a need is something else. And that's your ultimate goal.

Safe Conduct

The Consumer Product Safety Commission (CPSC) has jurisdiction over about 15,000 types of consumer products, from automatic-drip coffee makers to toys to lawn mowers.

If you are working on an invention in a particular category of consumer product, and you have health and safety questions about certain materials, or you want to be sure you are meeting current voluntary industry standards and/or mandatory government standards, a query to the CPSC can save you time and money.

You can call the CPSC's main number at (301) 504-0400 and be directed to the appropriate department.

Making Prototypes

If you cannot make a prototype, there are plenty of places to get help. In some cities prototype makers are listed in the yellow pages. Universities and engineering schools often have workshops where you can make connections to get products prototyped. Local inventor groups are also excellent sources of assistance.

Many invention marketing companies offer prototype making services, but be very careful. Know what you are getting yourself into when contracting prototype work. I would not engage anyone to do such work without first having inspected the person's shop and checked references.

If you cannot find a prototype maker, there is a list of people who bill themselves as prototype development assistance providers at the end of this chapter. I have not personally checked out the quality of their work or their credentials. Please be sure to obtain references or referrals from others before hiring one of these outfits or, in the case of inventor organizations, anyone they recommend.

Multiple Submissions

If you have more than one prototype, and the situation is appropriate, you may wish to consider making submissions to more than one manufacturer at the same time. I have no set rule about this and take it case by case, guided by experience.

If a company asks you to hold off further presentations until it has an opportunity to review the item at greater length, try to set guidelines. In all fairness, some products require a reasonable number of days to be properly considered. However, if you feel the company is asking for an unreasonable period of time, seek some earnest money to hold the product out of circulation. The amount of time and money is negotiable. Also insist that the product not be shown to anyone outside the company. See page 239 for a sample Option Agreement.

Mutual Dependency

You need the company or you would not be there. Show yourself as being independently creative, while at the same time taking the "we approach" and not the "I approach."

In order for your product to sustain itself through the review and development process, it will need a champion. Typically this standard-bearer will come from among those attending your first meeting. Get others involved. Turn "your idea" into "our idea."

If You Sign My Paper, I'll Sign Yours

Many companies ask that an agreement be signed before they'll accept the submission of outside ideas. This may surface when you first approach the company or on the day you make the formal presentation. I have never had a problem with such requests. I always know with whom I am dealing and feel confident in the relationship. If I did not, I wouldn't be there in the first place.

A suspicious attitude may seriously inhibit your progress. Put your time and energies into creating concepts versus overprotecting them. Become paranoid over this and no one will ever see your ideas.

Outside submission agreements take many forms. The two samples included on pages 218-223 are commonly used formats. Most companies use variations of these forms. Regard these forms as examples to familiarize you with the types of documents in case you are presented with such a document by a prospective licensee, or if you want to use one to protect yourself before looking at another individual's ideas.

Protecting Your Invention Submission

In some instances, it is appropriate for the inventor to have the company sign an agreement to hold a product secret and confidential. This agreement can be modified for use with your prototype provider as well. A sample agreement can be found on page 224.

First Impressions Are Lasting Impressions

The inventor is always selling two things: the concept and the inventor. Your personal credibility is often more important than the credibility of any particular single creative concept.

It is critical that a corporate executive buy the inventor as much as the invention. You may be capable of dreaming up numerous innovative products for a company to consider, and will want to be invited back again and again. Without respect from corporate executives, however, your products will never be taken seriously. And you cannot put a dollar value on the ability to make an encore.

When to Make the Pitch: Avoid Cold Calls

Presentations are best when carefully choreographed and staged. Nothing is usually gained by ambushing executives outside of their offices. As Agesilaus II, King of Sparta, said, "It is circumstance and proper timing that give action its character and make it either good or bad."

For every rule ever told about when to sell, another rule proves it wrong. The best rule on when to sell ideas is whenever possible. I operate under the principal that when you have something hot, burn it. When it gets cold, sell it for ice.

I licensed our game, Adver*teasing,* to Cadaco in late spring, and the manufacturer shipped it to stores that August. Cadaco put its resources on the line to make it happen on very short notice. By early December it had racked up sales of over 250,000 units.

While this example is the exception to the rule, had I waited until what I thought would be the "best" time to make the pitch, opportunity might have been delayed or the product may never have been licensed. The marketplace is temperamental and erratic.

Curtain Up. Light the Lights.

Once you have received an invitation to display your concept to a manufacturer, it's show time. And if you thought inventing was tough, you haven't yet experienced hardship. The moment you walk into a company's conference room with your invention at the ready, you pass into an eerie twilight zone. You are the hero in a video game and the corporate executives are the cast of characters you meet in your quest to win the game. Some pray for you to succeed. Others are gremlins out to gobble you up, shoot you down, and otherwise obliterate your ideas.

My Thoughts Exactly

Baa, baa, black sheep, have you any thoughts?

Yes sir, yes sir, to chuck what I've been taught

about channels and protocol and toeing the line.

You've got to make waves to do something fine.

—Richard Saunders

Thomas Alva Edison once observed, "Society is never prepared to receive any inventions. Every new thing is resisted, and it takes years for the inventor to get people to listen to him before it can be introduced."

Another inventor said, "Creation is a stone thrown uphill against the downward rush of habit."

World history is rich in stories about people's resistance to new ideas. Some seem unbelievable in retrospect. I find it helpful to recall some of these stories before I begin a presentation.

Impressionist art that sells for millions of dollars today was widely criticized when it was first introduced. The innovative Stravinsky was once labeled "cynically hell-bent to destroy music as an art form." Today he is regarded as a genius.

When railroads were established, farmers protested that the "iron horse" would scare their cattle to death and stop hens from laying eggs. The British Association for the Advancement of Science insisted the automobile would fail because a human driver "has not the advantage of the intelligence of the horse in shaping his path."

Inflexibility Is Inherent in the System

Remember that in order to operate, companies must have rules and controls. Loose cannons do not last long in corporate environments. Any organization must understandably have something of an established routine to survive. Therefore, executives, especially in larger companies, often fall into predictable and ordered routines. Your job is to break this routine, make people believe in you, and interest the company in buying your concept.

Idea Submission Agreement I

While _____ wishes to take every opportunity to improve its products and add profitable ones to its line, it has found certain precautions necessary in accepting disclosures from persons not in its employ. For an idea to be considered, this form must be completed in full, signed and returned with any disclosure of an idea or invention.

(Date)_____ 19___

To _____

I am submitting to you, for your evaluation and permanent record, copies of certain ideas, suggestions or other materials having to do with:

The information I am submitting to you consists of the following:

_____ 1) Description

_____ 2) Drawing or sketches

_____ 3) Samples

_____ 4) Copy of a patent application(s)

_____ 5) Other

(Please check the appropriate blank(s).)

In doing so, I agree to the conditions printed on the back of this sheet and further agree that such conditions shall apply to any additional disclosures made incidental to the original material submitted.

Signature _____

Name Printed _____

Address _____

Conditions of Submission

1) All submissions or disclosures of ideas are voluntary on the part of the submitter. No confidential relationship is established by submission or implied from receipt or consideration of the submitted material.

2) Patented ideas and ideas covered by pending applications for patent are considered only with the understanding that the submitter agrees to rely for his/her protection solely on such rights as he/she may have under the patent laws of the United States.

3) Ideas which have not been covered by a patent or a pending application for patent are considered only with the understanding that the use to be made of such ideas and the compensation, if any, to be paid for them are matters resting solely in the discretion of the Company.

4) If the subject matter offered the Company is a proposed trademark, advertising slogan, or merchandising plan, susceptible to trademark or copyright protection, the Company will examine it only under the terms set forth in this Agreement. The submitter shall rely for his/her protection solely on such rights as he/she may have under the Copyright and Trademark Laws of the United States.

5) The foregoing conditions may not be modified or waived.

Idea Submission Agreement II

AGREEMENT made and entered into this _____ day of _____, 19___, by and between _____ ("Disclosers" full name and address) and _____, a corporation organized under the laws of the State of _____, with offices located at _____.

WHEREAS Discloser is a developer of, or has licensing rights to, concepts for _____, and

WHEREAS, Discloser represents that he/she has developed a certain concept, device or other proprietary subject matter more specifically described at the end of this Agreement and on the attachment hereto (hereinafter referred to as the "Item"), and

WHEREAS, _____ desires to evaluate the commercial utility of the Item, and

WHEREAS, in order to make this evaluation possible, it will be necessary for Discloser to disclose confidential information concerning the Item to

_____.

NOW THEREFORE, in consideration of the mutual promises hereinafter contained, and for other good and valuable consideration, the parties agree as follows:

1) Discloser shall make full disclosure with respect to the Item to employees of _____ or one of its affiliates (collectively, _____) and shall submit to_____ all relevant data in connection therewith. The disclosure by Discloser to _____ is solely to enable _____ to evaluate the Item in order to determine its commercial utility. _____ _____ is under no obligation to market or produce the Item, unless and until a formal written agreement is entered into, and the obligations of _____ shall be only those which are set forth in any such agreement.

2) Discloser hereby represents to _____ that the Item is his own individual creation and wholly and solely the property

of Discloser and that Discloser has not assigned, sold, licensed, mortgaged, pledged, or otherwise transferred or encumbered the Item or entered into any agreement to do any of the foregoing with respect to the Item. The execution and performance of this Agreement by Discloser does not violate any contract, agreement or other restriction to which Discloser is a party or by which it is bound or any rights of any third party.

3) The disclosure of the Item and all information incidental thereto is confidential and shall be received by _____ in confidence. _____ shall not disclose such confidential information to others and shall take reasonable steps to prevent such disclosure. _____ agrees to use the same degree of care in protecting and safeguarding the confidentiality of the concepts and information disclosed hereunder as it uses for its own information of like importance. _____ shall not be liable for inadvertent disclosure or use of the Item by persons who are or have been in its employ, unless _____ fails to exercise the degree of care set forth above.

4) It is understood that _____'s willingness to evaluate the Item is not to be construed as an admission of the Item's novelty, priority, or originality. Discloser understands that _____ may have rights to the Item or particular elements thereof, due to prior access to information similar to the Item or elements thereof including, by way of illustration and not limitation, prior patents, prior publication, prior submissions to _____ by others, prior development by _____'s personnel or representatives, prior use by _____, prior knowledge, or prior sale. Accordingly, consideration of the Item by _____ shall not deprive _____ of its existing rights, if any, with respect to the Item or any element thereof.

5) Without limiting the generality of the provisions of paragraph 4 hereof, the obligations of _____ hereunder are not applicable to such information which:

   a) prior to disclosure by Discloser, was already known to _____ as evidenced by records kept in the ordinary course of business of _____ or by proof of actual use by _____.

b)   was known to the public or generally available to the public prior to the date of disclosure.

c)   becomes known to the public or is generally available to the public subsequent to the date of said disclosure through no act of _____ contrary to the obligations imposed by this Agreement.

d)   is disclosed by Discloser to an unrelated third party without restriction.

e)   is approved for public release by Discloser.

f)   is rightfully received from a third party without similar restriction and without breach of this Agreement.

g)   is independently developed by _____ without breach of this Agreement.

h)   is required to be disclosed by judicial or government action.

i)   is disclosed in a judicial or governmental proceeding subject to a protective order.

_____ shall be free of any obligations restricting disclosure and use of the information provided by Discloser hereunder, subject to Discloser's patent rights, if any of the provisions of a) through i) of this paragraph 5 are applicable to the information disclosed.

6)   Upon submission of the Item to _____, _____ shall consider the Item and as promptly as practicable advise Discloser of _____ 's interest or lack of interest therein, all subject to the terms, conditions and provisions of this Agreement.

7)   _____ shall not be obligated to take any action with regard to the Item other than pursuant to paragraphs 3 and 6 hereof.

8)   _____ will, upon request, return any letters, drawings, descriptions, specifications or other materials submitted to it in connection with the Item.

9) The provisions of this Agreement shall apply to any additional or supplemental information pertaining to the Item provided by Discloser to

_____ .

10) This writing reflects the entire agreement between the parties concerning the Item, and no modification, amendment, waiver or cancellation of this Agreement or any provision hereof shall have any validity or effect whatsoever unless in writing and signed by both parties hereto. Without limiting the generality of the foregoing, no agreement relating to the purchase or use of the Item by _____ or any of its affiliates, or relating to the terms of or consideration of such purchase or use, or relating to any compensation to, or reimbursement or any expenses of Discloser, shall be binding upon either party hereto unless in writing and signed by both parties hereto.

11) This Agreement shall be governed by, construed and enforced in accordance with the internal laws of the State of _____ without reference to principles or conflict of laws.

12) This Agreement shall be binding upon, and inure to the benefit of, the Discloser _____ and (and the affiliates of _____) and their respective heirs, executors, administrators, successors, and assigns.

IN WITNESS WHEREOF, the parties have signed this Agreement on the respective dates hereinafter written.

The Item is generally described as follows:

SEE ATTACHED

_____ (Company)

By: _____ (Discloser)

Date: _____

Agreement to Hold Secret and Confidential

The below described invention, idea or concept (hereinafter referred to as INVENTION) is being submitted to _____ of _____ (hereinafter referred to as COMPANY) by _____ of _____ on _____, 19___ (hereinafter referred to as INVENTOR) who is the inventor of record. The undersigned, in consideration of examining said INVENTION, with a purpose to opening negotiations to obtain a license to manufacture and sell said INVENTION, hereby agrees on behalf of himself/herself and said COMPANY that he/she represents, that:

1) He/she (during or after the termination of employment with said COMPANY) and said COMPANY, will keep said INVENTION, and any information pertaining to it, in confidence.

2) He/she will not disclose said INVENTION or data related thereto to anyone save for employees of said COMPANY, sufficient information about said INVENTION to enable said COMPANY to continue with negotiations for said license, and that anyone in said COMPANY to whom said INVENTION is revealed, shall be informed of the confidential nature of the disclosure and shall agree to hold confidential the information, and be bound by the terms hereof, to the same extent as if they had signed this Agreement.

3) Neither he/she nor said COMPANY shall use any of the information provided to produce said INVENTION until agreement is reached with INVENTOR.

4) He/she has the authority to make this Agreement on behalf of said COMPANY.

It is understood, nevertheless, that the undersigned and said COMPANY shall not be prevented by the Agreement from selling any product heretofore sold by said COMPANY, or any product in the development or planning stage, as of the date first above written, or any product disclosed in any heretofore issued U.S. Letters Patent or otherwise known to the general public.

The terms of the preceding section releasing, under certain conditions, the obligation to hold the disclosure in confidence does not however, constitute a waiver of any patent, copyright or other rights which said Inventor or any licensee thereof may have against the undersigned or said COMPANY.

PROTOTYPE DEVELOPMENT ASSISTANCE PROVIDERS

ALABAMA

Alabama Small Business Development Consortium
University of Alabama/Birmingham (205) 934-7260
1717 11th Avenue South, Suite 419
Birmingham, AL 35294

ARIZONA

Synergetics
P.O. Box 809 (602) 428-4073
3860 W. First
Thatcher, AZ 85552

ARKANSAS

Arkansas Science & Technology Authority
100 Main Street, Suite 450 (501) 324-9006
Little Rock, AR 72201

GENESIS Technology Incubator
University of Arkansas (501) 575-7227
Engineering Research Center
Fayetteville, AR 72701-6832

CALIFORNIA

Antelope Valley IWIEF Chapter
4444 E. Avenue R, Sp. 127 (805) 273-0144
Palmdale, CA 93550

The Dream Merchant
2309 Torrance Boulevard, Suite 201 (310) 328-1925
Torrance, CA 90501

DTSC/PPPRA
P.O. Box 806 (916) 324-1807
Sacramento, CA 95812-0806

Education Foundation
Inventors Workshop International (805) 484-9786
3201 Corte Malpaso, Suite 304-A
Camarillo, CA 93012

Inventors Helper Industries
4480 Treat Boulevard (415) 676-4975
Suite 310
Concord, CA 94521

The James F. Riordan Co.
3110 Camerosa Circle (916) 676-4729
Cameron Park, CA 95682

Orbic Controls
Box 23827 (510) 944-4987
Pleasant Hill, CA 94523

Perris IWIEF Chapter
1051 Davids Road (714) 657-2822
Perris, CA 92370

San Jose IWIEF Chapter
340 Rosewood Avenue (408) 248-1059
San Jose, CA 95117

COLORADO

Colorado Inventors Council, Inc.
P.O. Box 88 (303) 854-3851
Holyoke, CO 80734

Important!

It is prudent to require that any prototype maker sign a Confidential Disclosure Agreement before you show or discuss your product. A sample agreement is shown on page 224. Also, remember to always check the company's references before you hire them.

**Colorado University
Business Advancement
Centers**

335 S. 43rd Street   (303) 499-8114
Boulder, CO 80303

**Rocky Mountain
Inventors Congress**

P.O. Box 4365   (303) 758-6757
Denver, CO 80204

CONNECTICUT

SIMCO, Inc.

61 River Road   (203) 227-0041
Weston, CT 06883

FLORIDA

BranLabs, Inc.

5714 Oakhurst Drive   (813) 393-3552
Seminole, FL 34642

**Center for Health
Technologies, Inc.**

1150 N.W. 14th Street   (305) 325-2733
Suite 105
Miami, FL 33136-2112

**Data Safe Distributed
Systems**

1321 Woodgate Way   (904) 425-5029
Tallahassee, FL 32312

**Florida Product
Innovation Center**

2622 NW 43rd Street   (904) 334-1680
Suite B3
Gainesville, FL 32606-7428

The Inventor's Club

WSRE-TV   (904) 484-1224
1000 College Boulevard
Pensacola, FL 32504

GEORGIA

CAN DO Industries, Inc.

7610 Ball Mill Road   (404) 396-1401
Atlanta, GA 30350

HAWAII

Inventors Council of Hawaii

Box 27844   (808) 595-4296
Honolulu, HI 9628

IDAHO

Planet X

2211 Heron Street   (208) 336-7340
Boise, ID 83702

Boise State University

Technical & Industrial Extension Serv.   (208) 385-3767
1910 University Drive
Boise, ID 83752

ILLINOIS

Bradley University

Business Technology Incubator   (309) 677-2852
Peoria, IL 61625

**Center for Advanced
Manufacturing & Production**

Southern Illinois University   (618) 692-2166
Box 1108
Edwardsville, IL 62026

**Innovation Development
Corporation**

P.O. Box 1185   (312) 618-8129
Calumet City, IL 60409

Northern Illinois University

Technology Commercialization Office   (815) 753-1238
DeKalb, IL 60115-2874

Tech. Commercialization Center

Illinois State University　　　　　(309) 438-7127
Room 215, Media Center
Normal, IL 61761

University of Illinois at Chicago

Software Tech. Research Center　　(312) 413-7453
1033 W. Van Buren, Suite 700N, M/C 346
Chicago, IL 60607

Western Illinois University

212 Seal Hall　　　　　　　　　(309) 298-2211
Macomb, IL 61455

Solomon Zaromb

9S 706 William Drive　　　　　　(708) 654-8214
Hinsdale, IL 60521

KANSAS

Kansas Association of Inventors

2015 Lakin　　　　　　　　　　(316) 792-1374
Great Bend, KS 67530

Pittsburg State University

Center for Tech. Transfer　　　　(316) 235-4114
Shirk Hall
Pittsburg, KS 66762

KENTUCKY

Bluegrass Inventors Guild

P.O. Box 43610　　　　　　　　(502) 423-9850
Louisville, KY 40253-0610

LOUISIANA

University of Southwest Louisiana

Tech. Innovation Center　　　　　(318) 231-6767
P.O. Box 44172
Lafayette, LA 70505

A Fish Tale

The art of product presentation is not unlike the sport of fly fishing. It takes time and patience. In casting, the lure is presented and then pulled back. The lure's movement, aided by twitches, pauses, and jerks of the rod by the angler, entices the fish to strike. In the case of both fishing and selling your product, the object is to hook a big one.

MARYLAND

USDA-ARS-OTT

National Patent Program　　　　(301) 504-6786
Room 401, Building 005, BARC-W
Beltsville, MD 20705

MASSACHUSETTS

Cape Cod Thinkers Association

1600 Falmouth Road　　　　　　(509) 711-2622
Suite 123-MBE
Centerville, MA 02632

HighTech Design

26 Murray Street　　　　　　　(508) 535-2543
West Peabody, MA 01960

Innovative Products Research & Services

P.O. Box 335　　　　　　　　　(617) 862-5008
Lexington, MA 02173

Northeastern University

Division of Research Management　(617) 437-4587
360 Huntington Avenue, 423 Lake Hall
Boston, MA 02115

Worcester Inventors

65 Windsor Street　　　　　　　(508) 757-6178
Worcester, MA 01605

MICHIGAN

Bohning Co.
Lake City, MI 49651   (616) 229-4247

Inventors Center of Michigan
Ferris State University   (616) 592-3774
1020 E. Maple
Big Rapids, MI 49307

Michigan Technological University
Bureau of Industrial Development   (906) 487-2470
1400 Townsend Drive
Houghton, MI 49931

MINNESOTA

Excel Development
7007 Dakota Avenue (612) 934-1200
Chanhassen, MN 55317

Nordic Track
Product Planning & Development   (612) 368-2777
104 Peavey Road
Chaska, MN 55318

Northeast Business Innovation & Technology Resource Center
820 N. 9th Street   (218) 741-4241
Suite 140
Virginia, MN 55729

MISSOURI

Center for Technology
Grinstead 80 CMSU   (816) 543-4402
Warrensburg, MO 65093

Entech Engineering, Inc.
111 Marine Lane   (314) 434-5255
St. Louis, MO 63146

Central Missouri State University
Grubstead 80   (816) 543-4402
Warrensberg, MO 64093

Mid-America Inventors
2018 Baltimore   (816) 221-2442
Kansas City, MO 64108

Missouri Enterprize
Business Assistance Center   (314) 364-8570
800 West 14th Street, Suite 111
Rolla, MO 65401

Missouri Innovation Center
T-16 Research Park   (314) 882-2822
Columbia, MO 65211

University of Missouri—Rolla
Center for Technology Development   (314) 341-4559
Building 1, Nagogami Terrace
Rolla, MO 65401

MISSISSIPPI

Society of Mississippi Inventors
P.O. Box 13004   (601) 977-8799
Jackson, MS 39236-3004

MONTANA

Kings Tool
5350 Love Lane   (406) 586-1541
Bozeman, MT 59715

NEVADA

Concept Support and Development Corp.
121 Woodland Avenue #100   (702) 746-0700
Reno, NV 89523

NEW JERSEY

**New Jersey Institute
of Technology**
240 Dr. Martin Luther King Jr. Blvd.   (201) 596-3430
Newark, NJ 07102

NEW YORK

**Center for Innovative
Technology Transfer**
SUNY—College at Oswego   (315) 341-2128
209 Park Hall
Oswego, NY 13126

**New York Society of
Professional Inventors**
116 Steuart Avenue   (516) 598-3228
Amityville, NY 11701

New York State Energy R&D
Two Rockefeller Plaza   (518) 465-6251
Albany, NY 12223

NORTH CAROLINA

North Carolina SBDC
4509 Creedmoor Road, #201   (919) 571-4154
Raleigh, NC 27612

NORTH DAKOTA

Technology Transfer, Inc.
1833 East Bismarck Expressway   (701) 221-5346
Bismarck, ND 58504

University of North Dakota
ERIP/SBIR Program   (701) 777-3132
Box 8103 UND Station
Grand Forks, ND 58202

OHIO

Akron/Youngstown Inventors
1225 W. Market Street   (216) 864-5550
Akron, OH 44313

Innovation Alliance
1445 Summit Street   (614) 421-7163
Columbus, OH 43201

Invention Engineering, Inc.
27 Aberdeen Avenue   (513) 294-7447
Dayton, OH 45419-3101

**Inventors Council of
Greater Lorain County**
1005 North Abbe Road   (216) 365-4191
Elyria, OH 44305

Medical College of Ohio
P.O. Box 10008   (419) 381-4250
Toledo, OH 43699-0008

Ohio's Thomas Edison Program
77 S. High Street, 26th Floor   (614) 466-3887
Columbus, OH 43266-0330

**Ohio Technology
Transfer Organization**
Ohio State University   (614) 292-5485
Room 216 Bevis Hall
1080 Carmack Road
Columbus, OH 43210

Try, Try Again

Resistance, then success. Back in the 1930s, Charles B. Darrow invented a game called Monopoly. It was rejected by Parker Brothers and six or seven other companies. Parker Brothers eventually saw the light and published Monopoly, which annually sells more than one million units. Darrow became the first millionaire game inventor.

OKLAHOMA

**Invention Development
Society, Inc.**
8230 Southwest Eighth Street   (405) 376-2362
Oklahoma City, OK 73128

**Oklahoma Department
of Commerce**
Inventors Assistance Program   1-800-TRY-OKLA
1700 Sandra Drive
Midwest City, OK 73110

**Oklahoma Department
of Commerce**
Inventors Assistance Program   (405) 841-5161
P.O. Box 26980
Oklahoma City, OK 73126

Oklahoma Inventors Congress
P.O. Box 27291   (918) 245-6465
Tulsa, OK 74149-0291

OREGON

**Oregon Resource/Technology
Development Co.**
1934 Northeast Broadway   (503) 282-4462
Portland, OR 97232

PENNSYLVANIA

**Northwestern
Inventors Council**
Gannon University   (814) 871-7619
Erie, PA 16541

RHODE ISLAND

**Rhode Island Solar
Energy Association**
42 Tremont Street   (401) 942-6691
Crandston, RI 02920

SOUTH CAROLINA

Clemson University
Emerging Technology Center   (803) 646-4020
511 Westinghouse Road
Pendleton, SC 29670

**Enterprise Development Inc.
of South Carolina**
P.O. Box 1149   (803) 737-0843
Columbia, SC 29202

SOUTH DAKOTA

College of Engineering
South Dakota State University   (605) 688-4161
CEH 201 Box 2219
Brookings, SD 57007-0096

TENNESSEE

**Tennessee Inventors
Association**
P.O. Box 11225   (615) 483-0151
Knoxville, TN 37939

TEXAS

**Bill J. Priest Institute
for Economic Development**
Technology Transfer Center   (214) 565-5860
1402 Corinth
Dallas, TX 75215

**Inventors Information
Systems**
P.O. Box 927   (915) 698-3318
Abilene, TX 79604

**Network of American
Inventors and Entrepreneurs**
11371 Walters Road   (713) 537-8277
Houston, TX 77067

Office of Technology Development
M.D. Anderson Cancer Center   (713) 792-7598
1515 Holcombe Boulevard, Box 510
Houston, TX 77030

Technology Transfer Center
1402 Counth Street   (214) 565-5852
Dallas, TX 75215

Toy & Game Inventors of America
5813 McCart Avenue   (817) 292-9021
Ft. Worth, TX 76133

Ultimate Concepts, Inc.
Houston Inventors Society   (713) 686-7676
P.O. Box 740304
Houston, TX 77274-304

UTAH

National Congress of Inventor Organizations
710 N. 600 West   (801) 753-4700
P.O. Box 268
Logan, UT 84321

Science, Technology & Innovation
State Office   (801) 569-2973
P.O. Box 11
West Jordan, UT 84084

Utah Division of Energy
8764 S. Russell Park Road   (801) 538-5432
Salt Lake City, UT 84121

Weber State University
Technology Assistance Center   (801) 626-6309
Ogden, UT 84408-1801

VIRGINIA

Virginia Center for Innovative Tech.
Technology Commercialization   (703) 689-3000
CIT Building, Suite 600
2214 Rock Hill Road
Herndon, VA 22070

VERMONT

The Inventors' Ally
P.O. Box 1527   (802) 464-8918
Willington, VT 05363

WASHINGTON

Critical Data, Inc.
West 1100 Sixth Avenue, Suite B   (509) 838-4989
Spokane, WA 99204

Washington State University
Small Business Development Consortium   (509) 335-1576
245 Todd Hall
Pullman, WA 99164-4727

WISCONSIN

Center for Innovation & Development
University of Wisconsin—Stout   (715) 232-2565
103 First Avenue West
Menomonie, WI 54751

WEST VIRGINIA

West Virginia Small Business Development Consortium
West Virginia Development Office   (304) 558-2960
1115 Virginia Street East
Charleston, WV 25301-2406

WYOMING

Wyoming Small Business Development Consortium
Casper College   (307) 237-0621
125 College Drive
Casper, WY 82601

CANADA

Canadian Centre for Industrial Innovation

Evaluation and Technology Transfer   (514) 383-7712
75 Port Royal East, Suite 600
Montreal, Quebec H3L 3T1
Canada

Canadian Industrial Innovation Centre

156 Columbia Street West   (519) 885-5870
Waterloo, Ontario N2L 3L3
Canada

Technology Transfer Office

1800 Argyle, P.O. Box 519   (902) 424-7382
Halifax, Nova Scotia B3J 2R7
Canada

University of British Columbia

2194 Health Services   (604) 822-8580
Room 331, IRC Building
Vancouver, British Columbia V6T 1W5
Canada

GUAM

Guam Energy Office

P.O. Box 1167   (671) 472-8131
Agana, 96910
Guam

CHAPTER

13

How to Prepare Proposals

Ideas are such funny things; they never work unless you do.

—Herbert V. Prochnow

If It's Not on the Page, It's Not on the Stage

I believe every product submission should be accompanied by its own information package. This can take many forms, but typically I prepare a written proposal, the elements of which are outlined below, plus several colored-marker renderings and a video.

Elements of a Proposal

Every proposal should begin with an executive summary—a paragraph that allows corporate executives to get a quick read on the product and do a gut check. This is a simple paragraph, nothing elaborate: one that paints an image of the product and/or your objective.

I like to present all written materials in three-ring binders or folders, the kind that have cover and spine pockets. We design a cover page and use tabs to separate sections. These notebooks become one-stop information banks for ourselves and the executives. I purchase the binders in volume at large office supply outlets; they offer a large selection of colors and sizes.

Here are the categories (as appropriate) that I like to include in these binders:

1. **Operating instructions:** Take nothing for granted. Nothing! The worst thing that can happen is an executive's inability to use a product after you depart. Don't let the simplicity of your item draw you into a false sense of security. If I were to submit something as simple as a ball-point pen, I would prepare written, illustrated, and video instructions.

2. **Marketing Plan:** Highlight and detail your item's unique features and advantages over existing products. Define its appeal and target audience. Suggest follow-ups, including second generations and line extensions, if appropriate. Manufacturers like products that have a future, especially if they are required to spend beaucoup start-up dollars in the development and launch phases.

Good for the Long Haul

It is important to make clear to your prospective licensee that you are not just capable of delivering the item under consideration, but that a lot more can come from the same source: you.

A Helping Hand

Inventor and business organizations, as well as state and university assistance programs, can be of great value to the inventor who is preparing presentation packages for potential licensees.

For more information, see Chapter 17: University and Independent Innovation Research Centers, and Chapter 18: Inventor Organizations.

You do not need a graduate degree in marketing to work up this kind of information. Whatever you can do to define the market and positioning of your concept, the better.

3. **Trademark:** Offer possible trademarks and tag lines. I like to suggest everything from word marks to logotypes. If you have conducted a trademark search, include its results or the status of any applications you may have in play. The right mark or slogan can go a long way towards securing a sale. For information on trademark searches, see Chapter 9.

4. **Patents:** Before you submit any concept to a manufacturer (or investor), a patent search should be conducted. Include the results of any search, PTO actions, and so on. If a patent has been issued to you already, submit a copy. For information on patent searches, see Chapter 5.

5. **Advertisement / Publicity Copy:** Suggest advertising and publicity hooks or approaches. This information, like a good trademark, helps make presentations more persuasive and polished. It is also available to the manufacturer for focus groups.

6. **Videos:** Include a video demonstrating your product. We do not let much out of our studio without a demo video. If a picture is worth a thousand words, moving images are worth a million. Playback machines are found in every company. Other than a face-to-face presentation, nothing beats a video. By the way, you can purchase varying lengths of tapes from wholesale jobbers. Most of our tapes are pre-cut to 10 minutes. Shorter tapes also cost less to mail because they weigh less.

   In addition to doing video instructions, we also use videos to do technical briefings and to show people testing our prototypes. Nothing like showing a satisfied consumer. If we have a technical briefing and ideas for a focus group, we might consider submitting two separate tapes so that two departments do not have to share one tape.

7. **Technical:** In writing and on video you should address everything from design, engineering, and manufacturing issues to component sourcing and costing. Include exploded views, parts lists, and anything else that will help a corporate research and development type understand your item's technical and manufacturing profile.

8. **Personal Resume:** If you are unknown to the company, provide a background sheet on your capabilities. Depending upon the nature of the submission, a green light may depend upon the manufacturer's confidence in your ability to make the product happen, albeit under the company's guidance.

We approach every product submission in such a way that we could, if contracted by the manufacturer, take the item outside and do it ourselves. More and more manufacturers are relying on us to take products to tooling stage. This was not the case years ago, but with personnel cutbacks and increased workloads, often the only way a product gets done is through the inventor's being able to manage a development program.

When possible include with your proposal (in the "Technical" section):

- A sheet listing all components with respective prices from various sources. Pricing from three different sources is a good bet. When possible, a mix of domestic and off-shore numbers is best. Do not forget to include the volume the quotes are based upon, plus vendor contacts.

- Note the type of material(s) desirable, for example, polyethylene, wood, board, and so on. Provide substitutions and options for consideration.

- When you calculate the item's cost, do not forget to consider the price of assembly (if any). The quoting vendors will be helpful here.

- If your item requires retail packaging, add an extra 15-30 percent.

- Add an extra 20 percent for modifications and losses. At this point you have the item's hard cost.

- To arrive at the manufacturer's selling price, add in your royalty, an amount of money for promotion (if appropriate), and a gross profit margin for the manufacturer (65 percent). You may wish to estimate the mark-up at the retail end (if appropriate). A good estimate is the hard cost multiplied by three or four.

A manufacturer surely will do its own costing. If you feel confident providing figures or sources, it will be appreciated and may help your case. However, be aware that improper information can damage your credibility and may derail your submissions.

Negotiating a Licensing Deal

On Second Thought . . .

I have a close friend who beat a large manufacturer in court to the tune of a million dollars. After splitting the award with his partner and attorney, and paying taxes and costs, he came away with about $250,000. It sounded good at first. But he soon realized that every door was closed to him in his industry. None of the major companies wanted to see his concepts. They still don't. It was a high price to pay.

Do You Need a Lawyer for Contracts?

Maybe yes. Maybe no. This decision will depend upon your own experience in hammering out intellectual property deals. In my own case, I used a law firm for my first licensing agreement and then went it alone.

Lawyering, once a respected profession, has evolved into just another enterprise, one involving self-promotion, boredom, greed, and billable hours.

"I call it the Twilight Zone factor," said one lawyer in an interview with *Philadelphia* magazine. "Nothing we do as lawyers is rooted in reality. The fees we charge have no economic basis in the work actually done. It's whatever the market will bear. The issues we dispute are increasingly not real world issues but artificial conflicts that we created and that we prolong. And the worst part is that the expectations we have of ourselves all call for Superman in a three-piece suit. What I hate about being a lawyer is always reaching out to touch something—but it's never there."

Now that I have put lawyering into perspective, let me add that there are many very good lawyers. However, while a lawyer may know more than you about the fine points of agreements, few practicing lawyers have proven themselves to be sharp in business.

Negotiating contracts is a skill that can be learned. People tend to make it complex. It is not.

Do It Your Way

The most basic rule is to conduct your business in your style. Set the pace. Do not get caught up in your prospective licensee's timetables and priorities. Things get worse under pressure.

Lawyers tend to intimidate most people. If you let them, they'll confound you with facts, blind you with Latin, and plague you with precedents. Whether you take a lawyer with you or negotiate yourself, be sure that everything is spelled out to your satisfaction—even if it is not the way something is taught at Harvard Law. Always insist upon clarity over form.

Winning

Getting what you want does not always have to be at another person's expense. It is possible to get what you want and still let your opponent have something. After all, you are entering into what you both hope will be a long and mutually beneficial relationship. As our political process demonstrates, societies thrive best not on triumph in domestic debates, but on reconciliations. Nothing would ever be accomplished if every technical disagreement turned into a civil war.

A good deal is one in which the two sides both meet their needs. Needs can be reconciled. Compromise is okay. Unfortunately, not every person you meet at the bargaining table believes in this theory. Often you'll encounter a slick customer. These characters never appear to be the killers that they are.

Other people will give off signals that they are not trustworthy from the start. Terms aside, my own rule of thumb is that unless I am totally comfortable with the executives and the company, I don't even sit down to deal. **No deal is better than a bad deal.**

Agree to Agree

Prior to discussing the nuts and bolts of a specific licensing contract, I want to know two things up front:

1. Is the company willing to pay me a standard royalty on the net sales of each unit sold, and

2. Is the company willing to pay me a to-be-negotiated advance against future royalties. No strings attached.

Once I establish a basis for negotiation, contractual terms typically do not stand in the way. And they should not so long as I want to sell and the manufacturer is serious about licensing. Everything usually shakes out. But, if a question arises about wanting to pay an advance and royalty, I refuse to deal until we settle on those points.

Advances

The advance is important because it can help you recoup a portion of your outlay for research and development, and it serves as a barometer for gauging the amount of interest at the company. It is not always possible to earn an advance that will cover all of your expenses.

A good way to calculate an advance is to base it on one quarter or one third of the royalties the manufacturer would have to pay you during the first year of sales, assuming the product sells so many units at a predetermined price point. For example, if the company tells you it estimates 100,000 units in year one, and that it will sell at an average net wholesale price of $10, your royalty, if 5 percent, would amount to $50,000. Twenty-five percent of $50,000 is $12,500. At one third the numbers work out to an advance of $16,667.

Up-front Money

How much should you ask for as an advance? While many times this figure will depend upon the product and its estimated sales, ask the company what it typically pays. At the same time, doublecheck with others who may have licensed products to the same manufacturer. Inventors tend to share information with each other quite freely.

This advance could be paid as a lump sum upon signing, or in stages during the first year of the agreement. This is all open to negotiation, but the more you can get up front the better.

Royalties

Earned

Most industries that license concepts from outside inventors have royalty structures worked out. In publishing there is a sliding scale that climbs from 10 percent to 12.5 percent and, after a certain number of copies sold, levels off at 15 percent of the cover price. In character licensing and toy invention the royalty rate ranges from 4 percent to 15 percent, although E.T. in its heyday captured 20 percent. Most character and toy invention licenses hit somewhere between 5 and 8 percent.

Guaranteed

Guaranteed minimum royalties can also be negotiated. In this case, the licensee of your item would guarantee you no less than a certain amount of money within a predetermined time frame. Again, this amount would be based upon what the licensee projects your item will earn on an annual basis.

Many agreements have clauses for both earned royalties and minimum guaranteed royalties.

The Option Agreement

There are times when a manufacturer is not quite ready to do a licensing agreement, but wants exclusive rights to study or research your invention for a limited period of time. In this case, an Option Agreement is appropriate.

The amount of money a manufacturer is willing to pay for an option is open to negotiation. I have received options as high as $20,000 for a 30-day hold on a product. The money is usually against future royalties should a deal materialize. If no deal is struck, you keep the money.

I am not sure that there is a formula for negotiating option money. The fact that a manufacturer is willing to pay an option does show a certain level of interest and commitment. From there it follows that the greater the interest and commitment, the greater the amount of money the manufacturer is willing to risk.

I have received option money and then seen the products sit in a closet until the option period expired. Some companies eat like elephants and pass food like sparrows.

However, options are not always desired. You may not want to do an option if it will throw off your timing in terms of opportunities to present the same product to other potential licensees. Further, you must be sure that the option is not offered as a ploy to take your invention out of play for a period of time.

Here is a draft of an Option Agreement that you may wish to use as it is written or as the basis for something you are working up.

Be aware: Most manufacturers will not expect you to do everything for free once they license a product. Typically they will fund additional R&D design work. However, they may not offer funding to you unless you ask for it.

Option Agreement

THIS AGREEMENT made as of this _____ day of _____, 199__ between _____, located at _____ (hereinafter "LICENSOR"), and _____, located at _____ (hereinafter "LICENSEE").

WHEREAS, LICENSOR has invented a _____, (hereinafter "Item"), and

WHEREAS, LICENSOR has presented the Item to LICENSEE for evaluation and possible licensing; and

WHEREAS, LICENSEE wishes to review and evaluate the Item;

It is therefore agreed between the parties as follows:

LICENSOR agrees that LICENSEE may examine and evaluate the Item for a period commencing on the date of this Agreement and ending _____ ("Review Period"). LICENSOR represents that it has such rights in and title to the Item as to enable it to grant LICENSEE an exclusive license for its manufacture and sale. LICENSOR agrees that it will not license or disclose the Item or similar items during the Review Period to any other person, firm, corporation, or other entity in that would compete with LICENSEE.

LICENSOR agrees that should LICENSEE wish to license the Item, LICENSOR will enter into a mutually satisfactory licensing agreement for the exclusive use in the (define territory, e.g., US, Europe, worldwide, etc.) of the Item with LICENSEE or a subsidiary or affiliate designated by LICENSEE.

In consideration of the foregoing, LICENSEE agrees to pay to LICENSOR the sum of US$_____, along with other good and valuable consideration, the receipt of which is hereby acknowledged. If LICENSEE decides to license the Item, it will so notify LICENSOR, by a written confirmation sent to LICENSOR at the address specified above, mailed no later than the last day of the Review Period, and both parties agree to negotiate a licensing agreement within thirty (30) days thereafter. In that event, LICENSEE may apply the above-referenced paid consideration against any royalties payable under the executed license agreement. In the event that LICENSEE does not elect to use the Item, it is agreed that LICENSOR shall be entitled to retain the entire sum payable hereunder.

IN WITNESS WHEREOF, the parties have executed this Agreement as of the date first written above.

LICENSOR: _____   LICENSEE: _____

Title: _____   Title: _____

A String of Pearls

Before you head into the boardroom to negotiate a contract, put these pearls of wisdom in your briefcase.

- Stop. Look. And listen.

- Say no, then negotiate.

- Honesty is the best policy. Always.

- Never swing at the first pitch.

- Trust everybody, but cut the cards.

- Don't wish for luck; prepare for luck.

- If it looks like a duck and quacks like a duck, expect a duck.

- Don't get mad; don't get even; get ahead.

- Never murder a person who is committing suicide.

- Bite the bullet only if you can stand the taste of gunpowder.

Levy's Rules of Contract Negotiation

1. Negotiate yourself. In choppy seas, the captain should be on the deck. No one will do it better than you. No one has more to gain or lose.

2. Thou shalt not committee. Any simple problem can be made insoluble if enough people discuss it.

3. Don't deal with lawyers. It is always best to negotiate with an executive who is in a decision-making position. Lawyers are paid not to make executive decisions but to set rules and follow them. They see themselves as protectors, saving the executives from themselves. Yet I have found that the most successful executives break rules all the time.

4. Never respond to pygmies chewing at your toe nails. Don't roll over just because a lawyer says that without x, y, or z the project will not be approved. The company wants to do the project with you or you would not be in negotiation. Executives, not lawyers, are responsible for profits. If your invention can boost revenues, executives will shine.

5. Two plus two is never four. Exceptions always outnumber rules. Established exceptions have their exceptions. By the time one learns the exceptions, no one remembers the rules to which they correspond.

6. Written words live. Spoken words die. As they say in the theater, if it ain't on the page, it ain't on the stage. During negotiations, confirm every conversation with a memorandum to eliminate any misunderstanding about who agreed to what.

7. When in doubt, ask. Asking dumb questions is far easier than correcting dumb mistakes.

8. Keep it short and to the point. The length of a business contract is inversely proportional to the amount of business.

9. Do not accept standard contracts. In any so-called standard contract, boilerplate terms should be treated as variables. Not until a contract has been in force for six months will its most harmful terms be discovered. Nothing is as temporary as that which is called permanent.

10. **Have fun.** The moment I stop enjoying a negotiation, I pick up my marbles and go home. An agreement is a form of marriage and both parties must be compatible for it to succeed.

The Licensing Agreement

Terms of Endearment

If you get to the point that a company wishes to cut a licensing deal for your property, the two sample licensing agreements on the following pages will be invaluable.

Some manufacturers have standard licensing agreements which they prefer to use. Others will ask you to submit the paperwork. In either case, you will find that most issues are negotiable.

These licensing agreements are extremely important documents to you as an inventor faced with the opportunity to license an invention, trademark, design, or copyright. They are worth many thousands of dollars in terms of the time and effort put into crafting them.

You or your lawyer may use them as the basis for your agreements, or go through them cherry-picking clauses and terms that you find appropriate to your situation.

It is my hope that these agreements serve to alert you and your legal counsel to many points that are either omitted entirely from manufacturers' boilerplate contracts or are written entirely from the manufacturer perspective.

My dad, a former assistant attorney general for the Commonwealth of Pennsylvania, taught me that contracts are as good as the people that sign them. Remember this. Do not make a deal with people you do not trust, no matter how much you feel the agreement is in your favor.

He also instructed me that every contract has its own spirit and that I should always be clear about the spirit of any agreement I execute. Often I will articulate such an understanding and put it on record through letters to the manufacturer. While not part of the formal agreement, should push ever come to shove in a court of law, a judge and jury would probably not disregard the letters.

License Agreement I

AGREEMENT made this _____ day of _____, 19___ by and between _____, located at _____ (hereinafter referred to as LICENSOR) and _____, located at _____ (hereinafter referred to as LICENSEE).

Witnesseth:

WHEREAS, LICENSOR represents and warrants that it is the creator of

_____

_____

_____herein after referred to as the ITEM.

WHEREAS, LICENSOR hereby warrants that it is the sole and exclusive owner of all rights in the ITEM, that it has the sole and exclusive right to grant the license herein, and that it is not engaged in litigation or conflict of any nature whatsoever involving the ITEM; and

WHEREAS, LICENSEE is in the business of making and selling (<u>enter type of product, e.g. shoes, faucets, toys, etc.</u>); and

WHEREAS, LICENSEE is desirous of obtaining the sole and exclusive rights to manufacture and sell the ITEM (<u>enter territories, e.g. worldwide, USA and its possessions, Europe, etc.</u>).

NOW, THEREFORE, for and in consideration of the sum of One Dollar (US$1.00) and other good and valuable consideration, receipt of which is hereby acknowledged, and for the performance of the mutual covenants hereinafter to be performed, it is agreed as follows:

DEFINITIONS:

- Item: As used in this Agreement, the term ITEM refers to the ITEM, described hereinabove, and any Improvements, Accessories or Extensions thereto, whether developed by or for LICENSEE.

- Improvements: As used in this Agreement, the term Improvements means any design or technical refinements or advances made by or for LICENSEE and reflected in the ITEM as marketed.

- Accessories: As used in this Agreement, the term Accessories means any products making use of the ITEM, as well as equipment developed by or for LICENSEE designated for use with the ITEM.

- Extensions: As used in this Agreement, the term Extensions means any products that are sold independently by LICENSEE under the ITEM's trademark, i.e., products that trade on the name of the ITEM, but are not necessarily marketed as accessories to the ITEM.

- Collateral Merchandise: As used in this Agreement, the term Collateral Merchandise means products that are sold under the ITEM's trademark or trade on its good will. (Note: You may not want to grant these rights. If not, omit this reference. Or, you may wish to allow for certain collateral merchandise and not others. If so, spell it out here. Collateral Merchandise could range from t-shirts to tennis rackets.)

1. (a) LICENSOR hereby grants to LICENSEE the sole and exclusive right, privilege and license to make, reproduce, modify, use and/or sell, to have made, reproduced, modified, used and/or sold and to sublicense others to make, reproduce, modify, use and/or sell the ITEM and any images, representations, and material associated with the ITEM as well as the subject matter of a patent application which is to be filed on the ITEM pursuant to Paragraph 2(a) hereof.

    (b) All rights and licenses not herein specifically granted to LICENSEE are reserved by LICENSOR and, as between the parties, are the sole and exclusive property of LICENSOR and may be used or exercised solely by LICENSOR. Included within this understanding is the right of LICENSOR to use whatever trademark LICENSEE markets the ITEM under; however, it is understood and agreed between the parties that LICENSOR will not license the concept to a third party for a product that would compete with the ITEM as marketed by LICENSEE.

    (c) This Agreement shall continue for as long as the ITEM upon which royalties would be payable to LICENSOR, under provisions of this Agreement, shall continue to be manufactured or sold by LICENSEE.

2. (a) LICENSOR agrees to use its best efforts and to bear the expenses of obtaining patent protection in the USA, and any and all such

patents will be in the name of and remain the property of LICEN-SOR during and following any termination or cancellation hereof.

(b) To the best of LICENSOR's knowledge, the ITEM does not infringe any patent rights.

3. (a) LICENSEE shall pay LICENSOR ___ percent of the "Net Sales" of the ITEM, its Improvements, Accessories or Extensions. As used in this Agreement, "Net Sales" are defined as sales computed on prices charged by LICENSEE to its customers for the ITEM, less a deduction not to exceed seven and a half (7.5%) percent which provides for freight allowances, sales allowances actually credited, customary trade discounts (but not cash discounts), volume discounts (not to exceed usual industry practices), to the extent taken, directly applicable to the sale of licensed products, and less returns (but not for exchange) which are accepted and credited by LICENSEE. No deduction shall be made for non-collectible accounts. No costs incurred in the manufacture, sales, distribution, exploitation or promotion of the ITEM, its improvements, accessories or any adaptations thereof shall be deducted from any royalties payable by LICENSEE to LICENSOR, nor shall any deductions from due royalties be made for taxes of any nature. LICENSEE cannot barter or trade or do so-called "charge-backs" on the ITEM without computation of full royalties.

(b) In event LICENSEE sells FOB direct from a foreign manufacturing location, LICENSEE shall pay LICENSOR ___ percent of all Net Sales, as defined hereinabove, of the ITEM.

(c) LICENSEE may grant foreign manufacturing sublicenses on the ITEM upon any terms and conditions which it wishes to grant and establish so long as said terms and conditions are competitive and market prevailing; and provided that in the event LICENSEE does grant such sublicenses to manufacture and market the ITEM, LICENSEE shall pay to LICENSOR fifty (50%) percent of any and all moneys received, including, but not limited to royalties (which shall be no less than 2% of the sales of the ITEM), advances, guar-antees, mold or pattern lease or rental fees, etc. as received by LICENSEE from any such sublicense or grant. LICENSEE agrees to send a copy of any sublicense agreement, or shall otherwise promptly inform LICENSOR in writing about the terms of the sublicense to manufacture the ITEM.

(d) In the event LICENSEE sells the ITEM as a premium, the royalty rate shall be _____ percent of the revenue derived by LICENSEE from the sale of the ITEM as a premium. For purposes of this Agreement, the term "premium" shall be defined as including, but not necessarily limited to, combination sales, free or self-liquidat-ing items offered to the public in conjunction with the sale or promotion of a product or service other than the ITEM, including traffic building or continuity visits by the consumer, customer, or any similar scheme or device, whose primary purpose in regard to each of the sales described above is not directed at the sale of the ITEM itself and the prime intent of which is to use the ITEM in such a way as to promote, publicize, and/or sell the products, services, or business image of the user of such ITEM rather than the ITEM itself.

(e)   All royalties payable by LICENSEE to LICENSOR based upon LICENSEE's sales of the ITEM shall accrue upon LICENSEE's shipment and invoicing of the item.

(f)   If payments are made to LICENSOR from a foreign country, LICENSEE assumes sole responsibility for procuring any permits and documents needed to make all payments under any exchange regulations, and all such royalties shall be made by LICENSEE to LICENSOR in U.S. dollars at the then prevailing rate of exchange as used by Chase Bank.

4.   (a)   Upon execution of this Agreement, LICENSEE shall pay to LICENSOR the sum of US$_____, as a non-refundable, guaranteed advance against royalties. In no event shall such advance be repaid to LICENSEE other than in the form of deductions from payments due under Paragraphs 3 and 9 hereof.

(b)   Should LICENSEE, for any reason, and at any time, decide not to manufacture the ITEM, or if in any calendar year the ITEM fails to generate US$_____ in royalties for LICENSOR, LICENSOR shall have the right to terminate this Agreement. And should LICENSEE cease to manufacture said ITEM, or should this Agreement be terminated for a breach of conditions by LICENSEE, then all rights to the development work done on the ITEM on behalf of and/or paid for by LICENSEE (e.g. breadboards, models, etc.) shall belong to LICENSOR free-and-clear and LICENSEE shall have no further claim to the ITEM.

(c)   The receipt or acceptance by LICENSOR of any written statements furnished pursuant to this Agreement or of any payments made hereunder (or cashing of any checks paid hereunder) shall not preclude LICENSOR from questioning the correctness thereof at any time.

5.   LICENSEE agrees to introduce the ITEM on or before (date). LICENSEE agrees that during the term of this Agreement it will diligently and continuously manufacture, sell, distribute and promote the ITEM and that it will make and maintain adequate arrangements for the distribution of the ITEM and satisfy demand for the ITEM.

6.   LICENSEE has the right to change the form of the ITEM as submitted by LICENSOR, and to produce and sell it under new form(s); provided, however, that all the provisions of this Agreement shall apply to said new form(s) of the ITEM.

7.   (a)   LICENSEE shall mark the ITEM and its packages, containers and display cards with the words, "Patent Pending," if advised that a patent application is pending and until advised that a patent has been issued on the ITEM, at which later time LICENSEE shall mark the ITEM, containers and display cards with the specific patent number.

8.   (a)   LICENSEE shall annually furnish LICENSOR, free-of-charge, (enter a number) samples of each ITEM and its packaging prior to its availability at retail for purposes of quality control. In the case of the ITEM as made by LICENSEE's foreign sublicensees, LICENSEE shall annually furnish, free-of-charge, to LICENSOR (enter a number) samples of each and its packaging.

9.   (a)   The rights to sublicense the ITEM in foreign countries covered by this Agreement are conditioned upon introduction of the ITEM on or before (enter a date). In the event LICENSEE has not entered into a fully executed sublicense agreement or made formal, documented arrangements to sell the ITEM through a distributor in a foreign country prior to (enter a date, usually one year after domestic release), all rights with respect to such foreign country shall automatically revert to LICENSOR. However, LICENSEE shall be entitled, after expiration of any manufacturing sublicense agreement or distributor arrangement in any foreign country entered into prior to (enter a cut-off date) to enter into a new agreement or arrangement in such foreign country.

     (b)   When either party considers it necessary or desirable to obtain patent protection in a foreign country covered by this Agreement (other than the USA), it shall notify the other party of that decision and upon the agreement of said other party, LICENSOR shall, if obtaining the foreign patent protection is not barred by prior use or publication, promptly file such foreign application and the expenses thereof shall be shared equally by both parties. In the event the parties do not agree to share the expense of obtaining such patent protection, the party requesting the foreign patent protection may proceed alone, at its own expense, and all rights obtained shall belong to that party alone, notwithstanding any other provisions of this Agreement. If LICENSEE has not timely requested the filing of a patent application in a foreign country, LICENSOR is under no obligation to file such foreign application, and in the event that neither party has filed a patent application in a particular foreign country, or the parties have agreed not to file an application or have agreed to discontinue an application or patent, then the royalties for such foreign country shall be payable to LICENSOR as if the patent protection had been obtained in that country and the expenses shared equally by the parties.

(c) LICENSEE may sublicense its rights hereunder to use the ITEM on Collateral Merchandise under the terms and conditions it wishes to grant and establish as long as they are competitive and market prevailing; and provided that in the event LICENSEE grants such sublicenses, LICENSEE shall pay to LICENSOR the sum of fifty percent (50%) of any and all moneys received, including, but not limited to royalties, advances, guarantees and consultation fees it may receive.

(d) In the case of sublicenses of Collateral Merchandise, LICENSEE shall ensure that sublicensees cause the above trademark notices, if one or both trademarks are used by LICENSEE, to appear on all collateral merchandise, its packaging, containers, advertising, etc.

10. (a) LICENSOR agrees to indemnify, defend and save harmless LICENSEE against actions brought against LICENSEE with respect to any claim or suit that LICENSOR is not the originator of the Item.

(b) LICENSOR agrees to indemnify LICENSEE against losses, claims and expenses with respect to losses, claims or expenses which arise only from an act or omission by LICENSOR which is done in bad faith.

(c) LICENSOR shall not indemnify LICENSEE on claims whereby LICENSEE has been said to have been previously shown an ITEM similar to the ITEM by the claimant. LICENSEE warrants that it has never seen an item similar to the subject ITEM.

(d) LICENSOR agrees to indemnify and hold harmless LICENSEE from and against any claim of infringement of trade secrets arising out of LICENSEE's sale of the ITEM.

(e) LICENSEE will conduct its own patent, copyright and trademark searches and satisfy itself that the ITEM does not infringe anything in any of these fields, and once satisfied LICENSEE agrees to indemnify, defend and save harmless LICENSOR from and against all damages, costs and attorney fees resulting from all claims, demands, actions, suits or prosecutions for patent, copyright, trademark based upon use of the ITEM or its components and all forms of the ITEM as produced and sold by LICENSEE, its sub-sidiaries, affiliates and sublicensees. LICENSEE shall be given prompt notice of any claim against LICENSOR and shall have the right to defend such claim with counsel selected by LICENSEE. LICENSOR agrees to co-operate with LICENSEE in connection with the defense of any such claim.

(f) LICENSEE agrees to indemnify, defend and save harmless LICENSOR from and against all damages, costs and attorney fees resulting from all claims, demands, actions, suits or prosecutions for personal injury or property damage. LICENSEE agrees to cover LICENSOR under its product liability insurance, with an insurance company providing protection for itself and LICENSOR against any such claims or suits relating to personal injury, product manufacture, property damage, or materials failure but in no event in amounts less than (enter a number) million dollars or the limits of its policy, whichever is greater, and within thirty (30) days before manufacture of the ITEM, LICENSEE will submit to LICENSOR a certificate of insurance naming LICENSOR as an insured party, and covering LICENSOR requiring that the insurer shall not terminate or materially modify such without written notice to LICENSOR at least twenty (20) days in advance thereof.

(g) In the event of infringement of any patent that may be issued to LICENSOR on the ITEM and upon notice thereof from LICENSEE, LICENSOR shall, within thirty (30) days, notify LICENSEE of its election to prosecute or not prosecute a suit for infringement. If LICENSOR prosecutes said suit, it may select legal counsel and pay legal fees and costs of prosecution subject to being reimbursed therefor from any recovery in said suit. The balance of any recovery shall be divided equally between LICENSOR and LICENSEE. If LICENSOR elects not to prosecute any infringement suit, LICENSEE may do so after notice to LICENSOR of that intention. LICENSEE may then select legal counsel and shall bear all the legal fees and costs subject to reimbursement therefor from any recovery in said suit. The balance of any recovery shall be distributed as follows: One-fourth (1/4) to LICENSOR and three-fourths (3/4) to LICENSEE.

11. LICENSEE shall, within thirty (30) days following the end of each calendar quarter, starting with the month following the quarter in which sales of the ITEM commence, submit to LICENSOR a report covering the sales of the ITEM during the preceding quarter, and LICENSEE shall therewith send to LICENSOR payment of the amount due under Paragraphs 3 and 9 hereof. Such quarterly statements shall be submitted whether or not they reflect any sales.

12. (a) LICENSEE agrees to keep full and accurate books of account, records, data and memoranda respecting the manufacture and sales of the ITEM in sufficient detail to enable the payments here

under to LICENSOR to be determined, and LICENSEE gives LICENSOR the right, upon notice, at its own expense, to examine said books and records, only insofar as they concern the ITEM and not more often than once in any calendar year, for the purpose of verifying the reports provided for in this Agreement. In the event LICENSOR shall examine the records, documents and materials in the possession or under the control of LICENSEE with respect to the subject matter, such examination shall be conducted in such a manner as to not unduly interfere with the business of LICENSEE. LICENSOR and its representative shall not disclose to any other person, firm or corporation any information acquired as a result of any such examination; provided, however, that nothing herein contained shall be construed to prevent LICENSOR and/or its duly authorized representative from testifying, in any court of competent jurisdiction, with respect to the information obtained as a result of such examination in any action instituted to enforce the rights of LICENSEE under the terms of this Agreement.

(b) In the event that LICENSEE has understated Net Shipments or underpaid royalties by 5% or more for any contract year, LICENSEE shall forthwith and upon written demand also pay to LICENSOR all reasonable costs, fees and expenses incurred by LICENSOR in conducting such audit.

(c) Payments found to be due LICENSOR as a result of a delay or an examination shall be paid immediately at the prime rate quoted by Chase Bank at the close of business on the due date plus 1% per annum until paid.

13. (a) LICENSEE agrees to send all payments and reports due hereunder to LICENSOR at address noted in this Agreement's preamble.

(b) LICENSOR's Tax ID (*or Social Security No.*) is _____.

14. (a) If LICENSEE shall at any time default by failing to make any payment hereunder, or by failing to make any report required under this Agreement, or by making a false report, or for cause, and LICENSEE shall fail to remedy such default within ten (10) days for money, and thirty (30) days for reports, after notice thereof by LICENSOR, LICENSOR may, at its option, terminate this Agreement and the license granted herein by notice to that effect, but such act by LICENSOR shall not relieve LICENSEE of its liabilities accruing up to the time of termination. In the case of subsequent default, the time period which to remedy the default shall be reduced to fifteen (15) days.

(b) Should a third default take place, LICENSOR may, at its option, terminate this Agreement.

15. It is understood and agreed that if LICENSEE does not introduce the ITEM on or before (enter a date) or does not sell the ITEM for a period of 90 consecutive days or more except as provided in Paragraph 16 hereof, LICENSOR may give notice to LICENSEE of its desire to terminate this Agreement for that reason and if LICENSEE does not within thirty (30) days resume producing and selling of the ITEM, this Agreement and the license granted herein shall terminate as of the end of that thirty (30) day period.

16. It is understood and agreed that in the event an act of government, or war conditions, or fire, flood or labor trouble in the factory of LICENSEE, or in the factory of those manufacturing parts necessary for the manufacture of the ITEM, prevents the performance by LICENSEE of the provisions of this Agreement, then such non-performance by LICENSEE shall not be considered a breach of this Agreement and such non-performance shall be excused, but for no longer than a period of six (6) months on any single occurrence.

17. This Agreement shall continue for as long as the ITEM covered by this Agreement shall continue to be manufactured by LICENSEE, or unless sooner terminated under the provisions of this Agreement.

18. LICENSEE agrees that if this Agreement is terminated under any of its provisions, LICENSEE will not itself, or through others, thereafter manufacture and sell the ITEM and all rights to the ITEM and to any patents filed hereunder shall revert to LICENSOR.

19. (a) LICENSOR agrees that LICENSEE may assign this Agreement to any affiliate corporation; provided, however, that such assignee shall thereafter be bound by the provisions of this Agreement.

    (b) LICENSEE may not assign this Agreement, or any part thereof, without the expressed written permission of LICENSOR, which shall not be unreasonably withheld, unless it is selling its entire business as a going concern, and the same restriction shall be binding upon successors and assigns of LICENSEE.

    (c) Should LICENSEE wish to sell its rights in the ITEM or transfer this Agreement to any person or corporation that does not plan to purchase LICENSEE's entire business as a going concern, but has an interest only in said ITEM, then said prospective buyer will have to strike a separate deal with LICENSOR for its consent to the sale.

20. (a) In case of the Receivership or Bankruptcy of LICENSEE, by reason of which LICENSEE is prevented from carrying out the spirit of this Agreement, after written notice thereof by LICENSOR, LICENSOR may, at its option, terminate this Agreement and the license granted herein by notice to that

effect, but such act shall not relieve LICENSEE of its liabilities accruing up to the time of termination.

(b)   If LICENSEE, at any time after the execution of this Agreement and prior to and during the preparation of said ITEM for production, display and offering for sale, shall elect not to produce, display, offer or produce said ITEM, which election shall be in writing sent by Registered or Certified or Express Mail to LICENSOR, then LICENSOR's sole and exclusive remedy shall be to keep the advance against royalties, as provided for in Paragraph 4 hereof, for breach of this Agreement, and such Agreement shall thereafter be of no further force and effect, and the license shall be deemed canceled and neither party shall have claim against the other.

(c)   Immediately upon expiration or termination of this Agreement, for any reason whatsoever, all the rights granted to LICENSEE hereunder shall cease and revert to LICENSOR, who shall be free to license others to use any or all of the rights granted herein effective on and after such date of expiration or termination, and to this end, LICENSEE will be deemed to have automatically assigned to LICENSOR upon such expiration or termination, all copyrights, trademarks and service mark rights, equities, good will, titles, designs and concepts, and other rights in or to the ITEM. LICENSEE will upon the expiration or termination of this license execute any instruments requested by LICENSOR to accomplish or confirm the foregoing. Any assignments shall be without consideration other than mutual covenants and considerations of this Agreement. In addition, for whatever reasons, LICENSEE will forthwith refrain from any further use of the trademarks or copyrights of any further reference to any of them, direct or indirect.

(d)   In the event of termination of this Agreement, for any reason other than for failure to pay or make reports due hereunder to LICENSOR, LICENSEE shall have the right to dispose of its existing inventory for a period of 60 days. Further, upon termination of this Agreement, LICENSEE agrees to assign to LICENSOR the right to receive directly any royalties due to LICENSOR from any collateral merchandise sublicensee of the ITEM.

21.   All notices wherever required in this Agreement shall be in writing and sent by Certified Mail, Registered or Express Mail to the addresses firstabove written.

22.   If any provisions of this Agreement are for any reason declared to be invalid, the validity of the remaining provisions shall not be affected thereby.

23.	This Agreement shall be binding upon and inure to the benefit of the parties hereto and their successors and assigns as herein provided and said successors and assigns shall be libel hereunder. LICENSOR may assign its rights to receive royalties under this Agreement.

24.	It is expressly agreed that LICENSOR is in no way the legal representative of LICENSEE and has no authority, expressed or implied, on behalf of LICENSEE to bind LICENSEE or to pledge its credit.

25.	(Optional) It is a condition of this license that, if LICENSEE decides to market the ITEM under LICENSOR's trademark, _____. LICENSEE shall cause the following notice to appear on the ITEM and its advertising, promotional, packaging and display materials therefoer:

_____ is a TM of

_____

	Used with permission. All Rights Reserved.

26.	This Agreement shall be construed in accordance with the laws of the State of _____.

IN WITNESS WHEREOF, the parties have executed this Agreement in duplicate originals the day/year first hereinabove written.

LICENSOR _____

By _____

LICENSEE _____

By _____

License Agreement II

AGREEMENT made this _____ day of _____, 19___, between _____ (hereinafter called "Licensor"), and _____ (hereinafter called "Licensee"):

Witnesseth:

WHEREAS Licensor has rights to the name, character, symbol, design, likeness and visual representation of

_____ (which name, character, symbol, design, likeness and visual representation and/or each of the individual components thereof shall hereinafter be called the "Name"), said Name having been used over the facilities of numerous stations in radio and/or television broadcasting and in allied fields, and in promotional and advertising material in different businesses and being well known and recognized by the general public and associated in the public mind with Licensor, and

WHEREAS Licensee desires to utilize the Name upon and in connection with the manufacture, sale and distribution of articles hereinafter described,

NOW, THEREFORE, in consideration of the mutual promises herein contained, it is hereby agreed:

1. **Grant of License.**

    (a) Articles. Upon the terms and conditions hereinafter set forth, Licensor hereby grants to Licensee and Licensee hereby accepts the right, license and privilege of utilizing the Name solely and only upon and in connection with the manufacture, sale and distribution of the following articles:

    (b) Territory. The license hereby granted extends only to
    _____. Licensee agrees that it will not make, or authorize, any use, direct or indirect, of the Name in any other area, and that it will not knowingly sell articles covered by this agreement to persons who intend or are likely to resell them in any other area.

    (c) Term. The term of the license hereby granted shall be effective on the ____ day of _____, 19____ and shall continue until the ____ day of _____, 19____, unless sooner terminated in accordance with the provisions hereof.

2. **Terms of Payment.**

    (a) *Rate.* Licensee agrees to pay to Licensor as royalty a sum equal to ____ percent (____%) of all net sales by Licensee or any of its affiliated, associated or subsidiary companies of the articles covered by this agreement. The term "net sales" shall mean gross sales less quantity discounts and returns, but no deduction shall be made for cash or other discounts or uncollectible accounts. No costs incurred in the manufacture, sale, distribution or exploitation of the articles shall be deducted from any royalty payable by Licensee.

    (b) *Minimum Royalties.* Licensee agrees to pay to Licensor a minimum royalty of _____ dollars ($_____) as a minimum guarantee against royalties to be paid to Licensor during the first contract year, said minimum royalty to be paid on or before _____. The advance sum of _____ dollars ($_____) paid on the signing hereof shall be applied against such guarantee. No part of such minimum royalty shall in any event be repayable to Licensee.

    (c) *Periodic Statements.* Within thirty (30) days after the initial shipment of the articles covered by this agreement, and promptly on the 15th of each calendar month thereafter, Licensee shall furnish to Licensor complete and accurate statements certified to be accurate by Licensee showing the number, description and gross sales price, itemized deductions from gross sales price and net sales price of the articles covered by this agreement distributed and/or sold by Licensee during the preceding calendar month, together with any returns made during the preceding calendar month. For this purpose, Licensee shall use the statement form attached hereto, copies of which form may be obtained by Licensee from Licensor. Such statements shall be furnished to Licensor whether or not any of the articles have been sold during the preceding calendar month.

(d) *Royalty Payments.* Royalties in excess of the aforementioned minimum royalty shall be due on the 15th day of the month following the calendar month in which earned, and payment shall accompany the statements furnished as required above. The receipt or acceptance by Licensor of any of the statements furnished pursuant to this agreement or of any royalties paid hereunder (or the cashing of any royalty checks paid hereunder) shall not preclude Licensor from questioning the correctness thereof at any time, and in the event that any inconsistencies or mistakes are discovered in such statements or payments, they shall immediately be rectified and the appropriate payment made by Licensee.

3. Exclusivity.

(a) Nothing in this agreement shall be construed to prevent Licensor from granting any other licenses for the use of the Name or from utilizing the Name in any manner whatsoever, except that Licensor agrees that except as provided herein it will grant no other licenses for the territory to which this license extends effective during the term of this agreement, for the use of the Name in connection with the sale of the articles described in paragraph 1.

(b) It is agreed that if Licensor should convey an offer to Licensee to purchase any of the articles listed in paragraph 1, in connection with a premium, giveaway or other promotional arrangement, Licensee shall have ten (10) days within which to accept or reject such an offer. In the event that Licensee fails to accept such offer within the specified ten (10) days, Licensor shall have the right to enter into the proposed premium, giveaway or promotional arrangement using the services of another manufacturer, provided, however, that in such event Licensee shall have a three (3) day period within which to meet the best offer of such manufacturer for the production of such articles if the price of such manufacturer is higher than the price offered to Licensee by Licensor.

4. **Good Will, Etc.** Licensee recognizes the great value of the good will associated with the Name, and acknowledges that the Name and all rights therein and good will pertaining thereto belong exclusively to Licensor, and that the Name has a secondary meaning in the mind of the public.

5. **Licensor's Title and Protection of Licensor's Rights.**

   (a) Licensee agrees that it will not during the term of this agreement, or thereafter, attack the title or any rights of Licensor in and to the Name or attack the validity of this license. Licensor hereby indemnifies Licensee and undertakes to hold it harmless against any claims or suits arising solely out of the use by Licensee of the Name as authorized in this agreement, provided that prompt notice is given to Licensor of any such claim or suit and provided, further, that Licensor shall have the option to undertake and conduct the defense of any suit so brought and no settlement of any such claim or suit is made without the prior written consent of Licensor.

   (b) Licensee agrees to assist Licensor to the extent necessary in the procurement of any protection or to protect any of Licensor's rights to the Name, and Licensor, if it so desires, may commence or prosecute any claims or suits in its own name or in the name of Licensee or join Licensee as a party thereto. Licensee shall notify Licensor in writing of any infringements or imitations by others of the Name on articles similar to those covered by this agreement which may come to Licensee's attention, and Licensor shall have the sole right to determine whether or not any action shall be taken on account of any such infringements or imitations. Licensee shall not institute any suit or take any action on account of any such infringements or imitations without first obtaining the written consent of the Licensor so to do.

6. **Indemnification by Licensee and Product Liability Insurance.** Licensee hereby indemnifies Licensor and undertakes to defend Licensee and/or Licensor against and hold Licensor harmless from any claims, suits, loss and damage arising out of any allegedly unauthorized use of any patent, process, idea, method or device by Licensee in connection with the articles covered by this agreement or any other alleged action by Licensee and also from any claims, suits, loss and damage arising out of alleged defects in the articles. Licensee agrees that it will obtain, at its own expense, product liability insurance from a recognized insurance company which is qualified to do business in the State of _____, providing adequate protection (at least in the amount of $100,000/$300,000) for Licensor (as well as for Licensee) against any claims, suits, loss or damage arising out of any alleged defects in the articles. As proof of such insurance, a fully paid certificate of insurance naming Licensor as an insured party will be submitted to Licensor by Licensee for Licensor's prior approval before any article is distributed or

sold, and at the latest within thirty (30) days after the date first written above; any proposed change in certificates of insurance shall be submitted to Licensor for its prior approval. Licensor shall be entitled to a copy of the then prevailing certificate of insurance, which shall be furnished Licensor by Licensee. As used in the first two sentences of this paragraph 6, "Licensor" shall also include the officers, directors, agents, and employees of the Licensor, or any of its subsidiaries or affiliates, any person(s) the use of whose name may be licensed hereunder, the package producer and the cast of the radio and/or television program whose name may be licensed hereunder, the stations over which the programs are transmitted, any sponsor of said programs and its advertising agency, and their respective officers, directors, agents and employees.

7. **Quality of Merchandise.** Licensee agrees that the articles covered by this agreement shall be of high standard and of such style, appearance and quality as to be adequate and suited to their exploitation to the best advantage and to the protection and enhancement of the Name and the good will pertaining thereto, that such articles will be manufactured, sold and distributed in accordance with all applicable Federal, State and local laws, and that the policy of sale, distribution, and/or exploitation by Licensee shall be of high standard and to the best advantage and that the same shall in no manner reflect adversely upon the good name of Licensor or any of its programs or the Name. To this end Licensee shall, before selling or distributing any of the articles, furnish to Licensor free of cost, for its written approval, a reasonable number of samples of each article, its cartons, containers and packing and wrapping material. The quality and style of such articles as well as of any carton, container or packing or wrapping material shall be subject to the approval of Licensor. Any item submitted to Licensor shall not be deemed approved unless and until the same shall be approved by Licensor in writing. After samples have been approved pursuant to this paragraph, Licensee shall not depart therefrom in any material respect without Licensor's prior written consent, and Licensor shall not withdraw its approval of the approved samples except on sixty (60) days' prior written notice to Licensee. From time to time after Licensee has commenced selling the articles and upon Licensor's written request, Licensee shall furnish without cost to Licensor not more than ten (10) additional random samples of each article being manufactured and sold by Licensee hereunder, together with any cartons, containers and packing and wrapping material used in connection therewith.

8. **Labeling.**

    (a) Licensee agrees that it will cause to appear on or within each article sold by it under this license and on or within all advertising, promotional or display material bearing the Name the notice "Copyright © (year) _____" and any other notice desired by Licensor and, where such article or advertising, promotional or display material bears a trademark or service mark, appropriate statutory notice of registration or application for registration thereof. In the event that any article is marketed in a carton, container and/or packing or wrapping material bearing the Name, such notice shall also appear upon the said carton, container and/or packing or wrapping material. Each and every tag, label, imprint or other device containing any such notice and all advertising, promotional or display material bearing the Name shall be submitted by Licensee to Licensor for its written approval prior to use by Licensee. Approval by Licensor shall not constitute waiver of Licensor's rights or Licensee's duties under any provision of this agreement.

(b) Licensee agrees to cooperate fully and in good faith with Licensor for the purpose of securing and preserving Licensor's (or any grantor of Licensor's) rights in and to the Name. In the event there has been no previous registration of the Name and/or articles and/or any material relating thereto, Licensee shall, at Licensor's request and expense, register such as a copyright, trademark and/or service mark in the appropriate class in the name of Licensor or, if Licensor so requests, in Licensee's own name. However, it is agreed that nothing contained in this agreement shall be construed as an assignment or grant to the Licensee of any right, title or interest in or to the Name, it being understood that all rights relating thereto are reserved by Licensor, except for the license hereunder to Licensee of the right to use and utilize the Name only as specifically and expressly provided in this agreement. Licensee hereby agrees that at the termination or expiration of this agreement Licensee will be deemed to have assigned, transferred and conveyed to Licensor any trade rights, equities, good will, titles or other rights in and to the Name which may have been obtained by Licensee or which may have vested in Licensee in pursuance of any endeavors covered hereby, and that Licensee will execute any instruments requested by Licensor to accomplish or confirm the foregoing. Any such assignment, transfer or conveyance shall be without other consideration than the mutual covenants and considerations of this agreement.

(c) Licensee hereby agrees that its every use of such Name shall inure to the benefit of Licensor and that Licensee shall not at any time acquire any rights in such Name by virtue of any use it may make of such Name.

9. **Promotional Material.**

(a) In all cases where Licensee desires artwork involving articles which are the subject of this license to be executed, the cost of such artwork and the time for the production thereof shall be borne by Licensee. All artwork and designs involving the Name, or any reproduction thereof, shall, notwithstanding their invention or use by Licensee, be and remain the property of Licensor and Licensor shall be entitled to use the same and to license the use of the same by others.

(b) Licensor shall have the right, but shall not be under any obligation, to use the Name and/or the name of Licensee so as to give the Name, Licensee, Licensor and/or Licensor's programs full and favorable prominence and publicity. Licensor shall not be under any obligation whatsoever to continue broadcasting any radio or television program or use the Name or any person, character, symbol, design or likeness or visual representation thereof in any radio or television program.

(c) Licensee agrees not to offer for sale or advertise or publicize any of the articles licensed hereunder on radio or television without the prior written approval of Licensor, which approval Licensor may grant or withhold in its unfettered discretion.

10. **Distribution.**

(a) Licensee agrees that during the term of this license it will diligently and continuously manufacture, distribute and sell the articles covered by this agreement and that it will make and maintain adequate arrangement for the distribution of the articles.

(b) Licensee agrees that it will sell and distribute the articles covered by this agreement outright at a competitive price and at not more than the price generally and customarily charged the trade by Licensee and not on an approval, consignment or sale or return basis, and only to jobbers, wholesalers and distributors for sale and distribution to retail stores and merchants, and to retail stores and merchants for sale and distribution direct to the public. Licensee shall not, without prior written consent of Licensor, sell or distribute such articles to jobbers, wholesalers, distributors, retail stores or merchants whose sales or distribution are or will be made for publicity or promotional tie-in purposes, combination sales, premiums, giveaways, or similar methods of merchandising, or whose business methods are questionable. In the event any sale is made at a special price to any of Licensee's subsidiaries or to any other person, firm or corporation related in any manner to Licensee or its officers, directors or major stockholders, there shall be a royalty paid on such sales based upon the price generally charged the trade by Licensee.

(c) Licensee agrees to sell to Licensor such quantities of the articles at as low a rate and on as good terms as Licensee sells similar quantities of the articles to the general trade.

11. **Records.** Licensee agrees to keep accurate books of account and records covering all transactions relating to the license hereby granted, and Licensor and its duly authorized representatives shall have the right at all reasonable hours of the day to an examination of said books of account and records and of all other documents and materials in the possession or under the control of Licensee with respect to the subject matter and terms of this agreement, and shall have free and full access thereto for said purposes and for the purpose of making extracts therefrom. Upon demand of Licensor, Licensee shall at its own expense furnish to Licensor a detailed statement by an independent certified public accountant showing the number, description, gross sales price, itemized deductions from gross sales price and net sales price of the articles covered by this agreement distributed and/or sold by Licensee to the date of Licensor's demand. All books of account and records shall be kept available for at least two (2) years after the termination of this license.

12. **Bankruptcy, Violation, Etc.**

   (a) If Licensee shall not have commenced in good faith to manufacture and distribute in substantial quantities all the articles listed in paragraph 1 within three (3) months after the date of this agreement or if at any time thereafter in any calendar month Licensee fails to sell any of the articles (or any class or category of the articles), Licensor in addition to all other remedies available to it hereunder may terminate this license with respect to any articles or class or category thereof which have not been manufactured and distributed during such month, by giving written notice of termination to Licensee. Such notice shall be effective when mailed to Licensee.

   (b) If Licensee files a petition in bankruptcy or is adjudicated a bankrupt or if a petition in bankruptcy is filed against Licensee or if it becomes insolvent, or makes an assignment for the benefit of its creditors or an arrangement pursuant to any bankruptcy law, or if Licensee discontinues its business or if a receiver is appointed for it or its business, the license hereby granted shall automatically terminate forthwith without any notice whatsoever being necessary. In the event this license is so terminated, Licensee, its receivers, representatives, trustees, agents, administrators, successors and/or assigns shall have no right to sell, exploit or in any way deal with or in any articles covered by this agreement or any carton, container, packing or wrapping material, advertising, promotional or display material pertaining thereto, except with and under the special consent and instructions of Licensor in writing, which they shall be obligated to follow.

   (c) If Licensee shall violate any of its other obligations under the terms of this agreement, Licensor shall have the right to terminate the license hereby granted upon ten(10) days' notice in writing, and such notice of termination shall become effective unless Licensee shall completely remedy the violation within the ten-day period and satisfy Licensor that such violation has been remedied.

(d) Termination of the license under the provisions of paragraph 12 shall be without prejudice to any rights which Licensor may otherwise have against Licensee. Upon the termination of this license, notwithstanding anything to the contrary herein, all royalties on sales theretofore made shall become immediately due and payable and no minimum royalties shall be repayable.

13. **Sponsorship by Competitive Product.** In the event that any of the articles listed in paragraph 1 conflicts with any product of a present or future sponsor of a program on which the Name appears or is used, or with any product of a subsidiary or affiliate of such sponsor, then Licensor shall have the right to terminate this agreement as to such article or articles by written notice to Licensee effective not less than thirty (30) days after the date such notice is given. In the event of such termination, Licensee shall have sixty (60) days after the effective date of such termination to dispose of all of such articles on hand or in process of manufacture prior to such notice, in accordance with the provisions of paragraph 15. However, in the event such termination is effective as to all the articles subject to this agreement and the advance guarantee for the then current year has not been fully accounted for by actual royalties by the end of the 60-day disposal period, Licensor shall refund to Licensee the difference between the advance guarantee which has been paid for such contract year and the actual royalties. The refund provision contained in the preceding sentence pertains only to termination occurring pursuant to this paragraph 13, and shall not affect the applicability of any other paragraph to such termination except as expressly contradicted herein.

14. **Final Statement upon Termination or Expiration.** Sixty (60) days before the expiration of this license and, in the event of its termination, ten (10) days after receipt of notice of termination or the happening of the event which terminates this agreement where no notice is required, a statement showing the number and description of articles covered by this agreement on hand or in process shall be furnished by Licensee to Licensor. Licensor shall have the right to take a physical inventory to ascertain or verify such inventory and statement, and refusal by Licensee to submit to such physical inventory by Licensor shall forfeit Licensee's right to dispose of such inventory, Licensor retaining all other legal and equitable rights Licensor may have in the circumstances.

15. **Disposal of Stock upon Termination or Expiration.** After termination of the license under the provisions of paragraph 12, Licensee, except as otherwise provided in this agreement, may dispose of articles covered by this agreement which are on hand or in process at the time

notice of termination is received for a period of sixty (60) days after notice of termination, provided advances and royalties with respect to that period are paid and statements are furnished for that period in accordance with paragraph 2. Notwithstanding anything to the contrary herein, Licensee shall not manufacture, sell or dispose of any articles covered by this license after its expiration or its termination based on the failure of Licensee to affix notice of copyright, trademark or service mark registration or any other notice to the articles, cartons, containers, or packing or wrapping material or advertising, promotional or display material, or because of the departure by Licensee from the quality and style approved by Licensor pursuant to paragraph 7.

16. **Effect of Termination or Expiration.** Upon and after the expiration or termination of this license, all rights granted to Licensee hereunder shall forthwith revert to Licensor, who shall be free to license others to use the Name in connection with the manufacture, sale and distribution of the articles covered hereby and Licensee will refrain from further use of the Name or any further reference to it, direct or indirect, or anything deemed by Licensor to be similar to the Name in connection with the manufacture, sale or distribution of Licensee's products, except as provided in paragraph 15.

17. **Licensor's Remedies.**

    (a) Licensee acknowledges that its failure (except as otherwise provided herein) to commence in good faith to manufacture and distribute in substantial quantities any one or more of the articles listed in paragraph 1 within three (3) months after the date of this agreement and to continue during the term hereof to diligently and continuously manufacture, distribute and sell the articles covered by this agreement or any class or category thereof will result in immediate damages to Licensor.

    (b) Licensee acknowledges that its failure (except as otherwise provided herein) to cease the manufacture, sale or distribution of the articles covered by this agreement or any class or category thereof at the termination or expiration of this agreement will result in immediate and irremediable damage to Licensor and to the rights of any subsequent licensee. Licensee acknowledges and admits that there is no adequate remedy at law for such failure to cease manufacture, sale or distribution, and Licensee agrees that in the event of such failure Licensor shall be entitled to equitable relief by way of temporary and permanent injunctions and such other further relief as any court with jurisdiction may deem just and proper.

    (c) Resort to any remedies referred to herein shall not be construed as a waiver of any other rights and remedies to which Licensor is entitled under this agreement or otherwise.

18. **Excuse for Non-Performance.** Licensee shall be released from its obligations hereunder and this license shall terminate in the event that governmental regulations or other causes arising out of a state of national emergency or war or causes beyond the control of the parties render performance impossible and one party so informs the other in writing of such causes and its desire to be so released. In such events, all royalties on sales theretofore made shall become immediately due and payable and no minimum royalties shall be repayable.

19. **Notices.** All notices and statements to be given, and all payments to be made hereunder, shall be given or made at the respective addresses of the parties as set forth above unless notification of a change of address is given in writing, and the date of mailing shall be deemed the date the notice or statement is given.

20. **No Joint Venture.** Nothing herein contained shall be construed to place the parties in the relationship of partners or joint venturers, and Licensee shall have no power to obligate or bind Licensor in any manner whatsoever.

21. **No Assignment or Sublicense by Licensee.** This agreement and all rights and duties hereunder are personal to Licensee and shall not, without the written consent of Licensor, be assigned, mortgaged, sublicensed or otherwise encumbered by Licensee or by operation of law.

Licensor may assign but shall furnish written notice of assignment.

22. **No Waiver, Etc.** None of the terms of this agreement can be waived or modified except by an express agreement in writing signed by both parties. There are no representations, promises, warranties, covenants or undertakings other than those contained in this agreement, which represents the entire understanding of the parties. The failure of either party hereto to enforce, or the delay by either party in enforcing, any of its rights under this agreement shall not be deemed a continuing waiver or a modification thereof and either party may, within the time provided by applicable law, commence appropriate legal proceeding to enforce any or all of such rights. No person, firm, group or corporation (whether included in the Name or otherwise) other than Licensee and Licensor shall be deemed to have acquired any rights by reason of anything contained in this agreement, except as provided in paragraphs 6 and 21.

This agreement shall be construed in accordance with the internal laws of the State of _____.

IN WITNESS WHEREOF, the parties hereto have caused this instrument to be duly executed as of the day and year first above written.

By _____ ("Licensor")

Title: _____

By _____ ("Licensee")

Licensing Toys and Games

Like Father, Like Son

Lincoln Logs were invented in 1916 by John Lloyd Wright, son of architect Frank Lloyd Wright. John got his idea for the toy while in Tokyo with his father observing construction techniques for an "earthquake proof" hotel. Today, Lincoln Logs are produced from about four train carloads per month of Ponderosa pine from national forests in Oregon.

Practice staying in the lines, but occasionally go over them.
—MICHAEL SATTEN, TOY INVENTOR

The Toy Industry: An Overview

The toy industry is one of the last and most potentially lucrative frontiers for the independent inventor. If you want proof of this, ask the inventors of Teenage Mutant Ninja Turtles, Pictionary, Trivial Pursuit, Nerf, Twister, K'nex, Bumble Ball, Super Soaker, Koosh, Water Babies, Pound Puppies, and Micro Machines.

The United States is the largest market for toys in the world, followed by Japan, Germany, France, and the United Kingdom. American retail toy sales totaled an estimated $18 billion in 1994, according to Barry Alperin, chairman of Toy Manufacturers of America (TMA). In 1994 the U.S. toy industry shipped over 2.5 billion toys, comprising an estimated 120,000 individual products with 5,000 to 6,000 new items introduced annually.

Americans spend an average of $325 per child on toys each year, most of it in the fourth quarter. The toy manufacturers spent close to $800 million on advertising in 1994, with 90 percent of the total going to television spots, according to the TMA.

The United States also leads the world in toy development. According to the authors of *Inside Santa's Workshop: How Toy Inventors Develop, Sell and Cash In On Their Ideas*, a measure of professional independent inventors' productivity is the percentage of new products that come from them. Major toy companies get thousands of unsolicited ideas over the transom, and certain senior executives interviewed for *Inside Santa's Workshop* credit the inventing community with originating 50 to 75 percent of their lines every year.

The 70 professional inventors interviewed for *Inside Santa's Workshop* told the authors that they and/or their groups come up with an average of 100 to 150 original concepts each year.

Five of the largest toy makers—Mattel, Hasbro, Tyco, Lego Group, and Little Tykes—control 60 percent of the industry's business. Five retailers—Toys R Us, Kay-Bee, Wal-Mart, K-Mart and Target—sell 60 percent of all toys.

Success Stories

- Scrabble was invented in 1931 by Alfred M. Butts, an out-of-work architect who was a lifelong devotee of anagrams and crossword puzzles. The game was originally called "Crisscross Words." Renamed Scrabble in 1948, over 100 million Scrabble games have been sold worldwide to date.

- Trivial Pursuit was invented by three young Canadians in 1979. In the game's first year on the market, Selchow & Righter, the game's original U.S. manufacturer, sold 22 million sets at retail prices as high as $40. The next year it sold six million copies, and the next five million. Trivial Pursuit has sold more than 75 million games worldwide.

- Remember Dr. Erno Rubik, the Hungarian engineer and mathematician who invented the Rubik Cube? During its three-year hot streak, this innocent-looking, two-and-a-quarter-inch puzzle with more than forty-three quintillion possible combinations but only one true solution, sold an estimated 100 million authorized copies, plus another 50 million knock-offs and at least ten million books explaining how to solve it.

- Monopoly was brought to Parker Brothers in 1933 by the late Charles B. Darrow, who developed the game while he was unemployed during the Depression. It was initially rejected as having "52 fundamental errors," but was later published in 1935, and is now licensed in 33 countries and printed in 23 different languages. Over 100 million sets of Monopoly have been sold worldwide.

- Crayola Crayons celebrated its 90th birthday in 1993. Alice Stead Binney conceived the name Crayola for her husband Edwin's crayons in 1903. She derived it from the French word "craie" (stick of color) and the word "oleaginous" (oily). Each year the company Binney & Smith produces more than two billion Crayola Crayons. That would be enough crayons to circle the globe four and a half times or produce a giant crayon 35 feet wide and 400 feet tall—100 feet taller than the Statue of Liberty. Kids ages two to eight spend an average of 28 minutes a day coloring. The average child in the United States will wear down 730 crayons by age ten.

- Cincinnati barber Merle Robbins came up with the card game Uno and licensed it to International Games in 1972. Today Uno is sold in 26 countries and is available in 12 languages, with sales of over 80 million units worldwide.

- The Hula Hoop, introduced by Wham-O in 1958, set the standards by which all fads are measured. Within four months of its introduction, more than 25 million Hula Hoops were sold. The Hula Hoop was inspired by childrens' bamboo exercise hoops from Australia.

- A game invented by a wealthy Canadian couple to play aboard their yacht was so popular with their friends that they approached Edwin S. Lowe, most famous for publishing bingo games in the 1920s, to make samples of "The Yacht Game" for them to give as gifts. Lowe liked the game so much that he offered to buy the rights from the couple and they agreed. He eventually changed the name of the game to Yahtzee. In 1973, Milton Bradley acquired the E.S. Lowe Company.

- The world's best-selling top, one that can hit speeds of 10,000 rpms, the Wiz•z•zer, was invented by Paul L. Brown, the sixth child in a family of twelve. His parents were so poor that, when Paul was seven years old, they had to send him and four of their other children to the Clark County Children's Home in Springfield, Ohio for care. A high school dropout, Paul graduated first in his class from Fenn College Engineering Defense Training Program. In 1970, Mattel Toys licensed his top and named it the Wiz•z•zer. In 1995 the Wiz•z•zer will celebrate its 25th anniversary, and surpass 25 million units sold.

Oh, You Beautiful Doll!

Barbie, the world's best-selling and most widely recognized fashion doll in history, celebrated her 36th birthday in 1995. She is named after the daughter of her inventor, Ruth Handler. Since 1959, over 600 million Barbies and members of the Barbie clan (Ken is Barbie's real-life brother) have been sold in 67 countries throughout the world. Mattel sells over 20 million Barbie fashions annually, making the California toymaker the largest producer of "petite" women's wear in the world. Barbie generates over one billion dollars a year in sales.

Toy Fair: A Networking Mecca

If you want to see what's in and what's not and who's who in the toy industry, the best place is the annual American International Toy Fair. Here you'll be able to take the pulse of the industry and make contacts. Toy Fair, as it is more commonly known, is the largest toy trade show in the United States.

In 1994 Toy Fair featured 1,634 exhibitors, including 247 foreign companies from 27 countries, and was attended by 19,105 commercial retail buyers from 87 countries. It is also one of the most extensively covered trade shows in the United States; the 1994 Fair hosted over 700 national and international print and broadcast reporters.

Toy Fair is held in New York City for ten days, beginning on the second Monday in February. Exhibits are located in permanent showrooms in—or within short walking distance—of the two Toy Center buildings at 200 Fifth Avenue and 1107 Broadway, and concurrently for four days in display booths at the Jacob K. Javits Convention Center.

For more information on Toy Fair, its exact dates, seminars, and so on, contact: Toy Manufacturers of America, Inc., 200 Fifth Avenue, Suite 740, New York, New York 10010. The telephone number is (212) 675-1141; the fax number is (212) 633-1429.

Is It Easy to Get through the Door of Toy Companies?

No. Most major toy companies—where the big money is made on TV-promoted toys and games—depend upon the professional inventing community. You can be a part of it if you are smart about how you present yourself and your concept.

The best way to get started is through the offices of a professional toy broker. While not all toy manufacturers will recommend brokers, many—such as Milton Bradley, Hasbro, and Fisher-Price—will do so if asked.

Professional agents typically take between 30 and 50 percent of any advances and royalties; some may request as much as 60 percent. This assumes that you have an acceptable "looks like–works like" prototype. If the agent has to invest a lot of money to further develop your idea, you may have to give up even more of the pie.

There is no up-front money required by the best professional brokers. They not only have access to the toy companies' highest levels, but they are in the loop to know who's looking for what. Many brokers are professional inventors themselves.

Before signing up with a broker, you should have a face-to-face meeting. This will not be a problem for the pros. See the products the broker has had a hand in developing and/or licensing. Talk to the inventors behind these products and see what they think of the broker.

Lastly, insist that the broker arrange for advances and royalties to be paid directly from the toy manufacturer to you. Do not allow the broker to receive your money and then redistribute it. I say this not because I worry about the honesty of the broker, but because you want the path of least resistance for your advances and royalties. No toy company would deny a request to divide the money and send separate checks and reports. If the broker says the company won't do it, go to another broker.

The Idea Is Never Enough

No matter how terrific your concept may be, the manufacturer must make sure that the product can be manufactured and sold at the right price. Warren Bosch, a former director of product development at Tyco, told me, "It's one thing to make a piece of something, and another to make two and a half million of them. That kind of development work is often just as creative as the idea itself."

Inventor Michael Satten perhaps summed it up best when he told me, "Things have to pass through so many people. Imagine. You have to show it to a guy who has to like it; he has to take it back and show it to a group that has to like it; the engineering has to work; it has to work at a cost; it has to be tested with kids and tested with parents; the trade has to see it; the trade has to like it; the agency has to do the commercials; the commercials have to be tested with kids and parents who have to like what they see; then the trade has to see the product again and like the commercial. And after all that, there is a problem in China. It's mind-boggling."

Questions to Ask Yourself

Before you spend money on a prototype and approach a possible licensee, make sure that you have given thought to the following points. Toy companies do not license ideas or figments of your imagination, they license original, well-developed products that meet certain criteria.

- **Is your idea original?** Talk to toy store managers who have been around for a few years, attend Toy Fair, conduct a patent search (see Chapter 5). All too often inventors with no memory for product re-invent something that has been done before. This wastes everyone's time, and does nothing to forward your position.

- **Does your idea have visual appeal? Perceived value?** Hire an industrial designer if you do not possess the skills to design your concept. It's cheaper and wiser to begin on paper than in plastic. Toy executives are accustomed to a high level of professionalism when it comes to design. It must not only look good, it must have perceived value.

- **Does your item have play value? Repeat play value?** Child-test a breadboard or mock-up of the item. Make sure it works, and that it has the ability to sustain interest. In the case of games, it could take dozens of play sessions to fine-tune a concept.

- **Does your concept have wide market appeal?** Major companies have no interest whatsoever in small market segments. In order for a product to make the cut at the larger manufacturers, it must have very broad market appeal. In other words, it must sell to Toys R Us, Wal-Mart, K-Mart, Target, and other mass-market outlets. If the company does not feel it can sell hundreds of thousands of units the first year, it will not license the submission.

- **Is your product safe?** All products selected for manufacture by United States or foreign toymakers must meet one or more of the following safety standards:
  1. CPSC Regulations
  2. ASTM F963-92
  3. DEHP Content
  4. CONEG
  5. EN-71 (1988)
  6. BS 5665 (1989)
  7. HD271 (for electrical toys)
  8. Perspiration and saliva test (Germany)
  9. Phenol Content (Germany)

There is also a battery of functional abuse exercises which, depending upon the product, can include: transit test, aging test, humidity test, drop test, torque and tension test, compression test, abrasion test, adhesion test, and life test.

The K'nex Step

When plastics manufacturer Joel Glickman began bending drinking straws into geometric shapes to pass the time at a wedding reception, he knew it was more than a temporary diversion. In just its second complete year on the market, his Connector Set Toy Company has sold almost four million K'nex sets, won more than two dozen consumer and educational awards, and begun distribution in Europe and Asia. When I visited his factory, Joel told me that he was molding 18 million K'nex components daily to fill orders.

Sweet Thoughts

While Eleanor Abbott of San Diego, California, was recuperating from polio in the 1940s, she occupied herself with devising games and activities for youngsters who had polio. One of her inventions was called "Candy Land." Her young friends liked it so much that she submitted it to Milton Bradley Company, where it was immediately accepted. Candy Land sells over one million copies per year.

- **Can your product be manufactured? At what cost?** You may have to build or hire someone to build a full-scale, fully operational model to determine production and fabrication requirements and see if anything would render the concept infeasible. Play value and technical elegance mean little if your concept cannot be made for the right price.

  In costing toys, we use this rule-of-thumb: hard cost times four for non-promoted items; hard cost times five for TV-promoted items.

- **What competitive products are on the market?** Be able to tell your target licensee the category in which you envision your item as well as the competitive atmosphere. Buy competitive products, and tear them apart. Know what makes them tick.

- **What "wow factors" make your concept unique?** Demonstrate to yourself what makes your item unique compared to product in its category.

  Once you have satisfied these issues, then and only then embark on building a looks like–works like prototype.

A Degree in Toy Design

The Fashion Institute of Technology in New York City began offering a Bachelor of Fine Arts degree in Toy Design, the first known program of its kind, in 1989.

The FIT toy design curriculum emphasizes conceptual and technical design development, supported by a knowledge of safety and regulatory requirements, child psychology and anatomy, production, packaging, marketing and consumer motivation fundamentals, plus a general overview of the toy industry. A strong liberal arts component is included to broaden the designer's perspective as an interpreter of cultural trends. Industry internships with toy companies and working with children as teacher assistants are also important elements of this program.

FIT graduated its fourth class of toy designers in 1994 and within six months 95 percent of the students were placed in jobs at toy companies and inventor groups.

For more information, contact Professor Judy Ellis, Fashion Institute of Technology, Toy Design Department, 227 West 27th Street, New York, New York 10001.

TMA Freebies

Toy Manufacturers of America (TMA) publishes numerous guides that may be of interest to the independent inventor who is considering the toy industry.

1. *Toy Inventor/Designer Guide:* For a copy of this brochure, which describes ways to sell your invention/design or manufacture it yourself, please write to TMA's Communications Department *Toy Inventor/Designer Guide,* Toy Manufacturers of America, 200 Fifth Avenue, Room 740, New York, New York 10010.

2. The TMA *Guide to Toys & Play:* For a copy of this safety booklet (revised July 1993), please send a postcard with your name and address to Toy Booklet, P.O. Box 866, Madison Square Station, New York, New York 10159-0866. Or you may call the TMA, toll-free: 1-800-851-9955.

3. *Catalogue of Toys for the Blind:* For a copy of this publication, which lists toys appropriate for blind and visually impaired children, please send your name and address to Toy Manufacturers of America, 200 Fifth Avenue, Room 740, New York, New York 10010.

TOY MANUFACTURERS

The toy industry's greatest natural resource is the inventor who, after all, is responsible for its greatest products. In an era of intense global competition, the inventor becomes even more important.

The following list represents some of the toy makers who entertain outside submissions.

Artin Industrial Co. Ltd.
21-25 Sze Mei St.   (852) 325-5215
Sanpokong, Kowloon, Hong Kong
FAX: (852) 352-3497

Bandai America, Inc.
12851 E. 166th St.   (310) 926-0947
Cerritos, CA 90701
FAX: (310) 926-8030

Blue Box Toys, Inc.
200 Fifth Ave.   (212) 255-8388
New York, NY 10010
FAX: (212) 255-8520

Cadaco
4300 W. 47th St.   (312) 927-1500
Chicago, IL 60632
FAX: (312) 927-3937

Cap Toys, Inc.
26201 Richmond Rd.   (216) 292-6363
Bedford Heights, OH 44146
FAX: (216) 292-4815

Cardinal Industries, Inc.
21-01 51st Ave.   (718) 784-3000
Long Island City, NY 11101
FAX: (718) 482-7877

Child Guidance
A Division of Azrak/Hamway Int'l, Inc.   (212) 675-3427
1107 Broadway
New York, NY 10010
FAX: (212) 243-4271

Colorforms
133 Williams Dr.   (201) 327-2600
Ramsey, NJ 07446
FAX: (201) 327-2506

Commonwealth Toy & Novelty Co.
45 W. 25th St.   (212) 242-4070
New York, NY 10010
FAX: (212) 645-4279

Craft House Corp.
328 N. Westwood Ave.   (419) 536-8351
Toledo, OH 43607
FAX: (419) 536-4159

Dakin, Inc.
20969 Ventura Blvd. #25   (818) 340-5800
Executive Center
Woodland Hills, CA 91364
FAX: (415) 259-2525

Decipher Inc.
253 Granby St.   (804) 623-3600
Norfolk, VA 23510
FAX: (804) 623-3630

Diversified Specialists, Inc.
1100 W. Sam Houston Parkway N   (713) 526-2888
Houston, TX 77043
FAX: (713) 526-5049

Duncan Toys Co.
15981 Valplast Rd.   (216) 632-1631
P.O. Box 97
Middlefield, OH 44062
FAX: (216) 632-1581

Empire
P.O. Box 427   (919) 823-4111
Daniel St. Extension
Tarboro, NC 27886
FAX: (919) 823-8609

The Ertl Co., Inc.
Highways 136 & 20   (800) 553-4886
P.O. Box 500
Dyersville, IA 52040
FAX: (319) 875-8263

In spite of what you hear every Christmas from certain litigious product liability lawyers who grandstand for the media by pointing out defective toys—in hopes of getting clients—the major toy companies take health and safety issues quite seriously.

Fisher-Price, Inc.

636 Grand Ave.
East Aurora, NY 14052
FAX: (716) 687-3667

(716) 687-3000

Funrise Toy Corp.

6115 Variel Ave.
Woodland Hills, CA 91367
FAX: (818) 883-3809

(818) 883-2400

Goetz Dolls, Inc.

8257 Loop Road—Radisson
Baldwinsville, NY 13027
FAX: (315) 638-4320

(315) 635-1055

Hasbro Games Group

Milton Bradley and Parker Brothers
443 Shaker Rd.
E. Longmeadow, MA 01028
FAX: (413) 525-1767

(413) 525-6411

Hasbro Toy Group

(Playskool, Kenner, Kid Dimension)
1027 Newport Ave.
Pawtucket, RI 02860
FAX: (401) 727-5779

(401) 431-TOYS

Happiness Express Inc.

50 W. 23rd St.
New York City, NY 10010
FAX: (212) 675-9271

(212) 675-0461

Imperial Toy Corp.

2060 E. 7th St.
Los Angles, CA 90021
FAX: (213) 489-4467

(213) 489-2100

The United States is the largest market for toys in the world, followed by Japan and Western Europe. American retail toy sales totaled an estimated $18.7 billion in 1994, up from $17.5 billion in 1993, according to the TMA.

The TMA reports that the strongest performers were male action toys (up 33.9%), ride-ons (up 10.2%), vehicles (up 9.5%), and activity toys (up 8.2%).

Irwin Toy Ltd.

43 Hanna Ave.
Toronto, ON M6K 1X6
Canada
FAX: (416) 533-3257

(416) 533-3521

Just Toys, Inc.

50 W. 23rd, 7th Fl.
New York, NY 10010
FAX: (212) 741-8793

(212) 645-6335

Kenner Products

615 Elsinore Pl.
Cincinnati, OH 45202
FAX: (513) 579-4704

(513) 579-4000

Kid Dimension, Inc.

A Division of Hasbro, Inc.
1027 Newport Ave.
Pawtucket, RI 02862
FAX: (401) 727-5252

(401) 727-5548

Larami Corp.

303 Fellowship Rd. Ste #110
Mt. Laurel, NJ 08054
FAX: (609) 439-9732

(609) 439-1717

Learning Curve Toys

311 W. Superior #416
Chicago, IL 60610
FAX: (312) 654-8227

(312) 654-5960

Lego Systems, Inc.

555 Taylor Rd.
Enfield, CT 06083-1600
FAX: (203) 749-6077

(800) 243-4870

Lewis Galoob Toys, Inc.

500 Forbes Blvd.
S. San Francisco, CA 94080
FAX: (415) 952-7084

(415) 952-1678

Little Tikes Co.

2180 Barlow Rd.
Hudson, OH 44236
FAX: (216) 650-3067

(216) 650-3000

Mattel Toys

333 Continental Blvd.
El Segundo, CA 90245-5012
FAX: (310) 463-0571

(310) 524-2000

Meritus Industries, Inc.
511-5N Regent St.   (201) 740-8887
Livingston, NJ 07039
FAX: (201) 740-0140

Micro Games of America
16730 Schoenborn St.   (818) 894-2525
North Hills, CA 91343-6122
FAX: (818) 894-1666

Milton Bradley Co.
see Hasbro Games Group

National Latex Products Co.
246 E. 4th St.   (419) 289-3300
Ashland, OH 44805
FAX: (419) 289-7118

Natural Science Industries, Ltd.
50-01 Rockaway Beach Blvd.   (718) 945-5400
Far Rockaway, NY 11691
FAX: (718) 318-1194

Nylint Corp.
1800 16th Ave.   (815) 397-2800
Rockford, IL 61104
FAX: (815) 397-7845

OddzOn Products
1696 Dell Ave.   (408) 866-2966
Campbell, CA 95008
FAX: (408) 866-2972

The Ohio Art Co.
One Toy St.   (419) 636-3141
Bryan, OH 43506
FAX: (419) 636-7614

Parker Brothers
see Hasbro Games Group

Playmates Toys, Inc.
16200 Trojan Way   (714) 739-1929
La Mirada, CA 90638
FAX: (714) 739-2912

Playskool
A Divison of Hasbro, Inc.   (401) 431-TOYS
1027 Newport Ave.
Pawtucket, RI 02860
FAX: (401) 727-5779

Child's Play
Children have the greatest sales potential of any age or demographic group, controlling more than $90 billion themselves and directly influencing at least $130 billion of additional spending by adults, according to Dr. James McNeal, a Texas A&M marketing expert.

Pressman Toy Corp.
200 Fifth Ave.   (212) 675-7910
New York, NY 10010
FAX: (212) 645-8512

Remco Toys
A Division of Azrak/Hamway Int'l Inc. (212) 675-3427
1107 Broadway
New York, NY 10010
FAX: (212) 243-4271

Roadmaster Corp.
P.O. Box 344   (618) 393-2991
Olney, IL 62450
FAX: (618) 395-1057

Rose Art Industries, Inc.
6 Regent St.   (201) 535-1313
Livingston, NJ 07039
FAX: (201) 533-9447

Shelcore, Inc.
347 Elizabeth Ave.   (908) 764-9000
Somerset, NJ 08873
FAX: (908) 764-9001

Spectra Star
350 E.18th St.   (602) 782-2541
Yuma, AZ 85364
FAX: (602) 783-9534

Talicor, Inc.
190 Gentry St.   (909) 593-5877
Pomona, CA 91767
FAX: (909) 596-6586

Tiger Electronics Inc.
980 Woodlands Parkway   (708) 913-8100
Vernon Hills, IL 60061
FAX: (708) 913-8118

Tomy America Inc.
450 Delta Ave. (714) 256-4990
Brea, CA 92621
FAX: (714) 256-4835

Tootsietoy—Strombecker Corp.
600 N. Pulaski Rd. (312) 638-1000
Chicago, IL 60624
FAX: (312) 638-3679

Totsy Manufacturing Co., Inc.
P.O. Box 509 (413) 536-0510
One Bigelow St.
Holyoke, MA 01040
FAX: (413) 532-9804

Toy Biz. Inc.
333 E. 38th St. (212) 682-4700
New York, NY 10016
FAX: (212) 682-5317

Toymax Inc.
200 Hicks St. (516) 997-9060
Westbury, NY 11590
FAX: (516) 997-9084

TSR, Inc.
P.O. Box 756 (414) 248-3625
Lake Geneva, WI 53147
FAX: (414) 248-0389

Tyco Playtime
1107 Broadway (212) 741-7222
New York, NY 10010
FAX: (212) 243-5125

Tyco Toys, Inc.
6000 Midlantic Dr. (609) 234-7400
Mt. Laurel, NJ 08054
FAX: (609) 722-0431

Uneeda Doll Co., Inc.
200 Fifth Ave. (212) 675-3313
New York, NY 10010
FAX: (212) 929-6494

University Games Corp.
1633 Adrian Rd. (415) 692-2500
Burlingame, CA 94040
FAX: (415) 692-2770

Western Publishing Co., Inc.
1220 Mound Ave. (414) 633-2431
Racine, WI 53404
FAX: (414) 631-1854

Yes! Entertainment Corp.
3875 Hopyard Rd., Ste. 375 (212) 647-0250
Pleasanton, CA 94588
FAX: (212) 647-9639

Modeled to a "T"

The die cast car business made its debut in 1906 when the Dowst Brothers Company in Chicago made a miniature Model T Ford. More than 50 million of the toy cars were sold over the next several years. The early vehicles were called Tootsietoys, recalling the nickname of one of the Dowst brothers' granddaughters. Dowst eventually merged with the Cosmos Manufacturing Company in 1926, the original supplier of Cracker Jack prizes. The new firm was called the Strombecker Corporation. Today, according to the company, Strombecker produces over 40 million cars per year.

How to Benefit from Federally Funded Technology

This country must sustain world leadership in science, mathematics, and engineering if we are to meet the challenges of today . . . and of tomorrow.

—PRESIDENT BILL CLINTON

Science in the National Interest

The U.S. government offers mechanisms to access the knowledge and expertise of people working in federal research and development (R&D) labs and centers who may be able to help you as an independent inventor.

About $25 billion of federally funded R&D takes place annually at more than 700 federal research and development laboratories and centers. Over 100,000 scientists and engineers, one-sixth of the nation's total, work at these sites across a full range of science and technology, including optics, energy, material sciences, biotechnology, medicine, electronics, and environment, to name a few.

"The federal laboratories were established to provide the capability to meet national research goals and needs, complementary to those existing at universities and industry. Largely, they have been very successful in this task," says John Schiffer of Argonne National Laboratory and the University of Chicago.

U.S. government laboratories are looking to transfer their technologies and expertise to the private sector for commercialization that will contribute to the improvement of the American economy.

Fifty years ago, in Vannevar Bush's seminal report, *Science: The Endless Frontier*, he wrote: "The Government should accept new responsibilities for promoting the flow of new scientific knowledge and the development of scientific talent in our youth. These responsibilities are the proper concern of the Government, for they vitally affect our health, our jobs and our national security."

The bedrock wisdom of this statement has been demonstrated time and again in the intervening half century. The return from our public investment in fundamental science has been enormous, both through the knowledge generated and through the education of an unmatched scientific and technical workforce. Discoveries in mathematics, physics, chemistry, biology, and other fundamental sciences have contributed to and been driven by important advances in engineering, technology, and medicine.

Godfather of American Invention

Thomas Jefferson created a variety of practical devices, including:

- dumbwaiter
- lazy susan
- folding campstool
- folding ladder
- folding music stand
- portable copying press
- portable writing desk
- revolving chair
- cipher wheel for cryptography
- moldboard plow
- hemp-breaker for threshing machines

Many of his inventions can be seen today at Monticello, Jefferson's Palladian estate on a hilltop near Charlottesville, Virginia.

At the Forum on Science in the National Interest in February 1994, Vice President Al Gore said, "Science reveals new worlds to explore, and by implication new opportunities to seize and new futures to create."

This is a time of transition for the federal laboratories. The ending of the Cold War is profoundly affecting the defense laboratories and altering their mission. Today the federal laboratories are seeking to connect themselves with private industry and are willing to share their expertise, and occasionally their facilities and special services.

For example, the U.S. Navy reaches out through an advertisement in technical magazines. It reads, in part: "The same Navy resources that forged the world's finest defense systems are playing an even larger role—sharing research. And we're ready to share it with you. . . . We'd like to help you succeed." For information on how you might tap into the Navy's R&D, call: 1-800-NAVY-TEC.

An ad by NASA Marshall Space Flight Center has this headline: "Making America Stronger by Transferring Space Program Technology to America's Industries." Listed are areas of expertise such as microgravity research, coatings, composites, optical systems, and so on. For information call: 1-800-USA-NASA.

Martin Marietta, the $9 billion Bethesda, Maryland–based defense contractor, has worked for the Pentagon on everything from World War II bombers to Gulf War spy satellites. Recently, however, Martin modified its military technology for the commercial marketplace. Video game–maker Sega sought Martin's help in developing a new computer graphics board. The board combines Sega's 32-bit 3-D polygon graphics and Martin's real-time photo-texturizing technology, which represents a quantum leap in realism and response time compared to traditional video games.

"We don't know the entertainment world, but we have the technology. They know the entertainment industry, so when you team like that, you take the best of what both bring to the party," said Martin Marietta spokesperson Marlene Duval.

As a toy inventor, I have on many occasions looked to federal laboratories for assistance with my work. Since most toys are no more than a miniaturization of real world things, I frequently find myself looking for advice on how to transfer some costly and esoteric technology into an affordable, promotable feature for a toy.

I have always found the scientists and engineers on the other end of the line to be more than willing to answer my questions and even, at times, to conduct experiments or fax me critical background data. Two of the mandates under which federal labs operate are to expand secondary use of product and process technology and to assist small companies through incubators.

For an extensive list of more than 300 federal laboratories, visit your local library and consult the second edition of the *Inventing and Patenting Sourcebook*, published by Gale Research. For the complete listing of federal labs, you can purchase—for $65 plus handling—a guide published by the National Technical Information Service (NTIS). To order their *Directory of Federal Laboratory and Technology Resources: A Guide to Services, Facilities, and Expertise,* call NTIS at (703) 487-4650.

Technology Transfer Business Magazine

If you are interested in the transfer of federal technologies, there is a terrific quarterly magazine that covers the subject in depth. It is called *Technology Transfer*

Business, The Magazine for Profitable Partnerships, and it's free-of-charge to professionals in technology transfer. Non-professionals must pay $25 per year. For subscription information, call: (703) 848-2800.

Federal Sources of Information

Listed below are four federal organizations that can be valuable sources of information. If you can't find what you are looking for among these, chances are it does not exist.

- Federal Laboratory Consortium for Technology Transfer
- National Technical Information Service
- Commerce Clearinghouse
- Defense Technical Information Center

Federal Laboratory Consortium for Technology Transfer

The Federal Laboratory Consortium for Technology Transfer (FLC) consists of more than 600 member research laboratories and centers from 16 federal departments and agencies. In existence since 1974, it develops and tests transfer methods, addresses barriers to the process, provides training, highlights grass-roots transfer efforts, and emphasizes national initiatives where technology transfer has a role. The FLC also brings these laboratories together with potential users of government-developed technologies in the private sector as well as state and local governments. Through its locator network, it matches potential private sector R&D collaborators with laboratory resources.

The FLC mission is to promote the rapid movement of federal technology research and development into the mainstream of the U.S. economy. The Consortium accomplishes this mission through a coordinated program that supports the technology transfer needs of its members and those interested in research and technology results.

The FLC is a volunteer organization of federal research and development laboratories and centers that work together to maximize the transfer of technology. It provides a network for laboratory personnel who meet with other laboratory representatives to exchange information and experiences. Requests from private businesses or individuals for technical assistance are processed through the network of laboratory representatives. Each individual FLC representative links the science and technology of their federal laboratory to user communities as well as providing guidance to other laboratories.

For additional information, contact the FLC at this address: FLC Administration Office, P.O. Box 545, Sequim, WA 98382. Or call (206) 683-1005; the fax number is (206) 683-6654.

National Technical Information Service

The National Technical Information Service (NTIS) has as its mission the collection and dissemination of scientific, technical, engineering, and business information produced by the U.S. government and foreign sources in order to increase American competitiveness in the global economy.

NTIS is a federal agency within the technology administration of the U.S. Department of Commerce, but it receives no appropriated funds for its operation.

Clip and Save

Listed below are the names and phone numbers of four federal organizations that can be invaluable sources of technology research and development information.

- Federal Laboratory Consortium for Technology Transfer: (206) 683 1005

- National Technical Information Service: (703) 487-4650 (to order their free catalogue of products and services)

- Commerce Clearinghouse: (202) 377-8100

- Defense Technical Information Center: 1-800-225-3842

All costs (including salaries and office rental) associated with collecting, abstracting, indexing, archiving, reproducing, and disseminating the nation's collection of scientific, technical, engineering, and business information are covered by sales revenue. In fiscal year 1993, NTIS generated $31.4 million.

Where Does the Information Come From?

More than 200 U.S. government agencies contribute to the NTIS collection, including NASA, the Environmental Protection Agency (EPA), the National Institute of Standards and Technology (NIST), the National Institutes of Health (NIH), and the Departments of Agriculture, Defense, Energy, Commerce, Interior, Health and Human Services, and Transportation. NTIS adds approximately 90,000 new works to its collection annually.

NTIS is the leading U.S. government agency in international technical and business information exchange. It actively acquires and distributes information produced by 15 different foreign government departments and other organizations. Almost 30 percent of the NTIS collection is attributed to international sources.

From Joint Ventures

Legislation passed in 1988 authorized NTIS to work with private industry to build strategic alliances through the use of contracts or cooperative agreements with the private sector, individuals, firms, or other organizations. As a result of this legislation, NTIS established its Joint Ventures Program.

This unique program allows NTIS to enter into joint ventures with businesses to create new information products from U.S. government–produced data and software. In addition, NTIS is looking for partnerships to open new channels for the sales and distribution of government information products.

For more information about joint ventures with NTIS, call (703) 487-4674.

What Does NTIS Information Cover?

The NTIS collection of more than two million works covers current technologies, business and management studies, and international market research reports. Additional research results are available in the fields of environment, energy, health, and social sciences. Hundreds of other areas are covered. The material is not limited to printed reports and documents. NTIS also provides information in electronic formats including datafiles on tape, datafiles on diskette, software programs on tape, software programs on diskette, and titles on CD-ROM.

FedWorld Online Information Service

FedWorld is an online information service established by NTIS to provide the general public with a user-friendly central resource for government information. FedWorld provides both dial-up and Internet access.

Since FedWorld was established in the early 1990s, they have received close to a million calls from over 120,000 registered users who have downloaded files over 400,000 times from their system. They currently serve many in the policy community as one of the primary points of dissemination for White House information. For example, over a gigabyte's worth of copies of the President's Report to America and the Health Security Plan were downloaded from FedWorld within 48 hours of the President's address to Congress.

FedWorld also allows users to "gateway" through the system to over 130 other publicly available government information systems, effectively providing one-stop shopping for many types of government information. As a result of the FedWorld gateway, many of these systems are accessible from the Internet for the first time.

FedWorld does not assess any charges to the public to use the basic FedWorld services; they do set a time limit of three hours per day. They recover system development and maintenance costs through the sale of products online and through subscriptions to a small number of specific databases and files in the system. Currently, FedWorld offers about 50 products for sale on the system, as well as offering several subscription products, including the Department of Labor's Davis Bacon Wage Determination Database and the Federal Aviation Administration's Airworthiness Directives. FedWorld also provides information dissemination services on behalf of other agencies, for which they charge the agencies based on costs. They have found that the incremental cost of adding another agency's information, even in a highly customized way, is substantially lower than it would be for the other agency to create its own stand-alone system with comparable support and customer service.

NTIS Catalog

To receive the latest NTIS products and services catalog, call (703) 487-4650. This catalog is free of charge.

Getting in Touch with NTIS

NTIS is located at 5285 Port Royal Road, Springfield, Virginia 22161. To reach them by phone, consult the list of key numbers below.

Subscription orders	(703) 487-4630
Reports and computer product orders	(703) 487-4650
Rush orders (reports and computer products)	(800) 553-NTIS
Rush orders in Virginia	(703) 487-4700
Fax	(703) 321-8547
Research services	(703) 487-4780
Order tracing	(703) 487-4660
NTIS deposit accounts	(703) 487-4770
Federal Computer Products Center	(703) 487-4763

NTIS Annual Conference

Each year NTIS holds a two-day conference designed to bring business, academia, government, and the information sciences the latest news about what government information is available, and how to find it. For more information, call: (703) 487-4811.

Commerce Clearinghouse

A Bonanza of Free Information

Finding information about local, state, and federal technology development and commercialization programs used to be like finding that proverbial needle in a haystack. The situation changed in October, 1990, when the U.S. Department of Commerce formally inaugurated its Clearinghouse for State and Local Initiatives on Productivity, Technology, and Innovations. Now information and expert guidance is just a phone call away.

Established by Congress under the Omnibus Trade and Competitiveness Act of 1988, the Clearinghouse's primary mission is to help share the wealth.

State and local governments are in the forefront of efforts to improve productivity, technology, and innovation. Centers of excellence, seed and venture capital programs, and manufacturing technology extension services are only a sampling of the variety of approaches which states and localities have undertaken to stimulate technological innovation.

The Clearinghouse has a database with information from more than 700 local, state, and federal technology programs and from more than 400 individuals working in technology programs and policy.

Technology programs are classified by the Clearinghouse into 12 categories: research center, university research grant, technology transfer, university research park, tax policy, management assistance, technical assistance, incubator, business capital, manufacturing extension, training, and policy development.

"Business is great!" says Elizabeth M. Robertson, an analyst at the Clearinghouse. She reported that almost half of her calls are from individuals, for example, inventors, entrepreneurs, and businessmen.

"I tend to give people who call at least three or four numbers and tell them that what they are doing is getting into a network," Ms. Robertson explained. And while it can be very frustrating trying to get information from a large government agency, she sees her job, in part, as helping "to keep the frustration factor way down" for people who call for assistance.

There are three main groups that call the Clearinghouse for assistance. They are state and local officials, the private sector, and federal officials. Queries are wide ranging, but here are some examples of the kinds of questions the Clearinghouse responds to:

- The state of New Hampshire wanted to set up a research grant program and sought to learn about how other states had done it. The Clearinghouse was able to provide a list of people who had experience in designing and initiating such programs.

- An engineer called to obtain a list of who in the United States could help him with opto-electronic research. The Clearinghouse gave him a list of laboratories conducting R&D in the field.

- Some independent inventors have called to get something as simple as the telephone number for the nearest inventor organization.

- At the federal level, the assistant secretary for intergovernmental affairs might call to find out who's who in a specific field of technology for the purpose of organizing a conference.

No task is too large or too small for the Clearinghouse team. And it is **free** to anyone who calls or writes. It is one of the most valuable services the federal government provides to those involved in invention and technology transfer.

"It is indeed rare, but refreshing, to have one's needs met in such a timely manner. You are to be commended," wrote Myrta A. Mason of the Association for Manufacturing Technology, referring to her experience with the Clearinghouse.

Clearinghouse Guides

The Clearinghouse provides informational booklets, two of which are described below, free of charge.

- *A Guide to Professional Program Developers and Administrators.* This 40-page booklet describes 21 national clearinghouses that provide information on community and economic development programs, state and local government management, or technology and technology transfer.
- *A Guide to Directories.* This 28-page booklet summarizes 19 directories of programs and initiatives undertaken by federal, state, and local governments to stimulate business competitiveness. The directories are grouped into two broad categories—technology, R&D; and general business and economic development.

To order the guides write to: The Clearinghouse for State and Local Initiatives in Productivity, Technology and Innovation; Technology Administration; U.S. Department of Commerce, Rm H-4418; Washington, D.C. 20230. Or you can call (202) 377-8100; the fax number is (202) 377-0253.

Defense Technical Information Center

The Defense Technical Information Center (DTIC) is a great place to dig for research. It is the central storehouse of scientific and technical information resulting from and describing R&D projects that are funded by the Department of Defense (DOD). It has over 1.8 million documents on file.

"We exist to help independent inventors and small businesses cut through the red tape and bureaucracy," says the helpful and hospitable Richard Sparks, DTIC program manager and the pointman for outside queries. He even answers his own phone. Richard may be reached at (703) 274-6902 or 1-800-363-7247.

During fiscal year 1994, DTIC provided 4,427 small business requesters with approximately 28,937 technical information packages and 7,920 technical reports on the 754 DOD solicitation topics.

Most functions of DTIC can be accessed by calling 1-800-225-3842. This is the number for general DTIC information and information on becoming a registered user. Or write to the Defense Technical Information Center, Registration and Services Section (DTIC-FDRB), Building 5, Cameron Station, Alexandria, Virginia 22304-6145.

Fees for DTIC

- Online connect: $100 annual fee; $40 per connect hour.
- Documents: $6 for reports up to 100 pages,
  10¢ per page when over 100 pages.
- Microfiche: $4 per report.

Origins

The Guiness Book of World Records describes the following origins for major innovations:

- The phonograph was first conceived by Charles Cros (1842–1888), a French poet and scientist, who described his concept in sealed papers deposited with the French Academy of Sciences on April 30, 1877. However, the realization of a practical device was first achieved by Thomas Alva Edison (1847–1931).

- The earliest patent for wireless telegraphy was awarded to U.S. inventor Dr. Mahlon Loomis (1826–1886). Dated July 20, 1872, the patent was entitled "Improvement in Telegraphy." The first patent for a system of communication by means of electromagnetic waves was granted on June 2, 1896 to Marchese Guglielmo Marconi (1874–1937).

- War rockets, propelled by gunpowder, were described by Zeng Kung Liang of China in 1042. The first liquid-fueled rocket was patented on July 14, 1914 by U.S. inventor Dr. Robert H. Goddard (1882–1945). On March 16, 1926, he launched a rocket that reached an altitude of 41 feet and traveled a distance of 184 feet.

University and Independent Innovation Research Centers

Inventive Thinking: A Guide

Robert Eberle's "SCAMMPERR" technique is an excellent way to make sure you consider lots of options when you brainstorm.

- **S**ubstitute. Who else instead? What else instead? Other ingredient? Other material?

- **C**ombine. How about a blend, an alloy, an ensemble?

- **A**dapt. What else is like this? What could I copy?

- **M**inify. Odor, form, shape? What to add? More time?

- **M**agnify. Greater frequency? Higher? Longer? Thicker?

- **P**ut to Other Uses. New ways to use as is?

- **E**liminate. What to subtract? Smaller? Condensed? Shorter? Lighter? Streamline? Understate?

- **R**everse. Interchange components? Other pattern?

- **R**earrange. Other layout? Other sequence? Turn it backward? Turn it upside down?

Doing easily what others find difficult is talent; doing what is impossible for talent is genius.

—AMIEL

Many universities have formal programs specifically designed to assist inventors with ideas or patents. Nearly every state has at least one college or university that can provide research and development facilities to technically oriented companies or individuals.

State governments are heavily involved with university research through cooperative programs and support of research parks and technology transfer programs. Known at some universities as "Advanced Technology Centers" or "Centers of Excellence," these programs are designed to increase cooperation between academic institutions and state-based industries. They assist in the creation of new firms through the development of new technology, attracting new business to the state, and increasing competition.

Specifically, technology transfer programs facilitate the transmission of new technologies from the laboratory to the private sector. Such technologies can become the impetus for the creation of new businesses, new products, or revitalization of industries.

Research parks are planned groupings of technology companies, often near universities, that encourage university/private partnerships. They draw industry to a particular location and provide incubator services.

Other kinds of services that are available at many university research centers include innovation evaluation; invention testing; counseling on the start-up of new high-technology businesses; assistance in identifying potential research funding from the federal government under programs such as the Small Business Innovation Research (SBIR) Grants and the Energy-Related Inventions Program (ERIP); counseling regarding patents, trademarks, and copyrights; and assistance in identifying sources of government contracts for goods or services needed. Many centers also have excellent libraries of technical information and computer literature searching capabilities.

Whenever I require specialized engineering skills on a project, especially to build complicated prototypes, one of the first places I check is the talent pool available at the local university research center. The costs are reasonable and there is no better place to get fresh, young minds working on a resolution to a particular problem.

Universities do have specialties, so it is best to check in advance before going out to the campus. For example, some universities concentrate on biomedical and life sciences research, others handle aerospace engineering, electronic systems, mechanics, physical sciences, fabrication technologies, and so forth.

Idea evaluation is the first major step after a concrete, detailed idea has been developed. This is a critical phase since every subsequent phase requires the investment of more time and money. University research centers are an excellent place to have an invention evaluated to determine its overall technical and commercial feasibility.

The university research centers listed on the following pages may be useful in these evaluation activities. They have people to help determine whether the new invention is a marked improvement over its competition; whether it is likely to be commercially viable; what the probable demand for it will be; who will produce it; and how it will be distributed. These specialists can help an inventor arrive at the decision to go ahead to the commercialization stage, to redesign the invention, or to kill the project altogether.

The failure rate of technological innovators is estimated by one Michigan study to be one quarter that of all entrepreneurs during the first five years of business. One way to decrease the chances of failure is to ensure that all the resources are available when required. A good university research center program can help toward this end.

Listed on the following pages are principal university research institutes, technology transfer centers, and research parks whose purpose is to promote innovation, invention, and product development.

Advanced Science and Technology Institute

1239 University of Oregon   (503) 346-3189
Eugene, OR 97403-1239
FAX: (503) 346-1352
Dr. Robert S. McQuate, Executive Director

Joint technology transfer and economic development organization of Oregon State University, University of Oregon, Portland State University, and Oregon Health Sciences University. Bridges university research activities and resources with the corporate community. Encourages cooperative research projects with industry through its Industrial Associates Program. Serves as a broker for patenting and licensing arrangements between industry and parent universities. Sponsors technical seminars, workshops, and conferences. Holds an Annual Executives Conference. **Publications:** Newsletter.

Argonne National Laboratory, Industrial Technology Development Center

9700 S. Cass Ave.   800-627-2596
Argonne, IL 60439
FAX: (708) 252-5230
Paul Eichamer, Director

Research support organization at Argonne National Laboratory, a U.S. Department of Energy unit operated by University of Chicago. Facilitates the exchange of Argonne research resources and inventions with U.S. industry and develops partnerships with industry. **Publications:** Tech Transfer Highlights (newsletter).

Arizona State University Research Park

2049 E. ASU Circle   (602) 752-1000
Tempe, AZ 85284
FAX: (602) 491-2273
Dawn Dunn, Project Director

Independent inventor Shawn Tyler Brown turned to the University of Portland in Oregon when he needed to test his invention, Magna Ears, a strap-on hearing enhancer. "Through the university's testing, we discovered that Magna Ears really worked," he says. How much did this exercise cost Brown? "Nothing. The university did it free of charge because they were interested in the product and what it could do for their patients," Brown reports. Ultimately, through the university, the state of Oregon became involved and sponsored Dr. David Lilly, who is associated with NASA, to test the product.

Seeks to link technological results of university and individual research with private industry. The Park provides tenants with leased land for construction of research and development facilities, laboratories, offices, pilot plants, facilities for production or assembly of prototype products, university and government research facilities, and corporate regional or national headquarter facilities related to these functions.

Auburn University Center for International Commerce

415 W. Magnolia Ave., Ste. 109   (334) 844-2352
Auburn, AL 36849-5248
FAX: (334) 844-4092
Dr. William Boulton, Director

Integral unit of the College of Business at Auburn University. Investigates technology transfer activities in U.S., Japan, and European countries. Conducts Japanese management seminars.

Auburn University Economic Development Institute

3354 Haley Center   (334) 844-4704
Auburn, AL 36849-5252
FAX: (334) 844-4709
Robert S. Montoy, Ph.D., Director

Research support activity of the Office of Provost at Auburn University, operating under a 23-member executive advisory council. Facilitates technology transfer activities through interdisciplinary technology assistance teams and projects. Conducts industrial research and development activities. Serves as a resource for generating projects that link education and economic development by supporting faculty and graduate research opportunities. Assists in implementation of community economic development plans, and in contract and grant development. Also provides assistance to faculty, graduate students, professionals, business and industry, and local and state governments in proposal and project development. Holds advisory council board meeting (annually). Provides technical assistance to entrepreneurs. Encourages collaboration among economic development networks. Maintains an office at 401 Adams, Suite 170, Montgomery, AL 36130, phone (205) 242-2688, fax (205) 242-2324, and at Calhoun State Community College, 6250 U.S. Highway 31 North, Decatur, AL, 35609. Publications: Annual Report.

California University of Pennsylvania Mon Valley Renaissance

Box 62   (412) 938-5938
California, PA 15419
FAX: (412) 938-4575
Richard H. Webb, Executive Director

Research support activity at California University of Pennsylvania. Multiprogram consortium with a focus on applied research and economic development services to business and industry. Sponsors seminars to assist entrepreneurs and provides training workshops. Offers assistance in government procurement, robotics, international procurement technology, technical training, numerical control programming, and graphic communications.

Center for Education and Research with Industry
Marshall University

1050 4th Ave.   (304) 696-3093
Huntington, WV 25755
FAX: (304) 696-6280
Larry D. Kyle, Director

Statewide network of 26 public and private colleges and universities headquartered at Marshall University. Facilitates joint ventures between academia, business, and government by linking campus resources and faculty expertise with technology transfer activities around the state. Administers joint research programs between academia and industry. Sponsors research conferences and quarterly meetings. Offers training, educational, and consulting services. Provides employee and management training and development programs. Holds Quality Days (in October). **Databases:** Faculty inventory at participating institutions. **Publications:** Annual Report.

Center for Innovation & Business Development

Box 8372, UND Sta.   (701) 777-3132
Grand Forks, ND 58202
FAX: (701) 777-2339
Bruce Gjovig, Director

Separately incorporated organization affiliated with the University of North Dakota. Works with applied research personnel at the university and other institutions to facilitate the commercialization of new technologies by providing support services to entrepreneurs, inventors, and small manufacturers in the areas of invention evaluation, technology commercialization, SBIR applications, technical development, licensing, and business development, including market feasibility studies and marketing and business plans. Provides specialized services in rural manufacturing ventures and commercial evaluations for DOE Energy-Related Inventions Program (ERIP). **Publications:** Annual Report; Entrepreneur Kit; Business Plan for Start-ups; Marketing Plan: Step by Step.

Center for Research and Technology Development

1828 L St. NW, Ste. 906   (202) 785-3756
Washington, DC 20036-5104
FAX: (202) 429-9417
Dr. Howard E. Clark, Director

Research arm of the American Society of Mechanical Engineers. Organizes, plans, and manages cooperative, interdisciplinary activities in research and/or technology development; reviews and summarizes the state of knowledge in a specific technical area; identifies and encourages the development of emerging technologies; collects, evaluates, and disseminates research and development results; provides peer reviews of industrial or governmental research and development work. Offers a patent award. Publications: Newsletter (quarterly).

College of DuPage
Technology Development Center

Business & Professional Institute   (708) 858-2800
22nd & Lambert Rd.
Glen Ellyn, IL 60137-6599
FAX: (708) 790-1197
John J. Sygielski, Director

Technology transfer unit of Economic Development Center established by Illinois Department of Commerce and Community Affairs ITEC Network at College of DuPage. Links high-technology businesses to university and other resources to assist in the production and commercialization of new ideas and products and to enhance the transfer of technologies from university laboratories into the marketplace. Sponsors training seminars, workshops, and conferences. Offers monthly economic development breakfasts to the business community. **Publications:** Cutting Edge (monthly newsletter).

Columbia University
Columbia Innovation Enterprise

Engineering Terrace, Ste. 363   (212) 854-8444
116th St. & Broadway
New York, NY 10027
FAX: (212) 854-8463
Jack M. Granowitz, Director

Research support activity at Columbia University. Transfers technology from the university to business and industry, especially in the area of biotechnology.

Connecticut Innovations Incorporated

Technology Assistance Center   (203) 258-4305
845 Brook St.
Rocky Hill, CT 06067-3405
FAX: (203) 563-4877
Eric C. Ott, Director

Stimulates technological innovation and economic growth in Connecticut in the areas of advanced marine applications, aerospace, applied optics and micro-electronics, biotechnology, computer applications, energy and environmental systems, materials technology, medical technology, and telecommunications. Provides risk capital for new product development and loans to launch and market new products. Finances high-technology companies with early stage capital and small technology-based firms doing federal research and development under the Small Business Innovation Research (SBIR) grant program. Sponsors seminars and conferences annually. Sponsors an annual Connecticut Technology Dinner. Serves as clearinghouse for information on the technical, financial, research, and educational services, programs, and resources available in Connecticut. **Publications:** Connecticut Technology Resource Guide.

Cornell University
Cornell Research Foundation, Inc.

Cornell Business & Technology Park   (607) 257-1081
20 Thornwood Dr., Ste. 105
Ithaca, NY 14850
FAX: (607) 257-1015
H. Walter Haeussler, President

Wholly owned nonprofit subsidiary of Cornell University. Patent and technology marketing arm of the University. Activities include technology transfer, evaluating legal and business contracts, and assisting with patenting and licensing. Holds title to patents and promotes technology transfer to industry. **Publications:** Selected Technology Available for Licensing (semiannually).

Dandini Research Park

DRI Research Parks, Ltd.   (702) 673-7315
PO Box 60220
Reno, NV 89506
FAX: (702) 673-7421

470-acre site that links the research and development activities of Park tenants with the Desert Research Institute's technological equipment, personnel, laboratories, and training programs. Instrument design, environmental testing, and research and development services may be done in cooperation with DRI staff.

Fox Valley Technical College
Technical Research
Innovation Park

1825 N. Bluemound Dr.   (414) 735-5600
PO Box 2277
Appleton, WI 54913-2277
FAX: (414) 735-2582
R.F. Martin, Sr., Vice President

Seeks to attract and encourage liaisions of highly technical businesses and industries with the College. Encourages marketing ventures, supports inventors and entrepreneurs, fosters research and development for the paper and printing industries, and promotes new technology industries as tenants in the Park. Operates the D.J. Bordini Technological Innovation Center of the College that includes a technical library, quality and productivity improvement center, communications network area, product development service center, flexography laboratory, high technology demonstration laboratories, and facilities for conferences and classes.

Georgia Institute of Technology
Advanced Technology
Development Center

430 Tenth St. NW, Ste. N-116   (404) 894-3575
Atlanta, GA 30318
FAX: (404) 894-4545
Dwight Holter, Associate Director

Promotes the development of advanced technology-based companies throughout Georgia, including firms involved in advanced structural materials, electronic equipment, biotechnology, health and medical products, artificial intelligence, environmental sciences, telecommunications, aerospace systems, instrumentation and test equipment, robotics, and related technologies. Operates the Technology Business Center, providing an entrepreneurial working environment, including modern office and laboratory facilities. Provides business assistance to start-up technology companies and supports technology commercialization ventures and economic development efforts in key technology areas within the state; offers technical and business management services to entrepreneurs; provides general management consulting, corporate communication assistantce, mar-

ket research, staffing, guidance, and shared business services. **Publications:** Technology Partners (quarterly newsletter).

Harvard University
Office for Technology and
Trademark Licensing

Univ. Pl., Ste. 410   (617) 495-3067
124 Mt. Auburn St.
Cambridge, MA 02138
FAX: (617) 495-9568
Joyce Brinton, Director

Integral unit of Harvard University. Formerly known as Office for Patents, Copyrights, and Licensing. Facilitates university-industry relations through technology licensing in the areas of applied sciences, recombinant DNA, hybridoma technology, software and courseware, chemistry, therapeutics, vaccines, bioprocesses, and medical, veterinary, and agricultural diagnostics.

Idaho State University
Research Park

Box 8020   (208) 236-4004
Pocatello, ID 83209-8020
FAX: (208) 236-4367
Dr. William G. Phillips

Facilitates the interaction of University research community with Park tenants and other professional organizations, particularly in the areas of health professions, biological and physical sciences, pharmacy, nuclear engineering, hazardous waste management, and state of the art decision support systems. Collaborates on technology transfer programs with the Idaho National Engineering Laboratory. Provides assistance to early state companies. Facilitates economic development and community planning initiatives. **Publications:** An Environment for Innovation (brochure).

Illinois Institute of Technology
Center for Managing Technology

10 W. 35th St.   (312) 567-4800
Chicago, IL 60616
Bob Zacony, Director

Examines the interaction between the research, development, and engineering sections of companies with other company sections such as manufacturing, finance, and marketing. Offers consulting services. Works with individual corporations or groups of businesses to facilitate the transfer of information.

Industrial Research Institute, Inc.

1550 M St. NW   (202) 872-6350
Washington, DC 20005
FAX: (202) 872-6356
Charles F. Larson, Executive Director

Independent, nonprofit research organization, formed as a consortium of approximately 270 industrial companies with technical research departments; governed by a board of directors comprised of representatives of member corporations. Promotes, through cooperative efforts of its members, economic and effective techniques of organization, administration, and operation of industrial research. Seeks to generate industrial and academic research collaboration as well as industry/government cooperation in matters related to research. Also endeavors to stimulate and develop an understanding of research as a force in economic, industrial, and social activities and encourages high standards in the field of industrial research. **Publications:** Research Technology Management (bimonthly).

Iowa State University of Science and Technology Center for Industrial Research and Service

2501 N. Loop Dr., No. 500   (515) 294-3420
Ames, IA 50010
FAX: (515) 294-4925
Jerry Rounds, Director

Integral unit of University Extension at Iowa State University of Science and Technology. Deals in problem areas of business, manufacturing, technology transfer, productivity, new product design, manufacturing processes, marketing, and related topics. Acts as a problem-handling facility and a clearinghouse for efforts to help Iowa's industry grow through studies highlighting not only production and management problems but also markets, marketing, and profit potential of possible new developments. Cosponsors an annual manufacturing conference and workshops and briefings as needed. Also cosponsors specialized extension courses for various phases of industry in the state. Consults with industry on special problems, offering recommendations and guiding research and testing programs. Maintains a six-person field staff and provides an information retrieval service available to industry and professional groups within the state. Administers a statewide University Center Program for Iowa businesses.

Lehigh University Office of Research and Sponsored Programs

526 Brodhead Ave.   (610) 758-3020
Bethlehem, PA 18015-3046
FAX: (610) 758-5994
Thomas J. Neischeid, Director

Integral unit of Lehigh University. Administers and coordinates, as administrative agency of the University, sponsored and cooperative research supported by government agencies, industry, and technical associations, including studies in physical, natural, social, and engineering sciences, and the humanities. Assists faculty and students in unsponsored research and scholarly efforts. Provides consulting and sponsors industrial liaison programs. **Publications:** Research Notes.

Louisiana State University Office of Technology Transfer

S. Stadium Dr., Rm. 146A   (504) 388-6941
Baton Rouge, LA 70803-6100
FAX: (504) 388-4925
Mani Iyer, Director

Identifies, protects, and transfers technology that originates from University research activities. Activities include patent work, finding state or national licensees, and starting new entrepreneurial projects. Projects include work in such areas as agriculture, analytical, biotechnology, chemistry, electronics, environmental sciences, food science, health care, medical devices, petroleum engineering, sports and leisure studies, textiles, and veterinary medicine. Provides consulting services to independent inventors. **Publications:** Newsletter (occasionally).

Massachusetts Institute of Technology Technology Licensing Office

Rm. E32-300   (617) 253-6966
28 Carleton St.
Cambridge, MA 02142
FAX: (617) 258-6790
Lita Nelsen, Director

Research support activity of Massachusetts Institute of Technology. Commercializes technology from the Institute in the areas of biotechnology, biomedicine, ceramics, chemistry, computers, electro-optics, integrated circuits, and polymers. Markets inventions and software developed at Lincoln Laboratory.

Miami Valley Research Foundation

3155 Research Blvd.   (513) 252-5906
Dayton, OH 45420
FAX: (513) 252-9314
Bruce Pearson, Executive Director

Independent, nonprofit corporation, comprised of University of Dayton, Wright State University, Central State University, Sinclair Community College, and Air Force Institute of Technology. Formerly known as Miami Valley Research Institute.

Facilitates the transfer of basic and applied scientific technological research from Miami Valley Research Park tenants and member institutions to production, manufacture, and marketing of materials and services. Solicits public and private grants for research personnel and facilities of member institutions and recruits tenants for the Park.

Minnesota Technology Corridor

1200 Washington Ave. S   (612) 370-0111
Minneapolis, MN 55415
FAX: (612) 332-6306
Charles Neerland, Executive Director

128-acre site established to foster technology transfer, research and development, and prototype manufacturing activities between the university and related businesses.

Mississippi Research and Technology Park

PO Box 2720   (601) 324-7776
Starkville, MS 39759
FAX: (601) 323-3726
John Rucker, Executive Director

> Idea evaluation is the first major step after a concrete, detailed idea has been developed. This is a critical phase since every following phase requires the investment of more time and money. University research centers are excellent places to have an invention evaluated to determine its overall technical and commercial feasibility.

Operates research laboratories, incubator and multitenant facilities, professional and business services, and office space to foster high-technology research collaboration between the University and industry tenants. Provides meeting rooms for seminars and conferences.

Montana State University— Bozeman
Advanced Technology Park

1711 W. College   (406) 587-4480
Bozeman, MT 59715
FAX: (406) 587-4480
Roger N. Flair, President

Provides access for park tenants to the research community and facilities at the university. **Publications:** Montana State University Resources Catalog.

NKU Foundation
Research/Technology Park

Highland Heights, KY 41099-8005   (606) 572-5126
FAX: (606) 572-5566

Facilitates the exchange of research resources between the University and park tenants.

Northern Illinois University
Technology Commercialization
Office

DeKalb, IL 60115-2874   (815) 753-1238
FAX: (815) 753-1631
Dr. Larry Sill, Director

Research support activity at Northern Illinois University. Assists inventors, entrepreneurs, and small- and medium-sized businesses with product research and development, construction of prototypes, technical and commercial assessments, patent applications, licensing, and locating funding sources. Sponsors seminars on commercialization process, intellectual property matters, and federal Small Business Innovation Research funding.

Northwestern University
Evanston Business and
Technology Center

1840 Oak Ave.   (708) 864-0800
Evanston, IL 60201
FAX: (708) 866-1808

Integral unit of Northwestern University/Evanston Research Park. Formerly known as Technology Innovation Center. Matches university resources with the needs of businesses, including assisting faculty entrepreneurs to commercialize their new technologies. The center maintains three administrative offices: the incubator component leases space to technology-based small companies; the Small Business Development Center provides technical and business assistance companies in the Research Park and northern Cook County, manages the Evanston Business Investment Corporation (a seed capital fund), and links companies to campus expertise; and the Minority Business Development Office administers special training projects. **Publications:** Park Facts (newsletter).

Ohio University
Innovation Center and Technology
Transfer Office

20 E. Circle Drive—Ridges   (614) 593-1818
Athens, OH 45701-2979
FAX: (614) 593-0186
David N. Allen, Director

A 47,000-foot business incubator at Ohio University. Created to foster entrepreneurial activities and to provide technical and business assistance to new and expanding companies. Facilitates consulting, product testing, and technical assistance between tenants, the university, and the community at large. Undertakes analysis to determine patentability of invention disclosures and manages university patent portfolio. Conducts commercial negotiations with licensees. Offers informative business-related workshops for tenants and the general public. Provides business and marketing consultation, in-house bookkeeping and clerical services.

Oregon State University
Office of Vice Provost for Research
and International Programs

Corvallis, OR 97331-2140   (503) 737-3467
FAX: (503) 737-3093
Dr. George H. Keller, Vice Provost

Integral unit of Oregon State University. Formerly known as Office of Vice Provost for Research, and International Programs. Coordinates university research activities, including individual projects in various academic schools and special research centers and institutes. Also administers University's Technology Transfer Program and international education and technical assistance programs. Maintains a database on faculty expertise. Research results published in professional journals and technical reports. Sponsors Biology Colloquium, Ava Pauling Lecture, and Condon Lectures.

Pennsylvania State University Research and Technology Transfer

207 Old Main (814) 865-6331
University Park, PA 16802
FAX: (814) 863-9659
Dr. K. Jack Yost, Associate Vice President

Research support activity at Pennsylvania State University. Transfers technology from the university to business and industry in the areas of electronics, materials, structural ceramics, polymers, and manufacturing.

Progress Center: University of Florida Research and Technology Park

1 Progress Blvd. (904) 462-4040
Box 10
Alachua, FL 32615
FAX: (904) 462-3932
Sandra Burgess, Park Manager

200-acre research and technology park open to both public and private research and manufacturing organizations emphasizing high-technology development, including electronics, biotechnology, advanced materials, pharmacology, and agriculture. Center provides a link between university researchers and industry and transfers new technologies from the laboratory to the marketplace. Offers assistance in entrepreneurial development, commercialization of scientific and technological innovations, and international marketing.

Riverfront Research Park

University Planning Office (503) 346-5566
University of Oregon
Eugene, OR 97403-1276
FAX: (503) 346-2299
Diane K. Wiley, Manager

67-acre site adjacent to Willamette River and the university. Provides opportunities for research interaction between the university and Park tenants engaged in such activities as industrial research and development, biotechnology, materials science, environmental technology, advanced computer and software development, and business, educational, and governmental research and consulting services. Seeks to assist in the diversification of the economic base of the state, particularly the Eugene-Springfield metropolitan area, and provide additional financial opportunities for the university and its community.

Rutgers University Technology Management Research Center

Graduate School of Management (201) 648-5982
92 New St.
Newark, NJ 07102
FAX: (201) 648-1664
Prof. George F. Farris, Director

Integral unit of Graduate School of Management at Rutgers University. Applied research in technology management, especially technology development, implementation and adoption of new technology, trends in technology, and strategic management of technology. Supports ongoing dialogue between industry, government, and academia through applications-oriented seminars, broad-based conferences, and intensive short courses. Offers an M.B.A. program concentration in technology management and a Ph.D. in management. **Publications:** Working Paper Series.

Southwest Center for Manufacturing Technology

University of New Mexico (505) 277-5538
Dept. of Electrical Computer Eng.
Albuquerque, NM 87131
FAX: (505) 277-1439
Prof. Mo Jamshidi, Staff Director

Technology transfer of advanced manufacturing technology to small and medium-sized industries.

Stanford University Office of Technology Licensing

900 Welch Rd., Ste. 350 (415) 723-0651
Palo Alto, CA 94304
FAX: (415) 725-7295
Katharine Ku, Director

Intellectual property activity of Stanford University. Licenses technology in the fields of scientific and medical instruments, pharmaceuticals, chemicals, computer software and databases, integrated circuit technology, optics, and microbiology. Evaluates, markets, and negotiates licensing agreements with industry.

State University of New York at Stony Brook Center for Industrial Cooperation

Engineering Rm. 127 (516) 632-8518
Stony Brook, NY 11794-2200
FAX: (516) 632-8205
Yacor Shamash, Director

Integral unit of State University of New York at Stony Brook. Assists in establishing applied research on topics of interest to local industry by keeping business aware of ongoing research at the university and providing consulting arrangements with faculty through the Stony Brook Research Foundation. Also assists in providing access to technical information sources. Research results published in brochures, periodic newsletters, and summaries of research. Center active in undergraduate course work in engineering and managerial economics. Maintains a collection on engineering, science, and technology; Godlind Johnson, librarian.

State University of New York, Health Science Center at Brooklyn Office of Research Administration

450 Clarkson Ave. (718) 270-1178

Box 69
Brooklyn, NY 11203
FAX: (718) 270-1407
Antony Selvadurai, Director

Integral unit of State University of New York Health Science Center at Brooklyn. Administers programs, extramural funding, and contracts for research in medical and basic sciences. Also transfers technology from the University to business and industry.

Tampa Bay Area Research & Development Authority

University Technology Center   (813) 974-2890
3702 Spectrum Blvd., Ste. 155
Tampa, FL 33612-9421
FAX: (813) 974-2216
Han Cusick, Executive Director

University-affiliated research and development authority leasing land to industries needing facilities for research and related scientific manufacturing in medicine, engineering, and natural sciences. Tenants share university research facilities and services, use faculty as consultants, and utilize graduate students as part-time work force. Oversees development of university-related research parks in the Tampa Bay area, particularly University Technology Center.

Technology Enterprise Division New Mexico Economic Development Department

1100 St. Francis Dr.   (505) 827-0265
Santa Fe, NM 87503
FAX: (505) 827-0588
Ronald Tafoya, Director

Designated to promote and coordinate the transfer of defense technology between federal, state, and local government facilities and the private sector. Links university research resources with government laboratories and private research facilities, particularly in the areas of non-invasive diagnosis, materials, explosives technology, plant genetic engineering, computer research applications, and commercial product development. Emphasis is on technology transfer, defense conversion, and manufacturing extension. **Publications:** New Mexico Technology Enterprise Forum (periodic newsletter); New Mexico Statewide Directory of Technology-Related Organizations (annual).

Texas A&M University Research Park

1500 Research Pkwy.   (409) 845-7275
Ste. 200
College Station, TX 77845
FAX: (409) 845-9262
John Millhollon, Director

Located on 434 acres west of main campus, the Park serves to assist private utilization of university resources, to promote closer ties between industry engaged in research and the university, to improve the quality and productivity of university research activities, and to accelerate the dissemination of new knowledge and the transfer of new technologies. **Publications:** Parkview Newsletter.

Texas Research Park

Texas Research & Technology Foundation   (210) 677-6000
San Antonio, TX 78245
York Duncan, Park Manager

Supports basic and applied science and new advanced technology enterprises in the San Antonio area by providing technology assesment and evaluation services, company formation services, business and lab incubator facilities, and venture capital. Provides industry with access to the resources of affiliated organizations. Conducts inventory and analyses of scientific and technical resources in San Antonio.

University of Alabama in Huntsville Center for the Management of Science and Technology

Administrative Science Bldg.   (205) 895-6408
Ste. 126
Huntsville, AL 35899
FAX: (205) 895-6006
Dr. William E. Souder, Director

Formerly known as Technology Management Studies Institute, University of Pittsburgh (1990). Management of high technology projects and new product innovation processes, including research and development management, organization design, group dynamics, and strategic management. **Publications:** Annual Report.

University of Arkansas Arkansas Center for Technology Transfer

131 Engineering Research Center   (501) 575-3747
Fayetteville, AR 72701
FAX: (501) 575-6615
Marcus C. Langston, Director

Provides research and technical facilities to analyze factors such as production, workforce motivation and training, and technical problems. Facilitates research and technology transfer in the areas of manufacturing technology, interactive technology, productivity, and industrial efficiency. Specific projects include bar-code reading of trucks at weigh stations, scales to weigh-in-motion, tactile sensing, and advanced brakes, clutches, and transmissions. Provides entrepreneurs and small businesses statewide with technical assistance and technical seminars on topics such as automation, productivity, quality, and computer-aided design and computer-aided manufacturing.

University of California, Irvine Office of University/Industry Research and Technology

University Tower, Ste. 380   (714) 856-7295
4199 Campus Dr.
Irvine, CA 92717-7700
FAX: (714) 725-2899
David G. Schetter, Director

University and industry partnership and cooperative research development, including industrial contract development, review, and approval; science and technology research identification; consortia participation; new consortia initiation; and federal and state science and technology center participation. Provides technology transfer services on licensed and emerging technologies and other materials.

University of Colorado
Office of Intellectual Resources and Technology Transfer

Campus Box 51   (303) 492-4975
Boulder, CO 80309
FAX: (303) 492-5810
Michael G. Gabridge, Ph.D., Director

Research support organization of University of Colorado. Formerly known as University of Colorado Foundation Inc. Facilitates research and technology transfer activity at the university in the areas of medicine, optoelectronics, molecular biology, pharmacy, and chemistry. **Publications:** Patent Administration and Technology Transfer at the University of Colorado.

University of Florida Research Foundation, Inc.
Office of Research and Graduate Education

223 Grinter Hall   (904) 392-1582
Gainesville, FL 32611
FAX: (904) 392-9605
Karen A. Holbrook, Ph.D., President

Independent, nonprofit entity affiliated with University of Florida. Development and negotiation of licensing agreements. Facilitates the transfer of University inventions to the marketplace. Administers contracts and grants from private industry, foundations, and other nongovernmental agencies. **Publications:** Annual Report, Brochure.

University of Georgia Research Foundation, Inc.

Boyd Graduate Studies Research Center   (706) 542-5969
Ste. 609
Athens, GA 30602-7411
FAX: (706) 542-5978
Dr. Joe L. Key, Executive Vice President

Separately incorporated research foundation affiliated with the University of Georgia. Primarily life sciences, agriculture, veterinary medicine, pharmacy, chemistry, physics, and forestry resources. Subcontracts research projects to the university, handles all university technology transfer, and intellectual property issues. Research results published in applicable scientific journals and books.

University of Illinois at Champaign-Urbana
Office of the Dean of the Graduate School and Vice Chancellor for Research

417 Swanlund Bldg.   (217) 333-7862
601 E. John St.

Champaign, IL 61820
FAX: (217) 244-3716
Dillon E. Mapother, Associate Vice Chancellor

Integral unit of the University of Illinois at Urbana-Champaign. Transfers technology from the university to business and industry, particularly in the areas of engineering and computing. Activities include negotiating research contracts and technical testing agreements, licensing university-owned software and patents, and consulting on university policies on classified research. Responds to occasional external inquiries about intellectual property.

University of Illinois at Chicago
Software Technologies Research Center

851 S. Morgan St.   (312) 996-3002
Rm. 1120
Chicago, IL 60607-7053
FAX: (312) 413-7585
Thomas DeFanti, Director

Integral unit of the Office of the Vice Chancellor for Research at the University of Illinois at Chicago. Promotes technology transfer by applying scientific visualization and virtual reality to manufacturing, engineering, education and training, health care, and the environment. Arranges training assistance for qualified clients' requirements for testing, prototyping, demonstration, market surveys, and business planning. Research results published in graduate theses, feature articles, and reviewed articles.

University of Iowa
University of Iowa Research Foundation

109 TIC   (319) 335-4063
Oakdale Research Campus
Iowa City, IA 52242
FAX: (319) 335-4489
W. Bruce Wheaton, Ph.D.

Integral unit of University of Iowa. Serves as the university's agent in obtaining legal protection for property created at the university; negotiates licensing agreements with companies prepared to carry the results of academic research into the marketplace; licenses technology developed under grants from federal and state agencies; and seeks mutually acceptable licensing agreements with businesses that fund contractual research.

University of Manitoba
Institute for Technological Development

227 Engineering   (204) 474-6200
Winnipeg, Manitoba
R3T 5V6
Canada
FAX: (204) 261-3475
Ray Hoemsen, Director

Integral unit of University of Manitoba. Serves as a liaison between industry and the University by coordinating cooperative research and development activities, facilitating technology transfer, and encouraging communication between industry, university, and government organizations. Maintains the Academic Technology Inventory, a database which profiles science and technology researchers and facilities at the University of Manitoba, University of Winnipeg, Brandon University, and Red River Community College. Sponsors special-interest workshops, open to all, and participates in a technological entrepreneurship course. Bestows EASIAP Design Award. Assists industry by providing scientific and technical information through the NRC's Industrial Research Assistance Program; patent information and technology searching and forecasting services; and an investors assistance program. Services to small businesses include identifying research and development resources, sources of funding assistance, and recommending students for temporary and permanent employment opportunities. **Publications:** Connections (quarterly newsletter).

University of Maryland
Engineering Research Center

College of Engineering   (301) 405-3906
College Park, MD 20742
FAX: (301) 403-4105
Dr. David Barbe, Executive Director

Integral unit of College of Engineering at University of Maryland, affiliated with the Departments of Aerospace Engineering, Mechanical Engineering, Electrical Engineering, Civil Engineering, Chemical Engineering, Materials, and Nuclear Engineering. Activities include the Technology Extension Service (TES), which offers technical assistance to the Maryland business community; the Technology Advancement Program (TAP), an incubator for start-up companies engaging in the development of technically oriented products and services; the Technology Initiatives Program (TIP) to support technological capabilities within the university; and the Maryland Industrial Partnerships (MIPS) program, a matching fund for industry sponsored research. **Publications:** ERC Update (quarterly newsletter).

University of Maryland
Science and Technology Center

7505 Greenway Center Dr., 001   (301) 982-9400
Greenbelt, MD 20770
FAX: (301) 982-7963

466-acre research and development park focusing in the areas of math, physics, computer sciences, electrical engineering, and mechanical engineering. Researchers are provided access on a contract basis to University of Maryland faculty and equipment. The Supercomputing Research Center is housed at the Center.

University of Miami
Innovation and Entrepreneurship
Institute

PO Box 249117   (305) 284-4692
Coral Gables, FL 33124
Carl McKeney, Executive Director

Entrepreneurship and innovation in Florida, including studies on high-technology ventures and black and Cuban-American entrepreneurs. Promotes interaction of entrepreneurs and capital and ser-

vice providers. Sponsors the University of Miami Venture Council Forum, a series of programs to assist small entrepreneurial companies that include company case studies, seminars, workshops, and lectures. Administers the Miami Accelerated Technology Transfer (MATT) Program, which facilitates the commercialization of newly developed technology at the University. **Publications:** Research Report Series; Friends of the Forum Directory.

University of Minnesota
Office of Research and Technology
Transfer Administration

1100 Washington Ave. SE   (612) 624-1648
Ste. 201
Minneapolis, MN 55415-1226
FAX: (612) 624-4843
A.R. Potami, Associate Vice President

Integral unit of University of Minnesota. Serves as the research support unit for University of Minnesota faculty members by administering non-programmatic aspects of all research, training, and public service projects funded by external sources. Reviews and processes all proposals and awards for research, training, and public service projects and is responsible for financial management, cash receipt, and financial reporting of project funds. Transfers proprietary technologies to companies through exclusive or non-exclusive license agreements. Assists faculty in locating potential sources of support for research, training, and public service programs. Works with faculty to stimulate the disclosure of patentable discoveries resulting from university research. **Publications:** R&D Outreach Newsletter.

University of Missouri
Missouri Research Park

300 Chesterfield Ctr., No. 220   (314) 537-2030
Chesterfield, MO 63017
FAX: (314) 537-0599
Rick Finholt, Executive Director

A 750-acre research park designed to be a link between academia and industry as a center for research and development in such fields as advanced manufacturing, medical technology, and agriculture.

University of Missouri—Rolla
Center for Technology Transfer
and Economic Development

1 Nagogami Terrace   (314) 341-4555
Rolla, MO 65401
FAX: (314) 364-5117
H. Dean Keith, Director

Technological innovation, economic development, and manufacturing.

University of Oklahoma
University Research Park

1700 Lexington Ave.   (405) 325-7233
Norman, OK 73069

FAX: (405) 325-7339

George Hargett, Administrator

Provides on a 1,500-acre tract facilities and services for research administered by the University's Office of Research Administration, as well as U.S. and state government, industry, and University of Oklahoma laboratories located in the Park.

University of Pennsylvania
Center for Technology Transfer

3700 Market St., Ste 300　　　　　　　(215) 898-9585

Philadelphia, PA 19104-3147

FAX: (215) 898-9519

Dr. Solash, Director

Research support activity at University of Pennsylvania. Formerly known as Office of Corporate Programs and Technology. Transfers technology from the university to business and industry in the fields of biotechnology, chemistry and chemical engineering, dental medicine, diagnostics, electrical engineering, laboratory devices and reagents, medical devices, mechanical engineering, materials science, pharmaceuticals, robotics factory automation, software, veterinary medicine, physics and optics.

University of Pennsylvania
Office of Sponsored Programs

Mellon Bldg., Ste. 300　　　　　　　(215) 898-7293

133 S. 36th St.

Philadelphia, PA 19104-3246

FAX: (215) 898-9708

Anthony Merritt, Executive Director

Integral unit of University of Pennsylvania. Formerly known as Office of Project Research and Grants; Office of Research Administration (1986). Administers extramurally sponsored research for all departments and research units of the university and handles processing of research applications. Responsible for licensing of inventions and other technology transfer activities.

University of Southern California
Office of Patent & Copyright
##　　Administration

3716 S. Hope St.　　　　　　　(213) 743-4926

Ste. 313

Los Angeles, CA 90007

FAX: (213) 744-1832

Rosanne Dutton, Director

Research activity at University of Southern California. Transfers technology from the university to the private sector, including the areas of medicine, engineering, pharmacy, and gerontology. Sponsors seminars for faculty and students. Publications: Newsletter.

University of Texas
Center for Technology
##　　Development and Transfer

College of Engineering　　　　　　　(512) 471-3696

Ernest Cockrell Jr., Hall 10.340

Austin, TX 78712-1080

FAX: (512) 471-3955

Dr. Dale E. Klein, Director

Technology transfer unit established by the state legislature within the University of Texas System, under the direction of the Bureau of Engineering Research; the College of Engineering; University of Texas at Austin. Links university researchers with industrial needs and facilitates the commercialization of academic research, university-generated technology, scientific information, and other intellectual property. Activities range from serving as a clearinghouse for technology to product development and the formation of businesses. Also serves as the hub for a statewide technology development and transfer network.

University of Utah
Research Park

505 Wakara Way　　　　　　　(801) 581-8133

Salt Lake City, UT 84108

FAX: (801) 581-7195

Charles A. Evans, Director

300-acre tract designed to facilitate scientific research projects between government agencies, industrial organizations, and the university. Facilities include general and technical libraries, scientific and research equipment, and computer center.

University of Washington
Office of Technology Transfer

1107 N.E. 45th St., Ste. 200　　　　　　　(206) 543-3970

Seattle, WA 98105

FAX: (206) 685-4767

Margret Wagner Dahl, Acting Director

Research support activity at the University of Washington. Transfers technology from the university to business and industry, particularly in the areas of bioengineering, biotechnology, engineering, medicine, and software technologies. Sponsors staff development workshops, and other workshops, conferences and sympasia designed to educate faculty inventors and entrepreneurs about technology transfer issues and processes. Publications: Annual Report; Listing of Available Technologies and Information Pieces.

University of Wisconsin—Madison
University Research Park

1265 Wisconsin Alumni
　　Research Foundation Bldg.　　　　　　　(608) 262-3677

610 Walnut St.

Madison, WI 53705-2336

FAX: (608) 265-2886

Wayne McGown, Director

The 296-acre Park facilitates technology transfer between the research produced on the university campus and applied research of industry and provides a long-term endowment income to the university. Leases land to private companies and agencies and offers laboratory and office services to start-up companies. Provides access to the University-Industry Research Program for matching research interests.

University of Wisconsin—Madison
University-Industry Relations

1215 WARF Bldg.　　　　　　　　　　(608) 263-2840
610 Walnut St.
Madison, WI 53705
FAX: (608) 263-2841
Dr. Stephen C. Price, Acting Director

Industrial research liaison and technology transfer activity of University of Wisconsin-Madison Graduate School. Program is part of a statewide network of public and private organizations promoting economic development and commercialization of technical innovations. Formerly known as University-Industry Research Program (1994). Identifies scientific resources (faculty, research program information, and specialized facilities) at the university for application to needs of industry, commerce, and government. Facilitates technology transfer activities with industry and provides advisory services to faculty on intellectual property and university-industry consortia development. University research results made available to the public. Sponsors technology seminars, conferences, and executive briefings. Cooperates with State Department of Development to provide technical assistance in small business development and corporate relocation. Administers grants to faculty involved in projects having economic development potential for state industries. Publications: Activity Report (annually).

University of Wisconsin—Milwaukee
Office of Industrial Research
and Technology Transfer

PO Box 340　　　　　　　　　　　　(414) 229-5000
Milwaukee, WI 53201
FAX: (414) 229-6967
Irving D. Ross, Jr., Director

University/industry research supporting facility in Graduate School at University of Wisconsin-Milwaukee. Serves as a catalyst in developing university/industry research programs and facilitates the transfer of technology between the university and industry. Handles patents, copyrights, licensing and proprietary agreements.

University Research Park

2 First Union Center, Ste. 1980　　　(704) 375-6220
Charlotte, NC 28282
FAX: (704) 338-9539
Seddon Goode, Jr., President

2,800-acre site providing companies located in the park with research and educational interaction with University of North Carolina at Charlotte.

University Technologies
International, Inc.

HM 382, 3330 Hospital Dr. NW　　　(403) 220-3790
Calgary, Alberta
T2N 4N1
Canada
FAX: (403) 270-3236
Beverley A. Sheridan, President and CEO

Wholly owned subsidiary of University of Calgary. Formerly known as Office of Technology Transfer, University of Calgary. Offers contract research and tranfers new technologies, inventions, and products conceived by University hospital, and client institution researchers to industry. Coordinates university and hospital inventions, products, patents, expertise, facilities, and technologies with research and development needs of industry. Administers license and joint venture programs. Publications: UTI Ink.

University Technology
Incorporated

Research Bldg. I　　　　　　　　　　(216) 368-5514
11000 Cedar Ave.
Cleveland, OH 44106
FAX: (216) 721-2310
Dr. James Cobatch, Director of Technology Management

Established to strengthen regional business and industry, stimulate research, and provide a channel for the commercialization of campus technologies. Identifies, evaluates, and implements development strategies for technologies, including intellectual property and protection; patent strategy; defining business and marketing opportunities; designing business structures; and forming business, commercial, and financial relationships. Licenses preferentially to companies prepared to invest in further on-campus research and development of technology. Creates businesses based on University technology.

Utah State University
Utah State University Research
and Technology Park

1770 N. Research Pkwy., Ste. 120　　(801) 750-6924
North Logan, UT 84341
FAX: (801) 750-6925
Wayne Watkins, Director

Fosters the interaction between the University research community and Park tenants. Administers a technology transfer program and provides incubator services to start-up companies.

Virginia Center for
Innovative Technology

CIT Bldg., Ste. 600　　　　　　　　　(703) 689-3000
2214 Rock Hill Rd.
Herndon, VA 22070-4005
FAX: (703) 689-3041
Dr. Robert G. Templin, Jr., President

Facilitates the transfer of technology from commonwealth universities to industry in biotechnology, computer-aided engineering, information technology, materials science engineering, fiber optics, coal and minerals, space, semicustom integrated systems, wood and renewable resources, power electronics, electrochemical processes, magnetic bearings, advanced ceramics, and command, control, communications and intelligence. Sponsors technology transfer directors at community colleges to solve technology-related problems of small- and medium-sized businesses. Publications: CIT Brochure.

Virginia Polytechnic Institute and State University
Virginia Tech Corporate Research Center

1872 Pratt Dr.　　　　　　　　(703) 231-3600
Blacksburg, VA 24060
FAX: (703) 231-3327
Joe W. Meredith, President

120 acres adjacent to the main campus and university airport providing building sites and buildings for lease to companies interested in developing or expanding a research relationship with the university. Houses an innovation center and other buildings to provide facilities for emerging companies.

Washington Research Foundation

1107 NE 45th St., Ste. 205　　　　(206) 633-3569
Seattle, WA 98105
FAX: (206) 633-2981
Ronald S. Howell, President

Independent, nonprofit foundation. Provides professional intellectual property management services for the successful commercialization of new technologies generated by research institutions in Washington state. Deals with new technologies in a wide range of areas, including biological materials, biotechnology, engineering, electronics, and software and computer science. **Publications:** WRF Report.

Washington State University
Research and Technology Park

NE 1615 Eastgate Blvd.　　　　　(509) 335-5526
Pullman, WA 99164-1802
FAX: (509) 335-7237
Larry M. Simonsmeier, Director

Encourages research interaction between Washington State University and industry. Seeks to promote economic development in southeastern Washington. The Park leases land to industries engaged in research, development, and light manufacturing. Industrial tenants share university research facilities and services, use faculty as consultants, and utilize graduate students as a part-time work force. Major research areas include agriculture, forestry, veterinary medicine, plant and animal biotechnology, and engineering. **Publications:** Quarterly Newsletter.

Washington University
Industrial Contracts and Licensing

1 Brookings Dr.　　　　　　　　(314) 362-5830
Campus Box 8013
St. Louis, MO 63110
FAX: (314) 362-5872
Susan E. Cullen, Ph.D., Associate Vice Chancellor

Research support activity at Washington University. Transfers technology from the university to business and industry in the area of biotechnology.

Wayne State University
Manufacturing Technology Center

FAB Bldg., 3rd Fl.　　　　　　　(313) 577-2788
656 W. Kirby
Detroit, MI 48202
FAX: (313) 577-1274
Gary R. Shields, Director

Research support activity at Wayne State University. Formerly known as Technology Transfer Center. Assists manufacturers, entrepreneurs, and inventors in gaining access to technical resources at the university and throughout southeast Michigan.

Western Michigan University
Technology Transfer Center

350 E. Michigan Ave., Ste. 29　　(616) 387-2714
Kalamazoo, MI 49007
FAX: (616) 344-6510
William H. Cotton, Director

Research support activity at Western Michigan University. Transfers technology from the University to the business and industrial sectors. Research results published in WESTOPS quarterly report. Offers resource identification, technical assistance, testing, product and feasibility evaluations, and online searches.

Wheeling Jesuit College
National Technology Transfer Center

316 Washington Ave.　　　　　　(304) 243-2455
Wheeling, WV 26003
Lee W. Rivers, Executive Director

Operates under a cooperative agreement between the National Aeronautics and Space Administration and Wheeling Jesuit College. Facilitates transfer of federally sponsored research and technology to U.S. business and industry. Offers courses and seminars to the private and public sector. Publications: Technology Touchstone (quarterly).

Yale University
Office of Cooperative Research

246 Church St., Ste. 401　　　　　(203) 432-7240
New Haven, CT 06510
FAX: (203)432-7245
Dr. Robert K. Bickerton, Director

Research support activity at Yale University. Facilitates the transfer of technology from the university to business and industry. Patents and licenses university inventions in the areas of physical sciences and engineering, medicine and biotechnology, and computer sciences. Administers industrial liaison programs in computer science

Inventor Organizations

Advice on Organizing

John Moreland, publisher of *The Dream Merchant* magazine, had this to say about organizations: "Inventor organizations may prove to be an excellent means to meet or network with other inventors [and] learn of available resources.

"However, most clubs meet once a month or less frequently and **should not** be considered a substitute for diligence, persistence, and action.

"My biggest complaint about any organization is that many times the organization is more concerned about the welfare of the 'organization' than [about] individuals [in] the organization; this seems more true in larger national organizations. Also there seems to be some sort of rivalry between some clubs, much like the competitiveness of sports teams. Instead of working together to achieve a synergistic concept, they are more content with spinning their own wheels and trying to develop their membership."

An idea can turn to dust or magic, depending on the talent that rubs against it.

—William Bernbach

Inventors are not known for their business acumen or marketing expertise. Getting product from the workshop or drawing board to the consumer can be a long, tough, and sometimes lonely haul. Operating on the frontier of an emerging idea is difficult enough. Spearheading the development, licensing, and manufacture of an invention can leave even the most experienced players nonplused.

Inventors can receive guidance and support from myriad sources, including, but surely not limited to, patent attorneys, government agencies, universities, reference books, and technical resource centers. Often the most sincere and valuable assistance comes from other inventors, peers who are also engaged in the "quest for fire."

Inventors are no different than any other group of people with a common interest. They love to get together and explore professional issues and share experiences, success stories, heartbreaks, "insider" information, personal contacts, methods, techniques, and dreams.

I always remind the executives to whom I license concepts that although on one level all of us independent inventors compete against each other for the placement of product, we are in reality very friendly towards each other. We do not perceive each other as competition in the same way K-Mart would consider Wal-Mart competition. We help each other with everything from the sharing of executive contacts and insights to the terms in our licensing agreements.

Inventor organizations often take root where inventors practice their trade. Seeking to establish relationships among themselves, independent inventors have formed non-profit organizations throughout the nation, groups that provide professional and social forums. They offer members all kinds of product development support, guidance, and resources. A common objective is to stimulate self-fulfillment, creativity, and problem-solving.

Naturally, some inventor groups are more sophisticated than others; some more organized. But all of them have something positive to offer.

"Inventor organizations are the biggest bargain around for helping inventors to get their invention going, avoid scams, and meet people who have similar interests and have solved similar problems," says Ray Watts, former editor of the *Inventor-Assistance Program News*.

Chuck Mullen, chairman of the board, Houston Inventors Association, says, "At our regular meetings of approximately 200 members (HIA has 500 members total), when a newcomer stands up and asks for help, there will be at least three or four members that will have the information and experience that he or she needs."

And Penny Becker of the Minnesota Inventors Congress feels the best free guidance and networking opportunities inventors can obtain are through the non-profit inventor organizations.

Upon hearing that I was writing this second edition of *The Inventor's Desktop Companion,* numerous national inventor service organizations called to request that I make special mention of them in this chapter. I explained that I take very seriously any personal endorsements and do not make them lightly. It is in this spirit that I call to your attention one organization that did not lobby me for a mention: Intellectual Property Owners (IPO).

If you want to join a national organization, I feel very comfortable and confident recommending that you consider membership in this national organization if it fits your particular needs. I know many of the people behind IPO and they are sincere, honest, unselfish, and totally committed.

In my opinion, IPO is the Rolls Royce of all national intellectual property organizations. It is outstanding and without equal in the array of invaluable services it can offer inventors.

IPO is a non-profit association representing companies and inventors that own patents, trademarks, and copyrights. It was founded in 1972 by a group of individuals concerned about the lack of understanding of intellectual property rights in the United States. Members include nearly 100 large and medium-sized corporations (including Union Carbide, Monsanto, United Technologies, P&G, AT&T) and more than 300 small businesses, universities, independent inventors, authors, executives, and attorneys.

Herb Wamsley, IPO's executive director, says that applications for individual inventor memberships are on the rise. While IPO is definitely not for the one-time or casual inventor, it should be given consideration by professional inventors who have interest in up-to-the-minute information about intellectual property (for example, GATT, patent law harmonization, etc.). For a full list of IPO's benefits and services, request a fact sheet and other background material from the organization.

Membership runs $110 for individual members (inventor or author) who are not attorneys. To be eligible for these reduced dues, an individual member must have filed or be planning to file a patent application in which he or she is named as an inventor.

For more information contact Herbert C. Wamsley, Executive Director; Ada E. Barnes, Deputy Executive Director; Katelyn M. Feeney, Executive Assistant; at Intellectual Property Owners (IPO), 1255 23d Street NW, Suite 850, Washington, D.C. 20037. The telephone number is: (202) 466-2396; the fax number is: (202) 466-2893.

By the way, do not join a national inventor organization at the exclusion of a local one. Both have their place and complement each other in terms of what they can deliver. Beware that many national organizations exist to make money off products and services and do very little to forward the cause of the independent inventor. Study the brochures and make an educated decision.

Make Your Voice Heard

Send comments, queries, and protests about intellectual property laws to members of the Senate Judiciary Committee.

- Orrin G. Hatch (UT) (Chairman): (202) 224-5251
- Spencer Abraham (MI): (202) 224-4822
- Joseph R. Biden (DE): (202) 224-5042
- Hank Brown (CO): (202) 224-5941
- Mike DeWine (OH): (202) 224-2315
- Russell D. Feingold (WI): (202) 224-5323
- Dianne Feinstein (CA): (202) 224-3841
- Chuck Grassley (IA): (202) 224-3744
- Howell Heflin (AL): (202) 224-4124
- Edward M. Kennedy (MA): (202) 224-4543
- Herb Kohl (WI): (202) 224-5653
- Jon Kyl (AZ): (202) 224-4521
- Patrick J. Leahy (VT): (202) 224-4242
- Paul Simon (IL): (202) 224-2152
- Alan K. Simpson (WY): (202) 224-3424
- Arlen Specter (PA): (202) 224-4254
- Strom Thurmond (SC): (202) 224-5972

Help in Starting an Inventor Organization

Fred Hart, of the Department of Energy, advises, "Search out the nearest inventors' organization and join. If there isn't one, start one up. It's only through a mutual support group that you can safely get help."

If you need help in starting your own inventor group where none exists, you may wish to contact Chuck Mullen at (713) 326-1795. He is chairman of the Houston Inventors Association (HIA) and founder of the D.C. Capital Inventor's Society, a clone of HIA.

"This is a non-profit effort by the Houston Inventors Association to grow inventor organizations nationwide in areas where there are none," says Mullen. "We see this as the first step to a strong and healthy national network of inventors, one which will reach out to school children, motivate and teach them how to invent. At the same time, we'll stress the importance of staying in school and mastering math and science so that they have the opportunity to grow up to become rich and famous inventors."

For the nominal cost of materials and postage, Mullen says HIA will furnish written materials about how to start a club plus send videotapes of their best guest speakers, tapes which groups can show at their monthly meetings. The D.C. Capital Inventors Society was the first such clone organization; I have been a speaker there, and it is quite a large group.

For more information, write: Charles Mullen, 204 Yacht Club Lane, Seabrook, TX 77586. Phone/fax: (713) 326-1795.

New Inventors' Coalition

Founded in the summer of 1994 to fight the attempt to change the term of U.S. patents from 17 years from issue to 20 years from filing an application, Intellectual Property Creators is a national organization of inventors, product developers, and entrepreneurs that is, at the time of this printing, seeking non-profit status.

Among IPC's founders are Dr. Raymond Damadian, inventor of MRI, inductee into the National Inventors Hall of Fame, and recipient of the National Technology Medal; Dr. Wilson Greatbach, inventor of the implantable cardiac pacemaker, inductee into the National Inventors Hall of Fame, and founder and president of Greatbach Industries; and Dr. Robert Rines, inventor of high definition imaging, scanning radar and sonar, and inductee into the National Inventors Hall of Fame.

For information on IPC, contact Paul Heckel, Intellectual Property Creators, 146 Main Street, Suite 404, Los Altos, California 94022. The telephone number is: (415) 948-8350; the fax number is (415) 948-7319.

ORGANIZATIONS

Following is a list of local inventor organizations and Small Business Development Centers (SBDC), arranged by state. This information is reprinted here courtesy of the *Inventor-Assistance Source Directory*, published by Battelle Pacific Northwest Laboratories.

ALABAMA

Alabama SBDC

Medical Towers Building   (205) 934-7260
1717 11th Ave., Suite 419
Birmingham, AL 35294
FAX (205) 934-7645
John Sandefur, State Director

Alabama Technology Assistance Program

1717 11th Avenue S, Ste. 419   (205) 934-7260
Birmingham, AL 35294
FAX (205) 934-7645
David L. Day, Program Coordinator

U.S. Small Business Administration Business Development

2121 Eighth Ave. North, Suite 200   (205) 731-1338
Birmingham, AL 35203-2398
FAX (205) 731-1404
Raymond W. Hembree, Economic Development Chief

University of Alabama Office for the Advancement of Developing Industries

1075 13th St. South   (205) 934-2190
Birmingham, AL 35294-4440
FAX (205) 934-1037
Wilson L. Harrison, Director

ALASKA

Alaska Inventors & Entrepreneurs Association, Inc.

P.O. Box 241801   (907) 276-4337
Anchorage, AK 99524-1801
FAX (907) 278-2982
Eden Larson, Executive Director

Alaska Science/Technology Foundation

550 W. 7th Ave.   (907) 272-4333
Suite 360
Anchorage, AK 99501
FAX (907) 274-6228
John W. Sibert, Ph.D., Exec. Director

Inventors Institute of Alaska

P.O. Box 876154   (907) 376-5114
Wasilla, AK 99687
Al Jorgensen, President

Kenai Peninsula Borough Economic Development District, Inc.

110 S. Willow, Ste. 106   (907) 283-3335
Kenai, AK 99611
FAX (907) 283-3913
Mike Sims, Director Business Development

University of Alaska SBDC

430 W. 7th Ave.   (907) 274-7232
Suite 110
Anchorage, AK 99501
FAX (907) 274-9524
Jan Fredericks, State Director

Doing It by the Book

To illustrate the importance of record-keeping, consider this story. According to the PTO, long before Alexander Graham Bell filed a patent application in 1875, Daniel Drawbaugh claimed to have invented the telephone. However, as Drawbaugh had no notes, journal, or log, the U.S. Supreme Court rejected his claim by a vote of four to three. Bell, on the other hand, possessed complete written records and was awarded a patent on the telephone.

World Trade Center Alaska

421 W. First Avenue   (907) 278-7233
Ste. 300
Anchorage, AK 99501
FAX (907) 278-2982
Robin Zerbel, Manager

ARIZONA

Arizona Technology Incubator

1535 N. Hayden Road   (602) 990-0400
Scottsdale, AZ 85257-3773
FAX (602) 970-6355
Dr. Robert J. Calcaterra, President/CEO

Gateway Community College SBDC

108 N. 40th St.   (602) 392-5233
Phoenix, AZ 85034
FAX (602) 392-5329
Kathy Evans, Director

Innovation Development Corp.

408 W. Forest Ave.   (708) 891-0316
Flag Staff, AZ 86001

Pima Community College SBDC

4907 E. Broadway Blvd., #123   (602) 748-4906
Tucson, AZ 85709-1250
FAX (602) 748-4585
Linda Andrews, Director
Santina M. Baumeister, Vice Pres. Research

Synergetics

P.O. Box 809   (602) 428-4073
3860 W. First
Thatcher, AZ 85552
Don Lancaster

University of Arizona Office of Technology Transfer

1430 E. Fort Lowell Rd., #200   (602) 621-5000
Tucson, AZ 85719-2366
FAX (602) 322-0846
Dr. Rita C. Manak, Ph.D., Director

University of Arizona Engineering Professional Development

Box 9, Harvill Bldg.   (602) 621-5104
Tucson, AZ 85721
FAX (602) 621-1443
Paul A. Baltes, Director

ARKANSAS

East Arkansas Business Incubator System

5501 Krueger Dr.   (501) 935-8365
Jonesboro, AR 72401
FAX (501) 931-5133
Guy Enchelmayer, Director

Inventors Congress, Inc.

P.O. Box 411   (501) 229-4515
Dandanelle, AR 72834
Garland E. Bull, President

CALIFORNIA

American Innovation Association (A.I.A.)

P.O. Box 1107   (805) 649-1046
Oak View, CA 93022
FAX (805) 649-5264
John Lamar, President & Founder

American Inventor Network

6770 Depot St.   (707) 823-3865
Sebastopol, CA 95472
FAX (707) 823-0913
Jeff McGrew, President

Antelope Valley IWIEF Chapter

4444 E. Ave. R, Sp. 127   (805) 273-0144
Palmdale, CA 93550
Marvin Clark, President

**Asian Inventor
Entrepreneur Cooperative**
18325 Malden St.   (818) 882-4506
Northridge, CA 91325-3685
FAX (818) 882-4506
C.C. Alfarero, Chairman

**Black Inventors
Technical Resources**
P.O. Box 6402   (310) 323-4668
Torrance, CA 90504
FAX (310) 323-4668
Joe Edmonds, Director

California Inventors Council
P.O. Box 2036   (408) 732-4314
Sunnyvale, CA 94087
Barret Johnson

**California State University
Center for New
Venture Alliance**
Hayward, CA 94542-3066   (510) 881-3805
FAX (510) 727-2039
Lawrence J. Udell, Executive Director

**California State University—
Los Angeles
Institute of Entrepreneurship**
5151 State University   (213) 343-2971
Los Angeles, CA 90032
Dr. Marshal E. Reddick, Director

**Central Valley Inventor's
Association, Inc.**
P.O. Box 1551   (209) 239-5414
Manteca, CA 95336
John Christensen, President

**CONNECT UC/San Diego
Extension
Technology/Entrepreneurship**
U.C.S.D. Extension—0176   (619) 534-6114
La Jolla, CA 92093-0176
FAX (619) 552-0649
William W. Otterson, Director

Cossman International, Inc.
P.O. Box 4480   (619) 320-7717
Palm Springs, CA 92263
FAX (619) 320-9247
E. Joseph Cossman, President

Info/Masters Libraries, Inc.
Box 1092   (805) 497-4053
Thousand Oaks, CA 91358-0917
David Alan Foster, President

**International Education
Foundation
Inventors Workshop**
7332 Mason Avenue   (818) 340-4268
Canoga Park, CA 91306-2822
FAX (818) 884-8312
Alan A. Tratner, Internat'l. President

**Inventors Assistance
League**
345 West Cypress St.   (818) 246-6541
Glendale, CA 91204
Ted DeBoer, Executive Director

**Inventors Guild of
America**
1852 Kirby Road   (818) 507-0263
Glendale, CA 91208
George Schmidt, President/CEO

**Inventors Helper
Industries**
4480 Treat Blvd., Suite 310   (510) 676-4975
Concord, CA 94521
FAX (510) 535-0450
Robert Lewis

**Inventors New
Venture Alliance**
P.O. Box 371148   (510) 726-1945
Montara, CA 94037
FAX (510) 726-1945
Martha Regan, Vice President

ISTAR Inc.
406 Alta Ave.   (310) 394-7332
Santa Monica, CA 90402-2714
FAX (310) 394-7332
Dr. A. Wortman, President

Macro—Search Corp.
2082 Business Cntr. Dr. #225   (714) 253-9930
Irvine, CA 92715
FAX (714) 253-0951
Gene Scott, President

Margolin Development Company Inventive Product Development

P.O. Box 2846 (714) 645-5950
Newport Beach, CA 92663
FAX (714) 645-5974
George D. Margolin, Ph.D., President

NASA—Ames Technology Commercialization Center

155-A Moffett Park Dr. (408) 734-4700
Ste. 104
Sunnyvale, CA 94089
FAX (408) 734-4946
John Gee, Director

NASA Far West Regional Technology Transfer Center

3716 South Hope Street (213) 743-6132
Room 200
Los Angeles, CA 90007
FAX (213) 746-9043
Harriette Reid, Mgr. Outreach Srvcs.

NCIO

P.O. Box 6158 (510) 376-7541
Rheem Valley, CA 94570
FAX (510) 376-7762
Norman C. Parrish, President (retired)

Orbic Controls

Box 23827
Pleasant Hill, CA 94523
FAX (510) 944-4987
Newton E. Ball, Principal Engr.

San Diego Inventors Group

11190 Poblado Road (619) 673-4733
San Diego, CA 92127
FAX (619) 451-6154
Greg W. Lauren, Coordinator

Santa Cruz IWIEF Chapter

730 Encino Dr. (408) 662-1936
Aptos, CA 95003
FAX (408) 662-1936
Terry Chappell

Sawyer Inventive Resource Center

520 Mendocino Ave. Ste 210 (707) 524-1773
Santa Rosa, CA 95401
FAX (707) 524-1772
Charles Robbins, Director

Small Entity Patent Owners Association

295 Stevenson Dr. (510) 934-1331
Pleasant Hill, CA 94523
FAX (510) 934-1132
Sherm Fishman, Executive Director

Technology Access Report

16 Digital Dr., Ste. 250 (800) 733-1556
Novato, CA 94949
FAX (415) 883-6421
Michael Odza, Publisher

Technology Properties, Ltd.

4010 Moorpark Ave., Ste. 215 (408) 243-9898
San Jose, CA 95117
FAX (408) 296-6637
Daniel Leckrone, Chairman

Introduced by Alex Osborn in his book, *Applied Imagination*, brainstorming is a process of spontaneous thinking whereby one generates numerous alternative ideas while deferring judgment. Use these guidelines in your next brainstorming session.

• **No criticism permitted.** Do nothing to inhibit the free flow of thought. Record each idea as it is offered and move on.

• **Go for quantity.** Osborn said "quantity breeds quality."

• **Hitchhiking welcome.** Hitchhiking takes place when one person's idea produces a similar idea or an enhanced idea from another brainstormer.

• **Freewheeling encouraged.** The more "off-the-wall" the concept, the better. Every idea should be recognized. You'll be surprised how often a solution lies within the most wacky or seemingly unimportant idea.

**The James F. Riordan
Co., Inc.**
3110 Camerosa Circle   (916) 676-4729
Cameron Park, CA 95682
James F. Riordan, President

**University of Southern
California
Inventors Workshop**
Far West Tech Transfer Center   (213) 743-6132
3716 S. Hope St. Room 200
Los Angeles, CA 90007
FAX (213) 746-9043
Robert L. Stark

COLORADO

**Affiliated Inventors
Foundation, Inc.**
902 N. Circle Dr., #208   (719) 635-1234
Colorado Springs, CO 80909-5002
FAX (719) 635-1578
John T. Farady, President

**Colorado Biomedical
Venture Center**
1610 Pierce Street   (303) 237-3998
Denver, CO 80214
FAX (303) 237-4010
Lewis Kontnick, President

**Colorado Inventors
Council, Inc.**
P.O. Box 88   (303) 854-3851
Holyoke, CO 80734
Michael R. McKensie, Director

Colorado SBDC
1391 N. Speer Blvd., Ste. 600   (303) 620-4000
Denver, CO 80204

**Colorado University
Business Advancement
Centers**
335 S. 43rd St.   (303) 499-8114
Boulder, CO 80303
Teri Wenz

National Inventors Association
P.O. Box 2032   (303) 755-1456
Littleton, CO 80161
Morton Levand

**Rocky Mountain
Inventors Congress**
P.O. Box 4365   (303) 758-6757
Denver, CO 80204
Cornelia A. Snyder, President

CONNECTICUT

**Connecticut Innovations, Inc.
Technology Assistance Center**
845 Brook St.   (203) 258-4305
Rocky Hill, CT 06067
FAX (203) 563-4877
Eric Oh, Director

**Innovators Network
of Greater Danbury**
37 Seneca Road   (203) 797-8955
Danbury, CT 06811
Jack Lander, Program Director VP

InoNet, Inc.
187 South Salem Road   (203) 438-1661
Ridgefield, CT 06877
FAX (203) 221-7060
Barbara Burnes, Principal

**University of Connecticut
Connecticut SBDC**
Allyn Larabee Brown Bldg.   (203) 486-4135
U-9, Route 44
Storrs, CT 06269-5094
FAX (203) 486-1576
John P. O'Conner, State Director

DELAWARE

**Delaware Economic
Development Office**
99 Kings Highway   (302) 739-4271
P.O. Box 1401
Dover, DE 19903
FAX (302) 739-5749
Susan Rhoades, Tech. Specialist

University of Delaware

005 Purnell Hall
Newark, DE 19716
FAX (302) 831-1423
Clinton Tymes, SBDC State Director

(302) 831-2747

DISTRICT OF COLUMBIA

Alliance for American Innovation

1100 Connecticut Ave NW, Suite 1200
Washington, DC 20036-4101
FAX (202) 467-5591
Steven Michael Shore, President

(202) 293-1414

Howard University SBDC

2600 6th St. NW P.O. Box 748
Washington, DC 20059
FAX (202) 806-1777
Levi Lipscomb, Acting Director

Intellectual Property Owners Association

1255 23rd St. NW, Suite 850
Washington, DC 20037
FAX (202) 466-2893
Herbert C. Wamsley, Exec. Director

(202) 466-2396

U.S. Department of Energy

1000 Independence Ave. CE 521, 5E052
Washington, DC 20585
FAX (202) 586-8134
Terry Levinson

(202) 586-1478

U.S. Small Business Administration

2328 19th NW
Washington, DC 20009
FAX (202) 205-7064
Bill Scheirer, Economist

(202) 205-6977

FLORIDA

Center for Health Technologies, Inc.

1150 N.W. 14th Street, Suite 105
Miami, FL 33136-2112
FAX (305) 325-2698
Eugene H. Man, Ph.D., President/CEO

(305) 325-2733

Edison Inventors Association

5563 Trellis Lane
Ft. Myers, FL 33919
FAX (813) 939-4996
Bernard A. Cousino, PTE,
    Chairman & Founder

(813) 481-0098

Edison Inventors Association, Inc.

P.O. Box 07398
Ft. Myers, FL 33919
Dr Gary Nelson, Executive Vice Pres.

(813) 275-4332

Florida Department of Commerce

107 West Gaines Street
Tallahassee, FL 32399-2000
FAX (904) 487-6516
Brent Gregory, Research Economist

(904) 487-3134

Florida Product Innovation Center

2622 NW 43rd St., Suite B3
Gainesville, FL 32606-7428
FAX (904) 334-1682
Pamela H. Riddle, Director

(904) 334-1680

Florida SBDC Network

19 West Garden Street
Pensacola, FL 32501
FAX (904) 444-2070
Jerry Cartwright, State Director

(904) 444-2060

Inventors Council of Central Florida

822 E. Wallace Street
Orlando, FL 32809
David Flinchbaugh, President

(407) 859-4855

Inventors Society of South Florida

P.O. Box 4306
Boynton Beach, FL 33424
Betty White, Dir. Public Affairs

(407) 736-6594

Klenner International

P.O. Box 1748
Ormond Beach, FL 32175
FAX (904) 673-5911
Anne E. Klenner, CEO

(904) 673-4339

**Palm Beach Society of
American Inventors**
P.O. Box 26   (304) 655-0536
Palm Beach, FL 33480
Kiki Shapiro

Product Center Corporation
4001 NW 97th Ave.   (305) 592-7260
Miami, FL 33178
FAX (305) 592-7937
Edward D. Miller, President

PSPI, Inc.
15 Dahoon Court South   (904) 382-1535
Homosassa, FL 34446-8922
Jerry Zajic, Tech Specialist

**Small Business
Development Center**
P.O. Box 161530   (407) 823-5554
Orlando, FL 32816-1530
FAX (407) 823-3073
Jim Hahn, Associate Director

**Tampa Bay Inventors
Council**
P.O. Box 2254   (813) 391-0315
Largo, FL 34646
R.V. Purdy, President

**The Enterprise Corporation
of Tampa Bay**
1111 N. Westshore Blvd.   (813) 288-0445
Ste. 200-B
Tampa, FL 33607
FAX (813) 554-2356
M. Scott Faris, President

**Tutle/International
Technology Marketing**
2601 Seabreeze Court   (407) 423-8016
Orlando, FL 32805
FAX (407) 291-1176
Edward G. Tutle, President/Consultant

**WSRE-TV
The Inventor's Club**
1000 College Blvd.   (904) 484-1224
Pensacola, FL 32504
FAX (904) 484-1255
William R. Bowman, Vice President

GEORGIA

CAN DO Industries, Inc.
7610 Ball Mill Road   (404) 396-1401
Atlanta, GA 30350
FAX (404) 671-1070
W.C. Stillwagon, President

**Georgia Institute of
Technology
Library and Information
Center**
Atlanta, GA 30332-0900   (404) 894-4508
FAX (404) 894-8190
Mary-Frances Panettiere,
    Head, Tech Resources

**Institute of Industrial
Engineers**
25 Technology Park   (404) 449-0460
Norcross, GA 30092

**Inventors Clubs of
America, Inc.**
P.O. Box 450261   1-800-336-0169
Atlanta, GA 31145-0261
FAX (404) 355-8889
Alexander T. Marinaccio, Chairman

**U.S. Department of Energy
Atlanta Support Office**
730 Peachtree St. NE, Suite 876   (404) 347-7139
Atlanta, GA 30308
FAX (404) 347-3098
Ronald L. Henderson,
Director Grants Management

**University of Georgia
SBDC Connection**
Chicopee Complex   (404) 542-2737
Athens, GA 30602
FAX (404) 542-6776
Deborah Sommer, Director

University of Georgia SBDC
1180 E. Broad St.   (706) 542-6804
Athens, GA 30602-5412
FAX (706) 542-6776
Donna Brown, Program Specialist

HAWAII

High Technology Development Corporation

300 Kahelu Ave., Suite 35   (808) 625-5293
Honolulu, HI 96789
FAX (808) 625-6363
William M. Bass, Jr., Exec. Director/CEO

Inventors Council of Hawaii

Box 27844   (808) 595-4296
Honolulu, HI 96287
George K. C. Lee, President

University of Hawaii—OTTED

2800 Woodlawn Dr., Ste. 280   (808) 539-3831
Honolulu, HI 96822
FAX (808) 539-3833
Keith T. Matsunaga, Mgr. Special Projects

University of Hawaii at Hilo SBDC

523 West Lanikaula Street   (808) 933-3515
Hilo, HI 96720-4091
FAX (808) 933-3683
Janet Nye, State Director

IDAHO

Boise State University Technical & Industrial Extension Service

1910 University Dr.   (208) 385-3689
Boise, ID 83752
FAX (208) 385-3877
James R. Steinfort, Director

Boise State University Idaho SBDC

1910 University Drive   (208) 385-1640
Boise, ID 83725
FAX (208) 385-3877
Ronald R. Hall, State Director

Idaho Innovation Center

2300 N. Yellowstone   (208) 523-1026
Idaho Falls, ID 83401
FAX (208) 523-1049
Duane Abbott, Executive Director

Idaho SBDC

315 Falls Ave. P.O. Box 1238   (208) 733-9554
Twin Falls, ID 83303-1238
FAX (208) 733-9316
Cindy R. Bond, Director

Idaho State University SBDC

2300 North Yellowstone Ste. 121   (208) 523-1087
Idaho Falls, ID 83401
FAX (208) 523-1049
Betty Capps, Regional Director

ISBDC Technology Connection

1910 University Dr.   (208) 385-3870
Boise, ID 83725
FAX (208) 385-3877
Burt Knudson, Tech Srvcs. Consult.

Lewis-Clark State College Idaho SBDC

500 8th Avenue   (208) 799-2465
Lewiston, ID 83501
FAX (208) 799-2878
Helen M. LeBoeuf-Binninger,
          Regional Director

North Idaho College Idaho SBDC

1000 W. Garden   (208) 773-9807
Coeur d'Alene, ID 83814
FAX (208) 777-8123
John Lynn, Regional Director

State of Idaho Dept. of Water Resources

Boc 83720   (208) 327-7959
Boise, ID 83702-0098
FAX (208) 327-7866
Gerald Fleischman, P.E.,
          Bioenergy Specialist

ILLINOIS

American Electronics Association

625 North Court   (708) 358-2705
Palatine, IL 60067
FAX (708) 358-3246
Candy Renwall

American Inventors Council
P.O. Box 4304
Rockford, IL 61110
FAX (815) 962-1495
Nicholas G. Parnello, President/Founder

(815) 968-1040

American Society of Design Engineers
P.O. Box 931
Arlington Heights, IL 60006

(708) 259-7120

ARCH Development Corp.
9700 South Cass Ave. Bldg. 200
Argonne, IL 60439
FAX (708) 252-2876
Neil Wyant, Manager, Licensing

(708) 252-5904

Bradley University Business Technology Incubator
Peoria, IL 61625
FAX (309) 677-3386
Roger Luman, Director

(309) 677-2852

Evanston Business & Technology Center SBDC
1840 Oak Ave.
Evanston, IL 60201
FAX (708) 866-1807
Richard C. Holbrook, Director

(708) 866-1817

Innovation Development Corporation
P.O. Box 1185
Calumet City, IL 60409
FAX (708) 891-0316
Bruce R. Baumeister, President

(708) 891-0316

Inventors Council
53 W. Jackson, Ste. 1643
Chicago, IL 60604
Don Moyer, President

(312) 939-3329

Northern Illinois University Technology Commercialization Office
DeKalb, IL 60115-2874
FAX (815) 753-1631
Larry R. Sill, Director

(815) 753-1238

Those Inventive Americans

In 1850, English inventor James Nasmyth wrote the following about inventive Americans: "There is not a working boy of average ability in the New England States ... who has not an idea of some mechanical invention or improvement ... by which he hopes to better his position, or rise to fortune."

Northwestern University Technology Transfer Program
633 Clark St.
Evanston, IL 60208-1111
FAX (708) 491-4800
Linda Kawano, Administrator

(708) 491-3005

Southern Illinois University at Edwardsville Office of Technology & Commerce
Campus Box 1108
Edwardsville, IL 62026
FAX (618) 692-2555
Jim Mager, Director

(618) 692-2166

Technology Search International, Inc.
500 East Higgins Road
Elk Grove Village, IL 60007-1437
FAX (708) 593-2182
John W. Morehead, President

(708) 593-2111

Triple I Inventors
1851 16th St.
Moline, IL 61265
Cathy Mallary

(309) 764-4508

University of Illinois at Chicago Great Cities Coordinating Office
1737 West Polk Street, Ste. 310
Chicago, IL 60612-7227
FAX (312) 413-0238
David L. Gulley, PhD, Senior Associate

(312) 996-4995

International Association of Professional Inventors
818 Westminster   (317) 459-3555
Kokomo, IN 46901
Tom Nix

Inventor & Entrepreneur Society of Indiana
P.O. Box 2224   (219) 989-2354
Hammond, IN 46323
FAX (219) 989-2750
Daniel Yovich, Professor

IOWA

Drake University, SBDC INVENTURE Program
2507 University Ave.   (515) 271-2655
Des Moines, IA 50311
FAX (515) 271-4540
Benjamin C. Swartz, Director

Inventors, Innovators, Ideas, Inc.
304 West 2nd St.   (319) 322-4499
Davenport, IA 52801
FAX (319) 322-8241
James A. Graham, President

Iowa Lakes SBDC
Gateway North Highway 71 North   (712) 262-4213
Spencer, IA 51301
FAX (712) 262-4047
John Beneke, Director

KANSAS

Innovative Technology Enterprise Corp. (ITEC)
112 W 6th St., Ste. 408   (913) 233-9102
Topeka, KS 66603
FAX (913) 296-1160
Clyde C. Engert, President

Kansas Association of Inventors
2015 Lakin   (316) 792-1374
Great Bend, KS 67530
FAX (316) 792-3406
Clayton Williamson, President

Western Illinois University SBDC
Seal Hall 214   (309) 298-2211
Macomb, IL 61455
FAX (309) 298-2520
Daniel D. Voorhis, Director

INDIANA

Indiana University Indust. Res. Liaison Program
One City Centre, Suite 200   (812) 855-6294
Bloomington, IN 47404
FAX (812) 855-8270
W.S. Johnson, Director

Indiana Inventors Association
5514 South Adams   (317) 674-2845
Marion, IN 46953
Robert Humbert, President

Indiana Inventors Association, Inc.
P.O. Box 2388   (317) 745-5597
Indianapolis, IN 46206-2388
Randall N. Redelman, President

Indiana Small Business Development Network
One N. Capitol,, Ste. 420   (317) 264-6871
Indianapolis, IN 46204
Stephen Thrash, Executive Director

Pittsburg State University Business & Technology Institute

Pittsburg, KS 66762-7560   (316) 235-4920
FAX (316) 232-6440
C. Dale Lemons, Executive Director

Wichita State University Kansas SBDC

1845 Fairmount   (316) 689-3193
Wichita, KS 67260-0148
FAX (316) 689-3647
Tom Hull, State Director

KENTUCKY

Bluegrass Inventors Guild

P.O. Box 43610   (502) 423-9850
Louisville, KY 40253-0610
FAX (502) 423-1452
Vance A. Smith, Attorney

University of Kentucky SBDC

225 Bus. & Econ. Bldg.   (606) 257-7668
Lexington, KY 40506-0034
FAX (606) 258-1907
Janet Holloway, State Director

University of Louisville SBDC

Burhans Hall—Shelby Campus   (502) 588-7854
Louisville, KY 40292
FAX (502) 588-8573
Ken Blandford, Mgmt. Consultant

LOUISIANA

Northeast Louisiana University SBDC

700 University Ave.   (318) 342-5506
Monroe, LA 71209
FAX (318) 342-5510
John Baker, State Director

Louisiana Department of Economic Development

P.O. Box 94185   (504) 342-5371
Baton Rouge, LA 70804-9185
Laverne Jasek, Econ. Dev. Spec.

Louisiana State University Louisiana Business & Technology Center

South Stadium Drive   (504) 334-5555
Baton Rouge, LA 70803-6100
FAX (504) 388-3975
Charles D'Agostino, Executive Director

Southern Technology Development

109 Brownlee Ave.   (318) 837-4042
Broussard, LA 70518-3021
FAX (318) 837-5552
Robert Montgomery, Patent Agent & Pres.

MAINE

C.I.E.

University of Maine   (207) 581-1488
Maine Tech Center
16 Godfrey Drive
Orono, ME 04473

Maine SBDC

University of Southern Maine   (207) 780-4420
246 Deering Ave.
Portland, ME 04102

University of Maine Department of Industrial Cooperation

5711 Boardman Hall   (207) 581-1488
Orono, ME 04469-5711
FAX (207) 581-1484
James S. Ward

MARYLAND

Capital Inventors Society

2106 Salisbury Road   (301) 585-1885
Silver Springs, MD 20910
John Boucher, Chairman of Commun.

Technology Transfer Information Center National Agricultural Library

10301 Baltimore Blvd.   (301) 504-6875
Beltsville, MD 20705-2351
FAX (301) 504-7098
Kate Hayes, Coordinator

MASSACHUSETTS

Douglas Industries, Inc.
P.O. Box 581   (508) 420-2002
Osterville, MA 02655
FAX (508) 420-2002
Garrett D. Douglas, President

EKMS, Inc.
100 Inman St.   (617) 864-4706
Cambridge, MA 02139
FAX (617) 864-7956
Edward Kahn, President

Enbede Co.
P.O. Box 335   (508) 934-0035
Lexington, MA 02173
FAX (508) 934-0035
Donald Job, President

HighTech Design
26 Murray St.   (508) 535-2543
W. Peabody, MA 01960
Steve Chamuel, President

Innovative Products Research & Services
393 Beacon St.   (508) 934-0035
Lowell, MA 01850
FAX (508) 934-0035
Donald Job, President

Inventors Association of New England
115 Abbot Street   (508) 474-0488
Andover, MA 01810
FAX (509) 474-0488
Donald L. Gammon, President

Technology Capital Network
201 Vassar St.   (617) 253-8214
Cambridge, MA 02139
FAX (617) 258-7264
Bill Wolf, Executive Director

University of Massachusetts SBDC
205 School of Management   (413) 545-6301
Amherst, MA 01003
FAX (413) 545-1273
John Ciccarelli, State Director

Worcester Area Inventors
65 Windsor Street   (508) 757-6178
Worcester, MA 01605
Jack Brady, President

MICHIGAN

American Association of Inventors
2853 State St.   (517) 799-3444
Saginaw, MI 48602
Dennis Ray Martin

Economic Development Corporation
P.O. Box 387   (616) 946-1596
Traverse City, MI 49685
FAX (616) 946-2565
Charles Blankenship, President

Ferris State University Technology Transfer Center
1020 E. Maple St.   (616) 592-3774
Big Rapids, MI 49307
FAX (616) 796-1448
Thomas Schumann, Program Director

Industrial Development Center
311 Houghton St.   (906) 884-2795
Ontonagon, MI 49953
Joe Nies, Director

InnCom
P.O. Box 507   (313) 963-0616
Ypsilanti, MI 48197
FAX (313) 963-7606
Barbara B. Eldersveld

Manistee County Economic Development Office
375 River St., Ste. 205   (616) 723-4325
Manistee, MI 49660
FAX (616) 723-1488
Thomas Kubanek, Econ. Dev. Dir.

Michigan SBDC
2727 Second Ave.   (313) 964-1798
Detroit, MI 48201
FAX (313) 964-3648
Sarah McCue, Communications Mgr.

University of Michigan College of Engineering

2901 Hubbard (313) 747-0041
Ann Arbor, MI 48109-2016
FAX (313) 747-0036
J. Downs Herold, Dir. of Liaison

MINNESOTA

Accessible Technologies, Inc.

494 Curfew (612) 338-3280
St. Paul, MN 55104
FAX (612) 645-3675
Wade Van Valkenburg, President

Excel Development Group Inc.

1721 Mount Curve Ave. (612) 374-3233
Minneapolis, MN 55403-1017
FAX (612) 377-0865
Andrew Berton, President

INNO TECH, Inc.

7007 Dakota Avenue (612) 937-1413
Chanhassen, MN 55317
FAX (612) 934-1180
Michael Marra, President

Inventors & Designers Ed Assoc

P.O. Box 268 (612) 430-1116
Stillwater, MN 55082
Bill Baker, Director

Inventors Club of Minnesota (ICM)

818 Dunwoody Blvd. (612) 220-4277
Minneapolis, MN 55403
Lisa Olson, President
Bill Braddock, Vice President

Metro ECSU Young Inventors Fair

3499 Lexington Ave. No. (612) 490-0058
St. Paul, MN 55126
FAX (612) 490-1920
Janet Robb, Coordinator

Minnesota Entrepreneurs Club

1000 LaSelle Ave. (612) 533-1932
Minneapolis, MN 55403-2005
FAX (612) 533-1865
Mike Tikkanen, Past President

Minnesota Inventors Congress

1030 E. Bridge St. PO Box 71 (507) 637-2344
Redwood Falls, MN 56283
FAX (507) 637-5929
Penny Becker, Executive Director

Minnesota Project Innovation, Inc.

111 Third Ave. South, Ste. 100 (612) 338-3280
Minneapolis, MN 55401-2551
FAX (612) 338-3483
Randall Olson, Executive Director

Minnesota SBDC

College of St. Thomas (612) 448-8810
1107 Hazeltine Blvd., Ste. 245
Chaska, MN 55318

Minnesota Technology, Inc. Center for Economic Development—UMD

Ste.140, Olcott Plaza, 820 N. 9th St. (218) 741-4241
Virginia, MN 55792
FAX (218) 741-4249
Mark Mueller, Director

Society of Minnesota Inventors

20231 Basalt St. (612) 753-2766
Anoka, MN 55303
Paul G. Paris, President

Inventors on Inventing

- Most people look at what is; they never see what can be. —Albert Einstein
- When you can do the common things in life in an uncommon way, you will command the attention of the world. —George Washington Carver
- If we are industrious, we shall never starve, for at the working man's house hunger looks in, but dares not enter. —Benjamin Franklin

The Collaborative

10 S. 5th St., Ste. 415   (612) 338-3828
Minneapolis, MN 55402
FAX (612) 338-1876
Mallory Wingrove, Member Svcs. Coord.

MISSISSIPPI

American Association of Inventive Dentists

420 Magazine St.   (601) 842-1036
Tupelo, MS 38801
Charles E. Moore

Confederacy of Mississippi Inventors

4759 Nailor Rd.   (601) 636-6561
Vicksburg, MS 39180
R.E. Paine

Inventive Women

- Windshield wipers were patented by Mary Anderson in 1903.
- The Melita Automatic Drip Coffee Maker was invented by Melita Bentz in Germany in 1908.
- A dishwasher was patented by Josephine Cochrane in 1914.
- Hattie Alexander developed the cure for meningitis in the 1930s.
- Lise Meitner discovered and named nuclear fission in 1939.
- The first disposable diaper was patented by Marion Donovan in 1951.
- Helen G. Gonet invented an electronic Bible in 1984.

To read more about women inventors, pick up a copy of *Mothers of Invention*, written by Ethlie van Vare and Greg Ptacek, published in 1987.

Delta Inventors Society

P.O. Box 257   (601) 686-4041
Stoneville, MS 38776
FAX (601) 686-4045
Gordon Tupper, President

Meridian Community College SBDC

Meridian, MS 39307   (601) 482-7445
FAX (601) 482-5803
J.W. Lang, Director

Mississippi SBDC

Suite 216, Old Chemistry Bldg.   (601) 232-5001
University, MS 38677
FAX (601) 232-5001
Raleigh Byars, State Director

Mississippi State University SBDC

P.O. Box 5288   (601) 325-8684
Mississippi State, MS 39762
FAX (601) 325-8686
Noel Estel Wilson, Jr., Managing Dir.

Society of Mississippi Inventors

P.O. Box 13004   (601) 982-6229
Jackson, MS 39236-3004
FAX (601) 982-6610
Dr. William Blair, President

MISSOURI

Central Missouri State University Center for Technology

Grinstead 80   (816) 543-4402
Warrensburg, MO 64093
FAX (816) 543-8159
Bernard L. Sarbaugh, Coordinator for Tech

EnTech Engineering, Inc.

1846 Craig Park Court   (314) 434-5255
St. Louis, MO 63146
FAX (314) 434-3270
Gary J. Weil, Vice President

Innovation Institute

Route 2, Box 184
Everton, MO 65646
FAX (417) 836-6337
Jerry Udell

(417) 836-6302

Mid-America Inventors

2018 Baltimore
Kansas City, MO 64108
FAX (816) 221-3995
Ed Stout

(816) 221-2442

Missouri Innovation Center

5650 A S. Sinclair Road
Columbia, MO 65203
FAX (314) 443-3748
Lisa Sireno, Assistant Director

(314) 446-3100

Missouri SBDC

St. Louis University
O'Neil Hall—100
3674 Lindell Blvd., Room 007
St. Louis, MO 63108

(314) 534-7204

St. Louis Technology Center

P.O. Box 12405
St. Louis, MO 63132
FAX (314) 432-1250
Gene J. Boesch, Managing Director

(314) 432-4204

University of Missouri— Columbia SBDC Mid-Missouri Inventor Association

1800 University Place
Columbia, MO 65211
FAX (314) 882-9931
Nick Arends, Business Counselor

(314) 882-7096

University of Missouri—Rolla Center for Technology Development

Bldg. 1, Nagogami Terrace
Rolla, MO 65401
FAX (314) 341-4992
Dr. John Amos, Professor

(314) 341-4559

Washington University

Campus Box 1054
1 Brookings Drive
St. Louis, MO 63130
FAX (314) 935-5862
H.S. Duke Leahy

(314) 935-5825

MONTANA

Business Development Center

305 W. Mercury Street
Butte, MT 59701
Teri Foley

(406) 723-4061

Creativity, Innovation, Productivity, Inc.

RR #1, Box 37
Highwood, MT 59450
Fred E. Davison, President

(406) 733-5031

Montana Department of Natural Resources & Conservation Energy Division

1520 E. 6th Ave.
PO Box 202301
Helena, MT 59620-2301
FAX (406) 444-6721
Howard E. Haines, Bioenergy Eng. Spec.

(406) 444-6697

Montana Inventors Association President of King Tool

5350 Love Lane
Bozeman, MT 59715
FAX (406) 585-9028
Clarence A. Emerson, Acting President

(406) 586-1541

Montana Science & Technology Alliance

46 N. Last Chance Gulch, Ste. 2B
Helena, MT 59620
FAX (406) 442-0788
David P. Desch, Senior Investments Manager

(406) 449-2778

Montana State University Montana Entrepreneurship Center

Reid Hall
Bozeman, MT 59717
FAX (406) 994-4152
Ann Keenan, Sr. Reg. Director

(406) 994-2024

Montana State University Research & Development Institute

1711 West College
Bozeman, MT 59715
FAX (406) 587-4480
Roger N. Flair

(406) 587-4479

Montana State University—Billings
Montana Entrepreneurship Center

Cisel Hall   (406) 657-2813
Billings, MT 59101
FAX (406) 657-2327
Dave Krueger, Regional Director

Montana Tradeport Authority

2722 3rd Ave. N, Ste. 300 W   (406) 256-6871
Billings, MT 59101-1931
FAX (406) 256-6877
Bruce Hofmann, Business Development
     Coordinator

University of Montana
Montana Entrepreneurship Center

Suite 204, McGill Hall   (406) 243-4009
Missoula, MT 59812
FAX (406) 243-4030
John Balsam, Western Regional Director

Yellowstone Inventors Association

3 Carrie Lynn   (406) 259-9110
Billings, MT 59102
W. T. George, Pre-Patent Assistant

NEBRASKA

Association of SBDCs

1313 Farnam St., Suite 132   (402) 595-2387
Omaha, NE 68182-0472
FAX (402) 595-2388
Jackie Johnston, Member Services Director

The Development Council

1007 Second Ave.   (308) 237-9346
P.O. Box 607
Kearney, NE 68848
FAX (308) 237-3103
Ron Tillery, President

University of Nebraska—Lincoln
UN Engineering Extension

W191 Nebraska Hall   (402) 472-5600
Lincoln, NE 68588-0535
FAX (402) 472-2410
Herbert Hoover, Information Specialist

University of Nebraska at Omaha
Nebraska Business Development Centers

1313 Farnam, Ste. 132   (402) 554-2521
Omaha, NE 68182
FAX (402) 554-3747
Robert Bernier, State Director

NEVADA

Concept Support and Development Corp.

121 Woodland Ave., #100   (702) 746-0700
Reno, NV 89523
FAX (702) 746-0700
J.H. Christiansen, President

Nevada Inventors Association

P.O. Box 9905   (702) 322-9636
Reno, NV 89507-0905
FAX (702) 322-0147
Don Costar, Pte, Founder

Nevada Inventors Association

P.O. Box 3121   (702) 877-1161
Carson City, NV 89702-3121
FAX (702) 849-3342
Rolin Stutes, President

NITEC
TMCC Institute for Business & Industry

4001 S. Virginia St.   (702) 829-9000
Reno, NV 89502
FAX (702) 829-9009
John Kleppe, PhD, PE, President

South Nevada Certified Development Co.

2770 S. Maryland Pkwy., Ste. #212   (702) 732-3998
Las Vegas, NV 89109
FAX (702) 732-2705
Dr. Thomas Gutherie, President/CEO

University of Nevada—Las Vegas
Nevada SBDC

4505 Maryland Parkway   (702) 895-0852
Box 456011
Las Vegas, NV 89154-6011
FAX (702) 895-4095
Sharolyn Craft, Director

University of Nevada—Reno SBDC

College of Business Administration   (702) 784-1717
Reno, NV 89557
FAX (702) 784-4337
Sam Males, State Director

NEW HAMPSHIRE

I.D.E.A.

Star Route Box 98   (603) 469-3304
Meridan, NH 03770
FAX (603) 469-3305
Richard Thompson

Manchester SBDC & Training Programs Office

1000 Elm Street, 8th Floor   (603) 624-2000
Manchester, NH 03101-1730
FAX (603) 634-2449
Robert T. Ebberson

TechWorks

15 Kancamagus Estates   (603) 447-6331
P.O. Box 337
Conway, NH 03818-0337
FAX (603) 447-6331
David S. Urey, Licensing Consultant

NEW JERSEY

National Society of Inventors

P.O. Box 434   (201) 994-9282
Cranford, NJ 07016
Sheila Kalisher, President

New Jersey SBDC

180 University Ave.   (201) 648-5950
Newark, NJ 07102
FAX (201) 648-1110
Alyson Miller, Assistant State Director

Rutgers Graduate School of Management New Jersey SBDC

180 University Ave.   (201) 648-1597
Newark, NJ 07102
FAX (201) 648-1596
Randy Harmon, Help Desk Manager

Society for the Encouragement of Research and Invention

P.O. Box 412   (201) 273-1088
100 Summit Ave.
Summit, NJ 07901

NEW MEXICO

Los Alamos Economic Development Corp.

P.O. Box 715   (505) 662-0001
Los Alamos, NM 87544
FAX (505) 662-0099
James M. Greenwood, Executive Director

New Mexico State University

Box 30001, Dept 3RED   (505) 646-2022
Las Cruces, NM 88003
FAX (505) 646-6530
Dr. Averett Tombes, VP Research & Economic Development

New Mexico Tech Research & Economic Development

College Avenue   (505) 835-5646
Socorro, NM 87801
FAX (505) 835-5649
Allan Gutjahr

Technology Ventures Corporation

4919 Marble NE   (505) 246-2882
Albuquerque, NM 87110
FAX (505) 246-2891
Richard Reisinger, Director Product Development

**University of New Mexico
Anderson School of
Management**

Albuquerque, NM 87131-1221   (505) 277-6471
FAX (505) 277-7108
Howard L. Smith, Interim Dean

NEW YORK

Brooklyn SBDC

111 Livingston St., Room 208   (718) 596-7081
Brooklyn, NY 11201
FAX (718) 596-6989
Harvey Heit

InfoEd

453 New Karner Road   (518) 464-0691
Albany, NY 12205
FAX (518) 464-0695
Earl Wells

**Information Service and
Research**

P.O. Box 95   (315) 476-7359
Syracuse, NY 13210
Brian McLaughlin

**Institute of Electrical and
Electronics Engineers, Inc.**

345 E. 47th St.   (212) 705-7867
New York, NY 10017

**International Licensing
Industry Association**

350 Fifth Ave., Ste. 6210   (212) 244-1944
New York, NY 10118-6293
FAX (212) 563-6552
Murray Altchule

Inventors of Greece in the USA

Pan Hellenic Society   (516) 223-5958
2053 Madison Ave.
South Merrick, NY 11566
Kimon Louvaris

New York Public Library

5th Avenue & 42nd St.   (212) 930-0917
New York, NY 10018
Donna Hopkins, Patents Librarian

New York SBDC

State University of New York   (518) 473-5398
SUNY Central Administration, S-523
SUNY Plaza
Albany, NY 12246
Toll-free: 1-800-732-SBDC

**New York Society of
Professional Inventors**

116 Steuart Ave.   (516) 598-3228
Amityville, NY 11701
FAX (516) 598-3241
Phil Knapp

**New York State
Energy Authority
Technology Information Assoc.**

2 Empire State Plaza, Ste. 1901   (518) 465-6251
Albany, NY 12223-1253
FAX (518) 432-9474
Paul W. Larrabee

Smith Engineering

114-56 142nd St.   (718) 529-3434
Jamaica, NY 11436
Leo Smith, President

**SUNY Farmingdale
New York Society of
Professional Inventors**

Lupton Hall   (516) 420-2397
Farmingdale, NY 11735
J.E. Manuel

**SUNY Institute of Technology
SBDC**

P.O. Box 3050   (315) 792-7546
Utica, NY 13504
FAX (315) 792-7554
Tom Reynolds, Director

Syracuse IWIEF Chapter

205 S. Central Ave   (315) 656-9210
Minoa, NY 13116
Ray DiPietro, Chairman

**Technology Business Advisory
Group**

100 Kinloch Commons   (315) 682-8502
Manlius, NY 13104-2484
FAX (315) 682-8508
Richard Labs, Research Director/CL&B

Western New York Technology
Development Center
200 Harrison St.   (716) 661-3336
Jamestown, NY 14701
FAX (716) 483-4470
William Wild, Director

NORTH CAROLINA

North Carolina SBDC
4509 Creedmoor Rd., #201   (919) 571-4154
Raleigh, NC 27612
FAX (919) 571-4161
Scott R. Daugherty, State Director

North Carolina Technological
Development Authority, Inc.
2 Davis Drive   (919) 990-8558
P.O. Box 13169
Research Triangle Park, NC 27709
FAX (919) 990-8561
Brent Lane, President

Small Business Technical
Development Center
34 Wall Street   (704) 285-0021
Room 707
Asheville, NC 28801
FAX (704) 285-0021
Deborah McKenna, Technology Counselor

NORTH DAKOTA

Center for Innovation &
Business Development
Box 8372, UND Station   (701) 777-3132
Grand Forks, ND 58202
FAX (701) 777-2339
Bruce Gjovig, Director

Garrison Area Improvement
Association
P.O. Box 445   (701) 463-2631
Garrison, ND 58540
FAX (701) 463-7487

Technology Transfer, Inc.
1833 East Bismarck Expressway   (701) 221-5346
Bismarck, ND 58504
FAX (701) 221-5320
Warren Enyart, CEO

University of North Dakota
North Dakota SBDC
118 Gamble Hall, Box 7308   (701) 777-3700
Grand Forks, ND 58202
FAX (701) 777-3225
Wally Kearns, State Director

OHIO

Akron/Youngstown
Inventors
1225 W. Market St.   (216) 864-5550
Akron, OH 44313
FAX (216) 867-7986
Ned Oldham, President

Editon Industrial Systems
Center
1700 N. Westwood Ave.   (419) 531-8610
Suite 2286
Toledo, OH 43607-1207
FAX (419) 531-8465
Charles Alter, Director
    Business Development

Hopewell Coop., Inc.
73 Maplewood Dr.   (614) 594-5200
Athens, OH 45701-1910
FAX (614) 592-6452
Ron Docie, President

Innovation Alliance
1445 Summit Street   (614) 421-7163
Columbus, OH 43201
FAX (614) 294-1002
N. Kent Brooks, President

Invention Engineering, Inc.
27 Aberdeen Avenue   (513) 294-7447
Dayton, OH 45419-3101
FAX (513) 294-7448
George D. Pierce, Jr., President

Inventors Connection
of Cleveland

P.O. Box 360804   (216) 226-9681
Cleveland, OH 44136

Inventors Connection
of Greater Cleveland

P.O. Box 46-254   (216) 581-4546
Bedford, OH 44146
Bob Abernathy

Inventors Council of Dayton
Young Inventors Committee

P.O. Box 630   (513) 439-4497
Dayton, OH 45459-0630
FAX (513) 439-1704
Ronald J. Versic, Chairman

Inventors Network, Inc.

1275 Kinnear Rd.   (614) 291-7900
Columbus, OH 43219
Rick Hagle, President
Paul Simmons, First Vice President

Inventor's Society at
Northwest, Ohio

617 Croghan St.   (419) 332-2221
Fremont, OH 43420
FAX (419) 334-6164
Mike Kingsbourgh, President

National Business
Incubation Association

20 E. Circle Dr., #190   (614) 593-4331
Athens, OH 45701
FAX (614) 593-1996
Dinah Adkins, Executive Director
Terry Murphy, Director Member Services

National Inventors
Hall of Fame
Inventure Place

80 W. Bowery, Suite 201   (216) 762-4463
Akron, OH 44308
FAX (216) 762-6313
Thomas B. Hollingsworth

NORSAC
SBIR Assistance Center

58 W. Center St.   (216) 375-2173
Akron, OH 44308
FAX (216) 762-3657
Mike Lehere

Ohio's Thomas Edison
Program

77 S. High St.   (614) 466-3887
26th Floor
Columbus, OH 43266-0330
FAX (614) 644-5758
Kamal Alvi, Project Coordinator

Ohio Technology Transfer
Research Resource Program

Room 216 Bevis Hall   (614) 292-5485
1080 Carmack Rd.
Columbus, OH 43210
FAX (614) 292-1893
Robert E. Bailey, Director

University of Akron
Center for Economic Education

213 Crouse Hall   (216) 972-7762
Akron, OH 44325-4210
FAX (216) 972-6990
Fred Carr, Director

Yankee Ingenuity Programs

623 Grant St.   (216) 673-1875
Kent, OH 44240
FAX (216) 678-5743
Charlie Clark, President

OKLAHOMA

Invention Development
Society

8230 SW 8th St.   (405) 787-0145
Oklahoma City, OK 73128
FAX (405) 789-8198
Julian Taylor, President

Marketing/Business
Assistance

P.O. Box 1474   (918) 258-8420
Broken Arrow, OK 74013
Zelda J. Anderson, Owner/Manager

New Technology Ventures, Inc.

7549 Southwest 34th St.   (405) 745-7800
Oklahoma City, OK 73179
FAX (405) 745-2276
Dr. Floyd E. Farha, President

O.T. Autry Area Vocational Technology School

1201 West Willow
Enid, OK 73703
FAX (405) 233-8262
Terry Henneke, Bid Assistant Coordinator

(405) 242-2750

Oklahoma Inventors Congress

P.O. Box 27850
Tulsa, OK 74149-0850
FAX (918) 245-2947
Kenneith F. Addison, Jr.

(918) 245-6465

Oklahoma SBDC Inventors Resource and Technology Center

100 S. University Ave., Rm. 106
Enid, OK 73701
FAX (405) 237-2304
Bill Gregory, Coordinator

(405) 242-7989

Oklahoma Student Inventors Exposition

8230 S.W. 8th St.
Oklahoma City, OK 73128
Betty Wright,
Childrens Invention Coordinator

(405) 787-0145

Rural Enterprises, Inc.

P.O. Box 1335
Durant, OK 74701
FAX (405) 920-2745
James E. Benish, RTAT Marketing

(405) 924-5094

Zuzak Corporation

5460 S. Garnett Ste. A
Tulsa, OK 74146
FAX (918) 622-6755
Tom Mosley, Jr.,
Director Technology Transfer

(918) 663-0155

....

OREGON

....

Central Oregon Community College Business Development Center

2600 N.W. College Way
Bend, OR 97701-5998
FAX (503) 383-7503
Robert L. Newhart II, Director

(503) 383-7290

African American Invention

To learn more about African American inventors and scientists, contact the Black Inventions Museum at P.O. Box 76122, Los Angeles, California 90076. The telephone number is (310) 859-4602. This museum is a nonprofit corporation.

Oregon Graduate Institute Saturday Academy

P.O. Box 91000
Portland, OR 97291-1000
FAX (503) 690-1470
Pete Taylors, Program Director

(503) 639-8701

Oregon Resource and Technology Development Fund

1934 N.E. Broadway
Portland, OR 97232-1502
FAX (503) 280-6080
John A. Beaulieu, President

(503) 282-4462

Oregon State Library Patent & Trademark Depository

State Library Building
Salem, OR 97310
FAX (503) 588-7119
Barbara O'Neill

(503) 378-4239

Theodore Texas & Associates

P.O. Box 4633
Portland, OR 97208
FAX (503) 246-4859
Dr. Brecharr Hemmaplardh, President

(503) 244-6447

This Lit'l Inventor

10333 Leverman Road SE
Aumsville, OR 97325
FAX (503) 749-3375
Bill Nasset, Public Information Provider

(503) 749-1052

Unitor Corporation

P.O. Box 309
Silverton, OR 97381
FAX (503) 873-7746
Arthur Ferretti, President

(503) 873-7746

....

Inventor Organizations

315

PENNSYLVANIA

American Society of Inventors
P.O. Box 58426   (215) 546-6601
Philadelphia, PA 19102
Jay W. Cohen, President

**Bucknell University SBDC
Product Development Center**
126 Dana Engineering Building   (717) 524-1249
Lewisburg, PA 17837
FAX (717) 524-1768
Charles Coder, Director

**Domestic & International
Technology Ltd**
115 West Ave.   (215) 885-7670
Jenkintown, PA 19046
FAX (215) 884-1385
Richard Stollman, President

**Gannon University
Northwestern Inventors
Council**
Erie, PA 16541   (814) 871-7619
FAX (814) 455-2631
Robert K. Jordon, Founder

**Lehigh University Inventors
Assistance**
621 Taylor Street   (215) 758-3446
Bethlehem, PA 18015
FAX (215) 758-4499
Dr. Bruce M. Smackey

**Penn State Behrend College
PENN Tap**
Station Road
Erie, PA 16563-0101
Jay Schenck

Pennsylvania SBDC
423 Vance Hall   (215) 898-1219
3733 Spruce St.
Philadelphia, PA 19104-6374
FAX (215) 573-2135
Gregory L. Higgins, Jr., State Director

Techni-Research Association, Inc.
P.O. Box T   (215) 657-1753
Willow Grove, PA 19090-0922
FAX (215) 576-7924
Dr. Louis Schiffman, President

**University City Science
Center**
3624 Market Street   (215) 387-2255
Philadelphia, PA 19104
FAX (215) 382-0056
Robert S. Krutsick, Executive Vice President

RHODE ISLAND

Bryant College SBDC
1150 Douglas Pike   (401) 232-6111
Smithfield, RI 02917
FAX (401) 232-6416
Douglas Jobling, State Director

**Rhode Island Partnership
for Science & Technology**
7 Jackson Walkway   (401) 277-2601
Providence, RI 02903
FAX (401) 277-2102
Claudia Terra, Executive Director

**Rhode Island Solar Energy
Association**
42 Tremont Street   (401) 942-6691
Cranston, RI 02920-2543
Domenic Bucci

SOUTH CAROLINA

Carolina Inventors Council
2960 Dacusville High Way   (803) 859-0066
Easley, SC 29640
Johnny Sheppard, President

**Clemson University
Battelle Liasion**
300 Brackett Hall   (803) 656-1296
Clemson, SC 29634-5705
FAX (803) 656-0202
William Chard, Associate Vice President Research

Enterprise Development Inc. of South Carolina

P.O. Box 1149   (803) 252-8806
Columbia, SC 29202
FAX (803) 252-0053
Ronald A. Young, Vice President

Southeast Manufacturing Tech Center

1201 Main St., Ste. 2010   (803) 252-6976
PO Box 1149
Columbis, SC 29202
FAX (803) 252-0056
Dallas Garrett

University of South Carolina— Aiken
O'Connell Economic Enterprise Institute

171 University Parkway   (803) 648-6851
Aiken, SC 29801
FAX (803) 641-3362
Dennis Rogers, Director

University of South Carolina SBDC
College of Business Administration

Columbia, SC 29208   (803) 777-4907
FAX (803) 777-4403
John Lenti, State Director

SOUTH DAKOTA

South Dakota State University
College of Engineering

CEH 201 Box 2219   (605) 688-4161
Brookings, SD 57007-0096
FAX (605) 688-5878
Duane Sander, Dean

South Dakota Public Utility Commission

State Capitol Bldg.   (605) 773-3201
500 E. Capital
Pierre, SD 57501
FAX (605) 773-3809
Steven Wegman, Engineer

TENNESSEE

Martin Marietta Energy Systems, Inc.
Office of Technology Transfer

P.O. Box 2009   (615) 483-0151
Oak Ridge, TN 37831-8242
FAX (615) 574-9241
Dennis A. Grahl

Tennessee Inventors Association

P.O. Box 11225   (615) 483-0151
Knoxville, TN 37939
FAX (615) 974-5492
Dewey Feezell, President

Tennessee SBDC/SW

320 S. Dudley   (901) 527-1041
Memphis, TN 38104
FAX (901) 527-1047
Sharon Taylor-McKinney, Business Consultant
Tom Bowie, Consultant

University of Tennessee Research Corporation

415 Communication Bldg.   (615) 974-1882
Knoxville, TN 37996-0344
FAX (615) 974-2803

TEXAS

Amarillo Inventors Association

P.O. Box 15023   (806) 376-8726
Amarillo, TX 79105
FAX (806) 376-7753
Worth Hefley, President

Austin Entrepreneurs and Inventors Organization

1418 Fairwood Rd.   (512) 451-8871
Austin, TX 78722
Tim Bigham

Austin Technology Incubator

8920 Business Park Dr.   (512) 794-9994
Austin, TX 78759
FAX (512) 794-9997
Jamin Patrick, Director

Baylor University
Center for Entrepreneurship
P.O. Box 98011   (817) 755-2265
Waco, TX 76798
FAX (817) 755-2271
Fadene Shirley, Program Coordinator

Bill J. Priest Institute for
**   Economic Development**
Technology Transfer Center
1402 Corinth   (214) 565-5852
Dallas, TX 75215
FAX (214) 565-5881
Pamela Speraw, Director

Entrepreneurs Online
10550 Richmond Ave.   (713) 784-8822
Suite 200
Houston, TX 77042
FAX (712) 735-2900
Mike Mead, President & Founder

Forensic & Small Business
**   Services**
Venture Assistance
5807 S. Braeswood Blvd.   (713) 729-1129
Houston, TX 77096
FAX (713) 729-1129
Daniel Altman, President

Houston Inventors
**   Association**
204 Yacht Club Lane   (713) 326-1795
Seabrook, TX 77586
FAX (713) 326-1795
Charles Mullen, Chairman

Information Insights
1418 Fairwood Rd.   (512) 451-8871
Austin, TX 78722
FAX (512) 451-1885
Tim Bigham, Principle Researcher

Inventors Information
**   Systems**
P.O. Box 927   (915) 698-3318
Abilene, TX 79604
FAX (915) 673-4855
Scottie D. Williams, Owner

NASA—Johnson Technology
**   Commercialization Center**
2200 Space Park Dr.   (713) 335-1250
Ste. 200
Houston, TX 77058
FAX (713) 333-9285
Jill Fabricant, Ph.D., Director

Network of American
**   Inventors and Entrepreneurs**
11371 Walters Rd.   (713) 537-8277
Houston, TX 77067
FAX (713) 537-1548
Wessie Cramer, Executive Director

Texas Innovation Network
Infomart
1950 Stemmons Fwy. Ste. 5037E   (214) 746-5140
Dallas, TX 75207
FAX (714) 746-3091
Dan Morrison, Executive Director

Texas Inventors Association
4000 Rock Creek Drive #100   (817) 265-1540
Dallas, TX 75204
Mrs. Eloyd Murphy, President

Texas Tech University SBDC
2579 S. Loop 289   (806) 745-1637
Lubbock, TX 79423
FAX (806) 745-6207
David F. Montgomery, Director

The Capital Network
8920 Business Park Dr.   (512) 794-9398
Ste. 275
Austin, TX 78759
FAX (512) 794-0448
David Gerhardt, Director

Toy & Game Inventors of America
5813 McCart Ave.   (817) 292-9021
Ft. Worth, TX 76133
FAX (817) 346-8697
Bruce Davis

U.S. Department of Energy
Dallas Support Office
1420 W. Mockingbird Lane, Suite 400   (214) 767-7245
Dallas, TX 75247
FAX (214) 767-7231
Curtis E. Carlson, Jr., Director

Ultimate Concepts, Inc.
P.O. Box 740304   (713) 873-6338
Houston, TX 77274-304
FAX (713) 778-9092
Brian P. Shannon, President

University of Houston SBDC
Texas Product Development
  Center
1100 Louisiana, Ste. 500   (713) 752-8440
Houston, TX 77002
FAX (713) 756-1515
Susan Macy, Director

UTAH

Business Creation
324 S. State Street   (801) 538-8770
Suite 500
Salt Lake City, UT 84114-7380
FAX (801) 538-8773
G. Michael Alder, Director

CEDO
777 S. State St.   (801) 226-1521
Orem, UT 84058
FAX (801) 226-2678
Karen Stewart, Marketing Director

Center for Entrepreneurship
  Training
P.O. Box 30808   (801) 967-4558
Salt Lake City, UT 84130-8080
FAX (801) 967-4017
Sterling Francom

Intermountain Society of
  Inventors and Designers
9888 Darin Drive   (801) 571-2617
Sandy, UT 84070
John Winder, President

Small Business
  Administration
125 South State Street   (801) 524-3209
Salt Lake City, UT 84138
FAX (801) 524-4160
Josie Valdez

Science, Technology &
  Innovation
State Office
P.O. Box 11   (801) 569-2973
West Jordan, UT 84084
Jake Vandermeide, Chairman

S.E.U.A.L.G.
P.O. Box 1106   (801) 637-5444
Price, UT 84501
FAX (801) 637-5448
Dennis Rigby, Proc. Outreach Spec.

University of Utah
Marriott Library
Salt Lake City, UT 84112   (801) 581-8394
FAX (801) 585-3464
David Morrison, Patents Librarian

Utah SBDC
102 West 500 South, #315   (801) 581-7905
Salt Lake City, UT 84101
FAX (801) 581-7814
Mike Warren
David Nimkin

Weber State University
Technology Assistance Center
Ogden, UT 84408-1801   (801) 626-6309
FAX (801) 626-6987
Stephen Reed, Director

VERMONT

State of Vermont
Agency of Development &
  Community Affairs
Montpelier, VT 05609   (802) 828-3221
FAX (802) 828-3258
Curt Carter,
    Development Programs Coordinator

The Catalyst Group
P.O. Box 1200   (802) 254-3645
Brattleboro, VT 05302-1200
FAX (802) 254-3645
Hervey Scudder, Sales Development

Yankee Know-How

"'Yankee know-how' and 'American ingenuity' are two vernacular phrases that reflect the fact that U.S. creativity and innovation conquered a continent. The United States continues to invent and create, and as was the case in the past, U.S. inventions and creations induce investment, beget businesses, and generate jobs. The world looks to the United States for leadership in agriculture and engineering, in business and science, in communication and transportation, and in entertainment and recreation."

—Bruce A. Lehman,
Commissioner of Patents and Trademarks

VIRGINIA

**American Intellectual Property
Law Association**
2001 Jefferson Davis Hwy.   (703) 415-0780
Suite 203
Arlington, VA 22202
FAX (703) 415-0786
Martha Morales, Associate Executive Director

**Center for Innovative Technology
Technology Commercialization**
CIT Bldg, Ste. 600   (703) 689-3000
2214 Rock Hill Rd.
Herndon, VA 22070
FAX (703) 689-3041
Cathy Renault,
Director Entrepreneur Development

**Clearinghouse for State &
Local Initiatives in
Production,
Technology, & Innovation**
304 F 8001 Forbes Pl.   (703) 487-4968
Port Royal Rd.
Springfield, VA 22151
FAX (703) 321-8199
Elizabeth M. Robertson, Program Analyst

N.C.I./E.R.E.C.
8260 Greensboro Dr. Ste. 325   (703) 903-0325
Mclean, VA 22102
FAX (703) 903-9750
Bill Martin

**NJ Committee on Science
and Technology**
4156 Elizabeth Lane   (703) 425-4825
Annadale, VA 22003
FAX (703) 425-6736
Mike Miller

**Renewable Energy Clearinghouse
EREC Energy Efficiency**
P.O. Box 3048   (800) 523-2929
Merrifield, VA 22110
Paul Hesse, Senior Technical Specialist

**Small Business Development
Center**
918 Emmet Street North   (804) 295-8198
Ste. 200
Charlottesville, VA 22903-4878
FAX (804) 295-7066
Charles A. Kulp, Director

Virginia Tech Graduate Center
2990 Telestar Court   (703) 698-6016
Falls Church, VA 22042
FAX (703) 698-6062
Wanda Hylton

**Wheeling Jesuit College
National Technology
Transfer Center**
2121 Eisenhower Ave.   (703) 518-8800
Ste. 400
Alexandria, VA 22314
FAX (703) 518-8986
Sally Rood, Associate Director

WASHINGTON

Critical Data, Inc.
906 South Cowley   (509) 838-2917
Spokane, WA 99202
FAX (509) 838-4989
Bill Hockett

**Federal Laboratory
    Consortium
Management Support Office**

P.O. Box 545   (206) 683-1005
Sequim, WA 98382-0545
FAX (206) 683-6654
George Linsteadt

**Innovation Assessment Center
    Program, SBDC Specialist**

135 Kruegel Hall   (509) 335-1576
Pullman, WA 99164-4727
FAX (509) 335-0949
JoAnna Slaybaugh-Taylor
Peter Stroosma

Innovators International

P.O. Box 4636   (206) 842-7833
Rolling Bay, WA 98061

NASA—Northwest Office RTTC

12318 N.E. 100th Place   (206) 827-5136
Kirkland, WA 98033
FAX (206) 827-5430
Barbara Campbell, Director

**Seattle University
Albers School of Business
    & Economics**

Broadway and Madison   (206) 296-5702
Seattle, WA 98122-4460
FAX (206) 296-5795
Harriet B. Stephenson, Ph.D.

Skagit Valley College, SBDC

2405 College Way   (206) 428-1282
Mount Vernon, WA 98273
FAX (206) 336-6116

Technology Targeting, Inc.

7454 S. 118th Pl.   (206) 227-8088
Seattle, WA 98178-3039
FAX (206) 227-8068
Dr. Norman Brown

The Business Consortium, Inc.

P.O. Box 599   (509) 647-2000
Wilbur, WA 99185
FAX (509) 647-2001
James M. Canode, Project Director

**Tri-Cities Enterprise
    Association**

2000 Logston Blvd.   (509) 375-3268
Richland, WA 99352
FAX (509) 375-4838
Dallas Breamer, President

**TRIDEC: Tri-City Industrial
    Development Council**

901 N. Colorado   (509) 735-1000
Kennewick, WA 99336
FAX (509) 735-6609
John Lindsay, President/CEO

**University of Washington
Office of Technology Transfer**

1107 NE 45th Street, Suite 200   (206) 543-3970
Seattle, WA 98105
FAX (206) 685-4767
Margaret Wagner-Dahl, Acting Director

**Washington State University
Innovation Assessment Center**

180 Nickerson St., #207   (206) 464-5450
Seattle, WA 98109
FAX (206) 464-6357
Jim Van Orsow, Director

**Washington State University,
    Tri-Cities**

100 Sprout Road   (509) 375-9207
Richland, WA 99352-1643
FAX (509) 375-5337
Wiboon Arunthanes, Ph.D.,
Assistant Professor Marketing

**Western Washington
    University SBDC**

309 Parks Hall   (206) 650-3899
Bellingham, WA 98225-9073
FAX (206) 650-4844
Lynn Trzynka

WEST VIRGINIA

West Virginia SBDC

950 Kanawha Blvd. East   (304) 558-2960
Charleston, WV 25301
FAX (304) 558-0127
Hazel Kroesser, State Director

WISCONSIN

**University of Wisconsin—
   Whitewater
Wisconsin Innovation Service
   Center**

402 McCutchan Hall   (414) 472-1365
Whitewater, WI 52190
FAX (414) 472-1600
Debra Malewicki, Program Manager

**Wisconsin Department of
   Development
Technology Deployment Fund**

P.O. Box 7970   (608) 267-9383
Madison, WI 53707
FAX (608) 267-2829

WYOMING

Small Business Administration

P.O. Box 2839   (307) 261-5761
Casper, WY 82602
FAX (307) 261-5499
Kay Stucker

Thinkers on Creativity

- It is axiomatic that to think intelligently is to think creatively. —Alex F. Osborn

- I suspect that almost all creativity is really the result of play, in the higher meaning of the word. —Sidney J. Parnes

- The more we observe and are aware of, the more mental connections we can make that will result in new and relevant ideas. —Melissa Strickland

- The empires of the future are the empires of the mind. —Winston Churchill

- Every healthy and creative individual resists engulfment by custom and rigid habits. —Herbert Bonner

**West Inventors Council
Discovery Lab**

109 Engineering Research   (304) 293-3612
P.O. 6101
Morgantown, WY 26506-6101
FAX (304) 293-3472
Scott Warner, Assistant Manager/Org. Coordinator

CANADA

British Columbia Inventors Society

P.O. Box 5086
Vancouver, British Columbia V6B 4A9
Canada
Jeff Galenzoski, President

**Canadian Centre for Industrial
   Innovation Montreal
Evaluation and Technology Transfer**

75 Port Royal East   (514) 383-7712
Suite 600
Montreal, Quebec H3L 3T1
Canada
FAX (514) 383-7040
Luc E. Morisset, Director

**Canadian Industrial Innovation
   Centre**

156 Columbia St. West   (519) 885-5870
Waterloo, Ontario N2L 3L3
Canada
FAX (519) 885-5729
Gary Svoboda, Manager
Bob Huehn, Senior Analyst

**Canadian Young Inventors' Fair
   Society**

c/o Box 12151   (604) 689-3626
1220—808 Nelson Street
Vancouver, British Columbia V6Z 2H2
Canada
FAX (604) 684-4589
Chris Webb, Director

**Copyrights, Inventions &
   Patents Association of Canada**

1518 Dome Tower   (403) 265-3011
333-7 Ave. SW
Calgary, Alberta T2P-2Z1
Canada
FAX (403) 266-4091
Ralf Koechling, President

Guerin Science & Technology

40 Fountainhead Road #2004　　　　(416) 661-3127
Downsview, Ontario M3J 2V1
Canada
FAX (416) 661-3127
Stephen Guerin, Inventor/Owner

IDEAS Digest Newsmagazine

Box 12151　　　　　　　　　　(604) 689-3626
1220—808 Nelson St.
Vancouver, British Columbia V6Z 2H2
Canada
FAX (604) 684-4589
Chris Webb, Editor/Owner/Publisher

Inventors' Alliance of Canada

47 Kenneth Avenue　　　　　　　(416) 762-3453
Toronto, Ontario M6P 1J1
Canada
FAX (416) 762-3301
Mark Ellwood, President

Inventors' Society of
Nova Scotia

1046 Barrington St.　　　　　　(902) 421-1250
Ste. 106
Halifax, Nova Scotia B3H 2R1
Canada
FAX (902) 429-9983
Richard Valee, President

Nova Scotia Department of
Economic Development

1800 Argyle　　　　　　　　　(902) 424-7382
P.O. Box 519
Halifax, Nova Scotia B3J 2R7
Canada
FAX (902) 424-5739
Terry Collins, Consultant

Simon Fraser University
University/Industry Liaison

Burnby, British Columbia V5A 1S6　(604) 291-5844
Canada
FAX (604) 291-3477
Ms. Teri Lydiard, Technology Transfer Officer

The Women Inventors
Project, Inc.

1 Greensboro Drive　　　　　　(416) 243-0668
Suite 302
Etobicoke, Ontario M9W1C8
Canada
FAX (416) 243-0688
Susan Best, Project Director
Mrs. Chips Klein, Co-Director

University of Western Ontario

London, Ontario N6A 5B8　　　(519) 661-2161
Canada
FAX (519) 661-3907
Gregor Reid, Director Res. Services

OTHER

American Samoa Government
Economic Development
Planning Office

Pago Pago, American Samoa 96799　(684) 633-5155
FAX (684) 633-4195
Dr. Meki Solomona, Director

Puerto Rico Products
Association

P.O. Box 363631　　　　　　　(809) 753-8484
San Juan, PR 00936-3631
FAX (809) 753-0855
Juan Rivera Bigas, Executive Director

Puerto Rico SBDC

Box 5253　　　　　　　　　　(809) 834-5556
College Station
Mayaguez, PR 00681
FAX (809) 832-5550
José Romaguera, Executive Director

Solarville

P.O. Box 1167　　　　　　　　(671) 472-8131
Agana, Guam 96910
FAX (671) 472-8131
Joseph E. La Ville, Owner/Manage

Federal Programs for Inventors

SBIR Participants

Eleven federal agencies with an extramural budget for research or research and development presently participate in the Small Business Innovation Research (SBIR) Program. They are:

- Department of Agriculture
- Department of Commerce
- Department of Defense
- Department of Education
- Department of Energy
- Department of Health and Human Services
- Department of Transportation
- Environmental Protection Agency
- National Aeronautics & Space Administration
- National Science Foundation
- Nuclear Regulatory Commission

A man's judgment cannot be better than the information on which he has based it.

—ARTHUR HAYS SULZBERGER

Uncle Sam Has Money for Inventors

There are many federal programs available to inventors, but the following are those which I feel have the most potential for independent inventors, if certain criteria are met.

Understand that these are government programs and, therefore, involve all kinds of paperwork. However, if you are able to tolerate working with a large bureauracy, and your invention meets Uncle Sam's needs, one of these programs may hold rewards for you.

Small Business Administration

The U.S. Small Business Administration (SBA) was created by Congress in 1953 to help America's entrepreneurs form successful small enterprises.

Small businesses are the backbone of the American economy. They create two of every three new jobs, produce 40 percent of the gross national product, and develop more than one half the nation's technological innovations. Today some 20 million small companies provide job opportunities in the United States.

Inventors are small businesses, whether they manufacture their concepts themselves or license them to larger manufacturers. As small business owners and operators, inventors must, therefore, know how to manage and finance their enterprises. This is where the SBA may be of assistance.

Through workshops, individual counseling, publications and videotapes, the SBA helps entrepreneurs understand and meet the challenges of operating businesses—challenges such as financing, marketing, and management.

There are two programs that the SBA administers that could be useful to inventors: SBIR and STTR.

Small Business Innovation Research (SBIR) Program

In 1982 Congress passed the Small Business Innovation Development Act creating the federal Small Business Innovation Research (SBIR) Program. The purpose of the program is to increase the opportunity for small firms to participate in federal research and development. In addition to encouraging the participation of small businesses, the program is designed to stimulate the conversion of research findings into commercial application.

The Act designated the Small Business Administration (SBA) to run the program, govern its policy, monitor its progress, and analyze its results.

The SBIR grant program has given out about $4 billion since it began in 1982.

Eleven federal agencies with an extramural budget for research or research and development presently participate in the SBIR Program. They are:

- Department of Agriculture
- Department of Commerce
- Department of Defense
- Department of Education
- Department of Energy
- Department of Health and Human Services
- Department of Transportation
- Environmental Protection Agency
- National Aeronautics & Space Administration
- National Science Foundation
- Nuclear Regulatory Commission

How SBIR Works

Under the SBIR Program, the involved federal agencies request highly competitive proposals from small businesses in response to solicitations outlining their R&D requirements. After evaluating the proposals, each agency awards funding agreements for determining the technical feasibility of the research and development concepts proposed. These awards are as follows:

❑ **Phase I.** Awards up to $100,000 are made for research projects to evaluate the scientific and technical merit and feasibility of an idea. Time frame: six months. Two-thirds of this work must be done by the small business.

Let's say that you have an idea for a device that could, if successful, solve a problem posed by one of the SBIR agencies. There just might be $100,000 in the agency's budget to help you prove out the concept.

❑ **Phase II.** The Phase I projects with the most potential are funded to further develop the proposed idea for up to two years. Phase II awards can be as high as $750,000. Time frame: two years, and this can be exceeded with justification. One half of this work must be done by the small business.

If you are successful in realizing the first stage of your R&D effort, and the sponsoring agency thinks you are onto something, you just might qualify for Phase II funding.

❑ **Phase III.** Once you get into the final stage, or the commercialization process, there are no more federal SBIR funds available.

At this point, the federal government encourages you to seek private sector investment and support. While the government may extend follow-up production contracts for your technology, it no longer wants to be your partner. Ideally the federal seed money has been enough to get you off the ground.

State-Supported SBIR Programs

State governments, anxious to build their own industrial bases, have actively supported the SBIR Program by (1) promoting SBIR to small businesses, (2) providing information and technical assistance to SBIR applicants, (3) providing matching funds to SBIR Phase I and II recipients, and (4) helping firms to obtain Phase III funding from both private and state sources.

Why do the states do this? They see independent inventors and small businesses as a good investment because chances are technologies developed in a particular state will stay there once commercialized. Innovation leads to hard goods, goods create jobs, jobs employ people, people pay taxes, and so forth.

How to Contact SBIR Agencies

Here are the contacts on an agency by agency basis. If you would like to know about SBIR at a particular agency or be put on the mailing lists for SBIR solicitations, contact the appropriate pointperson.

Department of Agriculture

Dr. Charles F. Cleland
Director, SBIR Program
U.S. Department of Agriculture
Cooperative State Research Service
Ag. P.O. Box 2243
Washington, DC 20250-2243
(202) 401-4002

Department of Commerce

Dr. Joseph Bishop
U.S. Department of Commerce
SBIR Program Office
1315 East-West Highway
Room 15342
Silver Springs, MD 20910
(301) 713-3565

Mr. James P. Maruca
Director, Office of Small and Disadvantaged
Business Utilization
U.S. Department of Commerce
14th and Constitution Avenue, NW
HCHB, Room 6411
Washington, DC 20230
(202) 482-1472

Mr. Norman Taylor
SBIR Program
NIST
U.S. Department of Commerce
Physics Building, Rm. A343
Gaithersburg, MD 20899
(301) 975-4517

Department of Defense

Mr. Robert Wrenn
SBIR Program Manager
OSD/SADBU
U.S. Department of Defense
The Pentagon—Room 2A340
Washington, DC 20301-3061
(703) 697-1481

Department of Education

Mr. John Christensen
SBIR Program Coordinator
U.S. Department of Education
Room 602D
555 New Jersey Avenue, N.W.
Washington, DC 20208
(202) 219-2050

Department of Energy

Dr. Samuel J. Barish
SBIR Program Manager—ER-16
U.S. Department of Energy
Washington, DC 20585
(301) 903-0569

Department of Health and Human Services

Mr. Verl Zanders
SBIR Program Manager
Office of the Secretary
U.S. Department of Health and Human
Services
Washington, DC 20201
(202) 690-7300

Department of Transportation

Dr. George Kovatch
DOT SBIR Program Director, DTS-22
U.S. Department of Transportation
Research and Special Programs
Administration
Volpe National Transportation Systems
Center
55 Broadway, Kendall Square
Cambridge, MA 02142-1093
(617) 494-2051

Environmental Protection Agency

Mr. Donald F. Carey
SBIR Program Manager
Research Grants Staff (8703)
Office of Research and Development
U.S. Environmental Protection Agency
401 M Street, SW
Washington, DC 20460
(202) 260-7899

National Aeronautics and Space Administration

Mr. Michael F. Battaglia
Program Manager, SBIR
U.S. National Aeronautics and Space
Administration Headquarters

300 E. Street, SW
Washington, DC 20546-0001
(202) 358-4658

National Science Foundation

Mr. Tony Centodocati
Mr. Richie Coryell
Mr. Michael Crowley
Mr. Daryl G. Gorman
Mr. Charles Hauer
Dr. Sara Nerlove
Mr. Roland Tibbetts
SBIR Program Managers
U.S. National Science Foundation
SBIR Program, Room 590
4201 Wilson Boulevard
Arlington, VA 22230
(703) 306-1391

Nuclear Regulatory Commission

Ms. Marianne M. Riggs
SBIR Program Representative
Financial Management, Procurement and
Administrative Staff
U.S. Nuclear Regulatory Commission
Mail Stop T-10 D5
Washington, DC 20555
(301) 415-5822

SBIR Pre-Solicitation Mailing List

On a quarterly basis, SBA publishes an SBIR Pre-Solicitation Announcement (PSA) which contains pertinent information on SBIR solicitations about to be released by the participating federal agencies. An automated phone system allows new firms to record their request for PSAs by calling: (202) 205-7777.

Toll-Free SBIR Assistance

For specific information on SBIR programs, call 1-800-827-5722. This number will connect you with the SBA's Small Business Answer Desk.

Small Business Technology Transfer (STTR) Program

There is another competitive SBA program that you may wish to consider. It is the Small Business Technology Transfer (STTR) program.

The main difference between SBIR and STTR is that all research and development in the STTR pilot program must be conducted jointly by the small business concern and a non-profit research institution (includes Federally Funded Research and Development Centers. See pages: 328-29). Not less than 40 percent of the work conducted under an STTR program award is to be performed by the small business concern and not less than 30 percent of the work is to be performed by the non-profit research institution.

❏ **Phase I.** Awards may be as much as $100,000 for up to a one-year effort.

SBIR Success Stories

- NASA gave Electronic Imagery, Inc. $486,600 for work on high definition full color virtual image processing.

- The U.S. Army awarded $48,488 to Coleman Research Corp. for its work on indirect fire weapon simulation.

- The Department of Agriculture funded Artificial Intelligence Atlanta, Inc. $49,963 for a system for forest fire detection.

- Health and Human Services paid Martek Corp. $49,920 for a new source of arachidonic acid for infant formula.

For a complete and current listing of SBIR awardees, call the SBA, Office of Innovation Research and Technology: (202) 205-7777. The report will include the identity of the federal agency that made the SBIR award, the topic of the SBIR effort, the applicable SBIR phase, and the dollar amount of the SBIR award.

STTR Participants

Five federal agencies with an extramural budget for research or research and development presently participate in the Small Business Technology Transfer (STTR) Program. They are:

- Department of Defense
- Department of Energy
- Department of Health and Human Services
- National Aeronautics & Space Administration
- National Science Foundation

❑ **Phase II.** Awards may be as high as $500,000 for a two-year effort.

Five federal agencies with an extramural budget for research or research and development presently participate in the STTR Program. They are:

- Department of Defense
- Department of Energy
- Department of Health and Human Services
- National Aeronautics & Space Administration
- National Science Foundation

If you wish to be placed in the STTR source file, phone: (703) 756-4261.

To receive STTR solicitations, contact the participating agency and brochures and forms will be sent to you. The SBIR office contacts listed on pages 326-327 also handle STTR requests.

Federally Funded Research and Development Centers (FFRDC'S)

Department of Defense

- Institute of Defense Analysis (Alexandria, VA)
- Logistics Management Institute (Bethesda, MD)
- National Defense Research Institute (Santa Monica, CA)
- Software Engineering Institute (Pittsburgh, PA)
- Center for Naval Analyses (Alexandria, VA)
- Lincoln Laboratory (Lexington, MA)
- Aerospace Corporation (El Segundo, CA)
- Project Air Force (Santa Monica, CA)
- Arroyo Center (Santa Monica, CA)

National Aeronautics and Space Administration

- Jet Propulsion Laboratory (Pasadena, CA)

Department of Health and Human Service

- NCI Frederick Cancer Research and Development Center, (Frederick, MD)

National Science Foundation

- National Astronomy & Ionosphere Center (Areolbo, PR)
- National Center for Atmospheric Research (Boulder, CO)
- National Optical Astronomy Observatories (Tucson, AZ)
- National Radio Astronomy Observatory (Green Bank, WV)

Nuclear Regulatory Commission

- Center for Nuclear Waste Regulatory Analyses (San Antonio, TX)

Federal Aviation Administration

- Center for Advanced Aviation System Development (McLean, VA)

Department of Energy

- Idaho National Engineering Laboratory (Idaho Falls, ID)
- Energy Technology Engineering Center (Canoga Park, CA)

- Oak Ridge National Laboratory (Oak Ridge, TN)
- Sandia National Laboratories (Albuquerque, NM)
- Ames Laboratory (Ames, IA)
- Argonne National Laboratory (Argonne, IL)
- Brookhaven National Laboratory (Upton, Long Island, NY)
- Continuous Electron Beam Accelerator Facility (Newport News, VA)
- Lawrence Berkeley Laboratory (Berkeley, CA)
- Lawrence Livermore National Laboratory (Livermore, CA)
- Fermi National Accelerator Laboratory (Batavia, IL)
- Los Alamos National Laboratory (Los Alamos, NM)
- Princeton Plasma Physics Laboratory (Princeton, NJ)
- Stanford Linear Accelerator Center (Stanford, CA)
- Pacific Northwest Laboratory (Richland, WA)
- Inhalation Toxicology Research Institute (Albuquerque, NM)
- Oak Ridge Institute for Science and Education, (Oak Ridge, TN)

National Institute of Standards and Technology

Established by Congress as the Bureau of Standards in 1901, the organization was renamed National Institute of Standards and Technology (NIST) in 1988 with passage of the Omnibus Trade and Competitiveness Act. NIST's mandate is to enhance the nation's technological competitiveness through fast and effective transfer of new technologies to U.S. industries.

NIST employs about 3,200 scientists, engineers, technicians, and support personnel, and hosts about 1,200 visiting researchers each year. Fiscal year 1995 operating resources from all sources total about $1 billion.

The Institute maintains research facilities at its Gaithersburg, Maryland and Boulder, Colorado sites. Major facilities include a 20-megawatt research reactor, a metals processing facility, a synchrotron radiation source, and computer networking and security research laboratories.

NIST laboratories perform research across a wide range of disciplines, affecting virtually every industry. Primary fields of NIST research include chemical science and technology, physics, material science and engineering, electronics and electrical engineering, manufacturing engineering, computer systems, building technology, fire safety, computing, and applied mathematics.

Reflecting its role as the only federal laboratory exclusively dedicated to serving the needs of U.S. industry, NIST offers more than 300 types of calibrations, 1,000 standard reference materials for calibrating instruments and evaluating test methods, 24 standard reference data centers, laboratory accreditation programs, and free evaluation of energy-related inventions.

To contact NIST with general inquiries about their programs, call: (301) 975-3058; or send a message via e-mail to inquiries@nist.gov.

NIST has developed several programs that may interest the independent inventor.

- Manufacturing Extension Partnership (MEP).
- Advanced Technology Program (ATP)

- Energy-Related Inventions Program (ERIP)
- Non-Energy-Related Inventions Program (N-ERIP)

NIST Programs Contact Info

To contact the National Institute of Standards and Technology (NIST) with general inquiries about their programs, call: (301) 975-3058; or send a message via e-mail to inquiries@nist.gov.

Listed below are several NIST programs that may interest the independent inventor.

- **Manufacturing Extension Partnership (MEP):** with general inquiries about MEP, call: (301) 975-3593.

- **Advanced Technology Program (ATP):** for ATP information call: 1-800-ATP-FUND (1-800-287-3863).

- **Energy Related Inventions Program (ERIP):** For more information, call the Office of Technology Innovation, NIST, at (301) 975-5500; the fax number is (301) 975-3839. Or you can send inquiries via e-mail: innovate@enh.nist.gov.

- **Non-Energy-Related Inventions Program (N-ERIP):** George Lewett and Dr. Gilbert M. Ugiansky can provide information on opportunities available under this program. Call (301) 975-5506, or fax (301) 975-3839.

Manufacturing Extension Partnership (MEP)

The MEP assists small and medium-sized companies in adopting new technologies. It is a combination of two earlier programs: Regional Centers for the Transfer of Manufacturing Technology and the State Technology Extension Program.

There are more than 370,000 small U.S. companies—those with fewer than 500 employees—accounting for about 95 percent of all U.S. manufacturing plants.

MEP helps these companies through a nationwide network of affiliated technology extension centers that provide hands-on technical assistance. The program also links smaller manufacturers with federal agencies, such as the Small Business Administration (SBA) and the Environmental Protection Agency (EPA), and with other technology-related resources at the state and federal levels.

Examples of the type of grassroots assistance provided to companies through MEP Centers include help in implementing computer-aided design and manufacturing systems, redesign of the factory floor to improve production efficiency, and adoption of improved process technologies and controls.With general inquiries about MEP, call: (301) 975-3593.

Listed below are the 28 Manufacturing Extension Partnership Centers.

A.L. Philpott Manufacturing Center
PO Box 5311
Martinsville, VA 24115
(703) 638-8777, FAX: (703) 638-6469
John D. Hudson, Jr.

Arizona Applied Manufacturing Center
c/o Gateway College
108 N. 40th St.
Phoenix, AZ 85034
(602) 392-5184, FAX: (602) 392-5329
Charles Klement

California Manufacturing Technology
    Center
13430 Hawthorne Blvd.
Hawthorne, CA 90250
(310) 355-3060, FAX: (310) 676-8630
Joan Carvell

Chicago Manufacturing Technology
    Extension Center
Homan Square
3333 W. Arthington
Chicago, IL 60624
(312) 265-2020, FAX: (312) 467-0615
Rheal Turcotte

Connecticut State Technology Extension
    Program
368 Fairfield
Storrs, CT 06269-2041

(203) 486-2585, FAX: (203) 486-5246
Peter Laplaca

Georgia Manufacturing Extension Alliance
Georgia Institute of Technology
223 O'Keefe Bldg.
Atlanta, GA 30332
(404) 894-8989, FAX: (404) 853-9172
Charles Estes

Great Lakes Manufacturing Technology
    Center
4600 Prospect Ave.
Cleveland, OH 44103-4314
(216) 432-5322, FAX: (216) 361-2900
George Sutherland

Hudson Valley Manufacturing Outreach
    Center
Hudson Valley Tech. Development
300 Westage
Business Center, Ste. 140
Fishkill, NY 12524
(914) 896-6934, FAX: (914) 896-7006
Douglas Koop

Iowa Manufacturing Technology Center
Des Moines Area Community College
2006 S. Ankeny Blvd., Building 18 EDG
Ankeny, IA 50021
(515) 965-7040, FAX: (515) 964-6206
Del Shepard

Kentucky Technology Service
PO Box 1049
Lexington, KY 40588
(606) 233-3502, FAX: (606) 259-0986
Kris Kimel

MAMTC Colorado Regional Office
Rockwell Hall
Colorado State University
Fort Collins, CO 80523
(303) 224-3744, FAX: (303) 224-3715
Craig Carlile

Manufacturing Outreach Center of NY
Southern Tier
UniPEG
1310 North St.
Endicott, NY 13760
(607) 748-9214, FAX: (607) 785-0026
Mike Dziak

Massachusetts Manufacturing Partnership
Bay State Skills Corp.
101 Summer St., 4th Fl.
Boston, MA 02110
(617) 292-5100, ext. 271,
    FAX: (617) 292-5105
Jan Pounds

Mid-America Manufacturing Technology
    Center
10561 Barkley, Ste. 602
Overland Park, KS 66212
(913) 649-4333, FAX: (913) 649-4498
Paul Clay

Midwest Manufacturing Technology Center
Industrial Technology Institute
PO Box 1485
2901 Hubbard Rd.
Ann Arbor, MI 48106
(800) 292-4484, FAX: (313) 769-4064
John Cattier

New Mexico Manufacturing Extension
    Program
New Mexico, Inc.
1601 Randolph Rd., SE
Ste. 210
Albuquerque, NM 87106
(505) 272-7800, FAX: (505) 272-7810
Bill Rector

New York City Manufacturing Outreach
    Center
NY ITAC
253 Broadway, Rm. 302
New York, NY 10007
(212) 240-6920, FAX: (212) 240-4889
Sara Garretson

New York Manufacturing Extension
    Partnership
Rensselaer Technology Park

385 Jordan Rd.
Troy, NY 12180-8347
(518) 283-1010, FAX: (518) 283-1112
Mark Tebbano

Oklahoma Alliance for Manufacturing
    Excellence, Inc.
616 S. Boston, Ste. 402
Tulsa, OK 74119-1216
(918) 592-0722, FAX: (918) 599-9514
Edmund J. Farrell

Pennsylvania Manufacturing Extension
    Program
North/East Region
Manufacturers Resource Center
125 Goodman Dr.
Bethlehem, PA 18015
(610) 758-5599, FAX: (215) 758-4716
Edith Ritter

Plastics Technology Deployment Center
GLMTC Manufacturing Outreach Program
Prospect Park Bldg.
4600 Prospect Ave.
Cleveland, OH 44103
(216) 432-5300, FAX: (216) 361-2088
David Thomas-Greaves

Pollution Prevention Center
Institute for Research & Technical Assistance
2800 Olympic Blvd., Ste. 101
Santa Monica, CA 90404
(310) 453-0450, FAX: (310) 453-2660
Katy Wolf

Southeast Manufacturing Technology
    Center
PO Box 1149
Columbia, SC 29202-1149
(803) 252-6976, FAX: (803) 252-0056
James Bishop

The Delaware Manufacturing Alliance
Delaware Development Office
99 Kings Hwy.
PO Box 1401
Dover, DE 19903
(302) 739-4271, FAX: (302) 739-5749
John J. Shwed

Upper Midwest Manufacturing Technology
    Center
Minnesota Technology, Inc.
111 Third Ave. S., Ste. 400
Minneapolis, MN 55401
(612) 654-5201, FAX: (612) 654-5207
Sandy Voigt

Washington Alliance for Manufacturing
657 N. 34th St.
Seattle, WA 98103
(206) 633-5252, FAX: (206) 632-1272
Peggy Flynn

Western New York Manufacturing
    Outreach Center
Western NY Tech. Develop. Center
1576 Sweet Home Rd.
Amherst, NY 14228
(716) 636-3626, FAX: (716) 636-3630
Betsy Poole

Western Pennsylvania Manufacturing
    Extension Program
4516 Henry St.
Pittsburgh, PA 15213
(412) 687-0200, ext. 234, FAX:(412) 687-
    5232
Ray Christman

Advanced Technology Program (ATP)

The ATP provides cost-shared awards on a competitive basis to industry for development of high-risk, enabling technologies that can form the basis for new and improved products, manufacturing processes, and services. The program provides funding to both individual companies and industry-led joint ventures.

Examples of current focused competition program areas include: tools for DNA diagnostics, information infrastructure for health-care, manufacturing of composite structures, component-based software, and computer-integrated manufacturing for electronics.

The ATP relies on ideas supplied and co-funded by industry. By providing funding at an early stage of technology development, the ATP accelerates research progress for promising technologies that otherwise might not be developed quickly enough to be successful in a rapidly changing marketplace.

For ATP information call: 1-800-ATP-FUND (1-800-287-3863).

Energy-Related Inventions Program

"The NIH (Not Invented Here) Syndrome does not apply here," says George Lewett, director, Office of Technology Innovation (OTI) at NIST, under whose direction are the Energy-Related Inventions Program (ERIP) and the Non-Energy-Related Inventions Program (N-ERIP).

NIST and the Department of Energy (DOE), under provisions of the Federal Non-Nuclear Energy Research and Development Act of 1974, have combined to offer a marvelous opportunity to inventors of energy-related concepts, devices, products, materials, or industrial processes. ERIP is designed as a process to discover and assist the development of worthy inventions that might otherwise never be commercialized.

If you have an idea that can make money for you while saving the country energy, ERIP may be for you. It is a free technical evaluation and market assessment for you plus the possibility for financial support. Since the program began in 1974, more than 31,000 inventions have been submitted for evaluation. More than 625 of these submissions have been recommended to DOE with more than 450 receiving funding. During the past few years, the average grant or contract award has been about $89,000, with the maximum award being just under $100,000.

Dr. Gilbert M. Ugiansky, deputy director, Office of Technology Innovation, says that the evaluations are a terrific deal for the inventor. "By the time we recommend a project to DOE, we have spent about $15,000 evaluating the inventor's submission." He explains that even if a submission is kicked back to the inventor for more work, or rejected altogether, his department has spent a minimum of $400 and a maximum of $4,000 on initial reviews by outside consultants.

But financial grants are not all that ERIP is about. "We give inventions and their inventors a stamp of approval through our endorsement," explains Lewett. "This is often the kind of credibility a savvy inventor can turn into venture capital or perhaps a venture with a manufacturer. Our evaluation can carry a lot of weight."

How to Get a Free Evaluation of Your Invention

Anyone can submit an invention at any stage of development for a free confidential evaluation. See page 341 for a copy of Evaluation Form 1019.

For more information, or if you have questions, write the Office of Technology Innovation, National Institute of Standards and Technology, Gaithersburg, Maryland 20899-0001. The phone number is (301) 975-5500; the fax number is (301) 975-3839. Or you can send inquiries via e-mail: innovate@enh.nist.gov.

Guidance for Preparing Your Submission

Fill out Evaluation Form 1019 in English and return it with a detailed written description of your invention with illustrations or drawings, if possible. No other language will be accepted. The drawings need not be done professionally. **Don't submit models or prototypes.**

A properly prepared and complete submission (as described below), submitted in duplicate, will reduce the time it takes to evaluate your invention. Note that the requirements differ from the submission requirements of the Patent and Trademark Office or other grant programs.

The submission should be typed on metric A4 (or 8 x 11 inch) paper using 2.5 cm (1 inch) margins. It should be no more than 25 pages with the use of tabbed inserts to separate sections. An easy-to-read, letter quality font of no more than five characters per cm (a fixed pitch font of 12 or fewer characters per inch or a proportional font of point size 10 or larger) should be used. Check the material for accuracy, clarity, and completeness before submitting it for evaluation. Content requirements are in the following sections.

Invention Evaluation Request Form (NIST-1019)

Please use the form NIST-1019 as the cover sheet of your submission. Make sure that you have signed the back page of the form.

Executive Summary

This section should summarize, in one or two pages, the material contained in the body of the submission. Six paragraphs should be used to do this, one paragraph for each of the required sections.

- The invention's title is at the top of the first page of the Executive Summary. The title should be chosen carefully since it will be the first impression that the reader gets of your technology. Simple phrases and key words highlighting your invention are preferred with a maximum title length of 12 words.

- The first paragraph describes the invention and the problem or need the invention is addressing.

- The second paragraph summarizes the technical advantages of the invention and why it is better than what is currently available or will be available in the near feature.

- The third paragraph summarizes the energy impact of the invention at the national level.
- The fourth paragraph summarizes economic considerations, such as the estimated price of the product, manufacturing costs, and unusual development or manufacturing requirements.
- The fifth paragraph summarizes why the invention is considered commercially viable and identifies any potential problems or anticipated barriers associated with commercializing this technology.
- The final paragraph summarizes what is required to continue the development of the invention so that a commercial product could result from your efforts.

Body of Submission

Section 1. Technical Description. The material you include should define what problem your invention is solving, what the innovation is, and how the innovation solves the identified problem. The description of your invention can include patents, technical drawings, photographs, theoretical analyses, experimental data, computer simulations, calculations, and so on. The following comments are provided as guidance:

- This section should show, at a minimum, that the invention is well developed on paper. Thus, adequate technical and engineering detail (such as energy and mass balances, description of thermodynamic cycles, and mathematical descriptions) must be provided for a thorough evaluation of the invention's technical feasibility. Supporting calculations, data, or experimental test results should be neatly organized.
- While a patent (when available) could adequately describe the operation of the invention, approximately 20 years of experience evaluating inventions has shown that those inventions that are recommended usually include technical material besides the patent.
- Press releases, newspaper clippings, testimonials, and so on, usually do not add to the technical description.

Section 2. Technical Advantages. This section should describe the current technical state of art and why your invention is considered an advancement or improvement. It should consider not only competing technology that is currently available but also that in development. It should identify the competition even if the competition addresses the problem in a different way; for example, a solar heating system competes not only with other solar heating systems but also with conventional fossil-fueled systems. After reading this section, the reader should be able to identify distinct advantages for the proposed product. Appropriate material should:

- identify features and operating conditions that are new and innovative;
- discuss how the invention has overcome technical problems that impeded the commercialization of prior or competing products or designs;
- provide drawings that highlight innovations. Written material in this section should be keyed to these highlights. Drawings should be neat and clearly labeled;
- clearly distinguish between test results and expected performance, between facts and conjecture;
- refer to calculations and data from Section 1, as appropriate: don't duplicate the information here.

Section 3. Energy Impact. This section should attempt to quantify the energy impact of the invention. Energy impact may be direct or indirect. In particular, the significance of the invention's impact on energy production and/or conservation at the national level should be evident. Inclusion of calculations, experimental results, and other data supporting claims of energy efficiency is strongly encouraged. Assumptions should be clearly stated and a basis for their validity provided.

Section 4. Cost and Economic Considerations. The section should discuss, and detail if possible, the costs associated with the invention. Unusual development or manufacturing requirements, construction techniques, and so on, should be identified. The estimated selling price of the eventual product should be compared with existing competitive products; if the eventual product will be more expensive than current products, you should discuss why the end-user would buy your product.

Section 5. Commercial Potential and Market Considerations. This section is not intended to be a market survey. As appropriate, this section should discuss any (industrial, consumer, regional, etc.) bias against the use of a particular product, why past attempts at commercializing either the proposed product or competing products have failed, who will use it, where will it be used, and so on. Any safety issues, standards requirements, and so on, should be identified and discussed. The discussion should also state how the proposed product will overcome these commercialization problems. The importance of this section is to show who and what is competing with the proposed product and what other considerations will affect the eventual commercialization of this invention.

Section 6. Developments and/or Commercialization Requirements. This section should describe what needs to be done to continue the development of the proposed product. You should also indicate the support you need. However, this should not be a formal proposal (which will be requested by DOE in the case of a recommendation). This information is required in order to increase the evaluators' understanding of the technology and the requirements for technical and commercial feasibility. Appropriate topics for this section include a description of engineering and development work, required materials and subcontractors to be used, estimates of funding to complete the tasks, and so on.

ERIP—A Success Story for Everyone

In a report entitled *The Economic, Energy, and Environmental Impacts of the Energy-Related Inventions Program,* published in July of 1994 by DOE, it states that as of October 1991, a total of 557 inventions were recommended to DOE by NIST, which screens all submitted inventions for technical merit, potential for commercial success, and potential energy impact. By the end of 1992, at least 129 of those inventions had entered the market, generating total cumulative sales of $763 million.

The success of ERIP inventors is also shown in their licensing income. It is estimated that in 1992 ERIP inventors earned royalties of $1 million, and over the lifetime of the program, royalties total $18.6 million.

The commercial progress of spin-off technologies is also documented. Altogether, 36 spin-off technologies have generated sales of $63 million. Further, it is estimated that at least 668 full-time employees were working on ERIP technologies in 1992, and that this resulted in a return of approximately $2.7 million in individual income taxes to the U.S. Treasury. Finally, more than $531 million of energy expenditures have been saved over the past decade as a result of the commercial success of just three ERIP projects.

OTI's Lewett reports that he currently receives about 200 inventions a month for evaluation and that he recommends, fairly steadily, three to four inventions per month. "The system works," he proudly says.

ERIP Success Stories

Here are some ERIP success stories: inventors who made it through the ERIP process and came out with substantial grants or endorsements that enabled them to develop their inventions. These three inventions have accounted for $144.4 million in cumulative sales through 1990.

- Ronald E. Brandon, of Schenectady, New York, inventor of steam turbine packing rings, received $51,900.

- John A. McDougal developed the Electronic Octane® controls for automobile engines with help from DOE in the form of a positive product evaluation. McDougal did not accept any grant money; instead he turned the government's stamp of approval into cash in private sector support. He then inked license agreements with Ford and Chrysler as a result of patent infringement litigation.

- Harry E. Wood, of New Orleans, Louisiana, invented the Thermefficient-100®, an industrial water heater that is highly efficient and gas fired. DOE financed his work to the tune of $72,600.

In addition to the direct financial support provided by ERIP grants, the program can indirectly help meet the inventor's need for financing. Inventors often use their NIST evaluation and ERIP award as a source of credibility to aid them in attracting additional resources to further develop their technologies. ERIP support can make an inventor more credible in the eyes of potential investors. For example, DOE reports that one inventor parlayed a $50,000 ERIP grant into a $1 million award from a private industrial research institute. Another inventor used his $75,000 award to garner $10 million in funding from a multinational corporation.

Rights to the Invention

Procedures for handling invention disclosures have been established by NIST to safeguard the proprietary rights of the inventors. During the NIST evaluation, the disclosures are kept under strict control, with access generally restricted to personnel of the OTEA and to those selected by the office to assist in evaluation of the disclosures. All personnel of the OTEA and other government evaluators are required to sign statements that advise them of the procedures and regulations of 18 U.S.C. 1906, which provides for criminal penalties that may be imposed on a government employee for unauthorized release of confidential information, including trade secrets. Special provisions are made for outside experts that require them to adhere to confidential provisions established by OTEA. These provisions provide safeguards against experts reviewing an invention where there is or may be a conflict of interest.

The DOE patent policy is geared to provide patent incentives to individual inventors and small businesses under the inventors' program. One way of supporting this policy is by DOE waiving all rights, if any, in NIST-recommended inventions. A special class waiver was obtained by DOE for the program in 1980 to apply where individual inventors or small companies receive grants under $100,000. Requests for waivers from other support recipients will be considered individually. When waivers are granted, the patent provisions do not normally include any background patent rights provisions.

In summary, DOE will not require anyone to relinquish any rights or interest in their invention in exchange for support under $100,000. For sums in excess of that amount the terms of the support will be negotiable.

Inventions Covered by ERIP

Invention disclosures evaluated by ERIP are categorized as follows:

- Fossil Fuel Production
- Direct Solar
- Other Natural Sources
- Combustion Engines and Components
- Transportation Systems, Vehicles, and Components
- Building, Structures, and Components
- Industrial Processes
- Miscellaneous

Non Energy-Related Invention Program

This is a relatively new program at NIST, one which is also managed by George Lewett, director of the Office of Technology Innovation. At the time of this book's publication, it was still being formalized and funding is small by government standards—approximately $150,000. Nevertheless, George Lewett and his colleague, Dr. Gilbert M. Ugiansky, can provide information on opportunities that may be available to you under this program. They may be reached by phone (301) 975-5506, or fax (301) 975-3839.

Department of Energy

We now move from the NIST campus in Gaithersburg, Maryland to the Forrestal Building at 1000 Independence Avenue SW in Washington, D.C. It is here that George Lewett and his Technology Innovation team send the submissions they have found to be technically and commercially feasible. It will be up to the Department of Energy's Inventions and Innovations Division to carry on with your invention.

What happens to your submission once it reaches DOE?

Your invention will be assigned to an ERIP invention coordinator. He or she will ask you to submit a preliminary proposal describing the support you require. In the proposal, you should describe who will do the work, how much it will cost, and what you hope to accomplish as a result. The coordinator will commission a market assessment of your invention and set up a date for you to attend a **Commercialization Planning Workshop**. Then the coordinator will weigh the NIST analysis, your proposal, the market assessment, and the availability of funds to decide how best to support your invention. You will then be asked to prepare a work statement regarding what you will do with the ERIP support.

What is a Commercialization Planning Workshop?

The Commercialization Planning Workshops are held quarterly and bring together NIST-recommended inventors with a faculty of eight or nine experts from private industry. You will be given the opportunity for individual consultation on how to develop a focused strategy for moving your invention into the market. Classes cover patents, licensing, business planning, market analysis, financing, and other issues related to the successful commercialization of your technology. Workshops compress some 50 hours of classroom sessions, individual consultation, planning exercises, and inventor presentations into an intensive three-day format.

The Innovative Concepts Program

There is another DOE program that may be of interest to the independent inventor. This program does not involve NIST; it is pure DOE. Under the direction of the DOE's Inventions and Innovations Division (IID), the Innovative Concepts (InnConn) Program can provide seed money to allow inventors to determine if their ideas are technically and economically feasible.

Since InnConn began in 1983, over 70 projects have been funded, of which 55 have been completed. Of those completed, 45 percent have received funding from other sources to continue the work, leading to numerous journal articles, masters and Ph.D. theses, and patents. Many projects have found commercial applications.

If you have an idea for improving an industrial process and need money to launch an R&D effort, DOE's InnConn may be willing to invest. InnConn supports projects that:

- increase industrial output
- improve health conditions or the safety of industrial operations

- increase the useful life of a plant and equipment
- develop sensors and controls for difficult environments
- have a dual-purpose application to both defense and commercial industrial requirements
- address environmental issues in industrial operations
- can improve the overall energy efficiency of industrial processes and/or industry-related transportation requirements.

The InnConn Cycle

InnConn operates in cycles of roughly one to two years. Up to 15 innovators per cycle are awarded seed money—typically $15,000 to $20,000—to explore the technical and economic feasibility of their concepts. The InnConn program also seeks funding from other federal organizations that may benefit from your idea. The involvement of other organizations increases the number of grants provided, increases the visibility of the projects, and enhances the potential to yield commercial products.

InnConn Fairs

Each InnConn cycle leads to a technology fair, where the innovators present their work to representatives of businesses and industries that share an interest in the technology.

Rights to Inventions and Data

Because a major aspect of the Innovative Concepts Program is to promote the projects that it is funding and to draw the interest of potential sponsors, ICP retains unlimited rights to publish the data generated by program-funded projects. However, any proprietary data are kept confidential. Inventors are encouraged to document their inventions before accepting ICP funds in order to retain all patent rights to their inventions.

Over $13 Million in Follow-On Funding Awarded

The Innovative Concepts Program has far exceeded its original objectives. Seven cycles have been completed since 1983. In 1983, the DOE program manager expected that at least one concept in ten would receive follow-on funding from another sponsor. As of January 1994, according to Fred L. Hart, DOE's lead invention coordinator, 35 of the 70 innovators funded by ICP "seed money" have received more than $13 million in follow-on funding from other sponsors.

In addition to the follow-on funding received to date, two companies have been started; 18 patents have been issued with several more pending; and many masters theses and more than 80 journal articles have been written on these concepts.

Many of these innovators might have continued their R&D without the support of ICP; however, they credit the program with financial and commercialization assistance at a critical early stage in the development process, which might not have been available otherwise. Also, over half the Innovative Concepts awardees that have applied to the Energy-Related Inventions program have been recommended to DOE for further funding.

For More Information

InnConn maintains a mailing list of potential contributors, who are notified of opportunities to apply. To get your name on the list, get in touch with the

Battelle Pacific Northwest Laboratories, P.O. Box 999, K8-11, Richland, WA 99352. The telephone number is (509) 372-4328; the fax number is (509) 372-4369. The e-mail address is ipn_sil@pnl.gov.

To get brochures and more program information, contact: Lisa Barnett, Program Manager, U.S. Department of Energy, Forrestal Building, CE-521, 1000 Independence Avenue, SW, Washington, D.C. 20585. Telephone: (202) 586-1478; fax: (202) 586-1605.

National Innovation Workshops

National Innovation Workshops are a cooperative effort of federal and regional programs to provide practical guidance and information to innovative businesses, entrepreneurs, and inventors. NIW sponsors include U.S. Department of Energy, U.S. Department of Commerce (NIST); National Technology Transfer Center, and American Intellectual Property Law Association.

There have been 77 NIWs since 1980, attracting and providing assistance to thousands of innovators, inventors, and entrepreneurs. The purpose of these workshops is to help the inventor, developer of new technologies, or entrepreneur bring your ideas to market. Each two-day seminar covers a range of topics: Patenting and Protection; Estimating the Worth of Your Invention; Licensing/Marketing; Business Plans; New Business Start-up; R&D/Venture Financing; DOE-NIST Energy-Related Invention Program; SBIR Program. A wide variety of free how-to and technical publications are distributed.

The attendance fee is less than $125, which includes all sessions, 9 a.m.–5 p.m. on a Friday and Saturday, plus two luncheons.

"It is difficult to attribute the success of any given invention to the fact that someone attended a workshop. The idea of the workshops is to assist inventors who are struggling with the process, and help them figure out what their next step should be," says Dr. Norris Bell, NIW national coordinator.

For details about these programs and future workshops, or to get your name on the mailing list, contact: Dr. Norris Bell, NIW National Coordinator, Virginia Tech, 2990 Telestar Court, Fall Church, Virginia 22042, or call (703) 698-6016; or contact Terry Levinson, Inventions and Innovation Division, U.S. Department of Energy at (202) 586-1478; fax: (202) 586-1605.

Would Your Group Like to Host an NIW?

If you would like to be considered as a host for a future NIW, please send a letter of request to Terry Levinson, Inventions and Innovation Division, U.S. Department of Energy, Mail Stop 5E-052, 1000 Independence Ave. SW, Washington, D.C. 20585.

Your letter of request should respond to the following questions, which incorporate factors that will be considered in selecting future sites.

- Are you geographically situated to attract inventors and entrepreneurs within a 300-mile radius (a day's drive)?

- Are there specific attributes your organization could provide to inventors attending the workshop?

- What level of corporate and other private sector support could you expect to be provided to keep the registration fees affordable (under $125 for the two-day event)?

- Has your organization had experience holding meetings for groups ranging from 85 to 300 attendees?
- Has your organization had experience meeting the needs of inventors, small businesses, and entrepreneurs?
- Will there be any special events in your region that would enhance attendance during the period you would like to hold the NIW?
- What dates do you have in mind? Why?
- If you are not selected at this time, would you be interested in future workshops?

According to Terry Levinson, it takes a minimum of six months to organize and hold a workshop. It takes time to find the right people whose skills are necessary to the success of the NIW, to arrange for speakers, and to find corporate sponsorship. "Please keep this in mind when considering your involvement," she requests.

National Inventors Conference

Once a year, usually in late spring-early summer, the PTO and Intellectual Property Owners, Inc., (IPO) a non-profit association representing corporations and individuals who own patents, trademarks, and copyrights, host hundreds of inventors and entrepreneurs in Washington, D.C. for the National Inventors Conference and Expo.

A tribute to the nation's inventors, the event marked its twenty-first year in July 1994. The first day is typically wall-to-wall with speakers. The 1994 program, for example, featured a keynote address by Bruce A. Lehman, Commissioner of Patents and Trademarks, which was followed by talks on everything from venture capital to invention licensing and marketing. I have spoken at this event on numerous occasions, attended it for many more years as an attendee, and find it always to be a worthwhile experience.

Speakers on the program with me at the 1994 event included Warren R. Bovee, assistant chief intellectual property counsel, 3M Company; Fred L. Hart, Lead Invention Coordinator, Inventions and Innovations Division, Department of Energy; John R. Kirk, Jr., chairman, National Council of Intellectual Property Law Associations; John Pfanstiehl, founder, ProMotorcar Products, Inc.; Frederick R. Schmidt, patent examining Group Director, PTO; Mark A. Spikell, co-founder, Entrepreneurship Center, George Mason University; and Gilbert M. Ugiansky, deputy director and chief evaluator, Office of Energy-Related Inventions, NIST.

Past corporations exhibiting at the Expo include: Allied Signal, Inc.; AMP, Inc.; Atari, Inc.; Bally Manufacturing Co.; Coca Cola; Dupont; GE; and Eastman Kodak.

The conference costs $125 per person and includes all sessions, resource materials, and lunch. The expo is free and it runs for two days.

For information or to be placed on the mailing list for future events, contact: Office of Public Affairs, PTO, Washington, D.C. 20231; (703) 305-8341. Or call IPO at (202) 466-2396.

NIST-1019
(REV. 5-94)

U.S. DEPARTMENT OF COMMERCE
NATIONAL INSTITUTE OF STANDARDS AND TECHNOLOGY

PUBLIC LAW 93-577, SECTION 14

ENERGY-RELATED INVENTIONS PROGRAM INVENTION EVALUATION REQUEST

Public reporting burden for this collection of information is estimated to average 10 hours per response including the time to read the instructions. Send comments regarding this burden estimate or any other aspect of this collection of information, including suggestions for reducing this burden, to the Office of Technology Innovation, National Institute of Standards and Technology, Gaithersburg, Maryland 20899-0001; and to the Office of Information and Regulatory Affairs, Office of Management and Budget, Washington, D.C. 20503.

Instructions for Requesting an Energy-Related Invention Evaluation

Be sure you have read the Program Description and Policy Statement provided with this Form. Read the "Memorandum of Understanding" on the back of this Form. Check the appropriate box, sign and date, then complete the front of this Form. Remember to make copies of this Form for your records. Prepare the material describing the invention in accordance with the instructions provided with this Form. Send this Form and the submission material describing the invention directly to:

Office of Technology Innovation : National Institute of Standards and Technology, Gaithersburg, Maryland 20899-0001.

NAME OR TITLE OF INVENTION

BREIF DESCRIPTION OF INVENTION

NAME AND ADDRESS OF INVENTOR	TELEPHONE NUMBER	FAX NUMBER
		E-MAIL ADDRESS

NAME AND ADDRESS OF OWNER (IF DIFFERENT FROM INVENTOR)	TELEPHONE NUMBER	FAX NUMBER
		E-MAIL ADDRESS

NAME AND ADDRESS OF SUBMITTER OR OTHER INTERESTED PARTY (IF DIFFERENT FROM INVENTOR OR OWNER)	TELEPHONE NUMBER	FAX NUMBER
		E-MAIL ADDRESS

	YES	NO
HAS INVENTION BEEN DESCRIBED TO OTHER AGENCIES OF THE GOVERNMENT? (IF YES, DISCUSS IN INVENTION DESCRIPTION)	☐	☐
HAS INVENTION BEEN DISCLOSED TO ANY PRIVATE COMPANIES, PATENT ATTORNEYS, ETC.? (IF YES, DISCUSS IN INVENTION DESCRIPTION)	☐	☐

OFFICIAL USE ONLY

ER NUMBER	ANALYST	DATE
DISPOSITION	TECHNICAL NUMBER	EVALUATOR
TECHNICAL CATAGORY	ENERGY CATAGORY	SERIAL NUMBER 146106

ELECTRONIC FORM

NAME OF COMPANY INVOLVED (IF ANY)	NUMBER OF EMPLOYEES	ANNUAL GROSS SALES

STATUS OF INVENTION DEVELOPMENT (CHECK APPROPRIATE BOX)
(USE ACCOMPANYING BOOKLET ON ENGINEERING STAGES OF DEVELOPMENT AS A GUIDE)

☐ CONCEPTUAL ☐ TECHNICAL FEASIBILITY ☐ DEVELOPMENT ☐ COMMERCIAL VALIDATION AND PRODUCTION PREPARATION ☐ FULL SCALE PRODUCTION ☐ PRODUCT SUPPORT

PATENT STATUS (CHECK APPROPRIATE BOX)

☐ NOT PATENTABLE ☐ NOT APPLIED FOR ☐ DISCLOSURE DOCUMENT PROGRAM ☐ PATENT APPLIED FOR ☐ PATENT GRANTED

PATENT NUMBER(S) _____

MEMORANDUM OF UNDERSTANDING

I have read the Program Description and Statement of Policy. As the owner, or with the authority from the owner who is listed on this Form, I have attached (or previously submitted) a disclosure of the identified invention for the purpose of evaluation by the National Institute of Standards and Technology pursuant to Section 14 of Public Law 93-577.

I understand that to help protect property rights in an unpatented invention, an appropriate statement or notation should be applied to the title page or first page of the invention description, and that if the description is so marked, the Government will consider all information that is in fact (a) trade secret or (b) commercial or financial information that is privileged or confidential, as coming within the exemption set out in U.S.C. 552(b)(4). Accordingly, I have checked directly below, the box which is applicable to this invention.

	YES	NO
An appropriate statement has been applied to the information I have submitted.	☐	☐
Please apply an appropriate statement to all material I have submitted describing the invention to which this request pertains. (Example: This material contains commercial or financial information which is confidential.)	☐	☐
No statement is required because the information submitted is not confidential.	☐	☐

I also understand that NIST will evaluate the invention described in the invention disclosure on the following conditions:

1. The Government will, in the evaluation process, restrict access to the description to those persons, within or without the Government, who have a need for purposes of administration or evaluation and will restrict their use of this information to such purposes.

2. The information submitted will not be returned and may be retained as a Government record.

3. The Government may make additional copies of the material submitted if required to facilitate the review process.

4. The acceptance of the information for evaluation does not, in itself, imply a promise to pay, a recognition of novelty or originality, or a contractual relationship such as would render the Government liable to pay for use of the information submitted.

5. The provisions of this Memorandum of Understanding shall also apply to additions to the disclosure made by me incidental to the evaluation of the invention.

IMPORTANT: THE PERSON SIGNING BELOW WILL BE THE ADDRESSEE (OTHERS WILL RECEIVE COPIES) FOR ALL CORRESPONDENCE CONCERNING THE EVALUATION. BE SURE THAT YOUR ADDRESS IS IN THE APPROPRIATE ADDRESS BLOCK ON THE FRONT OF THIS FORM.

REQUEST IS BEING SUBMITTED BY

☐ INVENTOR ☐ OWNER ☐ OTHER _____

SUBMITTER (SIGNATURE)	NAME (PRINTED OR TYPED)
ORGANIZATION (IF APPLICABLE)	**DATE**

NIST-1018 (REV. 5-94)
ELECTRONIC FORM

CHAPTER 20

Hot Press: Magazines, Newsletters, and Directories

The man with a new idea is a crank until the idea succeeds.

—MARK TWAIN

Publications can provide a wide variety of information to inventors. I cannot emphasize enough their value and the time you should dedicate to researching and reading them. I love prospecting through publications. Often when I need creative stimulation, I will go to a nearby library and spend hours leafing through a variety of publications. If nothing else, this exercise tends to focus my direction.

In the United States, more than 12,300 magazines are listed as commercial publications. If you add on smaller publications and newsletters, the total jumps to no less than 25,000. It is a sure bet that even the most general circulation magazines will run a few stories each year on inventions and creativity, and maybe even something on your particular field of interest. I keep up with what is being covered through the *Reader's Guide to Periodical Literature* at my local library. The *Reader's Guide* keeps track of hundreds of publications day-to-day.

Although the news is not as fresh as you'll find in a daily newspaper, industry trade magazines and newsletters are excellent sources for in-depth information. And they carefully track and report a very broad range of executive assignments, not just the upper echelon. It was through a toy trade magazine, for example, that I obtained the name of the Milton Bradley senior vice president of research and development who licensed our first toy, Starbird.

These kinds of esoteric business trade publications are found in many major libraries that offer comprehensive business reference rooms. University technical libraries also usually have a wide range of these kinds of magazines. These magazines make their money through the sale of advertising, not the sale of subscriptions. One of the best things about them is the abundance of trade advertisements.

There are a number of publications that I read on a regular basis. As you page through this chapter, look for the publications I have highlighted as must-reads; these selections can provide a good starting point for organizing your reading list.

As valuable as the following publications are, it is important to remember that there is no substitute for personal contacts. I recommend a phone call to authors and editors as the best way to get information on research and development opportunities from many publications.

Extra! Extra!

The Wall Street Journal provides the most current, fast-breaking information on industry. It was through a piece in this newspaper that I learned what was going on at Proctor & Gamble's Crest brand, information that led to our selling P&G what was to become a $12-million premium program—the Crest Fluorider, a big-wheel trike with a giant Crest tube as its mainframe.

MAGAZINES

Copyright Law Reports

CCH Incorporated   (312) 583-8500
4025 W. Peterson Ave.
Chicago, IL 60646
FAX: (800) 224-8299
Toll-free: (800) TELL-CCH
Ronald Miller

    Copyright law publication.

Idea Marketplace

PO Box 131758   (718) 761-5380
Staten Island, NY 10313
Glenda Pagan, Editor

    A magazine for inventors and entrepreneurs.

The Lightbulb

M & M Associates   (916) 468-2282
12424 Main St.
PO Box 1020
Fort Jones, CA 96032-1020
FAX: (916) 468-2238
Melvin L. Fuller
Maggie Weisberg
Stephanie D. Medvin

    Magazine of interest to inventors, creative people, and entrepreneurs.

Les Nouvelles

Licensing Executives Society Intl.   (216) 771-2600
1444 W. 10th St., Ste. 403
Cleveland, OH 44113
FAX: (216) 771-8478
Jack Stuart Ott

    Magazine on worldwide technology licensing.

Official Gazette—Patents

U.S. Government Printing Office   (202) 783-3238
Superintendent of Documents
Washington, DC 20402
FAX: (202) 512-2250

    Publication reporting on patents.

Official Gazette—Trademarks

U.S. Patent & Trademark Office   (703) 308-9400
Dept. of Commerce
Washington, DC 20231
FAX: (703) 308-9400

    Publication reporting on trademarks.

Toy Business

10711 Burnet Rd., Ste 305   (512) 873-7761
Austin, TX 78758-4459
Kathleen M. Carson, Editor

    Magazine with a fresh slant on the business of toys.

World Patent Information

Pergamon Press, Inc.   (914) 524-9200
660 White Plains Rd.
Tarrytown, NY 10591-5153
FAX: (914) 333-2444
V.S. Dodd
Susan Rosenthal

    Journal serving as worldwide forum for the exchange of information among professionals in the patent information and documentation field.

NEWSLETTERS

Advanced Coatings & Surface Technology

Technical Insights, Inc.   (201) 568-4744
PO Box 1304
Fort Lee, NJ 07024-9967
FAX: (201) 568-8247
Alan Brown

    Provides details of developments in coatings and surface modification in the context of a wide variety of industry applications. Focuses on such topics as traditional coatings processes, chemical vapor deposition (CVD), and ion beam methods. Discusses the commercial value of new developments. Includes colorado Expert Sources of Technical Information.

Advanced Manufacturing Technology

Technical Insights, Inc.   (201) 568-4744
PO Box 1304
Fort Lee, NJ 07024-9967
FAX: (201) 568-8247
Edward D. Flinn

    Reports on technological advances that are contributing to robotics, computer graphics, flexible automation, computer-integrated manufacturing, and new techniques for machining. Covers techniques for cutting costs, improving product quality, and increasing productivity. Includes supplements titled Test and Measurement Alert and Assembly Technologies.

Aeronautical Engineering

National Technical Information Service (NTIS)   (703) 487-4630
U.S. Department of Commerce
5285 Port Royal Rd.
Springfield, VA 22161

Covers documents on the engineering and theoretical aspects of design, construction, evaluation, testing, operation, and performance of aircraft and associated components, equipment, and systems.

ASTI Newsletter

Association for Science,   (703) 241-2850
Technology, and Innovation (ASTI)
PO Box 1242
Arlington, VA 22210
FAX: (703) 241-2850
Robert Adams

Publishes news of ASTI, which promotes the development, demonstration, and application of policies, standards, and techniques for improving management of innovation, and news of other organizations with similar goals. Recurring features include review.

The Authority Report

Arkansas Science & Technology Authority   (501) 324-9006
100 Main St., Ste. 450
Little Rock, AR 72201
FAX: (501) 324-9012
Courtney Johnson-Woods

Concentrates on issues related to the development of Arkansas' scientific and technological resources. Contains notices of publications available and related conferences as well as news of research, including research and grants funded by the Authority. Recurring features include interviews, reports of meetings, and news of educational opportunities.

Battelle Today

Battelle Memorial Institute   (614) 424-7818
505 King Ave.
Columbus, OH 43201-2693
FAX: (614) 424-3889
Harry R. Templeton

Reports on the broad interests and activities of Battelle worldwide in developing, managing, and commercializing technology. Recurring features include news of research, a collection, reports of meetings, review, and notices of publications available.

BNA's Patent, Trademark & Copyright Journal

Bureau of National Affairs, Inc. (BNA)   (202) 452-4500
1231 25th St. NW
Washington, DC 20037
FAX: (202) 822-8092
Toll-free: (800) 372-1033
James D. Crowne

Monitors developments in the intellectual property field, including patents, trademarks, and copyrights. Covers proposed and enacted legislation, litigation, Patent and Trademark Office decisions, Copyright Office practices, activities of professional associations, government contracting, and international developments.

Canadian Patent Reporter

Canada Law Book Company, Ltd.   (905) 841-6472
240 Edward St.
Aurora, ON L4P 2XG
Canada
FAX: (905) 841-5085
Toll-free: (800) 263-2037
T. Gary O'Neill
W. Frank H. Mulock

Reports on significant Canadian cases on patents, industrial design, copyright, and trademark law. Includes articles and editorials on current practices, new developments, and decisions of the Commissioner of Patents.

Category Reports

Marketing Intelligence Service, Ltd.   (716) 374-6326
33 Academy St.
Naples, NY 14512
Pat Peck

Provides product reporting services in the following areas: foods, beverages, snacks, health and beauty aids, household, and pets.

Center for Advanced Materials Processing

Center for Advanced   (315) 268-2336
Materials Processing (CAMP)
Box 5665
Potsdam, NY 13699-5665
FAX: (315) 268-7615
Dana M. Barry

Focuses on microcontamination controls and fine particle sciences for advanced materials processing. Recurring features include news of research, reports of meetings, notices of publications available, and a column titled From the Director's Desk.

Charles A. Lindbergh Fund, Inc.— Newsletter

Charles A. Lindbergh Fund, Inc.   (612) 338-1703
708 S. 3rd St., Ste. 110
Minneapolis, MN 55415
FAX: (612) 338-6826
Gene Bratsch

Contains news and features concerning individuals receiving Lindbergh Grants and the Lindbergh Award from The Charles A. Lindbergh Fund for work which contributes to a balance between technological advancement and environmental preservation; Lindbergh and aviation history and news; books and videos related to Lindbergh, the environment, and aviation; symposiums and other events sponsored by the organization.

Clean Coal Technologies

National Technical   (703) 487-4630
Information Service (NTIS)
U.S. Department of Commerce

5285 Port Royal Rd.
Springfield, VA 22161

Abstracts worldwide information on clean coal. Includes such topics as mechanical coal cleaning, desulfurization, coal gasification and liquefaction, flue gas cleanup, and advanced coal combustion.

Communications Product Reports

Management Information Corporation   (609) 428-1020
410 E. Rte. 70
PO Box 5062
Cherry Hill, NJ 08034
FAX: (609) 428-1683
Toll-free: (800) 678-4642
Don Stuart

Describes and evaluates communications products and services available in both the domestic and international marketplaces. Focuses on the voice and data communications marketplace, providing product/service price data and evaluation based upon comparison and user comments. Recurring features include a cumulative listing of previous issues and the products evaluated.

Computer Industry Litigation Reporter

Andrews Publications   (215) 399-6600
PO Box 1000
Westtown, PA 19395
FAX: (215) 399-6610
Toll-free: (800) 345-1101
Harry G. Armstrong

Tracks significant computer law developments relating to copyright, patent and trademark infringement, alleged misappropriation of trade secrets, antitrust allegations, and user-vendor breach of contracts. Publishes complete texts of key decisions in addition to "unbiased news summaries of cases around the country."

Extra! Extra!

Forbes, Fortune, Business Week: these publications are the best way to track the ins and outs of senior executives and the ups and downs of corporate earnings, trends, and competition. Time and Newsweek are also must reads. Inventors may also be interested in general science magazines such as Popular Mechanics, Popular Science, Discover, and Omni.

CyberEdge Journal

The Delaney Companies   (415) 331-3343
1 Gate Six Rd., Ste. G
Sausalito, CA 94965
FAX: (415) 331-3643
Ben Delaney

Seeks to provide an information channel for those involved in advancing the state of the art of human-computer interaction. Promotes the open and free exchange of ideas and information related to the role of cybernetics in the future. Encourages the synthesis and growth of new ideas and devices. Assists in the development of commercial products incorporating new concepts and technologies. Recurring features include letters to the editor, product news, interviews, news of research, news of business developments, a collection, reports of meetings, job listings, and review.

Department of Energy Reports

National Technical   (703) 487-4630
     Information Service (NTIS)
U.S. Department of Commerce
5285 Port Royal Rd.
Springfield, VA 22161

Provides unclassified technical reports, critical reviews, conference papers, and symposia papers produced by the Department of Energy.

Derivatives Engineering and Technology

Waters Information Services, Inc.   (607) 770-9242
PO Box 2248
Binghampton, NY 13902-2248
FAX: (800) 947-8963
Toll-free: (800) 947-7947

Concerned with derivatives dealers and their technology strategies, proprietary developments in pricing models and design analytics, risk management systems integration, technology suppliers and their solutions, and industry and regulatory environment issues.

Drug and Device Product Approval List

National Technical   (703) 487-4630
     Information Service (NTIS)
U.S. Department of Commerce
5285 Port Royal Rd.
Springfield, VA 22161

Lists the most recent new drug approvals, new animal and medical devices, and licenses issued for biological products.

East/West Business Package

Welt Publishing Company   (202) 371-0555
1413 K St. NW, Ste. 800
Washington, DC 20005-3405
FAX: (202) 682-5833

Disseminates information on east/west business and trade, business in China, and technology.

East/West Technology Digest

Welt Publishing Company   (202) 371-0555
1413 K St. NW, Ste. 800
Washington, DC 20005
FAX: (202) 408-9369
Walter Smith

Supplies information on new technology and license offers, with addresses for obtaining further information. Recurring features include news of research.

The Eco

Ecocenters Corporation   (216) 498-4900
31225 Bainbridge Rd.
Solon, OH 44139
FAX: (216) 498-4920
Michael T. Kovach

Provides case studies and industry news on technological innovations. Recurring features include news of research.

Electric Pages

Interactive Features, Inc.   (718) 499-1884
405 4th St.
Brooklyn, NY 11215
FAX: (718) 499-1970
Jack Powers

Covers new publishing and media technologies from a publisher's perspective. Discusses such topics as digital printing, fax publishing, CD-ROM audiotex, videotex, electronic advertising, telecommunications, multimedia, and other interactive publishing tools. Recurring features include letters to the editor, interviews, news of research, review, notices of publications available, and technology reviews.

Electro Manufacturing

Worldwide Videotex   (407) 738-2276
PO Box 3273
Boynton Beach, FL 33424-3273
FAX: (407) 738-2275

Covers research and development of products and services in manufacturing. Includes company profiles, studies in productivity, and new technology.

The Entrepreneur Network

Edward Zimmer   (313) 663-8000
1683 Plymouth Rd., No. K
Ann Arbor, MI 48105-1891
Toll-free: (800) 468-8871
Edward Zimmer

Provides information on private investors and other resources seeking growth-business opportunities.

ERC Update

Engineering Research Center (ERC)   (301) 405-3906
University of Maryland
College of Engineering
Wind Tunnel Bldg.
Room 2104
College Park, MD 20742
FAX: (301) 405-6707
Judith Mays

Features news concerning the Center's programs supporting research and assistance to businesses and industry.

Eureka! The Canadian Inventor's Newsletter

Canadian Industrial   (519) 885-5870
   Innovation Centre/Waterloo
156 Columbia St. W
Waterloo, ON N2L 3L3
Canada
FAX: (519) 885-5729
Toll-free: (800) 265-4559
Robin Dzijacky

Serves as a forum for Canadian inventors and innovators.

Federal Research Report

Business Publishers, Inc.   (301) 587-6300
951 Pershing Dr.
Silver Spring, MD 20910
FAX: (301) 587-1081
Toll-free: (800) 274-0122
Leonard A. Eiserer, Ph.D.

Provides information on research and development funds available from federal agencies and bureaus or associations that provide support money for research and development. Lists items in categories including environment/energy, transportation, medicine/health, education, and social sciences.

Foreign Trade Fairs New Products Newsletter

Printing Consultants   (908) 686-2382
Box 636, Federal Sq.
Newark, NJ 07101
John E. Felber

Provides descriptions of and manufacturer's addresses for new foreign products.

Foresight Update

Foresight Institute   (415) 324-2490
PO Box 61058
Palo Alto, CA 94306
FAX: (415) 324-2497
Christine L. Peterson

Dedicated to information exchange, knowledge transfer, and education in the fields of nanotechnology and molecular manufacturing. Recurring features include letters to the editor, interviews, news of research, a collection, reports of meetings, news of educational opportunities, review, notices of publications available, and a column titled Recent Events.

The Frame

Survey Sampling, Inc.   (203) 255-4200
1 Post Rd.
Fairfield, CT 06430
FAX: (203) 254-0372
Diane Chagares

Provides information on the survey research industry regarding sampling methodologies and technologies. Recurring features include letters to the editor, interviews, news of research, and reports of meetings.

Future Technology Intelligence Report

Future Technology Intelligence Report, Inc.
PO Box 423652
San Francisco, CA 94142
Anthony C. Sutton

Emphasizes "fundamentally new technology," such as weather engineering, free energy, and vibrational medicine. Recurring features include news of research, a collection, reports of meetings, review, and notices of publications available.

Government Inventions for Licensing: An Abstract Newsletter

National Technical   (703) 487-4630
    Information Service (NTIS)
U.S. Department of Commerce
5285 Port Royal Rd.
Springfield, VA 22161

Reports on mechanical devices and equipment and other government inventions in chemistry; nuclear technology, biology and medicine; metallurgy; electrotechnology; and optics and lasers. Discusses patent applications and miscellaneous instruments. Recurring features include a form for ordering reports from NTIS.

GRID

Gas Research Institute (GRID)   (312) 399-8100
8600 W. Bryn Mawr
Chicago, IL 60631
FAX: (312) 399-8170
C. Drugan

Reports on gas energy research and development sponsored by the Institute. Carries announcements of technical reports available.

Healthcare Technology & Business Opportunities

Biomedical Business International, Inc.   (714) 755-5757
1524 Brookhollow Dr.

Santa Ana, CA 92706
FAX: (714) 755-5707
Michael Gibb

Provides information on new technologies available for license or transfer, distributors seeking products, and suppliers seeking distributors. Reports on joint ventures, acquisitions sought, strategic partnering, product and corporate divestitures, and product and business development services. Recurring features include listings of U.S., European, and Japanese patents and their applications and colorado New Technology, Business Opportunities, Patents, and Information Resources.

Home Automation News

Home Automation Association   (202) 333-8579
808 17th NW, Ste. 200
Washington, DC 20006
FAX: (202) 337-3809
Nicholas A. Pyle

Carries Association news and articles on international trends in the integrated home control industry. Recurring features include new product announcements, product reviews, and news of members.

Industrial Bioprocessing

Technical Insights, Inc.   (201) 568-4744
PO Box 1304
Fort Lee, NJ 07024-9967
FAX: (201) 568-8247
Karen Dean

Focuses on industrial processes involving biological routes to produce chemicals and energy. Covers the conversion of biomaterials via fermentation, enzymatic conversion, biodegradation, advances in bio-reactor design, product separation, and process monitoring. Lists new patent introductions. Recurring features include news of research, review, notices of publications available, a collection, market forecasts, company profiles, and a supplement titled Forecast.

InKnowVation Newsletter

Innovation Development Institute   (617) 595-2920
45 Beach Bluff Ave., Ste. 300
Swampscott, MA 01907

Focuses on SBIR (Small Business Innovative Research) program opportunities for small businesses entering early-stage, high-risk research and development ventures. Discusses how SBIR programs can be used to obtain initial funding or to test promising ideas. Available only as part of SBIR service.

Innovation News

Aremco Products, Inc.   (914) 762-0685
PO Box 429
Ossining, NY 10562-0429
FAX: (914) 762-1663
Brenda T. Lyons

Features technical research, updates on process equipment, and photographs of ceramic materials. Covers high temperature design, including process information and case histories; and describes equipment used in ceramic processing.

Inside R&D

Technical Insights, Inc.   (201) 568-4744
PO Box 1304
Fort Lee, NJ 07024-9967
FAX: (201) 568-8247
Charles Joslin

Describes research and development breakthroughs in industry, government, and academic labs, with an emphasis on technology transfer. Emphasizes how the results of the research covered can be applied practically by industry. Recurring features include news of research and colorado Managing Innovation, Companies to Watch, Industrial/Technology Policy, Trends, and Forecasts and Analyses.

Inside the PTO

Questel Publishing
8000 Westpark Drive
McLean, VA 22102
Toll-free: (800) 326-1710

Covers developments at the U.S. Patent and Trademark Office.

Intelligence

Edward Rosenfeld   (212) 222-1123
PO Box 20008
New York, NY 10025-1510
FAX: (212) 222-1123
Toll-free: (800) 638-7257
Edward Rosenfeld

Covers technologies that affect the future of computing and offers viewpoints. Concentrates on business, research and government activities in neural networks, parallel processing, pattern recognition, natural language interfaces, voice and speech technologies, and art and graphics. Recurring features include editorials and news of research.

International Calculator Collector

International Association   (714) 730-6140
   of Calculator Collectors
14561 Livingston St.
Tostin, CA 92680
FAX: (714) 730-6140
Guy Ball

Promotes portable electronic calculator collecting. Documents the history of early calculator development. Features calculators of interest. Offers a buy/sell/trade section. Recurring features include letters to the editor, interviews, news of research, reprints of early model calculator advertisements, and colorado It all adds up . . . and Personal Remembrances.

Extra! Extra!

Government agencies must, by law, publish advance notice of R&D opportunities in the Commerce Business Daily. Published Monday through Saturday (except on federal holidays), CBD may be obtained at the Department of Commerce field offices and at most major urban and university libraries. If you want personal copies, contact the Superintendent of Documents, Government Printing Office, Washington, DC 20402; (202) 275-3054. The GPO takes Visa and Mastercard.

International New Product Newsletter

Box 1146   (508) 741-0224
Marblehead, MA 01945
FAX: (508) 741-0224
Pamela H. Michaelson

Provides "advance news of new products and processes, primarily from sources outside the United States." Emphasizes new products which can cut costs and improve efficiency. Recurring features include the column Special Licensing Opportunities which lists new products and processes that are available for manufacture under license, or are for sale or import.

International Product Alert

Marketing Intelligence Service, Ltd.   (716) 374-6326
33 Academy St.
Naples, NY 14512
FAX: (716) 374-5217
Toll-free: (800) 836-5710
Sherie Meeker-Barton

Provides concise reports on new products in 18 countries outside of the U.S. and Canada. Covers foods, beverages, non-prescription drugs, cosmetics, toiletries, pet products, and miscellaneous household items. Also lists products that are extensions of existing lines and lists packaging changes. Recurring features include occasional copies of advertising.

Inventors News

Inventors Clubs of America, Inc.   (404) 938-5089
PO Box 450261
Atlanta, GA 30345
Toll-free: (800) 336-0169
Alexander T. Marinaccio

Designed to keep inventors abreast with the latest information on patents and trademarks. Carries news of the Club and information on exhibits and inventions. Recurring features include editorials, news of research, letters to the editor, news of members, a collection, and a column titled Patents for Sale.

Japan Consumer Electronics Scan

Kyodo News International, Inc. (212) 397-3723
50 Rockefeller Plaza, Ste. 815
New York, NY 10020
FAX: (212) 397-3721
Toll-free: (800) 536-3510

Reports on new consumer products made available by Japanese companies. Includes information on price, target markets, and availability. Available online only.

Japanese Sensor Newsletter

Edison Sensor Technology Center (216) 368-2934
Case Western Reserve University
Bingham Bldg.
10900 Euclid Ave.
Cleveland, OH 44106-7200
FAX: (216) 368-8738
Dr. Akiya Kozawa

Relates developments in Japan regarding new sensors, applications, research, and technical meetings. Compiles all information from original Japanese sources.

Jet News

Water Jet Technology Association (314) 241-1445
818 Olive St., No. 918
St. Louis, MO 63101
FAX: (314) 241-1449
George A. Savanick, Ph.D.

Provides information on the latest research developments, equipment, products, and services available, membership news, and new applications relating to water jet technology. Recurring features include letters to the editor, news of research, collection, interviews, news of meetings, and news of educational opportunities.

The Journal of Proprietary Rights

Law & Business, Inc. (201) 894-8538
Prentice-Hall, Inc.
270 Sylvan Ave.
Englewood Cliffs, NJ 07632
Michael A. Epstein
Salem M. Katsh
Steven D. Glazer

Covers trends involving patent, trade secret, trademark, and intellectual property protection issues, including practical solutions.

Lookout Foods

Marketing Intelligence Service, Ltd. (716) 374-6326
33 Academy St.

Naples, NY 14512
FAX: (374) 5217
Toll-free: (800) 836-5710
Tom Vierhile

Carries photographs and detailed descriptions of the most innovative products, package design, line extensions, and marketing background in consumer goods categories. Also copies advertising support. Recurring features include product information: name of manufacturer, ingredients, nutritional information, background data, and marketing strategies.

Lookout Nonfoods

Marketing Intelligence Service, Ltd. (716) 374-6326
33 Academy St.
Naples, NY 14512
FAX: (716) 374-5217
Toll-free: (800) 836-5710
Tom Vierhile

Carries photographs and detailed descriptions of the most innovative products, package design, line extensions, and marketing background in consumer goods categories. Covers nonprescription drugs, cosmetics and toiletries, and miscellaneous household items. Also copies advertising support. Recurring features include product information: name of manufacturer, ingredients, nutritional information, background data, and marketing strategies.

Manufacturing Automation

Vital Information Publications (415) 389-8671
321 Carrera Dr.
Mill Valley, CA 94941-3995
FAX: (415) 345-7018
Peter Adrian

Reports on activities in the manufacturing automation field. Concentrates on market strategies, international joint ventures, licensing, and news of research and development. Gives company profiles.

Mealey's Litigation Report: Intellectual Property

Mealey Publications, Inc. (215) 688-6566
512 W. Lancaster Ave.
PO Box 446
Wayne, PA 19087
FAX: (215) 688-7552
Toll-free: (800) 632-5397

Covers all areas of intellectual property litigation, including copyright, patent, trademark, trade secrets, and unfair competition.

Merchant & Gould Computer Law Newsletter

Merchant, Gould, Smith, (612) 331-1099
Edell, Welter & Schmidt
3100 Norwest Center

90 S. 7th St.
Minneapolis, MN 55402
FAX: (612) 332-9081

Covers developments and provides information on computer law and intellectual property.

New from Europe

Prestwick Publications, Inc.   (305) 427-2924
390 N. Federal Hwy., No. 401
Deerfield Beach, FL 33441
Roy H. Roecker

Contains market forecasts, trends, and descriptions of new products and technologies from Europe. Descriptions include the developer's name and address and an explanation of why the new product is superior to existing products or processes. Provides an overview of the European economy, its research and new product emphasis, and governmental actions that will affect future market activity. Recurring features include news of research.

New from Japan

Prestwick Publications, Inc.   (305) 427-2924
390 N. Federal Hwy., No. 401
Deerfield Beach, FL 33441
Roy H. Roecker

Describes new Japanese products and technologies and explains why they are superior to existing products or processes. Covers consumer products, energy conserving processes and products, manufacturing methods, and electronic products. Recurring features include news of research.

New from U.S.

Prestwick Publications, Inc.   (305) 427-2924
390 N. Federal Hwy., No. 401
Deerfield Beach, FL 33441
Roy H. Roecker

Describes new products and technologies researched and developed in the United States. Examines a single product and its use and applications in depth in each issue.

New Mexico Technology Enterprise Forum

Economic Development   (505) 277-3661
    Communications Office
University of New Mexico
457 Washington SE, Ste. M
Albuquerque, NM 87108
FAX: (505) 277-6188
Richard W. Cole

Concerned with developments relating to technology in New Mexico. Reports on new projects being funded by the State of New Mexico, profiles technology firms and laboratories in New Mexico, and contains news briefs on pertinent legislation. Recurring features include information on workshops and useful publications, news of research, review, and a collection.

New Technology Week

King Communications Group, Inc.   (202) 662-9746
627 National Press Bldg.
Washington, DC 20045
FAX: (202) 662-9744
Ken Jacobson

Carries news on evolving technologies, especially those in defense-related fields. Follows legislation and government agency action affecting defense and high-tech industries. Lists recipients of foundation and research grants in the United States. Recurring features include a collection and news of employment opportunities.

NTIS Foreign Technology Newsletter

National Technical   (703) 487-4630
    Information Service (NTIS)
U.S. Department of Commerce
5285 Port Royal Rd.
Springfield, VA 22161

Carries abstracts of reports on the field of foreign technology. Covers biomedical technology; civil, construction, structural, and building engineering; communications; computer, electro, and optical technology; energy, manufacturing, and industrial engineering; and physical and materials sciences. Recurring features include notices of publications available and a form for ordering reports from NTIS.

NTT Topics

Kyodo News International, Inc.   (212) 397-3723
50 Rockefeller Plaza, Ste. 815
New York, NY 10020
FAX: (212) 397-3721
Toll-free: (800) 536-3510

Highlights research and development activities of Nippon Telegraph and Telephone. Focuses on marketing strategies. Available online only.

Points West

Center for the New West   (303) 592-5310
600 World Trade Center
1625 Broadway
Denver, CO 80202
FAX: (303) 572-5499
Joe Sullivan

Explores the recent economic transformations of American society. Focuses on business formation, entrepreneurship, technological innovation, changes in demography, family life, and the workplace.

PresentFutures Report

PresentFutures Group, Inc.   (703) 538-6181
282 N. Washington St.
Falls Church, VA 22046
Jeff Hallett
Kris Hefley

Examines the impact of the new information economy on industries, business markets, schools, and other Industry. Identifies important issues and developments in technology, organizational design, and public policy in terms of their impact on the future.

Product Alert

Marketing Intelligence Service, Ltd.   (716) 374-6326
33 Academy St.
Naples, NY 14512
FAX: (716) 374-5217
Toll-free: (800) 836-5710
Diane Beach
Pat Peck

Reports on new consumer goods launched in American retailing, including foods and beverages, non-prescription drugs, cosmetics and toiletries, and miscellaneous household items. Lists products that are an extension of an existing product line, package changes, and marketing plans. Recurring features include pictures as well as descriptions of the products.

Pump News & Patents

IMPACT Publications   (208) 726-2133
PO Box 3113
Ketchum, ID 83340
Mary Jo Helmeke

Examines new valve products, software, and technical developments. Includes literature reviews and manufacturers' addresses. Recurring features include news of research, a collection, reports of meetings, news of educational opportunities, review, notices of publications available, notices of upcoming seminars, and technical reports.

Remarks: Trademark News for Business

International Trademark Association   (212) 768-9887
1133 Avenue of the Americas
New York, NY 10036
FAX: (212) 768-7796
Mary McGrane

Features columns and news abstracts relating to trademarks.

Research Horizons

Research Communications Office   (404) 894-6987
Georgia Institute of Technology
223 Centennial Research Bldg.
Atlanta, GA 30332
FAX: (404) 894-6983
Mark Hodges

Reports highlights of engineering, scientific, and economic development, research conducted by Georgia Tech Research Institute and a broad range of research in Georgia Tech academic departments. Covers such diverse topics as robotics for the 21st century, natural gas heat pump, AIDS education, Bulgaria's eco-crisis, toughened ceramics, and ecological philosophy.

Research Money

Evert Communications Ltd.   (613) 728-4621
1996 Carling Ave.
Ottawa, ON K1Z 7K8
Canada
FAX: (613) 728-0385
Vincent Wright

Supplies reports and analyses of the forces driving science and technology investment in Canada, with special emphasis on government policies, granting programs, and other incentives for industry and universities. Tracks major expenditures on research and development and highlights areas where research monies are available. Recurring features include interviews, news of research, reports of meetings, news of educational opportunities, and a collection.

The Research Park Forum

Association of University   (602) 752-2002
    Related Research Parks
4500 S. Lakeshore Dr., Ste. 475
Tempe, AZ 85282
FAX: (602) 752-2003
Chris Boettcher
Doug McQueen

Covers Association news, research park activities, technology incubators, and technology development. Recurring features include a collection, reports of meetings, and job listings.

Research & Productivity Council—Alert

Research & Productivity Council   (506) 452-8994
921 College Hill Rd.
Fredericton, NB E3B 6C2
Canada
FAX: (506) 452-1395
Nancy MacLeod

Covers Council activities and projects. Published in English and French.

ResearchNews

Division of Research   (313) 763-5587
   Development and Administration
University of Michigan
3003 S. State St.
Ann Arbor, MI 48109-1274
Lee Katterman

Presents articles on research conducted at the University. Follows research developments in medicine, engineering, physical and natural sciences, social science, humanities, and the arts.

RISC Management Newsletter

Andrew Allison   (408) 626-4361
25420 Via Cincindela
Carmel, CA 93923-8412
FAX: (408) 626-4362
Andrew Allison

Analyzes strategic issues and trends in the computer industry.

RLE Currents

Research Laboratory   (617) 253-2556
   of Electronics (RLE)
Massachusetts Institute of Technology (MIT)
MIT Room 36-412
Cambridge, MA 02139-4307
FAX: (617) 258-7864
Dorothy A. Fleischer

Describes the intellectual concerns and activities of MIT's Research Laboratory of Electronics (RLE) investigators. Focuses on a specific area of research in each issue. Recurring features include interviews, news of research, reports of meetings, notices of publications available, news of RLE faculty and research staff honors, and alumni news.

Sandia Science News

Sandia National Laboratories   (505) 844-8066

Division 7161
PO Box 5800
Albuquerque, NM 87185
Ken Frazier

Reports on current research advances at Sandia National Laboratories.

Science Trends

Trends Publishing, Inc.   (202) 393-0031
National Press Bldg.
Washington, DC 20045
FAX: (202) 393-1732
Arthur Kranish

Reports on developments in general science, in education and throughout society. Covers research and development, current trends, information on scientific and technical publications, and the high-technology outlook. Recurring features include news of research, review, items on publications available, calls for papers, and notices of conferences, seminars, and symposia.

Sensor Business Digest

Vital Information Publications   (415) 389-8671
321 Carrera Dr.
Mill Valley, CA 94941-3995
FAX: (415) 345-7018
Peter Adrian

Covers sensors and other test and measurement, software, and analysis systems. Offers information on markets and developmental strategies, news of research and development, and new products.

SMT Trends

Vital Information Publications   (415) 389-8671
321 Carrera Dr.
Mill Valley, CA 94941-3995
FAX: (415) 345-7018
Sarah Collings

Covers news of developments in surface mount technology (SMT), printed circuit board manufacturing, component packaging, RCB inspection and testing, industry trends, markets, and issues such as ISO 9000 standards and elimination of toxics.

Superconductor Week

Atlantic Information Services, Inc.   (202) 775-9008
1050 17th St. NW, No. 480
Washington, DC 20036
FAX: (202) 331-9542
Toll-free: (800) 521-4323
Charles David Chaffee

Tracks news of the superconductor industry. Covers companies involved in applications, markets, products, and business developments.

Technologies Tomorrow

c/o F.A. Bick
PO Box 21897
Albuquerque, NM 87154

Covers a specific emerging technology in each issue, such as telecommunications, data bases, virtual corporations, microelectronics, virtual reality, and materials technology. Discusses what it is, what companies are developing it, and what products may be based on it.

Technology Access Report

University R&D Opportunities   (415) 883-7600
16 Digital Dr., Ste. 250
Novato, CA 94949
FAX: (415) 883-6421
Toll-free: (800) 733-1556
Michael Odza
John Gilles

Contains news, analysis, and notices of opportunities in research and development, resources, contracts and funding, joint ventures, licensing, and other information regarding technology, transfer and commercialization processes, and management of technology.

Technology Forecasts and Technology Surveys

PWG Publications   (310) 273-3486
205 S. Beverly Dr., Ste. 208
Beverly Hills, CA 90212
Irwin Stambler
Willard Wilks

Covers new developments in advanced technology and predicts future trends in areas such as sales volumes, consumer demand, new technological advances, and developments in the methodology for forecasting future trends. Concerned with a range of technologies, including electronics, computers, medical technology, chemicals, pulp and paper, food, and materials.

Technology NY

Anderson Research & Communications   (518) 283-8109
1223 Peoples Ave.
Troy, NY 12180
Olga K. Anderson

Provides a comprehensive analysis of all aspects of state-wide technology developments in companies and universities. Focuses on specific products, projects, and corporate ventures, including news briefs on capital availability and news concerning research and development issues. Recurring features include news of relevant legislative and regulatory activity, people and company news, reports on regional economic development, special reports, and a collection.

TechScan

Richmond Research, Inc.   (212) 741-0045
PO Box 537, Village Sta.
New York, NY 10014-0537
FAX: (212) 732-9617
Louis Giacalone

Provides information and insights on how various new technologies, products, and design techniques may be used to solve business problems. Recurring features include a collection, editorials, product reviews, and review.

United States Patents Quarterly

Bureau of National   (202) 452-4200
    Affairs, Inc. (BNA)
1231 25th St. NW
Washington, DC 20037
FAX: (202) 822-8092
Toll-free: (800) 372-1033
Cynthia J. Bolbach

Reports important decisions dealing with patents, trademarks, copyrights, unfair competition, trade secrets, and computer chip protection.

Valve News & Patents

IMPACT Publications   (208) 726-2133
PO Box 3113
Ketchum, ID 83340
Mary Jo Helmeke

Reviews new valve products, software, and current literature in the field. Recurring features include news of research, a collection, reports of meetings, news of educational opportunities, review, notices of publications available, notices of upcoming seminars, and technical reports.

Video Technology News

Phillips Business Information, Inc. (301) 424-3700
7811 Montrose Rd.
Potomac, MD 20854
FAX: (301) 309-3847
Toll-free: (800) 777-5006
Charlotte Wolter

Reports on video technologies from a business point of view. Provides industry analyses and forecasts, reports on new products and emerging media trends. Covers legal and regulatory developments. Incorporates the former *FutureHome Technology News*, merged December 1992.

Voice Technology News

Phillips Business Information, Inc. (301) 340-2100
7811 Montrose Rd.
Potomac, MD 20854
FAX: (301) 424-4297
Toll-free: (800) 777-5006
Ian C. McCaleb

Provides comprehensive news and analysis of the voice technology marketplace. Features business strategies, new product developments and enhanced technologies of voice processing and related markets. Recurring features include colorado Washington, New Services, Voice Messages, and Market Strategies.

Wind Energy Technology

National Technical (703) 487-4630
  Information Service (NTIS)
U.S. Department of Commerce
5285 Port Royal Rd.
Springfield, VA 22161

Abstracts worldwide information on all aspects of energy from the wind.

DIRECTORIES

ADWEEK Agency Directory

ADWEEK Directories (212) 536-6504
1515 Broadway, 12th Fl.
New York, NY 10036
FAX: (212) 536-5321
Michael Battaglia

Covers over 4,000 U.S. advertising agencies, public relations firms, media buying services, direct marketing and related organizations. Entries include: Agency name, address, phone, fax; names and titles of key personnel; major accounts; headquarters location; major subsidiaries and other operating units; year founded; number of employees; fee income; billings; percentage of billings by medium. Contains material formerly published in separate regional editions.

ADWEEK Client/Brand Directory

ADWEEK Directories (212) 536-6506
1515 Broadway, 12th Fl.
New York, NY 10036
FAX: (212) 536-5321
Toll-free: (800) 468-2395
Christopher B. O'Connell

Covers advertisers of over 5,000 brand-name products/services nationwide. Entries include: Client/brand name; company name, address, phone, fax; names and titles of key personnel; media expenditures; advertising agency.

Alphabetical Directory of French Trademarks

Compu-Mark 3 220-7211
St. Pietersvliet 7
Antwerp B-2000
Belgium

An index to over one million word trademarks and verbal elements of device trademarks applied for, registered, or renewed in France, and Italian trademarks with extension of protection for France. Entries include: Trademark, international classes of goods and services, application, registration or reference number, date published in *Bulletin Officiel de La Propriete Industrielle*. Volumes may be purchased separately.

Alphaphonetic Directory of International Trademarks

Compu-Mark 3 220-7211
St. Pietersvliet 7
Antwerp B-2000
Belgium

An index of over one million word trademarks and verbal elements of device trademarks filed under the Madrid Agreement at the World Intellectual Property Organization and published in *Les Marques Internationales*. Covers all international classes of goods and services. Entries include: Trademark, international classes of goods and services, registration number, publication date in *Les Marques Internationales,* owner, country of origin, renewal, cancellation or modifications. Individual volumes may be purchased separately.

Alphaphonetic Directory of International Trademarks—Classified by Suffix

Compu-Mark 3 220-7211
St. Pietersvliet 7
Antwerp B-2000
Belgium

An index that lists over 250,000 word trademarks and verbal elements of device trademarks filed under the Madrid Agreement at the World Intellectual Property Organization and published in *Les Marques Internationales*. Covers classes 1-5 for the chemical and pharmaceutical industries. Entries include: Trademark, international class of goods and services, registration number, renewal information, publication date in *Les Marques Internationales*.

American Educational Research Association—Biographical Membership Directory

American Educational   (202) 223-9485
   Research Association
1230 17th St. NW
Washington, DC 20036-3078
FAX: (202) 775-1824
Susan L. Wantland, Director of Publications

Covers over 21,000 individuals involved in educational research and development. Entries include: Name, address, phone, highest degree held, year received, affiliation, areas of occupational specialization, electronic mail address, professional interests.

Athletic Footwear Association—Trademark Directory

Athletic Footwear Association (AFA)   (407) 842-4100
200 Castlewood Dr.
North Palm Beach, FL 33408
FAX: (407) 863-8984

Covers Approximately 1,437 trademarks used by footwear companies.

Attorneys—Patent Directory

American Business Directories, Inc.   (402) 593-4600
American Business Information, Inc.
5711 S. 86th Circle
Omaha, NE 68127
FAX: (402) 331-1505

Number of listings: 4,461. Entries include: Name, address, phone (including area code), size of advertisement, year first in *Yellow Pages,* name of owner or manager, number of employees. Compiled from telephone company *Yellow Pages,* nationwide.

Binsted's Directory of Food Trade Marks and Brand Names

Food Trade Press Ltd.   (959) 63944
Station House
Hortons Way
Westerham, KT TN16 1B2
England
FAX: (959) 561285
Adrian M. Binsted

Covers about 7,000 United Kingdom manufacturers and importers, foreign manufacturers of food products, and their trademarks. Entries include: Brand name or trademark, company name, address, description of product, London agents.

BioScan: The Worldwide Biotech Industry Reporting Service

Oryx Press   (602) 265-2651
4041 N. Central, No. 700
Phoenix, AZ 85012-3397

FAX: (800) 279-4663
Toll-free: (800) 279-6799
Janet Woolum

Covers Over 1,000 companies currently doing product research and development in food processing, agriculture, medicine, and other fields in the biotechnology industry. Entries include: company name, address, phone, names and titles of key personnel, number of employees (including number of Ph.D.s), date founded, names and description of subsidiaries, names of principal investors and percentage of investment, name and description of agreements or contracts, description of research and development, products available, products in development, list of subject terms.

Bioscience Industries of the Southwest

Greater Houston Partnership   (713) 651-2235
1200 Smith, Ste. 700
Houston, TX 77002
FAX: (713) 651-2299

Covers Over 100 bioscience companies in the southwestern U.S. involved in production, manufacturing, scientific research, and technology development. Entries include: Company name, address, phone, description of products/research activities and scope; name and title of contact, corporate structure.

Brands and Their Companies

Gale Research Inc.   (313) 961-2242
835 Penobscot Bldg.
Detroit, MI 48226-4094
FAX: (313) 961-6083
Toll-free: (800) 877-GALE
Susan Edgar

Covers Approximately 282,000 trade names, trademarks, and brand names of consumer-oriented products and their 51,000 manufacturers, importers, marketers, or distributors. Entries include: For trade names—Name, description of product, company name, and source code. For companies—Name, address, source code, phone, fax, toll-free number.

BUSINESS

BUSINESS Datenbanken GmbH   (622) 1 166061
Kurfuersten-Anlage 6
Heidelberg D-69115
Germany
FAX: (622) 1 21536
D. Schumacher

Database covers about 20,000 manufacturers, importers and exporters, research establishments, chambers of commerce, trade promotion agencies, banks and investment companies offering or seeking business opportunities worldwide. Database includes: Name of company, address, type of opportunity or business, product and country codes.

Business Organizations, Agencies, and Publications Directory

Gale Research Inc.   (313) 961-2242
835 Penobscot Bldg.

Detroit, MI 48226-4094
Michael B. Huellmantel

Covers over 34,000 organizations and publications of all kinds that are helpful in business, including trade, business, commercial, and labor associations; government agencies and advisory organizations; commodity and stock exchanges; United States and foreign diplomatic offices; regional planning and development agencies; convention, fair, and trade organizations; franchise companies; banks and savings and loans; hotel/motel systems; publishers; newspapers; information centers; computerized information services; research centers; graduate schools of business; special libraries, periodicals, directories, etc. Entries include: Name, address, phone, fax, contact person; some entries include annotations. Contents are based in part on information selected from other Gale publications.

Business Week—R&D Scoreboard

McGraw-Hill, Inc.   (212) 512-2000
1221 Avenue of the Americas
New York, NY 10020
FAX: (212) 512-6322

Publication includes: List of more than 900 companies reporting expenditures of at least one million dollars for research and development in preceding year. Entries include: (In chart form) Company name, research and development expenses in preceding year, and several other statistical measures.

Canadian Corporate R&D Directory

Evert Communications Ltd.   (613) 728-4621
1296 Carling Ave.
Ottawa, ON K1Y 7K8
Canada
FAX: (613) 728-0385
Natalie Gallimore

Covers approximately 300 Canadian-based companies performing more than $750,000 worth of research and development annually; top 20 research and development intensive Canadian universities. Entries include: company or university name, address, phone, names and titles of key personnel, number of employees, financial data, description of activities.

Catalog of Government Inventions Available for Licensing to U.S. Businesses

U.S. Center for the Utilization   (703) 487-4650
    of Federal Technology
5285 Port Royal Rd.
Springfield, VA 22161
FAX: (703) 321-8547
Edward J. Lehman

Covers about 1,200 federal inventions, developed during the previous year, that are available for licensing by U.S. companies. Entries include: Invention name, inventor name, agency name, patent or patent application number, invention summary.

The China List

Asia Marketing & Management   (215) 735-7670

2014 Naudian St.
Philadelphia, PA 19146-1317
FAX: (215) 735-9661

Database covers over 2,500 scientific, technical, engineering, medical, research and development, educational, and professional companies and organizations in the People's Republic of China. Database includes: Organization or co. name, address, phone, fax, name and gender of decision-maker.

China's Research and Development Institutes Database

AsiaInfo Services, Inc.   (214) 351-3091
2474 Manana Dr., Ste. 121
Dallas, TX 75220
FAX: (214) 351-4861

Database covers approximately 8,000 research and development institutions in China. Entries include: Institute name, address, phone, fax, demographics, name and title of contact, scope of research and and achievements, new products.

Commercial Development Association— Membership Directory

Commercial Development   (202) 775-1849
    Association (CDA)
1330 Connecticut Ave. NW, Ste. 300
Washington, DC 20036
FAX: (202) 659-1699

Covers about 800 member executives involved in the development and promotion of new products. Entries include: Personal name, company name, address, phone, fax, biographical data, professional experience, photograph. Also available in a pocket-sized edition, listing member name, address, phone, published annually.

Companies and Their Brands

Gale Research Inc.   (313) 961-2242
835 Penobscot Bldg.
Detroit, MI 48226-4094
FAX: (313) 961-6083
Toll-free: (800) 877-GALE
    Susan Edgar

Extra! Extra!

There are no better sources than *Advertising Age* and *Ad Week* for the latest information on Fortune 500 new introductions, and who's spending how much to promote and introduce which products.

Covers over 51,000 companies that manufacture, distribute, import, or otherwise market their 282,000 consumer-oriented products. Entries include: Company name, address, phone, fax, trade names, description, and source of information. Based on *Brands and Their Companies*.

Compu-Mark Directory of Trademarks in the Benelux

Compu-Mark   3 220-7211
St. Pietersvliet 7
Antwerp B-2000
Belgium

An index to over 500,000 word trademarks and verbal elements of device trademarks registered or renewed at the Benelux Trademark Office and published in the *Benelux Trademark Gazette*. Entries include: Trademark, international classes of goods and services, registration number, renewals, modifications, publication date in *Benelux Trademark Gazette*.

Compu-Mark Directory of Trademarks in Germany

Compu-Mark   3 220-7211
St. Pietersvliet 7
Antwerp B-2000
Belgium

An index listing all pending applications and registrations of German trademarks and international trademarks claiming protection in Germany and in the International Register for trademarks designating Germany. Entries include: Trademark, international classes of goods and services, status, application or registration number, publication date in official register and *Les Marques Internationales*. Individual volumes may be purchased separately.

Compu-Mark Online

Thomson & Thomson   (617) 479-1600
500 Victory Rd.
North Quincy, MA 02171-3145
FAX: (617) 786-8273
Toll-free: (800) 692-8833

Database covers over 1.2 million federal trademark registrations and applications filed with the U.S. Patent & Trademark Office (USPTO); over 700,000 trademarks and service marks registered with the secretaries of state in individual states and Puerto Rico; over 450,000 Canadian trademark applications and registrations filed with the Canadian Intellectual Property Office. Database includes: Trademark name; ownership history; current status; application and/or registration number; description of product/service; international and/or U.S. classification; dates of first use, filing, registration, renewal, cancellation, and abandonment; USPTO Trademark Trial and Appeal Board data; etc. All active data records are to present, inactive U.S. federal records from 1975, inactive U.S. state records vary from state to state, and inactive Canadian records from 1978.

Conference Papers Index

Cambridge Scientific Abstracts   (301) 961-6700
7200 Wisconsin Ave., 6th Fl.

Bethesda, MD 20814
FAX: (301) 961-6720
Toll-free: (800) 843-7751
Cheryl Droffner

Covers authors of and sources for more than 72,000 scientific and technical papers in chemistry, geosciences, engineering, physical sciences, medicine, and life sciences presented at over 150 major regional, national, and international meetings each year. Entries include: Paper title, author name and address; conference title, location, date, and sponsors; availability, cost, ordering information.

Copyright Directory: Attorneys, Professors, Government Agencies, Searchers, Congressional Committees, Consultants and Clearinghouses

Copyright Information Services   (206) 378-2727
Association for Ed. Communications and Technology (AECT)
1025 Vermont Ave. NW, Ste. 820
Washington, DC 20005
Dr. Jerome K. Miller

Covers organizations and individuals concerned with copyright law, including copyright agencies, copyright clearinghouses, professional associations, consultants, congressional committees, and registration and title searchers. Entries include: Organization, individual, or agency name, address, phone, fax, contact.

Cosmetics & Toiletries—Who's Who in R&D Directory Issue

Allured Publishing Corp.   (708) 653-2155
362 S. Schmale Rd.
Carol Stream, IL 60188-2787
FAX: (708) 653-2192
Nancy Allured
Cynthia Champhey Urbano, Associate Publisher/Editor

Publication includes: List of cosmetic manufacturers and consultants for product development, legal, safety, and regulatory assistance; laboratories that do testing in toxicology, photobiology, microbiology, and safety in animal and human research. Entries include: For companies—Name, address, phone, product types, brand names, names of scientists or directors of product development laboratories. For consultants—Name, address, phone, areas of expertise, professional background, list of company facilities; some entries give association memberships. For testing laboratories—Company name, address, phone, date company founded, types of service, types of laboratory facilities, areas of expertise.

Current Research in Britain

Longman Information & Reference   (279) 442601
Longman Group U.K. Ltd.
Westgate House
The High
Harlow, EX CM20 1YR
England
FAX: (279) 444501
Lindsay Karpeta

Covers over 60,000 research projects at academic institutions, government agencies, and other organizations in the United Kingdom; also includes research by other institutions and by government departments. In 4 volumes: volume 1, physical sciences; volume 2, biological sciences; volume 3, social sciences; volume 4, humanities. Entries include: Institution and department name, address, phone, fax, department head, names of researchers, project descriptions, sponsoring bodies, dates of the work, descriptions of publications.

Directory: A Guide to Leading Independent Testing, Research, and Inspection Firms in America

American Council   (202) 887-5872
     of Independent Laboratories, Inc.
1629 K St. NW, Ste. 400
Washington, DC 20006
FAX: (202) 887-0021
Dana E. Marshall

Covers about 450 independent commercial laboratories worldwide offering research and development, testing and inspection services in the scientific and engineering fields. Entries include: Company name, address, phone, fax, names and titles of key personnel, subsidiary and branch names and locations, description of service, product/service.

Directory of American Research and Technology

R. R. Bowker Co.   (908) 464-6800
Reed Reference Publishing
121 Chanlon Rd.
New Providence, NJ 07974
FAX: (908) 665-6688
Toll-free: (800) 521-8110
Beverley McDonough

Covers over 13,000 nongovernmental facilities active in research with commercial applications. Entries include: Firm or organization name, address, phone, fax, electronic mail address, names and titles of key personnel, subsidiaries or divisions with R&D facilities or activities, description of research emphases, number of research staff with disciplines.

Directory of Associations of Inventors

World Intellectual Property   (22) 7309111
     Organization (WIPO)
34, chemin des Colombettes
Geneva CH-1211
Switzerland
FAX: (22) 7335428

Covers about 130 associations of inventors worldwide, primarily national groups; includes about 35 state and national organizations in the United States. Entries include: Association name, address, phone, name and title of principal officer, name of larger organization (if any) with which affiliated.

Directory of Contact R&D Companies

Delphi Marketing Services, Inc.   (212) 534-4868

400 E. 89th St., Ste. 2J
New York, NY 10128
FAX: (212) 369-6390

Covers companies offering contract research and development services. Entries include: Company name, address, phone, telex, name and title of contact, areas in which research is being done, areas of experience.

Directory of CSIRO Research Programs

Editorial and Publications Section   3 4190459
Commonwealth Scientific and Industrial Research
Organisation (CSIRO)
314 Albert St.
PO Box 89
East Melbourne, VI 3002
Australia
Tony Ermers

Covers about 220 research programs and 2,000 projects. Entries include: Program title, objectives, details, project titles, responsible division, name, title, and address of contact, staff, funds available.

Directory of European Professional & Learned Societies

CBD Research Ltd.   (81) 6507745
15 Wickham Rd.
Beckenham, KT BR3 2JS
England
FAX: (81) 6500768
Susan Greenslade

Covers approximately 6,000 associations and societies in natural sciences, technology, social sciences, business, law, and the humanities in Europe (excluding the United Kingdom and Ireland). Entries include: Association name, official English translation, acronym, address, phone, telex, year founded, subjects in which interested, activities, membership data, previous names, publications, membership in international organizations. This and *Directory of European Industrial and Trade Associations* were formerly combined in the *Directory of European Associations*.

Directory of Federal Laboratory & Technology Resources: A Guide to Services, Facilities, and Expertise

U.S. Center for the Utilization   (703) 487-4650
   of Federal Technology
5285 Port Royal Rd.
Springfield, VA 22161
FAX: (703) 321-8547
Edward J. Lehmann, Director, Office of Product Development

Covers more than 1,900 resources within federal laboratories, engineering, and technical information centers offering advice, specialized equipment, services, and facilities to assist nongovernmental technicians, researchers, and engineers. Entries include: Facility name, address, phone, contact person, capabilities.

Directory of Industrial Laboratories in Israel

National Center of Scientific   (32) 97781
   and Technological Information
84 Hachashmonaim St.
Tel Aviv 67011
Israel
FAX: (34) 92033
Geula Gilat

Covers 300 industrial laboratories engaged in research and development work in science, biotechnology, and the technology of engineering in Israel. Entries include: Laboratory name, address, year established, director, area of research.

Directory of Institutions, Consulting Organizations and Experts in Science and Technology in Africa

African Regional Centre for Technology
BP 2435
Avenue Sheikh Anta Diop
Dakar
Senegal

Covers about 6,000 researchers, consultants, and other experts at 600 institutions in the fields of science and technology, food, and energy in Africa. Entries include: Name of institution or expert, address, phone, names and titles of key personnel, biographical data, services provided, areas of expertise.

Directory of Low Temperature Research in Europe

Institute of Physics Publishing   (272) 297481
Techno House

Redcliffe Way
Bristol BS1 6NX
England
FAX: (802) 8647626
Toll-free: (800) 4882665
Paul C. MacDonald

Covers Approximately 500 laboratories, companies, institutions, and other organizations in Europe and the United Kingdom involved in low temperature research. Entries include: Organization name, address, phone, fax; names and titles of key personnel; geographical area served; description of research activities.

Directory of Manufacturing Research Centers

Manufacturing Technology   (312) 567-4730
   Information Analysis Center (MTIAC)
10 W. 35th St.
Chicago, IL 60616-3799
FAX: (312) 567-4736
Toll-free: (800) 421-0586
Michal Safar

Covers nearly 200 manufacturing research centers. Entries include: Center name, address, phone, names and titles of key personnel, number of employees, financial data, description of projects, publications, organization affiliations.

Directory of National and Regional Industrial Property Offices

World Intellectual Property   (22) 7309111
   Organization (WIPO)
34, chemin des Colombettes
Geneva CH-1211
Switzerland
FAX: (22) 7335428

Covers over 200 industrial property offices. Entries include: Administration name, address, phone, fax, telex, names and titles of key personnel.

Directory of Organizations & Institutes in Israel

National Center of Scientific   (32) 97781
   and Technological Information (COSTI)
84 Hachashmonaim St.
Tel Aviv 67011
Israel
FAX: (34) 92033
Toll-free: (03) 2332

Database covers over 700 scientific organizations, technical associations, industrial laboratories, and research institutes in Israel. Database includes: Name, address, phone, director, governing body, year founded, research fields, publications, goals, membership description.

Directory of Professional Representatives

European Patent Office   (89) 23990
Munich D-80298
Germany
FAX: (89) 23994465

Covers Professional representatives of the European Patent Office. Entries include: Name, address, phone, fax, telex, nationality.

Directory of R&D Institutions in the ROC Republic of China Database

National Science Council   (2 7) 377631
Science and Technology Information Center (STIC)
PO Box 91-37
Taipei
Taiwan
FAX: (2 7) 377663

Database covers research and development institutions in the Republic of China. Entries include: Institution name, address, phone, fax; year founded; contact person; major activities; sources of funding; major publications; cooperative efforts.

Directory of Research and Development Institutions: Non-Ferrous Metals Industry

United Nations Industrial   (222) 26310
    Development Organization (UNIDO)
Wagramerstrasse 5
Vienna A-1400
Austria
FAX: (222) 2307584

Covers research and development centers, institutes, laboratories, bureaus, data collection centers, and other facilities concerned with the production and processing of aluminium, copper, lead, nickel, tin, and zinc. Entries include: Institution name, address, phone, telex, cable address, names and titles of contact and other key personnel, financial data, information concerning affiliated organizations, description of on-going projects, anticipated projects, availability of pilot/installation equipment, technical cooperation and services offered, technical cooperation requested, and publications (if any).

Directory of Research Organizations and Facilities in South Africa

CSIR Information Services   (12) 8414408
Council for Scientific and Industrial Research (CSIR)
Box 395
Pretoria 0001
Republic of South Africa
FAX: (12) 862869
Ingrid de Bont

Covers 500 public, academic, and private organizations and/or departments within organizations conducting research in science and technology in the Republic of South Africa. Entries include: Organization name, street and mailing addresses, phone,

fax, activities. For research departments with the organization— Name, address, phone, number of staff, number of research staff, nature of research, special facilities.

Directory of Researchers for Manpower, Personnel Training and Human Factors

Manpower and Training   (619) 553-7000
    Research Information System
Defense Technical Information Center DTIC-AM
53355 Cole Rd.
San Diego, CA 92152-7213
FAX: (619) 553-7053
E. Byars Vicino

Covers over 750 researchers and research and development project managers; all listed are employed by an office of the Department of Defense, one of its contractors, or other government agency. Entries include: Name, phone and organization.

Directory of Sources of Funds for International R&D Cooperation

World Association of Industrial   (89) 438943
    and Technological Research Organizations
Danish Technological Institute
Teknologiparken
Arhus DK-8000
Denmark
FAX: (89) 438989
Moses D. Mengu

Covers research and development funding agencies, intergovernmental and nongovernmental agencies, and other organizations, including United Nations agencies worldwide, that provide funds for improving research facilities, training of personnel, and information exchange.

Directory of Soviet Medical Research Institutes

Flegon Press   (81) 7521296
37b New Cavendish St.
London W1M 8JR
England
FAX: (71) 4862094

Covers research institutes in the former Soviet Union. Entries include: Institute name, address, phone, director, type of institution.

The Directory of U.S. Trademarks

Thomson & Thomson   (617) 479-1600
500 Victory Rd.
North Quincy, MA 02171-3145
FAX: (617) 786-8273
Toll-free: (800) 692-8833
Lisa M. DePasquale

An index listing 1.2 million active and pending federal trademark registrations filed with the U.S. Patent & Trademark Office for all international classes of goods and services. Entries include: Trademark, registration/serial number, last reported owner name, Principal/Supplemental Register Indicator, *Official Gazette* publication date, intent-to-use information, design indicator, Section 8 & 15 filings, renewals.

Elsevier Business and Services Data Base Directory

Gordon Publications, Inc.   (201) 292-5100
Elsevier Business Press
301 Gibraltar Dr.
Box 650
Morris Plains, NJ 07950-0650
FAX: (201) 898-9281
Joseph Kenna

Covers over 550,000 manufacturers, sales organizations, distributors, wholesalers, research facilities, medical and health care organizations, transportation organizations, consultants, and related government educational facilities. Entries include: Individual name and title and/or function, company or facility name, address, phone, Standard Industrial Classification (SIC) code, number of employees.

Engineering and Electronics Research Centres

Cartermill International   (334) 477660
Technology Centre
St. Andrews
Fife KY16 9EA
Scotland
FAX: (334) 477180

Covers about 7,000 governmental laboratories, industrial research centers, and educational institutions with research and development activity in such fields as standards and metrology, aeronautics, civil engineering, computer theory and software, cybernetics, and transportation. Entries include: Center or organization name, English version of name (if needed), acronym, address, phone, fax, telex, name of parent body, names of director and senior technical section directors, activities, publications, products, annual research & development expenditure, size of graduate research staff, information services, special facilities, number of technical support staff.

Euromecum—European Higher Education and Research Institutions

Dr. Josef Raabe Verlags-GmbH   (228) 9702025
Konigswintererstrasse 418
Bonn D-53227
Germany
FAX: (288) 9702010

Covers over 60,000 higher education and research institutions in 27 European countries. Entries include: Institution name, address, phone, fax, funding status, name and phone of president; address, phone, and contact for public relations department, library, student office, and other departments and faculty; publications; equipment; services offered; governing body.

Europa World Year Book

Europa Publications Ltd.   (71) 5808236
18 Bedford Sq.
London WC1B 3JN
England
FAX: (71) 6361664

Covers national key organizations and firms and about 1,650 international organizations. Entries include: For international organizations—Name, address, principal officials, organization, function, activities, financial structure. For national organizations—Lists with names, addresses, officials, and key facts as appropriate, for the government, religious bodies, newspapers, radio and television stations, banks, trade associations, transport industry, tourism and cultural organizations, universities, research institutes.

European Centres of Development on Advanced Materials

Metra Martech Ltd.   (81) 5630666
Glenthorne House
Hammersmith Grove
London W6 0LG
England
FAX: (81) 5630040
P. Gorle

Covers over 500 organizations in Europe involved in research in advanced composite materials. Entries include: Organization name, address, phone, telex, name and title of contact, names and titles of key personnel, description of projects and products, facilities, technology and expertise available for transfer, requirements for transfers. Part of a series of directories that include coverage on advanced air quality, water and waste water treatment, sensors, and laser technology.

European Centres of Development on Water and Waste WaterTreatment

Metra Martech Ltd.   (81) 5630666
Glenthorne House
Hammersmith Grove
London W6 OLG
England
FAX: (81) 5630040
N. James

Covers over 200 organizations in Europe involved in water and waste water treatment. Entries include: Organization name, address, phone, telex, name and title of contact, names and titles of key personnel, description of projects and products, facilities, technology and expertise available for transfer, requirements for transfers.

European Directory of Consumer Brands and Their Owners

Euromonitor PLC (71) 2518024
87-88 Turnmill St.
London ECIM 5QU
England
FAX: (71) 6083149

Publication includes: List of 5,000 major brand-owning companies in Europe. Entries include: company name, address, phone, telex, financial data, brand names. Principal content of publication is index to 12,000 consumer brands in western Europe, classified by country and by market sector.

European Research Centres

Cartermill International (334) 477660
Technology Centre
St. Andrews
Fife KY16 9EA
Scotland
FAX: (334) 477180
Kevin Blyth

Covers over 13,000 major industrial laboratories, and research funding organizations that conduct, promote, or support research in science, technology, agriculture, and medicine; covers eastern and western Europe. Entries include: Organization name (with English translation), address, phone, telex, affiliation, name of research director, names of senior research staff and their departments, number of graduate research staff, annual budget, activities, major projects, contract work conducted, publications, information services, special facilities, number of technical support staff.

European Sources of Scientific and Technical Information

Cartermill International (334) 477660
Technology Centre
St. Andrews
Fife KY16 9EA
Scotland
FAX: (334) 477180
Lindsay Karpeta

Covers over 1,500 patents and standards offices, national offices of information, and organizations active in scientific fields in Europe, including former Soviet bloc nations. Provides English-language version of foreign terminology. Entries include: Organization name, address, phone, telex, year founded, name of contact, parent company, subject(s) covered, publications, library facilities, and information, consulting, and training services.

Expedition Planners' Handbook & Directory

Expedition Advisory Centre (71) 5812057
c/o Royal Geographical Society
I Kensington Gore
London SW7 2AR
England
FAX: (71) 5844447
Shane Winser
Nicholas McWilliam

Publication includes: Lists of funding organizations, equipment suppliers, travel services, and publishers of information for scientific expeditions and fieldwork outside the United Kingdom. Entries include: Name, address, phone, description of services. Principal content is articles on expedition and fieldwork planning.

Far East and Australasia

Europa Publications Ltd. (71) 5808236
18 Bedford Sq.
London WCIB 3JN
England
FAX: (71) 6361664
Lynn Daniel

Countries of the Far Eastern and Oceanic regions receive detailed physical, social, economic, and political analysis; includes lists of about 20,000 key organizations, firms, institutions; there is also an extensive general survey of the regions as a whole. Similar to the directories in the *Europa World Year Book*.

Federal Bio-Technology Transfer Directory

Biotechnology Information Institute (301) 424-0255
1700 Rockville Pike, Ste. 400
Rockville, MD 20852-1631
FAX: (301) 424-0257
Ronald A. Rader
Sally A. Young

Covers Over 2,800 federal (U.S. government) biomedical and basic biotechnology inventions and technology transfers from 1980-1993, including 1,200 U.S. patents, 900 patent applications, over 500 Collaborative Research & Development Agreements (CRADAs) and nearly 1,000 patent licenses. Entries include: project name, patent/application number, date, inventor name, abstract; contact name and phone begins each agency/laboratory section.

Federal Research in Progress (FEDRIP)

Office of Product Management (703) 487-4929
National Technical Information Service (NTIS)
5285 Port Royal Rd.
Springfield, VA 22161
FAX: (703) 487-4134

Database covers more than 150,000 federally-funded research projects currently in progress in the physical sciences, engineering, health, agriculture, and life sciences areas. Database includes: Project title, starting date, principal investigator, performing and sponsoring organization, detailed abstract, description of the research, objective, and findings (when available).

Federal Research Report

Business Publishers, Inc.   (301) 587-6300
951 Pershing Dr.
Silver Spring, MD 20910
Leonard Eiserer

Covers government and foundation grants and contracts for research, development, and education. Entries include: Grant title; name, address, and phone of contact; deadline; amount.

Fiber Optics Directory

Phillips Business Information, Inc.   (301) 340-2100
1201 Seven Locks Rd., Ste. 300
Potomac, MD 20854
FAX: (301) 309-3847
Toll-free: (800) 777-5006
Carol Eyler

Covers about 1,500 suppliers of cable, connectors, lasers, and other fiber optic systems, networks, and components; technical and engineering firms; consultants; research firms; and other technical and business resources; international coverage. Entries include: company name, address, phone, fax, number of employees, year fiber activities began, names and titles of key personnel, description of product/service.

FMARK

Institut National de la Propriete   (14) 2945260
   Industrielle (INPI)
26, bis rue de Leningrad
Paris F-75800
France
FAX: (14) 2940216

Covers 761,770 trademarks filed and published in France. Database includes: Trademark name, product, description, applicant name and address, agent, filing number and date, publication number.

Food Engineering Database

Chilton Co.   (215) 964-4000
Chilton Way
Radnor, PA 19089
FAX: (215) 964-2915
Jerry Clark

Covers more than 10,000 food and beverage plants with 20 or more employees; food and beverage research and development facilities; and company headquarters. Entries include: Company name, address, number of employees, phone, Standard Industrial Classification (SIC) codes.

Foundation Education Philanthropic Research Directory

American Business Directories, Inc.   (402) 593-4600
American Business Information, Inc.
5711 S. 86th Circle
Omaha, NE 68127
FAX: (402) 331-1505

Entries include: Name, address, phone, size of advertisement, name of owner or manager, number of employees, year first in *Yellow Pages*. Compiled from telephone company *Yellow Pages,* nationwide.

Futuretech

Technical Insights, Inc.   (201) 568-4744
PO Box 1304
Ft. Lee, NJ 07024-9967
FAX: (201) 568-8247
Peter Savage, Editorial Director

Publication includes: List of companies, laboratories, and researchers involved in advanced technology projects that will have broad applications in various industries, including high temperature resistant ceramic parts, plastic structural materials, and low cost dewater processes for solid-liquid systems (coal slurry, paper pulp, etc.); international coverage. Entries include: Company, laboratory, or researcher name, address, phone, background information, financial and technical data, analysis of market potential and investment opportunities. Principal content is articles and analyses of advanced technologies; focus is on one technology per issue.

Government Research Directory

Gale Research Inc.   (313) 961-2242
835 Penobscot Bldg.
Detroit, MI 48226-4094
FAX: (313) 961-6083
Toll-free: (800) 877-GALE
Monica M. Hubbard
Joseph M. Palmisano

Covers Over 4,200 research and development facilities operated or sponsored by the United States or Canadian governments, including research centers, bureaus, and institutes; testing and experiment stations; data collection and analysis centers; government-supported user facilities; cooperative research programs; and major research-supporting service units. Entries include: Unit name, address, phone, fax, mail address, name of director, staff, year founded, parent agencies, description of activities and fields of research, special research facilities, publications, public services, and library collections.

Imsmarq Canadian Trademarks Database

Imsmarq AG   (42) 362244
Dorfplatz 4
Cham CH-6330
Switzerland
FAX: (42) 360363

Database covers all trademark applications and registrations in Canada. Entries include: Trademark, status, description, current and previous owners, priorities, filing details, agents and representatives, international classification, index headings, wares and services description.

Imsmarq Danish Trademarks Database

Imsmarq AG   (42) 362244
Dorfplatz 4

Cham CH-6330
Switzerland
FAX: (42) 360363

Database covers all trademark registrations and applications in Denmark since 1880. Entries include: Trademark, international class, registration or application number, type.

Imsmarq Finnish Trademarks Database

Imsmarq AG   (42) 362244
Dorfplatz 4
Cham CH-6330
Switzerland
FAX: (42) 360363

Database covers all live trademarks for Finland since 1891. Entries include: Trademark, international class code, registration number; when known, status and type, trademark owner.

Imsmarq International Trademarks Database

Imsmarq AG   (42) 362244
Dorfplatz 4
Cham CH-6330
Switzerland
FAX: (42) 360363

Database covers over 280,000 valid international trademark registrations from more than 30 countries from the register maintained at the International Bureau of the World Intellectual Property Organization (WIPO). Entries include: Trademark, status, description, current and previous owner, filing details, any use refusals or limitations.

Imsmarq Norwegian Trademarks Database

Imsmarq AG   (42) 362244
Dorfplatz 4
Cham CH-6330
Switzerland
FAX: (42) 360363

Database covers all trademark registrations and applications in Norway since 1911. Entries include: Trademark, international class code, registration number; when known, status and type, owner.

Imsmarq Pharmaceutical Trademarks In-Use Database

Imsmarq AG   (42) 362244
Dorfplatz 4
Cham CH-6330
Switzerland
FAX: (42) 360363

Database covers over 300,000 registered and nonregistered trademarks currently in use in more than 50 countries. Entries include: Trademark, owner or distributor, recorded sales, product launch date, anatomical therapeutic class.

Industrial Research in the United Kingdom

Longman Group UK Ltd.   (279) 442601
Westgate House, 6th Fl.
The High
Harlow, EX CM20 1YR
England
FAX: (279) 444501
Lindsay Karpeta

Covers approximately 3,000 industrial laboratories in the U.K. with research and development facilities, government-controlled research facilities, research organizations, government departments concerned with R&D, universities and polytechnic institutes, trade and research associations, and professional organizations. Entries include: Most entries include facility, department, or agency name, research interest, address, phone, contact, telex, faculty or managers by specialty; many include annual R&D budget, details of publications produced, staff size, special facilities, name and product range of parent company; listings for associations usually include name, address, and phone only.

INPAMAR Trademarks, Spain Database

Departamento de Informacion   (13) 495300
    Technolgica
Oficina Espanola de Patentes y Marcas
Calle Panama 1
Madrid E-28071
Spain
FAX: (14) 572586

Database covers over 1.4 million trademarks, trade names, business signs, and international trademarks registered in Spain since 1876. Entries include: Trademark, registration number, date of registration, classification.

Inside R&D

Technical Insights, Inc.   (201) 568-4744
PO Box 1304
Ft. Lee, NJ 07024-9967
FAX: (201) 568-8247
Charles Joslin

Publication includes: Lists of companies and researchers developing new advanced technology products or processes, including metal glass film depositors, cell immobilization systems, and squeeze casting techniques. Entries include: Company or researcher name, address, phone, name of contact, description of product or process. Principal content is articles introducing or explaining new technologies.

Interamerican Copyright Institute— Directory of Specialists

Interamerican Copyright Institute   (11) 2119498
Faculty of Law of the University of Sao Paulo
Largo de Sao Francisco, Numero 95
Sao Paulo, SP 01005
Brazil

Covers national copyright institutes and individual members that work to augment the study and progress of the effective protection of copyright in North and South America.

International Brands and Their Companies

Gale Research Inc.   (313) 961-2242
835 Penobscot Bldg.
Detroit, MI 48226-4094
FAX: (313) 961-6083
Toll-free: (800) 877-GALE
Linda Irvin

Covers about 84,000 trademarks, trade names, and brands of consumer products from 28,000 international manufacturers, importers, and distributors. Entries include: For brands—Name, description, company name, source code. For companies—Name, address, phone, fax, toll-free phone, source code.

International Companies and Their Brands

Gale Research Inc.   (313) 961-2242
835 Penobscot Bldg.
Detroit, MI 48226-4094
FAX: (313) 961-6083
Toll-free: (800) 877-GALE
Linda Irvin

Covers Approximately 28,000 companies that manufacture, distribute, import, or otherwise market consumer products in countries other than the United States. Entries include: company name, address, phone, fax; trade names, description, source of information. Based on *International Brands and Their Companies,* which lists about 84,000 trade names and company names alphabetically; companies appear in a separate yellow pages section.

International Federation of Industrial Property Attorneys—Membership List

International Federation of Industrial Property Attorneys
Holbeinstrasse 36-38
Basel CH-4051
Switzerland

Covers 1,805 industrial property attorneys and patent lawyers in private practice.

International Federation of Institutes for Advanced Study—Annual Report

International Federation   (416) 926-7570
    of Institutes for Advanced Study (IFIAS)
39 Spadina Rd.
Toronto, ON M5R 2S9
Canada
FAX: (416) 926-9481
Dr. Jacob Spelt

Publication includes: List of about 40 member institutes, individuals serving as trustees, special advisors, and program administrators and directors; institutes are active in the humanities, natural and social sciences, and technology. Entries include: Name of institution or individual, location. Principal content of publication is essays on science and other subjects concerning world problems.

International Guide to Collective Administration Organizations

Little, Brown & Co.   (617) 859-5569
34 Beacon St.
Boston, MA 02108
FAX: (617) 859-0629
Toll-free: (800) 331-1664
David Sinacore-Guinn

Covers over 250 collective copyright administration organizations worldwide. Entries include: Organization name, address, phone, fax, telex, names and titles of key personnel, description, geographical area served, foreign administration, membership, licensing procedures.

International Intertrade Index of New Imported Products

International Intertrade Index   (908) 686-2382
Box 636, Federal Sq.
Newark, NJ 07101
John E. Felber

Covers manufacturers of new products that are announced at foreign trade fairs and available to United States importers. Entries include: Company name, address, description of new products, and prices. Subscription includes *Foreign Trade Fairs* newsletter.

International Invention Register

Catalyst   (619) 723-8064
PO Box 547

Fallbrook, CA 92028
Dudley Rosborough

An advertising newspaper devoted to classified ads describing inventions for sale or license; published approximately quarterly, but may vary to coincide with certain international trade shows; a typical issue may include 2,000 patents.

International New Product Newsletter

International New Product Newsletter   (508) 741-0224
Box 1146
Marblehead, MA 01945
FAX: (508) 741-0224
Pamela Hilton

Covers sources of new products, worldwide, but principally in the U.S.; includes all types of consumer, industrial, and scientific products; also lists special licensing opportunities; about 70-80 new products per issue, with no repetition. Entries include: For sources of new products—Description of product, rights available, organization name, address. (Some organizations are contacted only through the publisher of the newsletter, and listings for those show only a reference number.) For licensing opportunities—Description of product, name and address of contact.

International Research Centers Directory

Gale Research Inc.   (313) 961-2242
835 Penobscot Bldg.
Detroit, MI 48226-4094
FAX: (313) 961-6083
Toll-free: (800) 877-GALE
Anthony L. Gerring

Covers over 7,800 research and development facilities maintained outside the United States by governments, universities, or independent organizations, and concerned with all areas of physical, social, and life sciences, technology, business, military science, public policy, and the humanities. Entries include: Facility name, address, phone, fax, telex, e-mail, name of parent agency or other affiliation, date established, number of staff, type of activity and fields of research, special research facilities, publications, educational activities, services, and library holdings.

Inventing and Patenting Sourcebook

Gale Research Inc.   (313) 961-2242
835 Penobscot Bldg.
Detroit, MI 48226-4094
FAX: (313) 961-6083
Toll-free: (800) 877-GALE
Wendy Van de Sande

Covers agencies and organizations of interest to investors, including national and regional associations, business incubators, federal funding sources, publications, youth innovation programs, online resources, patent and trademark depository libraries, Small Business Administration centers and state innovation programs, university research centers, consultants and research services, 14,000 registered patent attorneys and agents, venture capitalist sources, federal laboratory assistance for inventors, and 200 leading corporations receiving patents. Entries include: Agency or organization name, address, phone, description of services.

Islamic Foundation for Science, Technology and Development— Directory of Universities and Institutes in OIC Member States

Islamic Foundation for Science,   (26) 322292
  Technology and Development
Box 9833
Jeddah 21423
Saudi Arabia
FAX: (26) 322274

OIC is the acronym for the Organization of the Islamic Conference.

Jewelers' Circular/Keystone—Brand Name and Trademark Guide

Chilton Co.   (610) 964-4470
Chilton Way
Radnor, PA 19089
FAX: (610) 964-4481
Kathleen Ellis

Covers over 5,000 manufacturers of jewelry store products. Entries include: Name of firm, address.

Kompass Industrial Trade Names

Kompass Publishers   (342) 326972
Reed Information Services Ltd.
Windsor Ct.
East Grinstead House
East Grinstead, WS RH19 1XD
England
FAX: (342) 335992
K. Viney, Product Manager

Covers 25,000 companies in the United Kingdom and their 100,000 active trade names and 18,000 lapsed trade names. Entries include: Company name, address, phone, fax, telex, trade names.

Laboratories—Research & Development Directory

American Business Directories, Inc.   (402) 593-4600
American Business Information, Inc.
5711 S. 86th Circle
Omaha, NE 68127
FAX: (402) 331-1505

Number of listings: 2,245. Entries include: Name, address, phone, size of advertisement, name of owner or manager, number of employees, year first in *Yellow Pages.* Compiled from telephone company *Yellow Pages,* nationwide.

Lawyers' List

Commercial Publishing Co., Inc.   (410) 820-4494
20 W. Dover St.
Easton, MD 21601
FAX: (410) 820-4474

Covers about 2,500 lawyers in general, corporate, trial, patent, trademark, and copywrite practices in the United States. Entries include: Firm name, address, phone, type of practice, names of representative clients, names of partners and associates. A general law list.

Licensing Executives Society—Membership Directory

Licensing Executives Society   (203) 232-4825
638 Prospect Ave.
Hartford, CT 06105-4298
FAX: (203) 232-0819
Peter J. Berry, CAE, Administrative Manager

Covers about 6,700 U.S. and foreign business executives, scientists, engineers, lawyers, and new-idea scouts having direct responsibility for the licensing of technology, patents, trade marks, and know-how. Entries include: Name, address.

Link-Up—New Lines Section

Learned Information, Inc.   (609) 654-6266
143 Old Marlton Pike
Medford, NJ 08055-8750
FAX: (609) 654-4309
Loraine Page

Publication includes: List of companies offering new online services, databases, and computer software and hardware. Entries include: Company name, address, phone, description of product, price.

Lockwood-Post's Directory of the Pulp, Paper and Allied Trades

Miller Freeman, Inc.   (212) 869-1300
1515 Broadway, 33rd Fl.

New York, NY 10036
FAX: (212) 869-0902
Harry Dyer

Covers almost 1,000 U.S. and Canadian pulp and paper companies and their mills; 4,000 paper converters; 3,000 paper merchants; and industry associations. Also lists suppliers of equipment, chemicals, technical services, and raw materials to the paper industry (SIC 26). Entries include: Company name, address, phone, personnel, and products and grades manufactured or handled; paper company listings also show divisions and subsidiaries, research centers; mill listings also show equipment and daily output; merchant lists also show nature of business (whether wholesaler, importer, etc.), conversion operations offered. Also available in a *Traveler's Edition* providing same data on pulp and paper companies and their mills with company name and personal name indexes.

MacRae's Industrial Directory Connecticut

MacRae's Blue Book, Inc.   (212) 475-1790
Business Research Publications, Inc.
817 Broadway
New York, NY 10003
FAX: (212) 475-1790
Toll-free: (800) MACRAES
Holli Savadove

Covers 3,900 firms in Connecticut. Entries include: Company name, address, phone, names of key executives, number of employees, plant size, Standard Industrial Classification (SIC) code, product or service provided, gross sales, bank references, year established; whether firm imports, exports, or has research facilities.

MacRae's Industrial Directory Maine/New Hampshire/Vermont

MacRae's Blue Book, Inc.   (212) 673-4700
Business Research Publications, Inc.
817 Broadway
New York, NY 10003
FAX: (212) 475-1790
Toll-free: (800) 622-7237
Holli Savadove

Covers 1,500 firms. Entries include: Company name, address, phone, names and titles of key executives, number of employees, plant size, Standard Industrial Classification (SIC) code, product or service provided, gross sales, bank references, year established, and whether firm imports or exports, and whether firm has research facilities.

MacRae's Industrial Directory Maryland/D.C./Delaware

MacRae's Blue Book, Inc.   (212) 673-4700
Business Research Publications, Inc.
817 Broadway
New York, NY 10003
FAX: (212) 475-1790
Toll-free: (800) 622-7237
Holli Savadove

Covers 3,700 industrial resources. Entries include: Company name, address, phone, names of key executives, number of employees, plant size, Standard Industrial Classification (SIC) code, product or service provided, gross sales, bank references, year established, whether firm imports or exports, whether firm has research facilities. Includes material formerly published in individual directories for Maryland and Delaware.

MacRae's Industrial Directory Massachusetts/Rhode Island

MacRae's Blue Book, Inc.   (212) 673-4700
Business Research Publications, Inc.
817 Broadway
New York, NY 10003
FAX: (212) 475-1790
Toll-free: (800) 622-7237
Holli Savadove

Covers about 6,000 firms in Massachusetts and Rhode Island. Entries include: company name, address, phone, names and titles of key personnel, number of employees, plant size, Standard Industrial Classification (SIC) code, product or service provided, gross sales, bank references, year established, whether firm imports or exports, whether firm has research facilities. Massachusetts and Rhode Island were covered in separate directories until 1984.

MacRae's Industrial Directory New Jersey

MacRae's Blue Book, Inc.   (212) 673-4700
Business Research Publications, Inc.
817 Broadway
New York, NY 10003
FAX: (212) 475-1790
Toll-free: (800) MAC-RAES
Holli Savadove

Covers about 8,269 firms with six or more employees in New Jersey. Entries include: Company name, address, phone, names of key executives, number of employees, plant size, Standard Industrial Classification (SIC) code, product or service provided, gross sales, year established, bank references, whether firm imports or exports, whether firm has research facilities.

MacRae's Industrial Directory New York State

MacRae's Blue Book, Inc.   (212) 673-4700
Business Research Publications, Inc.
817 Broadway
New York, NY 10003
FAX: (212) 475-1790
Toll-free: (800) MAC-RAES
Holli Savadove

Covers 10,752 companies in New York. Entries include: Company name, address, phone, names of key executives, number of employees, plant size, Standard Industrial Classification (SIC) code, product or service provided, gross sales, bank references, year established, whether firm imports or exports, whether firm has research facilities.

MacRae's Industrial Directory North Carolina/South Carolina/Virginia

MacRae's Blue Book, Inc.   (212) 673-4700
Business Research Publications, Inc.
817 Broadway
New York, NY 10003
FAX: (212) 475-1790
Toll-free: (800) 622-7237
Holli Savadove

Covers over 6,800 firms. Entries include: company name, address, phone, names and titles of key personnel, number of employees, plant size, 4-digit Standard Industrial Classification (SIC) code, product/service, gross sales, year established, bank references, whether firm imports or exports, whether firm has research facilities. North Carolina, South Carolina, and Virginia were covered in separate volumes until 1986.

MacRae's Industrial Directory Pennsylvania

MacRae's Blue Book, Inc.   (212) 673-4700
Business Research Publications, Inc.
817 Broadway
New York, NY 10003
FAX: (212) 475-1790
Toll-free: (800) MAC-RAES
Holli Savadove

Covers 7,581 companies. Entries include: Company name, address, phone, names of key executives, number of employees, plant size, Standard Industrial Classification (SIC) code, product or service provided, gross sales, year established, bank references, whether firm imports or exports, and whether firm has research facilities.

Martindale-Hubbell Law Directory

Martindale-Hubbell   (908) 464-6800
  Reed Reference Publishing

121 Chanlon Rd.
New Providence, NJ 07974
FAX: (908) 464-3553
Toll-free: (800) 526-4902
Drew Meyer, Publisher

Covers lawyers and law firms in the United States, its possessions, and Canada, plus leading law firms worldwide; includes a biographical section by firm, and a separate list of patent lawyers, attorneys in government service, in-house counsel, and services, suppliers, and consultants to the legal profession. Entries include: For non-subscribing lawyers—Name, year of birth and of first admission to bar, code indicating college and law school attended and first degree, firm name (or other affiliation, if any) and relationship to firm, whether practicing other than as individual or in partnership. For subscribing lawyers—Above information plus complete address, phone, fax, type of practice, clients, plus additional personal details (education, certifications, etc.). A general law list.

Middle East and North Africa

Europa Publications Ltd.   (71) 5808236
18 Bedford Sq.
London WC1B 3JN
England
FAX: (71) 6361664
Simon Chapman

Countries of the Middle East and North African regions receive detailed physical, social, economic, and political analyses; list of about 15,000 key organizations, firms, and institutions, similar to directories in the *Europa World Year Book*; there is also an extensive general survey of the regions as a whole.

National Council of University Research Administrators—Directory

National Council of University   (202) 466-3894
   Research Administrators
1 Dupont Circle NW, Ste. 220
Washington, DC 20036
Natalie Kirkman

Covers about 2,600 individuals engaged in the administration of sponsored research, training, and educational programs in colleges or universities, in organizations wholly organized and administered by colleges and universities, or in a consortium of colleges and universities. Entries include: Name, affiliation, address, phone.

New Products Monthly Reports

Berliner Research-Kaufman
29 Ironwood Dr.
Danbury, CT 06811-2703

Series includes over 50 separate reports, such as *New Container Products, New Pump Products,* etc., compiled from United States patent abstracts and various information from other countries. A reference to a new product may contain, as available, the product name, manufacturer name and address, uses, ingredients, description, unique features or innovative aspects, etc.

New Trade Names in the Rubber and Plastics Industries

Rapra Technology Ltd.   (939) 250383
Shawbury
Shrewsbury SY4 4NR
England
FAX: (939) 251118
Heather Rodenhurst

Covers about 1,000 manufacturers of thermoplastics, thermosetting resins, rubbers, adhesives, compounding materials and chemicals, and finished products and machinery, primarily in the United Kingdom and the United States; each edition covers about 4,500 new products. Entries include: Company name, address, trade name, graphic description, product description, citation, and product categories.

Nordres

Teknillisen Korkeakoulun Kirjasto   (04) 514112
Otaniementie 9
Espoo FIN-02150
Finland
FAX: (04) 514132

Database covers over 1,000 research units and institutes in Denmark, Finland, Iceland, Norway, and Sweden. Database includes: Organization name, address, field of activity, names and expertise of key personnel, special equipment, participation in Nordic and international cooperative projects.

Online Patents and Tradenames Databases

Aslib   (71) 2534488
   (Association for Information Management)
Information House
20-24 Old St.
London EC1V 9AP
England
FAX: (71) 4300514
J.F. Sibley

Covers more than 35 databases concerned with patents information in 15 countries, and about 20 hosts and producers. Entries include: For databases—Database name, frequency of updating, producer, host, content, costs, user aids. For hosts and producers—Company name, address, phone.

Pacific Research Centres: A Directory of Organizations in Science, Technology, Agriculture, and Medicine

Cartermill International   (334) 477660
Technology Centre
St. Andrews
Fife KY16 9EA
Scotland
FAX: (334) 477180
Kevin Blyth

Covers nearly 3,500 government, state, and private company laboratories, government and university research departments, industrial research centers, and hospital research laboratories in Australia, Brunei, China, Fiji, Hong Kong, Indonesia, Japan, Korea, Malaysia, New Caledonia, New Zealand, Papua New Guinea, Philippines, Samoa, Singapore, Solomon Islands, Taiwan, Thailand, Vanuatu, Vietnam, and some Pacific Islands. Entries include: Name of organization, name of parent company, English translation and acronym (where applicable), address, phone, fax, telex, names of key personnel, number of graduate research staff, description of product or service, summary of current projects, annual expenditure (in local monetary units) for research and development, status (whether unit of government, university, industrial company, or an independent), name, description of publications, special facilities, information services, number of technical support staff.

Patent Attorneys and Agents Registered to Practice before the United States Patent and Trademark Office

U.S. Patent and Trademark Office   (703) 305-8341
Washington, DC 20231

Covers about 11,000 attorneys and agents. Entries include: Name, address, phone, registration number.

Physics Today—Buyers' Guide Issue

American Institute of Physics   (301) 209-3100
1 Physics Ellipse
College Park, MO 20740

Publication includes: List of over 2,000 manufacturers and suppliers of more than 1,800 products and services used in physics research; directory of science buyer's guides to allied fields; coverage is worldwide. Entries include: For manufacturers and suppliers— Firm name, address, phone, toll-free phone, telex, fax, and TWX; sales and distributor offices, with addresses and phone numbers. For directory of related buyer's guides—Title, publisher name, address, frequency, price.

Polymer Engineering Group Directory

British Plastics Federation   (71) 4575000
6 Bath Pl.
London EC2A 3JE
England
FAX: (71) 4575045

Covers current research projects involving polymers at universities and polytechnics in the United Kingdom.

Poultry & Egg Marketing—Brand Name Marketing Issue

Poultry & Egg News, Inc.   (404) 536-2476
Gannett Publishing Co., Inc.
PO Box 1338
Gainesville, GA 30503
FAX: (404) 532-4894
Jim Mathis

Publication includes: List of over 100 companies that market poultry and egg products under specific brand names. Entries

include: company name, address, phone, name and title of contact, product line, brand names, and marketing data.

Prepared Foods—Source Book

Cahners Publishing Co.   (708) 635-8800
1350 E. Touhy Ave.
Des Plaines, IL 60017-5080
Dave Fusaro
Bob Swientek

Publication includes: About 600 food and beverage companies. Entries include: Company name, location, description of products. Principal content of publication is listings of ingredients and equipment.

PRODUCTSCAN

Marketing Intelligence Service Ltd.   (716) 374-6348
6473D Route 64
Naples, NY 14512
FAX: (716) 374-5217
Toll-free: (800) 836-5710
Tom Vierhile, Senior Editor

Database covers approximately 185,000 new consumer goods launched in U.S. and Canadian retail markets as well as in 18 countries outside North America. Compiled from the newsletters *Product Alert* and *International Product Alert*. Database includes: Brand name, product name; manufacturer name, address, phone; distributor name, corporate affiliates, distribution market, market region, industry; product category name and code, major flavors, product form, sample availability, package type, package material, shelving, package tags, innovation, Universal Product Code (UPC), description, varieties and extensions, stock keeping units; date and title of publication in which cited, previous reports, and secondary source citation.

Research Centers Directory

Gale Research Inc.   (313) 961-2242
835 Penobscot Bldg.
Detroit, MI 48226-4094
FAX: (313) 961-6083
Toll-free: (800) 877-GALE
Anthony L. Gerring

Covers over 13,400 university-related and other nonprofit research organizations established on a permanent basis to carry on continuing research programs in all areas of study; includes research institutes, laboratories, experiment stations, computing centers, research parks, technology transfer centers, and other facilities and activities; coverage includes Canada. Entries include: Unit name, name of parent institution, address, phone, fax, mail address, name of director, year founded, governance, staff, educational activities, public services, sources of support, annual volume of research, principal fields of research, publications, special library facilities, special research facilities.

Research & Development Product Source Telephone Directory

Cahners Publishing Co.   (708) 635-8800

1350 E. Touhy Ave.
PO Box 5080
Des Plaines, IL 60017-5080
FAX: (708) 390-2618
Nancy Rademacher

Covers about 4,000 manufacturers, distributors, and suppliers of products and equipment to industrial research facilities. Entries include: Company name, address, phone, and fax. Published as a thirteenth issue of R&D Magazine.

Research Services Directory

Gale Research Inc.　　　　　　　　(313) 961-2242
835 Penobscot Bldg.
Detroit, MI 48226-4094
FAX: (313) 961-6083
Toll-free: (800) 877-GALE
Anthony L. Gerring

Covers more than 4,500 commercial laboratories, consultants, firms, data collection and analysis centers, individuals, and facilities in the private sector that conduct contractual or proprietary research in all areas of business, government, humanities, social science, and science and technology. Entries include: Firm name, address, phone, fax, toll-free number, e-mail name of chief executive, name and title of contact, date founded, staff size and composition, rates charged, annual revenues, professional memberships, parent and/or affiliate organizations, description of research services and principal clients, affiliates, patents, licenses, special equipment.

Resources for Comparative Biomedical Research

National Center for Research Resources　　(301) 594-7933
National Institutes of Health (NIH)
Westwood Bldg., Rm. 857
Bethesda, MD 20892
FAX: (301) 480-2470
Dr. Ole Henriksen, Project Manager

Covers regional primate research centers, primate breeding and research projects, animal diagnostic and investigative laboratories, animal reference centers and information projects, special animal colonies and models, AIDS animal model projects and institutional training awards supported by the Comparative Medicine Program, National Center for Research Resources. Entries include: Resource or institution name, address, phone, names of principal investigators, principal area of research, and details on the resource.

Samir Husni's Guide to New Consumer Magazines

Cowles Business Media, Inc.　　　　(203) 358-9900
911 Hope St.
PO Box 4949
Stamford, CT 06907-0949
FAX: (203) 357-9014
Toll-free: (800) 795-5445
Dr. Samir Husni

Covers publishers of over 540 magazines issued for the first time during the previous year. Entries include: Magazine title, publisher name, address, names and titles of key personnel, price, number of editorial and advertising pages per issue, subject.

SASKTECH Directory

Communications Branch　　　　　　(306) 787-1619
Science and Technology Division
Saskatchewan Economic Development
1919 Saskatchewan Dr.
Regina, SK S4P 3V7
Canada
FAX: (306) 933-8244

Covers over 300 companies in Saskatchewan, Canada dealing in high technology industries, including microelectronics, automation, fibre optics, computer and software development, biotechnology, space and communication and related research; over 80 listings representing research organizations and Saskatchewan's two universities. Entries include: Company name, address, phone, fax, name and title of contact, line of business, activities, products, and services.

Science et Technologie au Quebec

Quebec Dans le Monde　　　　　　(418) 877-2728
CP 8503
Sainte-Foy, PQ G1V 4N5
Canada
FAX: (418) 877-3945
Denis Turcotte

Covers over 1,200 scientific associations, periodicals, research and development facilities, and research centers in Quebec. Entries include: Organization name, address, phone, fax, toll-free phone, description of services.

Science and Technology in Africa

Longman Group UK Ltd.　　　　　　(279) 442601
Westgate House
The High
Harlow, EX CM20 1YR
England
FAX: (279) 444501
Lindsay Karpeta

Part of a multivolume series; lists government agencies, educational institutions, and private organizations conducting research and development projects, and scientific and technological societies and associations in Africa. Volumes may vary somewhat in organization, but generally include overviews of science and technology in the area covered, detailed information on specific major projects, councils, etc., and briefer listings on other projects.

Science and Technology in Australasia, Antarctica, and the Pacific Islands

Longman Group UK Ltd.　　　　　　(279) 442601
Westgate House
The High
Harlow, EX CM20 1YR

England
FAX: (279) 444501
Jarlath Ronayne

Part of a multivolume series; lists government agencies, educational institutions, and private organizations conducting research and development projects; and scientific and technological societies and associations in Australasia, Antarctica, and the Pacific Islands. Volumes in series vary somewhat in organization, but generally include overviews of science and technology in the area covered, detailed information on specific major projects, councils, etc., and briefer listings on others.

Science and Technology in Canada

Longman Group UK Ltd.   (279) 442601
Westgate House
The High
Harlow, EX CM20 1YR
England
FAX: (279) 444501
Paul Dufour
John de la Mothe

Publication includes: Lists of Centres of Excellence, government information and funding sources, technology and research centers and programs, and technology transfer centers in Canada. Entries include: Name, address, phone, fax, description, name and title of contact. Principal contents are essays on the state of science and technology in Canada, including statistics and a bibliography. Part of the series *Longman Guide to World Science and Technology*.

Science and Technology in Eastern Europe

Longman Group UK Ltd.   (279) 442601
Westgate House
The High
Harlow, EX CM20 1YR
England
FAX: (279) 444501
Gyorgy Darvas

Part of a multivolume series; lists government agencies, educational institutions, and private organizations conducting research and development projects; and scientific and technological societies and associations in Eastern Europe. Volumes in series vary somewhat in organization, but generally include overviews of science and technology in the area covered, detailed information on specific major projects, councils, etc., and briefer listings on other projects.

Science and Technology in the Federal Republic of Germany

Longman Group UK Ltd.   (279) 442601
Westgate House
The High
Harlow, EX CM20 1YR
England
FAX: (279) 444501
F. Meyer-Krahmer

Extra! Extra!

Billing itself as the "Industry Newspaper for Engineers and Technical Management," *Electronic Engineering Times* is one of the better free publications in its field. It serves up the latest news on technologies, business, and new products. Its reporters cover major trade shows. For information on the *Electronic Engineering Times*, contact CMP Publications, 600 Community Dr., Manhasset, NY 11030; phone (516) 562-5882.

Part of a multivolume series; lists government agencies, educational institutions, and private organizations conducting research and development projects, and scientific and technological societies and associations in Germany. Volumes in series vary somewhat in organization, but generally include overviews of science and technology in the area covered, detailed information on specific projects, etc., and briefer listings on other projects.

Science and Technology in France and Belgium

Longman Group UK Ltd.   (279) 442601
Westgate House
The High
Harlow, EX CM20 1YR
England
FAX: (279) 444501
E. Walter Kellerman

Part of a multivolume series; lists government agencies, educational institutions and private organizations conducting research and development projects, and scientific and technological societies and associations in France and Belgium. Volumes in series may vary somewhat in organization, but generally include overviews of science and technology in the area covered, detailed information on specific major projects, councils, etc., and briefer listings on other projects.

Science and Technology in India, Pakistan, Bangladesh and Sri Lanka

Longman Group UK Ltd.   (279) 442601
Westgate House
The High
Harlow, EX CM20 1YR
England
FAX: (279) 444501
Abdur Rahman

Part of a multivolume series; lists government agencies, educational institutions, and private organizations conducting research and development projects, and scientific and technological societies and associations in the subcontinent. Volumes in series vary somewhat in organization, but generally include overviews of science and technology in the area covered, detailed information on specific major projects, councils, etc., and briefer listings on other projects.

Science and Technology in Japan

Longman Group UK Ltd.   (279) 442501
Westgate House
The High
Harlow, EX CM20 1YR
England
FAX: (279) 444501
Alun M. Anderson

Part of a multivolume series; lists government agencies, educational institutions and private organizations conducting research and development projects, and scientific and technological societies and associations in Japan. Volumes vary somewhat in organization, but generally include overviews of science and technology in the area covered, detailed information on specific major projects, councils, etc., and briefer listings on other projects.

Science and Technology in the Middle East

Longman Group UK Ltd.   (279) 442601
Westgate House
The High
Harlow, EX CM20 1YR
England
FAX: (279) 444501
Ziauddin Sardar

Part of a multivolume series *Longman Guide to World Science and Technology*; lists government agencies, educational institutions, and private organizations conducting research and development projects, and scientific and technological associations in the Middle East. Volumes in series vary somewhat in organization, but generally include overviews of science and technology in the area covered, detailed information on specific major projects, councils, etc., and briefer listings on other projects.

Science and Technology in Scandinavia

Longman Group UK Ltd.   (279) 442601
Westgate House
The High
Harlow, EX CM20 1YR
England
FAX: (279) 444501
George Ferne

Part of a multivolume series; lists government agencies, educational institutions, and private organizations conducting research and development projects, and scientific and technological associations in Scandinavia. Volumes vary somewhat in organization, but generally include overviews of science and technology in the area covered, detailed information on specific major projects, councils, etc., and briefer listings on other projects.

Science and Technology in the United Kingdom

Longman Group UK Ltd.   (279) 442601
Westgate House
The High
Harlow, EX CM20 1YR
England
FAX: (279) 444501
Anthony P. Harvey

Part of a multivolume series; lists government agencies, educational institutions, and private organizations conducting research and development projects, and scientific and technological societies and associations in the United Kingdom. Volumes in series vary somewhat in organization, but generally include overviews of science and technology in the area covered, detailed information on specific major projects, councils, etc., and briefer listings on other projects.

Science and Technology in the USA

Longman Group UK Ltd.   (279) 442601
Westgate House
The High
Harlow, EX CM20 1YR
England
FAX: (279) 444501
Albert H. Teich
Jill H. Pace

Part of a multivolume series; lists government agencies, educational institutions, and private organizations conducting research and developmental projects, and scientific and technological associations in the USA. Volumes in series vary somewhat in organization, but generally include overviews of science and technology in the area covered, detailed information on specific major projects, councils, etc., and briefer listings on other projects.

Science and Technology in the USSR

Longman Group UK Ltd.   (279) 442601
Westgate House
The High
Harlow, EX CM20 1YR
England
FAX: (279) 444501
Michael J. Berry

Part of a multivolume series; lists government agencies, educational institutions, and private organizations conducting research and development projects, and scientific and technological associations in the former Soviet Union. Volumes in series vary somewhat in organization, but generally include overviews of science and technology in the area covered, detailed information on specific projects, councils, etc., and briefer listings on other projects.

Scientific and Technical Organizations and Agencies Directory

Gale Research, Inc.   (313) 961-2242
835 Penobscot Bldg.
Detroit, MI 48226-4094

FAX: (313) 961-6083
Toll-free: (800) 877-GALE
Christine Jeryan, Managing Editor

Covers over 25,600 national and international organizations and agencies concerned with the physical and applied sciences, engineering, and technology, including associations, computer information services, consulting firms, educational institutions, foundations, government advisory organizations, federal government agencies, general grant and assistance programs, libraries and information centers, patent sources and services, research and development centers, scholarships, fellowships, and loans, science-technology centers, standards organizations, state academies of science, and state government agencies in the fields of aeronautics and space sciences, chemistry, computer science specialties, electronics, geography, geology, machinery, mathematics, metallurgy, meteorology, mineralogy, nuclear science, petroleum and gas, physics, plastics, transportation, water resources, and other areas. Entries include: Organization name, address, phone, and name of contact; additional descriptive text for most entries.

Society of Research Administrators—
Membership Directory
Society of Research Administrators   (312) 661-1700
500 N. Michigan Ave., Ste. 1400
Chicago, IL 60611
FAX: (312) 661-0769
Jan Schwarz

Covers 2,700 persons interested in the management of research of all types in all fields. Entries include: Name, office address and phone, highest degree held, areas of occupational specialization.

Sources Directory
Barrie Zwicker, Publisher   (416) 964-7799
4 Phipps St., Ste. 109
Toronto, ON M4Y 1J5
Canada
FAX: (416) 964-8763
Barrie Zwicker

Covers over 1,200 organizations, including industrial and other firms, trade associations, universities, colleges and other educational institutions, research organizations, public interest groups, churches, government agencies, and labor unions in Canada; all listings are paid. Also includes list of journalism awards, fellowships, prizes, and grants. Entries include: Organization name, address, phone, fax, contact name, names and titles of key personnel, description of activities, logo.

South Pacific Research Register
Pacific Information Centre   313 900
University of the South Pacific
GPO Box 1168
Suva
Fiji
FAX: 300 830
Jayshree Mamtora

Covers Approximately 600 researchers working in all disciplines in the South Pacific area. Entries include: Researcher name, home and business address, position, sponsor, research title, geographical area covered by research, date and duration of research.

State Government Research Directory
Gale Research Inc.   (313) 961-2242
835 Penobscot Bldg.
Detroit, MI 48226-4094
FAX: (313) 961-6083
Toll-free: (800) 877-GALE
Kay Gill
Susan E. Tufts

Covers about 850 state government-sponsored research programs, institutes, test centers, and facilities. Entries include: Program, institute, or facility name, address, phone, date established, name and title of director, parent agency, description and listing of staff, federal and university affiliation, fields of research, projects, scope of library, description of publication and seminar programs.

Status Report of the Energy-Related
Inventions Program
Office of Technology   (301) 975-5500
    Evaluation and Assessment
U.S. National Institute of Standards and Technology
Gaithersburg, MD 20899

Covers inventors of items recommended for possible Department of Energy support. Entries include: Project number, Department of Energy program coordinator and program office, invention name and patent number, inventor name, company affiliation, location, brief description of invention, evaluation status, disposition including award and explanation, relevant dates.

Extra! Extra!

In 1987, the Department of Energy launched the *Inventor-Assistance Program News*, the main information transfer mechanism of the States Inventors Initiative (SII). The purpose of the SII is to encourage inventor organizations, and any associations that assist inventors. For information, contact: Robin Conger, Editor, Battelle Pacific Northwest Laboratories, PO Box 999, K8-11, Richland, WA 99352; phone (509) 372-4328; fax (509) 372-4369.

Student Contact Book

Gale Research Inc. (313) 961-2242
835 Penobscot Bldg.
Detroit, MI 48226-4094
FAX: (313) 961-6083
Toll-free: (800) 877-GALE
Annette Novallo

Covers 800 organizations and 300 publications of interest to students from junior high school through community college levels who are preparing papers and speeches on topics of current interest or who are investigating career opportunities; includes associations, clearinghouses, government agencies, and companies that provide information free or for less than 25, as well as biographical sources. Entries include: For organizations—Name, address, phone, fax, description of purpose and activities, list of materials or assistance available. For publications—Title, description, frequency, publisher.

Think Tank Directory: A Guide to Nonprofit Public Policy Research Organizations

Government Research Service (913) 232-7720
701 Jackson, Ste. 304
Topeka, KS 66603
FAX: (913) 232-1615
Toll-free: (800) 346-6898
Lynn Hellebust

Covers Approximately 1,200 nonprofit organizations in the U.S. conducting research on public policy. Entries include: Organization name, address, phone, name and title of contact, financial data, organizational structure, governing body, publications, and description of activities.

Trademark World—Directory Section

Intellectual Property Publishing Ltd. (71) 7367111
Brigade House, 3rd Fl.
Parsons Green
London SW6 4TH
England
Kelly Wemmers

Publication includes: List of international trademark services and professionals in private practice and in industry. Entries include: Company, institution, or personal name, address, phone, telex, fax, names of key personnel. Principal content is articles on trademark practices and law, reviews of books in the field, conference reports, calendar of conferences and meetings.

TRADEMARKSCAN—Canada

Thomson & Thomson (617) 479-1600
500 Victory Rd.
North Quincy, MA 02171-3145
FAX: (617) 786-8273
Toll-free: (800) 692-8833

Database covers over 450,000 trademark registrations and applications filed with Canadian Intellectual Property Office. Database includes: Trademark; Thomson & Thomson-assigned international class; application and registration number; current and previous owners; description of product or service; current status; dates of first use, filing, registration, renewal; representative for service or agent.

TRADEMARKSCAN—U.K.

Thomson & Thomson (617) 479-1600
500 Victory Rd.
North Quincy, MA 02171-3145
FAX: (617) 786-8273
Toll-free: (800) 692-8833

Database covers trademark, international class number, application (serial) number, owner's name, description of product or service, current status, agent name, and any other pertinent information supplied by the UKPO. Basic data are obtained from the UKPO. Database includes: Information on over 650,000 trademark registrations and applications filed with the Trade Marks Branch of the Patent Office of the United Kingdom (UKPO). Includes active trademarks applications and registration from 1876, lapsed applications from July 1988, and lapsed registration from July 1983.

TRADEMARKSCAN—U.S. FEDERAL

Thomson & Thomson (617) 479-1600
500 Victory Rd.
North Quincy, MA 02171-3145
FAX: (617) 786-8273
Toll-free: (800) 692-8833

Database covers more than 1.2 million U.S. federal trademark registrations and applications filed with the United States Patent and Trademark Office (USPTO). Database includes: Trademark, serial and/or registration number, owner history including current owner, U.S. and international class(es), description of product or service, current status, dates of first use, filing, abandonment, publication, republication, registration, cancellation, and renewal; claims and disclaimers, Trademark Trial and Appeal Board data, filing correspondent. Approximately 40% of the records include images of the trademark logo or design elements. Basic data are obtained either directly from the USTPO or indirectly from *Official Gazette (Trademark Section) of the U.S. Patent and Trademark Office.*

TRADEMARKSCAN—U.S. STATE

Thomson & Thomson (617) 479-1600
500 Victory Rd.
North Quincy, MA 02171-3145
FAX: (617) 786-8273
Toll-free: (800) 692-8833

Database covers over 900,000 trademarks registered with the secretaries of state in the United States and Puerto Rico. Database includes: Trademark, state of registration, U.S. and international class(es), a description of the product or service, registration number, registration date, current status, date of first use, current owner. Basic data are obtained from secretaries of state and Puerto Rico.

UNESCO/ROSTLAC—Directory of Scientific and Technological Research Centres in Latin America and the Caribbean, Spain and Portugal

Regional Office for Science and   (598) 2 772023
    Technology for Latin America and the
    Caribbean (ROSTLAC)
United Nations Educational, Scientific and Cultural Organization
    (UNESCO)
Av. Brasil 2697
11300 Montevideo Uruguay
Montevideo
Uruguay
FAX: (598) 2 772140
Eduardo Martinez

Covers 2000 centers. Entries include: Institution name, address, phone, telex, fax, name and title of contact.

Unique 3-in-1 Research & Development Directory

Government Data Publications, Inc.   (718) 627-0819
1661 McDonald Ave.
Brooklyn, NY 11230
FAX: (718) 998-5960
Toll-free: (800) ASK-GOVT
Siegfried Lobel

Covers firms that received research and development contracts from the federal government during preceding fiscal year. Entries include: Awardee name, address, agency, description of work, dollar amount of contract, and other pertinent data. Additional contracts are listed in *R&D Contracts Monthly,* published in the same arrangement.

U.S. Source Book of R & D Spenders

Schonfeld & Associates Inc.   (708) 948-8080
1 Sherwood Dr.
Lincolnshire, IL 60069
FAX: (708) 948-8096
Carol Greenhut

Covers 5,700 public companies in the U.S. that spend money on research and development. Entries include: company name, address, phone, names and titles of key personnel, financial data, research and development budgets, fiscal year close, Standard Industrial Classification (SIC) code.

U.S. and World Advanced Technology Corporation Directory Service

U.S. Organization Chart Service, Inc.   (619) 443-6560
PO Box 15878
San Diego, CA 92115
Adriana F. Lippman

Covers approximately 12,000 high-technology companies, including defense related corporations, electronics, aerospace, communications, research and development corporations; international coverage. Entries include: Corporation name, address, phone, names and titles of key personnel; addresses and phone numbers of branch offices, subsidiaries, description of activities.

Universities in Italy

Editoriale Italiana S.N.C.   (63) 212653
Via Vigliena 10
Rome I-00192
Italy
FAX: (63) 211359

Covers about 8,000 universities, educational institutions, learned societies, and research establishments in Italy. Entries include: Name of institution or society, address, phone, key personnel or faculty.

Water Data Sources Directory

National Water Data   (703) 648-5672
    Exchange (NAWDEX)
Water Resources Division
U.S. Geological Survey
421 National Center
Reston, VA 22092
FAX: (703) 648-5704

Covers more than 2,000 organizations that are sources of water data; worldwide coverage. Entries include: Organization name, address, phone, data collection areas, type of water data collected and available, branch offices names and addresses, other organization activities.

Western Europe

Europa Publications Ltd.   (71) 5808236
18 Bedford Sq.
London WC1B 3JN
England
FAX: (71) 6361664

Publication includes: List of key organizations, firms, and institutions in western Europe. Entries include: Name and address. Principal content of publication is essays on the political, social, and economic conditions in each country. A select bibliography is also included. Style is similar to directories in the *Europa World Year Book.*

Who's Who in Technology

Gale Research Inc.   (313) 961-2242
835 Penobscot Bldg.
Detroit, MI 48226-4094
FAX: (313) 961-6083
Toll-free: (800) 877-GALE
Amy L. Unterburger

Covers 38,000 engineers, scientists, inventors, and researchers. Entries include: Name, title, affiliation, address; personal, education, and career data; publications, patents; technical field of activity; area of expertise.

World Buyers' Guide to Unusual & Innovative Products

Emir Publications
PO Box 134
Katong
Singapore 9143
Singapore 2951414 7433307
Ismeth Emir

Covers manufacturers and suppliers of unusual merchandise worldwide for mail order dealers, gift stores, novelty dealers, catalog businesses, importers, and opportunity seekers. Entries include: Name of firm, address, cable address, telex, fax, products.

World Directory of Sources of Patent Information

World Intellectual Property   (22) 7309111
   Organization (WIPO)
34, chemin des Colombettes
Geneva CH-1211
Switzerland
FAX: (22) 7335428

Covers about 150 information centers in 36 countries with collections of patent documents, periodicals specializing in patent related matters, and offering other information services concerning technological, legal, and economic data on industrial property and technology transfer. Entries include: Center or publication name, documentation available, other information sources, public services.

World Environmental Research Directory

Pira International   (372) 376161
Randalls Rd.
Leatherhead, SR KT22 7RU
England
FAX: (372) 377526
Susan Farrell, Associate Consultant

Covers industrial, academic, governmental, and private research organizations working on environmental research projects.

Entries include: Name, address, phone, fax; list of research projects with name of project leader, length of project, funding amount and source; list of published papers.

The World of Learning

Europa Publications Ltd.   (71) 5808236
18 Bedford Sq.
London WC1B 3JN
England
FAX: (71) 6361664
Michael Salzman

Covers over 26,000 academic institutions, including universities, colleges, libraries, museums, learned societies, academies, and research institutes and 150,000 staff and officials; 400 cultural, scientific, and educational organizations; international coverage. Entries include: For universities—Name, address, phone, founding date, names and titles of key personnel, and statistics on size of faculty, student body, and library; some listings include names of full professors. For other organizations and institutions—Similar information and detail as appropriate, listings may include number of members, publications, and subject interests; listings for international organizations include more detail.

World Technology/Patent Licensing Gazette

Techni Research Associates, Inc.   (215) 657-1753
York & Davisville Rds.
Willow Grove Plaza
Willow Grove, PA 19090
FAX: (215) 659-3818
Louis F. Schiffman, Editorial Director

Covers leading firms, private and government research laboratories, universities, inventors, consultants, and others that have new products, new process developments, and new technologies available for license or acquisition. Entries include: For products, processes, and technologies—Descriptor, description of item, name and address of contact. For seminars and meetings—Name, location, dates, description.

CHAPTER

21

Surfing the Inventor's Internet

The realization that there are other points of view is the beginning of wisdom.

—CHARLES M. CAMPBELL

Electronic Sources

BIZ. ORBIT. CompuServe. DIALOG. BYTE. NEXIS. These words look and sound like ITT cable addresses, but represent just six of the major players in an industry encompassing thousands of online databases. There is such an abundance of information available electronically that choosing the appropriate database can be a complicated business.

The publication *Gale Directory of Databases,* a complete guide to the worldwide electronic database industry, covers more than 9,000 databases and more than 800 online services. Check your local library for this comprehensive directory.

Years ago, electronic databases were used and understood only by specially trained librarians. Not anymore. Today online databases are available to anyone, and no prior knowledge of computers is required. The learning curve is short. A few online databases can be learned with less than 20 minutes of practice.

Online databases are not perfect by any means. A personal touch is often required to iron out the finer points and make sure that some original material has not been omitted. The data are, after all, input by human beings. But, online databases provide more information faster, easier, and more cost-effectively than manual searches.

The inventor is required to stay informed about myriad fields of research and development, many outside one's specialty. Before the advent of online databases this was a difficult, time-consuming task. But with the present network of electronic tracking systems, the inventor is able to monitor fast-breaking events as they occur. All one requires is a PC connected to a phone line, and entry to numerous online databases. And, of course, you can now maintain select databases on your PC utilizing CD-ROM technology.

Many of the publications cited in Chapter 20, Hot Press, are available via online databases. Your access to online databases may be as close as your own computer terminal or as far away as your nearest public library. For more information

Selecting the Right Database

The selection of an appropriate database can be difficult unless you understand the criteria used to define and evaluate individual patent databases, according to John T. Butler, a science and engineering reference librarian and bibliographer at the University of Minnesota, Twin Cities.

In *Successful Patents and Patenting for Engineers and Scientists,* Mr. Butler gives the following criteria:

- Functional orientation
- Scope
- Text elements
- Indexing and classification
- Graphical elements
- Timeliness

on databases for use by inventors, see the subheading Online Patent Searches located in Chapter 5, How to Conduct a Patent Search.

Internet—A Trip Through Cyberspace

Can the Internet be of value to inventors? You bet! But at the same time, it can be costly to access and use efficiently and a risky place at best.

If you want to do it right, according to Jim Ball in a piece entitled "Avoiding Potholes on the Information Superhighway" that ran in the February 1995 edition of the Department of Energy's *Inventor-Assistance Program News* (IPN) ". . . you'll need the right equipment, patience, a good service provider, patience, an understanding family, patience, and patience. And if you have the help of a 'nerd' friend, so much the better."

Internet is a worldwide maze of computer networks that began in the 1960s as a U.S. Defense Department sponsored project called ARPANET.

By 1984, according to the Internet Society of Reston, Virginia, a total of 1,024 computers were linked directly to the Internet. By October 1994 the number had grown to 3.8 million; and millions of additional unaccounted for computers were connected to Internet by way of access providers and online services (like America On-Line or CompuServe).

The Internet doubles in size every ten months. "The Internet grows like a mutant strain of info-kudzu no reference librarian can prune," writes Rob Pegoraro, self-proclaimed Cyber-Surfer for *The Washington Post*.

"It's safe to say that few people imagined that this enterprise would spawn the sprawling, hyperkinetic creature of today—a place where over a billion e-mail messages zap back and forth every month among 35 million users."

But is the information stored inside Internet-linked computer bases and transmitted on the Internet secure? This is an important issue for any inventor, so I did some research to find out what the experts think.

Alan Brill, who runs the global, high-tech security business for Kroll Associates in New York City told Mike Wallace of *60 Minutes,* "On the (information) superhighway there are car jackings, drive-by shootings, and some of the rest stops are pretty dangerous places to hang around. Until companies understand that, they are putting themselves at risk."

The Computer Emergency Response Team (CERT) at Carnegie Mellon University, Pittsburgh, is funded by the federal government to help prevent break-ins on the Internet. But that's a tough job because the Internet is a web of tens of thousands of different companies, universities, government agencies, and private individuals with no central control. Dane Gary, a former manager of CERT, told Mike Wallace of *60 Minutes,* ". . . there is no one in charge of the Internet. There is no network manager. There is no network administration at the Internet level."

Then he added a reality check. "You cannot make a computer secure. You can reduce the risk, but you can't guarantee security. It isn't going to happen." In other words, any computer can fall victim to hackers. So be careful about how you store and communicate what may be proprietary information on any of your inventions that do not have patent protection.

Having made this point, let's look at the benefits of the Internet. It is staggering to think about the number of resources an Internet account brings to your computer screen.

Edee Edwards, a technical librarian at Pacific Northwest Laboratory in Richland, Washington, writes in IPN that there are three benefits to inventors in the Internet.

(1) E-mail: The most basic level of Internet usage is for electronic mail. Two or more people can interchange mail. Note: No one can be included in your discussion unless added into the address list by you or a recipient. Or you can participate in larger open discussion groups (LISTSERV or USENET) on highly specialized topics.

(2) Search/Retrieve: You can connect yourself to remote computers and search for (telnet) or retrieve (FTP) information. For example, you can use many library catalogs through the Internet.

(3) Search/Browse: You can search and browse many computers' holdings. Tools and software, such as Gopher, Archie, and Mosaic, help with this task. To adequately describe all of these services would require a book. In fact, there are no less than 188 books about the Internet on the market.

Resources for Inventors

The following are some resources you may wish to explore.

• Telnet to fedworld.gov for loads of government information, including the U.S. Patent and Trademark Office Bulletin Board (#116). Contains the full ASCII text of 1994 patents.

• FTP to town.hall.org for the full text of 1994 U.S. patents.

• You can e-mail the word "help" in the body of a message to patents@world.std. com, or phone (617) 489-3927 to find out about an e-mail service for new patents issued.

• Using Mosaic, check out the home pages of the U.S. Patent and Trademark Office at http://www.uspto.gov or DOE at http://www.doe.gov.

• Internet Patent News Service is free of charge and may be reached through the Internet at http://sunsite. unc.edu/patents/intropat.html.

APIPAT

American Petroleum Institute (API)   (212) 366-4040
Central Abstracting & Information Services
275 Seventh Ave., 9th Floor
New York, NY 10001-6708
FAX: (212) 366-4298

Contains more than 230,000 citations, with abstracts (from 1980), to patents related to the petroleum refining and petrochemical industries issued in all industrial countries, as well as European patents issued under the European Patent Procedure (December 1978), and world patents issued under the Patent Cooperation Treaty. Beginning in January 1981, also covers patents related to oil field chemicals. Type of database: bibliographic.

ASIST

U.S. Patent and Trademark Office (PTO)   (703) 308-0322
Office of Information Products Development
Crystal Plaza 2, 9D30
Washington, DC 20231
FAX: (703) 308-0493

Contains information on U.S. patents. Comprises the following files: Patentee-Assignee File—contains assignments at time of issue for utility patents issued from 1969 to date, other patents issued from 1977 to date, and inventors from 1975 to date. Attorney Roster File—contains name, address, and telephone number of attorneys licensed to practice before the PTO. Corresponds to *Attorneys and Agents Registered to Practice before the U.S. Patent and Trademark Office.* Also includes subject terms from the *Index to the U.S. Patent Classification* and U.S. Class and International Patent Classification (IPC) numbers from the *IPC-USPC Concordance* as well as the *Manual of Classification* (two files), *Manual of Patent Examining Procedure, Classification Definitions,* and *Classification Orders Index.* Type of database: bibliographic; directory.

BNA Patent, Trademark & Copyright Law Daily

The Bureau of National Affairs, Inc. (BNA)   (202) 452-4132
1231 25th St., NW
Washington, DC 20037
FAX: (202) 452-4062
Toll-free: (800) 862-4636

Contains the complete text of news articles and analyses covering judicial decisions, legislation, and administrative actions relating to patent, trademark, and copyright law. Covers the state and federal courts, U.S. Congress and state legislatures, and federal regulatory agencies, including the U.S. Patent and Trademark Office and Copyright Office. Includes summaries of cases to be published in *United States Patents Quarterly* and articles which may later appear in *BNA's Patent, Trademark & Copyright Journal* (described in a separate entry). Type of database: full-text.

BNA's Patent, Trademark & Copyright Journal

The Bureau of National Affairs, Inc. (BNA)   (202) 452-4132
1231 25th St., NW
Washington, DC 20037
FAX: (202) 452-4062
Toll-free: (800) 862-4636

Contains the complete text of *BNA's Patent, Trademark & Copyright Journal,* covering legislation, committee reports, international developments (e.g., treaties, conventions), and court and federal agency rulings on patents, trademarks, copyright, and unfair competition. Primary source of information is the U.S. Patent and Trademark Office. Type of database: full-text.

Brands and Their Companies

Gale Research Inc.   (313) 961-2242
835 Penobscot Bldg.
Detroit, MI 48226-4094
FAX: (313) 961-6815
Toll-free: (800) 347-4253

Contains listings of more than 355,000 trade names, trademarks, and brand names of consumer goods available from more than 63,000 owners and distributors. For each trade name, provides name, brief product description, company name, U.S. Patent and Trademark Office (PTO) trademark code, source of information, and, as available, address and telephone, toll-free, and fax numbers of the manufacturer, distributor, importer, or information source. Information is obtained from some 150 published sources and by annual questionnaire mailings and telephone calls to listed companies. Corresponds to print editions *Brands and Their Companies, Companies and Their Brands, International Brands and Their Companies,* and *International Companies and Their Brands.* Type of database: directory.

British Library Catalogue: Science Reference and Information Service

British Library   071 323 7494
Science Reference and Information Service (SRIS)
25 Southampton Bldgs.
Chancery Lane
London WC2A 1AW
England
FAX: 071 323 7435
e-mail: 81: BLI404 (Telecom Gold)

Contains approximately 290,000 citations to books, periodicals, monographs, conference proceedings, and directories held by the British Library Science Reference and Information Service. Covers scientific and technical literature, literature on patents, trademarks, and registered designs, and some business publications. Includes title, brief notes (e.g., name of sponsoring body for conference proceedings, date of meeting), physical description (e.g., number of pages), publisher and/or distributor, and SRIS classification. Type of database: bibliographic.

BYTE

McGraw-Hill, Inc.   (212) 512-4686
1221 Avenue of the Americas
New York, NY 10020
FAX: (212) 512-4256

Contains the complete text of *BYTE,* a magazine covering news and information on microcomputer products and their applications. **Type of database:** full-text.

BYTE Information Exchange

Delphi Internet Services Corporation   (800) 227-2983
BYTE Information Exchange (BIX)
1030 Massachusetts Ave.
Cambridge, MA 02138
e-mail:Liberty (BIX)

An online electronic conferencing system offering a forum for the discussion and exchange of information on a variety of topics of interest to personal computer users. Covers personal computers (including specific brands), operating systems, computer applications, semiconductors and microprocessor chips, new technologies, microcomputing, telecommunications, artificial intelligence, and computer languages. Features industry news, downloadable software, previews of upcoming *BYTE* articles, the complete text of published issues of *BYTE* (described in a separate entry), and a wide variety of Exchanges covering special-interest topics. Each Exchange offers idea-sharing with other users, interviews with industry specialists, new product coverage, classified ad listings, vendor support conferences, the Microbytes Daily newswire providing daily updated computer industry information, public-domain software programs, and programs that accompany articles appearing in *BYTE* magazine.

Among the BIX Exchanges available are: Amiga Exchange—for users of Amiga personal computers, including programming and developer issues. IBM Exchange—for users of the IBM PC, AT series and workalikes, PS/2 series, OS/2 operating system, PC/DOS and MS/DOS, alternative 386 operating systems, utility software, new IBM products, and IBM-compatible hardware and peripherals. Mac Exchange—for users of Apple Macintosh personal computers, including new hardware and software products, office uses, and desktop publishing. Telecomm Exchange—covers computer telecommunications, including facsimile transmission, ISDN, and packet-switching networks. Tojerry Exchange—using personal computers in education, high-level mathematics, space exploration, and other high-tech areas. Writer's Exchange—covers desktop publishing, journalists using computers for reporting, word-processing programs, science fiction and poetry, and writers' conferences. **Type of database:** full-text; software; bulletin board.

Canadian Patent Index

University of British Columbia Library   (604) 822-5404
1956 Main Hall
Vancouver, BC V6T 1Z1
Canada
FAX: (604) 822-3989
e-mail: patscan (Envoy 100)

Contains more than 200,000 citations to Canadian patents. Enables the user to search by patent or application number, classifi-

cation, and keywords. Corresponds to the *Patent Office Record.* **Type of database:** bibliographic.

Canadian Patent Reporter

Canada Law Book Inc.   (905) 841-6472
240 Edward St.
Aurora, ON L4G 3S9
Canada
FAX: (905) 841-5085

Contains the complete text and headnotes of approximately 5900 decisions covered in *Canadian Patent Reporter.* Covers significant cases on patents, industrial design, copyrights, and trademarks from the Commissioner of Patents, Registrar of Trade Marks, and various Canadian courts. Also includes cases tried under the Combines Investigation Act and the Competition Act. Provides cross-references to other Canada Law Book databases. **Type of database:** bibliographic; full-text.

Canadian Trade Marks

Southam Electronic Publishing   (416) 445-6641
1450 Don Mills Rd.
Don Mills, ON M3B 2X7
Canada
FAX: (416) 445-3508

Contains descriptions of approximately 300,000 Canadian trademarks, covering all current registrations and applications. For each trademark, includes owner, agent, and type of trademark (e.g., word, design, certification); description of the trademark and the goods or services to which it applies; whether it is licensed to be used by others and associated claims (e.g., used in Canada since 1950); previous owner; registration and application dates; and other trademarks that are registered by the same owner. Data are gathered by the Register of Trade Marks, Consumer and Corporate Affairs Canada. **Type of database:** directory.

Canadian Trade Marks Opposition Board

Southam Electronic Publishing   (416) 445-6641
1450 Don Mills Rd.
Don Mills, ON M3B 2X7
Canada
FAX: (416) 445-3508

Contains trademark-related decisions from the Registrar of Trade Marks, from hearing officers designated by the Registrar, and from the Canada Trade Marks Opposition Board. Includes type of opposition, style of cause, application number, ruling officer, date of decision, and text of decision. **Type of database:** full-text.

Caribbean Patent Information Network

United Nations Economic Commission   0809 6237308
    for Latin America and the Caribbean (ECLAC)
Subregional Headquarters for the Caribbean
22 St. Vincent St.
P.O. Box 1113
Port of Spain
Trinidad and Tobago

FAX: 0809 6238485

e-mail: ECLAC@UNDP.ORG (Internet)

Contains more than 2300 citations, with abstracts, to patents issued primarily in English-speaking Caribbean countries. Data are obtained from Industrial Property Offices or Registrar Offices in each country. Corresponds to *Caribbean Patents Index*. Type of database: bibliographic.

CAS Registry File

Chemical Abstracts Service (CAS)   (614) 447-3731

2540 Olentangy River Rd.

P.O. Box 3012

Columbus, OH 43210-0012

FAX: (614) 447-3751

Toll-free: (800) 753-4227

e-mail: help@cas.org (Internet)

Contains structure, name, and formula information for nearly 13 million substances identified by Chemical Abstracts Service in journals and patents. Covers more than 16,000,000 substance names. Typical data elements include CAS Registry Number, index name, synonyms, complete molecular formula, class identifier, CAS Registry Number locator, and stereochemistry. Enables the user to retrieve full information on a particular substance covered in the various Chemical Abstracts Service databases. Data are derived from more than 12,000 journals; patents from 27 countries and two international organizations; and books, conference proceedings, dissertations, and technical reports. Type of database: numeric; full-text.

CASSIS/ASIGN

U.S. Patent and Trademark Office (PTO)   (703) 308-0322

Office of Information Products Development

Crystal Plaza 2, 9D30

Washington, DC 20231

FAX: (703) 308-0493

Contains patent reassignment information from the U.S. Patent and Trademark Office. Type of database: full-text.

CASSIS/CLASS

U.S. Patent and Trademark Office (PTO)   (703) 308-0322

Office of Information Products Development

Crystal Plaza 2, 9D30

Washington, DC 20231

FAX: (703) 308-0493

Contains U.S. Patent Classification System (PCS) categories assigned to more than 5.5 million utility, design, plant, reissue, and X-numbered patents, as well as defensive publications and statutory invention registrations. Type of database: full-text.

CD-ROM Electronic News

Ellis Enterprises, Inc. (EEI)   (405) 749-0273

4205 McAuley Blvd., Suite 315

Oklahoma City, OK 73120

FAX: (405) 751-5168

Toll-free: (800) 729-9500

Contains information on CD-ROM technology and national standards for CD-ROM products. Includes a forum enabling the user to interact with other participants. Information is provided by users. Type of database: full-text; bulletin board; directory.

Chemical PatentImages ™

MicroPatent   (203) 466-5055

250 Dodge Ave.

East Haven, CT 06512-3358

FAX: (203) 466-5054

Provides full-text images of U.S. patent documents. Includes chemical structures and formulas, drawings, and tables. Corresponds in part to the PatentImages™ CD-ROM. Also front-page searchable on 11 fields including title and abstract. Type of database: image; full-text.

Chinese Patent Abstracts in English Data Base

European Patent Office (EPO)   1 52126307

European Patent Information and Documentation System (EPIDOS)

Schottenfeldgasse 29

Postfach 82

Vienna A-1072

Austria

FAX: 1 521265492

Contains citations, with English-language abstracts, to nearly 50,000 patents issued by the Patent Office of the People's Republic of China. Includes patent numbers from equivalent patents published in other countries. Record items include standard bibliographic data and patent family information. Type of database: bibliographic.

CLAIMS ™/CITATION

IFI/Plenum Data Corporation   (910) 392-0068

102 Eastwood Rd., Suite D-6-F

Wilmington, NC 28403

FAX: (910) 392-0240

Toll-free: (800) 368-3093

Contains information on every U.S. and non-U.S. patent (over 5 million patent numbers) cited in U.S. patents. Each of the more than 3.8 million records identifies, by patent number, all later U.S. patents in which an earlier patent is cited. Each patent record contains the patent number of each later patent (both U.S. and non-U.S.) that cites it. Type of database: bibliographic.

CLAIMS ™ Compound Registry

IFI/Plenum Data Corporation   (910) 392-0068

102 Eastwood Rd., Suite D-6-F

Wilmington, NC 28403

FAX: (910) 392-0240

Toll-free: (800) 368-3093

Contains a listing of more than 15,000 chemical compounds referenced five or more times since 1950 in patents covered by the CLAIMS/UNITERM database. Each record includes the IFI compound term number, main compound name, available synonyms, molecular formula, element count, fragment codes, and fragment terms. Corresponds to the *Compound Term List Name/Number Order for the IFI Chemical Patent Databases* and *Compound Term List Molecular Formula Order for the IFI Chemical Patent Databases*. Type of **database**: full-text; numeric.

CLAIMS ™/Comprehensive Data Base

IFI/Plenum Data Corporation   (910) 392-0068
102 Eastwood Rd., Suite D-6-F
Wilmington, NC 28403
FAX: (910) 392-0240
Toll-free: (800) 368-3093

Contains enhanced indexing of the U.S. chemical and chemically related patents included in the CLAIMS/UNITERM database. Each patent record has the same set of general and compound terms selected from the controlled vocabulary (uniterms) used in CLAIMS/UNITERM. In addition, all fragment terms include roles to indicate the function (e.g., reactants, products, non-reactants) of each chemical substance. Also includes systematic indexing of polymers and the substructure fragmentation system used in indexing Markush structures. Type of **database**: bibliographic.

CLAIMS ™/Reassignment & Reexamination

IFI/Plenum Data Corporation   (910) 392-0068
102 Eastwood Rd., Suite D-6-F
Wilmington, NC 28403
FAX: (910) 392-0240
Toll-free: (800) 368-3093

Contains more than 280,000 citations to patents for which ownership has been transferred from one party to another (reassigned patents) or for which the patentability has been reviewed and rejected, reaffirmed, or extended by the U.S. Patent and Trademark Office (reexamined patents). Includes bibliographic data for all patents. For reassigned patents, also includes names of former and new assignees, date of reassignment, and type of reassignment (e.g., quarter interest). For reexamined patents, also includes name and location of reexamination requestor, request number and date, reexamination certificate date, and text from the certificate describing the results of the reexamination. Type of **database**: bibliographic.

CLAIMS ™/Reference

IFI/Plenum Data Corporation   (910) 392-0068
102 Eastwood Rd., Suite D-6-F
Wilmington, NC 28403
FAX: (910) 392-0240
Toll-free: (800) 368-3093

Contains classification codes and titles for classes and subclasses provided in the U.S. Patent Office *Manual of Classification* and the *Index to the U.S. Patent Classification*. Covers approximately 400 main classes and 105,000 subclasses that pertain to patents issued for mechanical, electrical, and chemical inventions. Also contains vocabulary and thesaurus terms that appear in the CLAIMS™/UNITERM database as assigned by IFI/Plenum Data Company. Type of **database**: dictionary.

CLAIMS ™/U.S. Patent Abstracts

IFI/Plenum Data Corporation   (910) 392-0068
102 Eastwood Rd., Suite D-6-F
Wilmington, NC 28403
FAX: (910) 392-0240
Toll-free: (800) 368-3093

Contains more than 2.3 million citations to granted U.S. utility patents, reissued patents, and defense publications announced in the U.S. Patent Office *Official Gazette*. Equivalent patents issued by Belgium, Germany, France, Great Britain, and The Netherlands are also included for chemical patents. Covers patents in these areas of science and technology: aerospace and aeronautical engineering, agricultural engineering, biomedical technology, chemistry, chemical engineering, civil engineering, electronics and electrical engineering, electromagnetic technology, mechanical engineering, medicine, nuclear science, and telecommunications. Also includes records previously contained in CLAIMS/CHEM. Abstracts are included for chemical patents since 1950, electrical and mechanical patents since 1963, and design patents since 1976. Full text claims and front page information for U.S. patents issued from 1971 forward. Type of **database**: bibliographic.

CLAIMS ™/UNITERM

IFI/Plenum Data Corporation   (910) 392-0068
102 Eastwood Rd., Suite D-6-F
Wilmington, NC 28403
FAX: (910) 392-0240
Toll-free: (800) 368-3093

Contains more than 2.3 million citations, with abstracts, to all granted U.S. utility patents, reissue patents, and defensive publications. Covers patents related to chemistry from 1950 to date, mechanical and electrical patents from 1963 to date, and design patents from 1976 to date, with additional subject indexing for chemical and chemically related patents. Each patent covered in the file has been assigned a set of at least 20 descriptors (uniterms) selected from a controlled vocabulary consisting of general terms, fragment terms, and compound terms. Citations to *Chemical Abstracts* as well as foreign patent equivalents from Belgium, Germany, France, Great Britain, and The Netherlands are also included. Covers the patents listed in the Chemical Section of the *Official Gazette* of the U.S. Patent Office. Type of **database**: bibliographic.

Classification and Search Support Information System

U.S. Patent and Trademark Office (PTO)   (703) 308-0322
Office of Information Products Development
Crystal Plaza 2, 9D30
Washington, DC 20231
FAX: (703) 308-0493

Contains current U.S. Patent Classification System categories and titles assigned to utility patents issued from 1969 to date and other patents issued from 1977 to date. For each patent, provides date of issue, title, assignee at the time of issue, state or country of residence of first inventor, and status. Also contains patent abstracts from the most recent two years, as disc space allows. Type of **database**: bibliographic; full-text.

COMLINE Japan News Service

COMLINE Business Data, Inc. 03 5401 4567
1-12-5 Hamamatsucho, Minto-ku
Tokyo 105
Japan
FAX: 813 54012345

Covers Japanese industry, with emphasis on new products, new technologies, ongoing research, and marketing strategies. Contains more than 100,000 citations, with English-language abstracts, to Japanese literature covering the electronics, biotechnology and medical technology, industrial automation, chemicals and materials, computers, telecommunications, and transportation industries in Japan as well as the Japanese economy and financial industry. Sources include more than 100 Japanese-language publications, as well as government reports, seminars, trade shows, and exhibits. Corresponds to the COMLINE Daily News series covering the individual industries; each Daily News service is described in a separate entry. Type of database: bibliographic.

Commission d'Opposition des Marques de Commerce

QL Systems Limited (613) 238-3499
901 St. Andrew's Tower
275 Sparks St.
Ottawa, ON K1R 7X9
Canada
FAX: (613) 238-7597
Toll-free: (800) 387-0899

Contains trademark-related decisions from the Registrar of Trademarks and the Canada Trade Marks Opposition Board. Includes all decisions released in French, and French translations of selected decisions released in English. Type of database: full-text.

Compu-Mark Online

Thomson & Thomson (617) 479-1600
500 Victory Rd.
North Quincy, MA 02171-3145
FAX: (617) 786-8273
Toll-free: (800) 692-8833

Contains information on more than 1.5 million U.S. federal trademark registrations and applications filed with the U.S. Patent and Trademark Office; over 900,000 trademarks and servicemarks registered in individual U.S. states, American Samoa, and Puerto Rico; and over 500,000 trademark applications and registrations filed with Canadian Intellectual Property Office. For each trademark, includes description of mark, description of goods or services, international classification, owner history, registration and/or application date and number, type of trademark (e.g., word or design), current status, disclaimers, limitations, priority claims, and related domestic and foreign registrations. Type of database: full-text.

Compu-Mark U.K. On-Line

Compu-Mark (UK) Ltd. 071 278 4646
New Premier House, Suite 3
150 Southampton Row
London WC1B 5AL

England
FAX: 071 278 5934

Contains all trademark registrations, applications, and pending applications filed with the British Patent Office. Sources include the *Official Trade Marks Journal of the UK,* pending applications, and official publications of new trademark applications. Type of database: full-text.

Copyright Monographs File

U.S. Library of Congress (202) 707-6100
Cataloging Distribution Service
Washington, DC 20541-5017
FAX: (202) 707-1334

Contains bibliographic information for nearly 8 million items registered or renewed for copyright protection. Covers literary works (except periodicals), works of performing and visual arts, and sound recordings submitted to the U.S. Copyright Office. For each item, provides author and copyright claimant, title of work, dates of creation and copyright filing, registration number, and related works. Type of database: bibliographic.

Current Patents Fast-Alert

Current Drugs Ltd. 071 580 8393
Middlesex House
34-42 Cleveland St.
London W1P 5FB
England
FAX: 071 580 5646

Contains citations, with abstracts, to U.S., British, European, and international pharmaceutical patents and applications. Covers anti-inflammatory, antiallergic, and gastrointestinal patents; antimicrobial patents, including antivirals, vaccines, antifungals, antibacterials, and antiparasitics; cardiovascular patents, including cardiovascular chemotherapy and blood products; and central nervous system (CNS) patents, including antidepressants, antipsychotics, analgesics, antiepileptics, antimigraines, and treatments for Alzheimer's Disease, Parkinson's Disease, and stroke. Abstracts include therapeutic classification and examples of preferred compounds, inventor name, and International Patent Classification (IPC). Type of database: bibliographic.

DYNIS

Control Data Systems Canada, Ltd. (800) 267-0444
Information Services
One Antares Dr., Ste. 400
Nepean, ON K2E 8C4
Canada
FAX: (613) 225-9211

Contains the complete text of registered and pending trademark applications in Canada. Type of database: full-text.

EDS Shadow Patent Office

Electronic Data Systems (EDS)
Toll-free: (800) 258-6739
e-mail: webmaster@www.spo.eds.com;
    spo_patent@ spo.eds.com.

Offers easy-to-use patentability and infringement searches against the full-text of 1.7 million U.S. patents issued since January 1972. Access through URL: http://www.eds.com/patent.html. **Type of database:** full-text.

EPAT

| France Institut National | (01) 42945260 |
| de la Propriete Industrielle (INPI) | |

26 bis, Rue de Leningrad
Paris F-75800
France
FAX: (01) 42940216

Lists patents applied for and published in the European Patent Office's printed European Patent Bulletin. Contains more than 560,000 citations, with abstracts in the original patent language, to claims in English and French, and all bibliographic and legal information. **Type of database:** full-text; bibliographic.

ESPACE EP-A

| European Patent Office (EPO) | 1 52126307 |

European Patent Information and Documentation System
(EPIDOS)
Schottenfeldgasse 29
Postfach 82
Vienna A-1072
Austria
FAX: 1 521265492

Contains full-text images of European patent applications, including illustrations (e.g., formulas, chemical structures, technical drawings) and claims. Covers 13 bibliographic data elements, including names of inventor, applicant, and agent; application title in English, French, and German; International Patent Classification (IPC) code; country of origin; countries designated; dates of application and publication; and opponents and dates of opposition. Corresponds to *European Patent Bulletin* and in part to the European Patents Register online database. Full-text images of the first pages of patent applications are available in ESPACE FIRST. **Type of database:** full-text; image.

F-D-C Reports

| FDC Reports, Inc. | (301) 657-9830 |

5550 Friendship Blvd., Suite 1
Chevy Chase, MD 20815
FAX: (301) 656-3094

Contains information related to the business, financial, and legal aspects of the health care and chemical industries. Comprises the following five files: The Gray Sheet—monitors regulatory developments for medical devices and diagnostics at the U.S. Food and Drug Administration's Center for Devices & Radiological Health, as well as policies and congressional reform initiatives concerning premarket approvals, 510(k) exemptions, and related policies. The Pink Sheet—provides news and analysis of developments affecting prescription and over-the-counter drugs. The Rose Sheet—monitors regulatory and legal for the cosmetics, toiletries, fragrances, and skin care industries. The Blue Sheet—covers developments in biomedical research and health policy, particularly as it applies to the U.S. National Institutes of Health. The Tan Sheet—tracks the U.S. Food and Drug Administration's new OTC Office of Drug Evaluation and its recently formed Nonprescription Drugs Advisory Committee; emphasis is on nutritional supplements, including vitamins and minerals. **Type of database:** full-text.

Federal Technology Report

| McGraw-Hill, Inc. | (212) 512-4686 |

1221 Avenue of the Americas
New York, NY 10020
FAX: (212) 512-4256

Contains the complete text of *Federal Technology Report,* a newsletter covering business opportunities at U.S. federal laboratories under the U.S. Departments of Energy, Defense, and Commerce. Covers technology transfer laws, licensing arrangements, and cooperative research and development agreements. **Type of database:** full-text.

Financial Times Business Reports: Technology

| FT Business Enterprises Ltd. | 0932 761444 |

Number One
Southwark Bridge
London SE1 9HL
England
FAX: 44932 781425

Contains the complete text of all articles appearing in 7 newsletters covering business and financial developments in the computer and telecommunications industries: *Advanced Manufacturing* (formerly *Automated Factory*); *Computer Product Update; Electronic Office; Mobile Communications; Personal Computer Markets; Software Markets; Telecom Markets.* **Type of database:** full-text.

Friday Memo

| Information Industry Association (IIA) | (202) 639-8262 |

555 New Jersey Ave., NW, Suite 800
Washington, DC 20001
FAX: (202) 638-4403

Contains the complete text of *Friday Memo,* a biweekly newsletter covering the information industry in the United States. Includes national policy positions; activities of executives in the information industry; major events of the IIA and other information groups; reports on new information services and products; innovative technologies; new publications; business operations; research; and industry trends. **Type of database:** full-text.

FullText

| MicroPatent | (203) 466-5055 |

250 Dodge Ave.
East Haven, CT 06512-3358
FAX: (203) 466-5054

Contains the complete text of all patents issued by the U.S. Patent and Trademark Office (USPTO) since 1975. For each patent, includes descriptions of drawings, background and summary of the invention, examples, and claims. Users can search and retrieve patents by up to 11 fields, and key words in title and abstract. Source of information is the complete text of patents issued electronically by the USPTO. Front-page patent information is available through U.S. Patent Search. Images of patents are available in PatentImages™; each product is described in a separate entry. **Type of database**: full-text.

GENESEQ

Derwent Information Ltd.	(071) 344 2800
Derwent House	
14 Great Queen St.	
London WC2B 5DF	
England	
FAX: (071) 344 2900	
e-mail: LONDON DERWENT (DIALMAIL)	

Contains descriptions of nucleic acid and protein sequences from patent applications and granted patents. Covers all nucleotide sequences greater than nine bases, all protein sequences greater than three amino acids, and probes of any length. Includes type of molecule (e.g., DNA, RNA, protein), length of molecule, accession number, description of the sequence, genus and species of the organism from which the sequence was obtained, table of sequence-specific features, patent number, publication date, priority details for the patent, patent assignee, patent inventors, patent title, and patent sequence. Nucleotide sequences are in IUPAC format, and protein sequences are in one-letter format. Data are derived from patent applications and granted patents published by 14 major patent issuing authorities, including the European Patent Office (EPO), Japan Patent Information Organization (JAPIO), and U.S. Patent and Trademark Office (USPTO). **Type of database**: bibliographic; full-text; numeric.

Genetic Technology News

Technical Insights, Inc.	(201) 568-4744
P.O. Box 1304	
Fort Lee, NJ 07024-9967	
FAX: (201) 568-8247	
e-mail: info@insights.com (Internet)	

Contains the complete text of *Genetic Technology News*, a monthly newsletter covering commercial applications in the chemical, energy, food, and pharmaceutical industries of products resulting from genetic engineering. Includes market forecasts, business news from leading biotechnology firms, announcements of investment opportunities, reports on research activities at universities and research institutions, and reports of new products. Also provides information on recently granted biotechnology patents, including number, title, assignee, and issue date. Many items include names, addresses, and telephone numbers of people who may be contacted for further information. **Type of database**: directory; full-text.

Government-Industry Data Exchange Program

U.S. Navy	(909) 273-4677
Naval Warfare Assessment Center	
Government-Industry Data Exchange Program (GIDEP)	
GIDEP Operations Center	

P.O. Box 8000
Corona, CA 91718-8000
FAX: (909) 273-5200

Contains non-classified technical data on design, development, production, and operation of all types of equipment and systems, from lawn mowers to computers and aircraft. Data are submitted by users in industry and government agencies, including approximately 1500 military, aerospace, and commercial industrial organizations. Comprises the following five files: The Engineering Data File—contains engineering evaluation and qualification test reports, non-standard parts justification data, parts and materials specifications, manufacturing processes, and other related engineering data on parts, components, materials, and processes. Also includes reports on specific engineering methodology and techniques, air and water pollution, alternative energy sources, and manufacturer-certified test results for commercial and military devices. Primary purpose of this data collection is to help prevent duplicate testing by contractors and government agencies. The Reliability-Maintainability Data File—contains failure rate/mode and replacement rate data on parts, components, and materials based on field performance information and/or reliability demonstration tests of equipment, subsystems, and systems. Also contains reports on theories, methods, techniques, and procedures related to reliability and maintainability practices. The Metrology Data File—contains metrology data, i.e., engineering data on test systems, calibration systems, measurement technology, more than 25,000 calibration procedures for test equipment, and more than 11,000 test equipment maintenance manuals. Includes National Institute of Standards and Technology (NIST) metrology-related data. The Failure Experience Data File—contains objective alerts, agency action notices, safe alerts, and failure information generated when significant problems, including safety and fire hazards, are identified for parts, components, processes, fluids, and materials. Also contains information on problem and failure analysis. The Product Manufacturing Data File—contains Diminishing Manufacturing Sources and Material Shortages (DMSMS) Notices and Product Change Notices (PCN). DMSMS Notices provide advance notice of product discontinuance to allow activities enough time to determine the need for an alternate source, conduct redesign, or make a life-of-type buy. PCNs notify the user of changes to technical characteristics or parameter (form, fit, or function) in an item or material. **Type of database**: numeric; full-text.

The Gray Sheet

FDC Reports, Inc.	(301) 657-9830
5550 Friendship Blvd., Suite 1	
Chevy Chase, MD 20815	
FAX: (301) 656-3094	

Monitors regulatory developments for medical devices and diagnostics at the Food and Drug Administration's Center for Devices & Radiological Health, as well as policies and congressional reform initiatives concerning pre-market approvals, 510(k) exemptions, and related policies. Also covers product innovations and industry news, including start-ups, financing deals, international developments, technology reimbursement, and other industry activities. **Type of database**: full-text.

The Green Sheet

FDC Reports, Inc.	(301) 657-9830
5550 Friendship Blvd., Suite 1	

Chevy Chase, MD 20815
FAX: (301) 656-3094

Provides news and information on the pharmacy profession and the pharmaceutical distribution system. **Type of database:** full-text.

HDTV Report

Phillips Business Information, Inc. (PBI)   (301) 340-2100
7811 Montrose Rd.
Potomac, MD 20854-3363
FAX: (301) 424-7261

Reports news and developments in the field of high-definition television (HDTV) technology. Coverage includes current developments in the United States, Europe, and Japan. **Type of database:** full-text.

Improved Recovery Week

Pasha Publications, Inc.   (703) 528-1244
1616 N. Fort Meyers Dr., Suite 1000
Arlington, VA 22209
FAX: (703) 528-1253
Toll-free: (800) 424-2908

Provides coverage of the improved oil and gas industry, including technology transfer information; news; advanced secondary techniques such as horizontal drilling, infill drilling, and profile modification; and enhanced oil recovery such as thermal, CO_2, steam, and microbial. **Type of database:** full-text.

IMSWorld New Product Launches

IMSWorld Publications Ltd.   (071) 393 5100
7 Harewood Ave.
London NW1 6JB
England
FAX: (071) 393 5900

Contains information on more than 25,000 new ethical pharmaceutical products introduced in 57 countries worldwide. For each product, provides product name, country of launch, manufacturer, corporate name, composition, number of components, therapeutic class, language of text, indications, launch date, year of launch, and pack information. Source is original data gathered by IMS offices worldwide. Corresponds to the *Compendium* section of the *New Product Launch Letter*. **Type of database:** full-text.

IMSWorld Patents International

IMSWorld Publications Ltd.   (071) 393 5100
7 Harewood Ave.
London NW1 6JB
England
FAX: (071) 393 5900

Contains product patent information on more than 1000 pharmaceutical compounds, either marketed or in active research and development. Provides pharmaceutical product family profiles. For each patent, provides drug name, synonyms, trade names, therapeutic class, compound name, Chemical Abstracts Service (CAS)

Registry Number, description, patent number, assignee, priority numbers and dates, publication date, patent country, estimated patent expiration dates, published application number by country, extensions to patent terms for the United States and Japan where granted, and U.S. marketing exclusivity information, if applicable. Corresponds to *Patents International*. **Type of database:** directory.

INNOVATION

Longman Cartermill Ltd.   0334 77660
Technology Centre
St. Andrews
Fife KY16 9EA
Scotland
FAX: 0334 77180

Contains information on new products, processes, ideas, and methodologies developed at public universities, polytechnics, and research institutions in the U.K. Covers both patented and unpatented research with commercial potential in the physical, engineering, and life sciences. Includes item description, stage of development (e.g., projected, early, intermediate, finished), patent status, and contact information. Corresponds to *Innovation*. **Type of database:** directory.

Inside R & D

Technical Insights, Inc.   (201) 568-4744
P.O. Box 1304
Fort Lee, NJ 07024-9967
FAX: (201) 568-8247
e-mail: info@insights.com (Internet)

Contains the complete text of *Inside R&D*, a weekly newsletter on developments with industrial potential available for licensing. Covers chemistry, electronics, materials engineering, instrumentation, biotechnology, robotics, and emerging technologies. Includes name, address, and telephone number of a contact person. Also provides articles on research and development management. Sources include technical publications, patents, and interviews with research scientists and administrators. **Type of database:** directory; full-text.

INSPEC™

Institution of Electrical Engineers (IEE)   01438 767297
Michael Faraday House
Six Hills Way
Stevenage SG1 2AY
England
FAX:01438 742840
e-mail: 11472 INSPEC UK (DIALMAIL)
e-mail: INSPEC (DATAMAIL)
e-mail: 315K (STNMAIL)
e-mail: INSPEC@IEE.ORG.UK (Internet)

Contains more than 4.7 million citations, with abstracts, to the worldwide literature in physics, electronics and electrical engineering, computers and control, and information technology. Primary coverage is of journal articles and papers presented at conferences, although significant books, technical reports, and dissertations are also included. Topics covered in physics include: mathematical and theoretical physics; electromagnetism and optics;

quantum field theory and elementary particle physics; nuclear physics; atomic and molecular physics; gases, fluid dynamics, and plasmas; structural, thermal, mechanical, electrical, magnetic, and optical properties of condensed matter; acoustics; geophysics; astronomy and astrophysics; instrumentation and measurement; and related interdisciplinary topics. Topics covered in electronics and electrical engineering include: circuits and components; electron devices and materials; electromagnetics and communication; energy and power systems and applications; instrumentation; and telecommunications. Topics covered in computers and control include: systems and control theory; control technology; computer programming and applications; and computer systems and equipment. Information technology topics include applications of modern communications and computing to the production, transmission, storage, and interpretation of visual, oral, and digitally encoded information. Hardware coverage includes microcomputers and related peripherals. Corresponds to these publications: *Physics Abstracts, Electrical and Electronics Abstracts,* and *Computer and Control Abstracts.* **Type of database:** bibliographic.

International Technology & Business Opportunities

Klenner International, Inc.   (904) 673-4339
PO Box 1748
Ormond Beach, FL 32175
FAX: (904) 673-5911

Contains information on more than 30,000 technology items available for licensing that have been collected from organizations around the world. Items are divided into 56 subject categories in these broad groups: chemicals, biologicals, mechanicals, electronics, and miscellaneous. For each item, provides the title of the technology, licensor, technical description, list of main uses and advantages, degree of development and, if applicable, the patent number. **Type of database:** directory.

Investigational Drugs Database

Current Drugs Ltd.   (071) 580 8393
Middlesex House
34-42 Cleveland St.
London W1P 5FB
England
FAX: (071) 580 5646

Provides access to information on the research and development activities within the pharmaceutical industry, especially pharmaceutical compounds, patents, and applications. Comprises four modules: Commercial—provides information on sales forecasts, licensing opportunities, financial data and corporate strategies. Sources include newsletters, financial reports, business databases, and company communications. Literature—provides citations to articles appearing in key biomedical and chemical journals. Meetings—provides information from scientific conferences. Patents—covers patents filed through the World, European, U.S., and Japanese patent systems; provides coverage of novel compounds, new indications, processes, formulations, diagnostic tests, and biotechnology discoveries. Searchable by company name, identification name, patent name, developmental status, indications and actions, Chemical Abstracts Service Registry Number, chemical name, synonyms, class, licensing, sales potential, patents, bibliography, pharmcast, biology, pharmaceutics, clinical data, evaluation, and current opin-

ion. Weekly updates are available on diskette or via the Internet on day of release. **Type of database:** bibliographic; full-text; image.

The Journal of Proprietary Rights

Prentice Hall Law & Business   (201) 894-8538
270 Sylvan Ave.
Englewood Cliffs, NJ 07632
FAX: (201) 894-8666
Toll-free: (800) 447-1717

Contains the complete text of *The Journal of Proprietary Rights,* a monthly publication covering substantive developments in the areas of copyright, patents, trademarks, and intellectual property in general. **Type of database:** full-text.

Lessons from the Future

21st Century Media Communications, Inc.   (604) 688-7103
548 Cardero St.
Vancouver, BC V6G 2W6
Canada

Contains the complete text of selected Dr. Tomorrow columns which provide predictions of technological developments and trends, including compact disc technology, robotics, computers, telecommunications, and satellites. **Type of database:** full-text.

LEXIS Federal Patent, Trademark, & Copyright Library

LEXIS-NEXIS   (513) 865-6800
9443 Springboro Pike
P.O. Box 933
Dayton, OH 45401-0933
FAX: (513) 865-6909
Toll-free: (800) 227-4908

Contains patent, trademark, and copyright case decisions from the Supreme Court since 1790, from the Courts of Appeals since 1789, from the Federal Circuit Court of Appeals since 1982, from the District Courts since 1789, from the Court of Claims from 1940 to 1982, and from the Claims Court since October 1982. Also contains patent and trademark case decisions from the U.S. Court of Customs and Patent Appeals from 1952 to 1982, from the Board of Patent Appeals and Interferences since 1981, from the Commissioner of Patents and Trademarks since 1981, from the Trademark Trial and Appeals Board since 1982, from the U.S. Court of International Trade since 1980, the U.S. Customs Court from 1962 to 1982, and the U.S. Bankruptcy Court since 1979. Also includes *Code of Federal Regulations* Titles 19 and 37, *United States Code Service* Titles 17 and 35, the complete text of the current developments section of *BNA's Patent, Trademark & Copyright Journal* and of the American Bar Association's *Patent, Trademark & Copyright Law* since 1981, and the complete text of all utility, plant, and design patents files in the United States. since 1975. **Type of database:** full-text.

LEXIS United Kingdom & Commonwealth Legal Libraries

LEXIS-NEXIS   (513) 865-6800
9443 Springboro Pike

P.O. Box 933
Dayton, OH 45401-0933
FAX: (513) 865-6909
Toll-free: (800) 227-4908

Consists of several libraries of United Kingdom, European, and Commonwealth statutes, statutory instruments, case decisions, and legal journals, including English General Library (ENGGEN)—contains *Patent Cases* since 1945. United Kingdom Intellectual Property Library—contains statutes and statutory instruments and case decisions reported in *Reports on Patent Design and Trademark Cases* since 1945 and unreported cases since 1980. **Type of database:** full-text.

LEXPAT

LEXIS-NEXIS   (513) 865-6800
9443 Springboro Pike
P.O. Box 933
Dayton, OH 45401-0933
FAX: (513) 865-6909
Toll-free: (800) 227-4908

Contains the complete text of more than one million U.S. patents, including all utility patents issued since 1975. Includes the complete specification, claims, and abstract, as well as changes resulting from certificates of correction, reclassification, and reassignment. Over 40 segments of information in each patent document are uniquely identified for searching, including patent number, title, inventor, issue date, examiner, filing date, assignee, claims, and the complete text of the specification. Patent documents are organized in four files: Util, containing all utility patents; Plant, containing all plant patents; Design, containing all design patents; and All, containing all utility, plant, and design patents. The following search-aid files are also available: Patent Classifications File (CLASS), containing classes, subclasses, and patent numbers; Manual of Classification (CLMNL), containing the classification schedule; and Index to Classification (INDEX), containing citations, cross-references, and scope notes to classes and subclasses. **Type of database:** full-text.

Library of Congress—Copyright Information Database

Library of Congress Information System   (202) 707-3000
101 Independence Avenue SE
Washington, DC 20559
FAX: (202) 707-9100

Contains work registered for copyright since 1978. It includes books, films, music, maps, sound recordings, software, multimedia kits, drawings, posters, sculptures, and other items. Database also includes documents relating to copyright ownership, such as name changes and transfers, and serials. Files include: Introduction to the Copyright Office, Copyright Basics, Copyright Information Circulars, Copyright Registration, Research in Copyright Office Files, Copyright Office Announcements, Other Copyright Topics, Advisory Committee on Copyright Registration and Deposit (ACCORD) CARP & Licensing Information, and What's New in the Copyright Office. **Access:** Telnet: locis.loc.gov; Gopher: Library of Congress; gopher.loc.gov; Choose: copyright. **Type of database:** full-text.

LitAlert

Research Publications International   (203) 397-2600

12 Lunar Dr., Drawer AB
Woodbridge, CT 06525
FAX: (203) 397-3893
Toll-free: (800) 336-5010

Contains more than 20,000 citations, with abstracts, to cases involving U.S. patent and trademark violations filed in U.S. District Courts and reported to the Commissioner of Patents and Trademarks. For each case includes description of action, names of parties in action, patent or trademark cited, and patent assignee or trademark owner. Corresponds to *LitAlert—The Patent and Trademark Litigation Alert*. **Type of database:** bibliographic.

Local Government Information Network

LOGIN Services Corp.   (612) 331-5672
125 Main St. SE, Ste. 341
Minneapolis, MN 55414
FAX: (612) 331-5532
Toll-free: (800) 328-1921
e-mail: loginservice.lgin@ksinet.com (Internet)
e-mail: 73304,702 (CompuServe)
e-mail: rosesk.lgin@ksinet.com (Internet)

Contains information on more than 60,000 innovative and efficient ways to carry out all aspects of local government. Topics covered include administration, communications, computers, corrections, cost reductions, courts, demographics, economic development, energy, environment, finance, health and human services, housing, public safety, public works, revenue, taxation, transportation, utilities, and wastes. In addition to a title and one- to two-page description of the item, records identify the person or organization to be contacted for further information. Bibliographic citations are provided for items from the literature on local government issues. Information is provided by users, local government agencies, information clearinghouses, and research centers. **Type of database:** directory; full-text.

MARPAT

Chemical Abstracts Service (CAS)   (614) 447-3731
2540 Olentangy River Rd.
P.O. Box 3012
Columbus, OH 43210-0012
FAX: (614) 447-3751
Toll-free: (800) 753-4227
e-mail: help@cas.org (Internet)

Contains more than 250,000 graphic representations of generic (Markush) structures from more than 72,000 patents published since 1988 and cited in CA File. All chemical subject areas are represented, with the exception of alloys, metal oxides, inorganic salts, intermetallics, and polymers. Records also contain bibliographic citation, abstract, and CAS Registry Numbers. **Type of database:** image; bibliographic.

MARPATpreviews

Chemical Abstracts Service (CAS)   (614) 447-3731
2540 Olentangy River Rd.
P.O. Box 3012
Columbus, OH 43210-0012

FAX: (614) 447-3751
Toll-free: (800) 753-4227
e-mail: help@cas.org (Internet)

Contains more than 3500 graphic representations of generic (Markush) structures from more than 1000 patents cited in the *CApreviews* database (described in a separate entry). All chemical subject areas are represented, with the exception of alloys, metal oxides, inorganic salts, intermetallics, and polymers. Records also contain bibliographic citation, abstract, and CAS Registry Numbers. Type of database: image; properties; numeric; bibliographic.

McGraw-Hill Publications Online

McGraw-Hill, Inc.   (609) 426-5000
Princeton-Hightstown Rd.
Hightstown, NJ 08520
FAX: (609) 426-7352
e-mail: 13957 (DIALMAIL)

Contains the complete text of 13 business magazines and 34 newsletters published by McGraw-Hill as follows: *Aerospace Daily* (since 1989). *Aerospace Propulsion* (since 1991). *Airports* (since 1989). *Aviation Daily* (since 1989). *Aviation Europe* (since October 1991). *Aviation Week & Space Technology* (since 1985). *Biotechnology Newswatch* (since September 1986). *Business Week* (since 1985). *BYTE* (since March 1987). *Chemical Engineering* (since 1985). *Clean-Coal/Synfuels Letter* (since 1988). *Coal Week* (since 1988). *Coal Week International* (since 1988). *Data Communications* (since 1985). *Electric Utility Week* (since September 1986). *Electrical World* (since 1992). *ENR: Engineering News-Record* (since 1985). *Federal Technology Report,* formerly *Tech Transfer Report* (since 1990). *Hazardous Waste Business* (since 1991). *Independent Power Report,* formerly *Cogeneration Report* (since 1988). *Industrial Energy Bulletin* (since 1989). *Inside Energy/with Federal Lands* (since 1988). *Inside F.E.R.C.* (since 1988). *Inside F.E.R.C.'s Gas Market Report* (since 1988). *Inside N.R.C.* (since 1985). *Integrated Waste Management,* formerly *Waste-to-Energy Report* (since 1988). *LAN Times* (since 1989). *Metals Week* (since 1988). *Modern Plastics* (since 1991). *Nuclear Fuel* (since July 1985). *Nucleonics Week* (since July 1985). *The Physician & Sportsmedicine* (since 1990). *Platt's International Petrochemical Report* (since 1988). *Platt's Oilgram News* (since September 1986). *Platt's Oilgram Price Report* (since September 1986). *Postgraduate Medicine* (since 1990). *Power* (since 1992). *Regional Aviation Weekly* (not updated, earliest issues date from 1989). *Securities Week* (since May 1986). *Standard & Poor's Emerging & Special Situations* (since 1991). *Standard & Poor's The Review of Banking & Financial Services* (since September 1991). *Standard & Poor's The Review of Securities & Commodities Regulations* (since July 1991). *UnixWorld* (since July 1991). *Utility Environment Report* (since 1991). *The Weekly of Business Aviation* (since 1989). NOTE: Most of these publications are available as separate files on several online services. Further details on the online accessibility of each individual publication may be found in individual entries under the publication title. Type of database: full-text.

Michigan Corporation Data Base

Michigan State Department of Commerce   (517) 334-6327
Corporations and Securities Bureau
P.O. Box 30054
Lansing, MI 48909

Contains information on 420,000 active and inactive corporations, 14,000 active and inactive limited partnerships, and approximately 6000 trade/service marks in the State of Michigan. Type of database: full-text.

Modern Intellectual Property

Prentice Hall Law & Business   (201) 894-8538
270 Sylvan Ave.
Englewood Cliffs, NJ 07632
FAX: (201) 894-8666
Toll-free: (800) 447-1717

Provides the complete text of *Modern Intellectual Property, Second Edition* by Michael A. Epstein. Contains information on trade secrets, copyright law, patent law, trademark law, the protection of ideas, biotechnology, and other areas of intellectual property law. Type of database: full-text.

New Product Announcements/Plus

Information Access Company (IAC)   (415) 378-5000
362 Lakeside Dr.
Foster City, CA 94404
FAX: (415) 358-4759
Toll-free: (800) 321-6388

Contains the complete text of approximately 200,000 press releases from more than 15,000 companies on new products and technologies and corporate activities. Covers announcements of new products and services, product modifications, and new technologies and processes from manufacturers, distributors, and service companies in nearly 60 industries (e.g., communications, medical and health services, textiles). Includes new product description, specifications, and applications; information on trade names, prices, model numbers, availability, and licensing agreements; and the name, address, and telephone number of a company contact. Also covers new facilities and expansions, distribution and licensing agreements, joint ventures, mergers and acquisitions, contract awards, litigation, corporate financial results, and market demographics. Information may be searched by product codes based on Standard Industrial Classification (SIC) codes, company name, product trade name, product uses and applications, location of company, and special feature codes indicating discussions of price or performance specifications. Also provides selected news releases from BusinessWire and PR Newswire. Type of database: full-text.

New Technology Week

King Communications Group, Inc.   (202) 638-4260
627 National Press Bldg.
Washington, DC 20045
FAX: (202) 662-9744

Contains the complete text of *New Technology Week,* a newsletter covering technological developments in such areas as superconductivity and energy storage, high definition television (HDTV), magnetic levitation transport, computers, semiconductors, and new materials. Provides list of National Science Foundation Grants, including recipient's name and institution, amount of award, and research topic, plus profiles of international industry leaders, government officials, and laboratories. Type of database: full-text.

NTIS FedWorld

U.S. National Technical Information Service   (703) 487-4650
5285 Port Royal Rd.
Springfield, VA 22161

Provides access to the results of U.S. and foreign government sponsored research and development and engineering activities. It provides the full technical reports for most of the results announced. FedWorld features include: document ordering with popular downloadable products; collection of files/documents on government information and other data; databases and subdatabases of government information provided by other agencies or info sources (Davis-Bacon, Patent Licenses, CALS, etc.); a gateway connection to other government systems and databases; FTP of Library of Files including text files and images; public mail conferences; White House press release and documents; and federal jobs. Special offerings include databases of abstracts of government patents available for license; abstracts of reports on international trade; and information on government grants and assistance programs. Also available in CD-ROM and print. **Access**: Telnet: fedworld.gov; Login: First/Last Name; Password: Chosen by user. Modem: 703321-8020. **Type of database**: full-text.

OG/PLUS

Research Publications International   (203) 397-2600
12 Lunar Dr., Drawer AB
Woodbridge, CT 06525
FAX: (203) 397-3893
Toll-free: (800) 336-5010

Contains information on patents issued in the current year by the U.S. Patent and Trademark Office. Comprises the following three files: Patents Issued—contains full-text images of the *Official Gazette,* including chemical formulas, technical drawings, and illustrations. For each patent, includes patent number, title, issue date, inventor name and state or country, assignee name and state or country, filing date, application serial number, related patents and applications, country, date, and application of priority, U.S. class and cross-reference classes, and International Patent Classification (IPC) class. Also includes exemplary claims from the *Official Gazette* and abstracts of newly issued patents (not covered in the Official Gazette). Patent Status File—contains information on post-issue actions affecting U.S. patents. Covers more than 20 types of actions, including corrections, disclaimers, litigation and adjudication, cancellations, reexaminations, reissues, reassignments, and withdrawals. Includes patent number, status change, and effective date of change. Corresponds to the monthly *Patent Status File* and in part to the Patent Status File online database. LitAlert—contains citations, with abstracts, to cases involving U.S. patent violations filed with the PTO by the U.S. District Courts during the current year. Covers patent infringement, false description or designation of origin, product dilution, and unfair competition. For each case, includes description of action, names of parties in action, patent cited, and patent assignee. Corresponds to the monthly *LitAlert—The Patent and Trademark Litigation Alert* and in part to the LitAlert online database. **Type of database**: bibliographic; full-text; image.

ORBPAT

ORBIT QUESTEL   (703) 442-0900
8000 Westpark Dr.
McLean, VA 22102

FAX: (703) 893-4632
Toll-free: (800) 456-7248

Enables the user to conduct crossfile searching of patents databases accessible online through the ORBIT Search Service. **Type of database**: directory.

PAPERCHEM

Institute of Paper Science and Technology   (404) 853-9500
500 Tenth St., NW
Atlanta, GA 30318
FAX: (404) 853-9510
e-mail: 25526 (DIALMAIL)

Contains more than 300,000 citations, most with abstracts, to the worldwide scientific and technical literature and patents on theories, resources, products, and processes of the pulp, paper, and board industries. Both manufacturing and use aspects are covered. Also provides author and inventor affiliation information, including organization name and location. Corresponds to the *Abstract Bulletin of The Institute of Paper Science and Technology (ABIPST)*. **Type of database**: bibliographic.

PATDATA

CD Plus Technologies   (212) 563-3006
CDP Online
333 7th Ave.
New York, NY 10001
FAX: (212) 563-3784
Toll-free: (800) 950-2035

Contains citations, with abstracts, to approximately 700,000 U.S. utility patents issued since 1971 and all reissue patents and defense publications issued by the U.S. Patent and Trademark Office since 1975. Each record includes patent number, date, title, inventor, inventor address, patent assignee, application data, foreign priority information, and codes of the United States and International Patent Classification Systems. Also includes cited references to U.S. and non-U.S. patents. **Type of database**: bibliographic.

Patent Explorer/Software

Research Publications International   (703) 413-5050
Rapid Patent Service
1921 Jefferson Davis Hwy., Ste. 1821-D
Arlington, VA 22202
FAX: (800) 457-0850
Toll-free: (800) 336-5010

Contains the complete text of more than 10,000 software patents. Information is grouped into 28 classes, and is searchable by keyword, abstract, claim, U.S. class, inventor, and assignee. **Type of database**: full-text.

Patent Licensing System Database

U.S. National Technical Information Service
5285 Port Royal Rd.
Springfield, VA 22161

Provides information on hundreds of new government research and development inventions available each year for licensing. The database represents the following agencies: Agriculture, Commerce, Transportation, Environmental Protection Agency, Health and Human Services, Interior, and Veteran Affairs. **Access:** Telnet: fedworld.gov.

Patent Status File

Research Publications International   (203) 397-2600
12 Lunar Dr., Drawer AB
Woodbridge, CT 06525
FAX: (203) 397-3893
Toll-free: (800) 336-5010

Contains more than 300,000 references to post-issue actions affecting U.S. patents. Covers more than 20 types of actions, including corrections, disclaimers, litigation and adjudication, cancellations, reexaminations, reissues, reassignments, and withdrawals. Includes patent number, type of action, status change, and effective date of change. Corresponds to the *Patent Status File*. **Type of database:** directory.

PatentBible

MicroPatent   (203) 466-5055
250 Dodge Ave.
East Haven, CT 06512-3358
FAX: (203) 466-5054

Contains patent classification information. Contains the complete text of the following six U.S. Patent and Trademark Office publications: *Manual of Classification, Index to the U.S. Patent Classification System, Classification Definitions, Patent-to-Classification, Patent-to-Class Index,* and *Concordance.* **Type of database:** full-text.

PatentImages ™

MicroPatent   (203) 466-5055
250 Dodge Ave.
East Haven, CT 06512-3358
FAX: (203) 466-5054

Contains full-text images of all U.S. patents, including drawings. Each disc covers some 1100 patents and is issued within two weeks of the U.S. Patent and Trademark Office (PTO) issue date. Approximately 100 discs per year will be issued. A chemical subset is also available comprising 30 discs per year and detailing only patents of a chemical nature. For each image, provides patent number, issue year, patent assignee, state/country, patent classification codes, title, text, and image. Enables the user to retrieve patents by searching by keyword in title or abstract. Also features 11 additional search fields. **Type of database:** image; bibliographic; full-text.

PatentView

Research Publications International   (203) 397-2600
12 Lunar Dr., Drawer AB
Woodbridge, CT 06525
FAX: (203) 397-3893
Toll-free: (800) 336-5010

Contains citations, abstracts, and images of all patents issued by the U.S. Patent and Trademark Office (PTO). For each patent, provides number, type, reissue patent number, title of invention, issue date, inventor name, inventor state and country, assignee name, assignee state and country, filing date, related filing date, serial number, related serial number, related patent number, priority country, priority date, priority application number, international classification, U.S. classification, U.S. classification cross-reference, abstract, exemplary claim, and image. Software enables the user to view patent drawings in four positions—upright, right-rotated 90 degrees, inverted, and left-rotated 90 degrees—as well as enlarge drawings to view details. Issued as the full Utility Patents set or as three subsets: Chemical, General/Mechanical, and Electrical. **Type of database:** bibliographic; image.

PATOSEP (Patent Online System Europe)

Wila Verlag Wilhelm Lampl GmbH   (089) 5795220
Landsberger Str. 191a
Munich D-80686
Germany
FAX: (089) 5795220

Contains more than 560,000 citations to patent applications and patents granted by the European Patent Office (EPO) since 1978 and current information on the legal status of European patents and published applications. Included are all changes, corrections, and amendments which occur during examination proceedings as well as any new event referring to the legal status, including date of grant, withdrawal or rejections, oppositions, licensee, and new proprietor. Covers patent title in German, English and French; patent and application number; document type; language of application, publication, and procedure; designated states; international patent classification; name and address of inventor, patent assignee and legal representative; dates of application, publication, and filing; priority information; related information referenced patent and non-patent literature; source; legal status information and alteration; complete text in German of the Main Claim for patents granted since the 31st week of 1990; and the complete text in German, English, and French for all patent applications published since 1992. **Type of database:** bibliographic; English; French.

PCT PatentSearch

MicroPatent   (203) 466-5055
250 Dodge Ave.
East Haven, CT 06512-3358
FAX: (203) 466-5054

Contains citations, with abstracts, to world patent applications published by the World Intellectual Property Organization. Software enables the user to search by title, patent number, date of issue, inventor(s), assignee, application serial number and filing date, classification, status, and keyword in title and abstract. Full-text images of patent applications are available in ESPACE World. **Type of database:** bibliographic.

Pharmaceutical Business News

FT Business Enterprises Ltd.   0932 761444
Number One
Southwark Bridge

London SE1 9HL
England
FAX: 44932 781425

Contains the complete text of *Pharmaceutical Business News,* a newsletter covering pharmaceutical product development worldwide. Covers research and pre-clinical trials of new drugs, legal and regulatory developments, and company news, including joint ventures, mergers and acquisitions, and financial performance. **Type of database:** full-text.

The Pink Sheet

FDC Reports, Inc.   (301) 657-9830
5550 Friendship Blvd., Suite 1
Chevy Chase, MD 20815
FAX: (301) 656-3094

Provides news and analysis of developments affecting prescription and over-the-counter drugs. Includes updates on regulatory policies and actions by the Food and Drug Administration (FDA), the Federal Trade Commission (FTC), Congress, the courts, and other federal and state agencies. Also covers the pharmaceutical industry, tracking mergers and acquisitions, new product introductions, drug reimbursement initiatives, research and development, biotechnology start-ups, international developments, and other industry activities. **Type of database:** full-text.

PROMT ™

Information Access Company (IAC)   (415) 378-5000
362 Lakeside Dr.
Foster City, CA 94404
FAX: (415) 358-4759
Toll-free: (800) 321-6388

Contains more than 2.8 million citations, with abstracts and selected full texts, to the worldwide business literature on companies, markets, products, and technologies for major international, national, and regional manufacturing and service industries. Covers new products and technologies, mergers and acquisitions, capital expenditures, market data, product sales, marketing strategies, foreign trade, and regulations. Sources include more than 1200 business, financial, and trade magazines, newspapers, newsletters, reports, BusinessWire news releases, and COMLINE News Service, providing abstracts of articles and news originally published in Japanese. Corresponds to *PROMT* (Predicasts Overview of Markets and Technology) and includes information from the following Information Access Company (IAC) databases: Aerospace/Defense Markets & Technology, Annual Reports Abstracts, Marketing and Advertising Reference Service, and New Product Announcements/Plus (described in separate entries). **Type of database:** bibliographic; full-text.

Public Hearings on Software Patents

U.S. Patent and Trademark Office
Box 4
Washington, DC 20231
e-mail: www@uspto.gov

Provides the complete transcripts of public hearings on software patents, prepared remarks from the hearings, and e-mail comments. **Access:** URL: http://www.uspto.gov/hearings.html. **Type of database:** full-text.

Rapra Trade Names

Rapra Technology Ltd.   0939 250383
Shawbury
Shrewsbury SY4 4NR
England
FAX: 0939 251118

Contains information on more than 40,000 trade names and trademarks of rubber and plastics products and materials. Covers rubber, thermoplastics, thermoplastic elastomers, and thermosets; additives and processing aids; finished and semi-finished products; and manufacturing and processing equipment. For each trade name or trademark, includes product category and description, company name, address and other contact information, and a citation to the registration document. Corresponds to *RAPRA New Tradenames.* **Type of database:** bibliographic.

The Rose Sheet

FDC Reports, Inc.   (301) 657-9830
5550 Friendship Blvd., Suite 1
Chevy Chase, MD 20815
FAX: (301) 656-3094

Monitors regulatory and legal developments for the cosmetics, toiletries, fragrances, and skin care industries, as well as scientific developments and testing methodologies. Also reports on cosmetics industry news, covering mergers and acquisitions, Europe 1992 developments, marketing strategies, new product introductions, and promotions and advertising at retail. **Type of database:** full-text.

Search Master® Intellectual Property Library

Matthew Bender & Company, Inc.   (518) 487-3000
1275 Broadway
Albany, NY 12204-2694
FAX: (518) 487-3584
Toll-free: (800) 223-5297

Contains 15 Matthew Bender publications that deal with intellectual property law: Computer Law—contains the complete text of *Computer Law,* by David Bender. Covers concepts and techniques of evidence and discovery for computer-based information and protection of computer software under intellectual property laws. Includes information on principles and characteristics of computer hardware and software, protection of computer software from misappropriation under copyright, patent, and trade secret laws, application of law of evidence and procedure to the use of computer data, international software protection, and how to introduce, oppose, buttress, or attack the probative value of, and interpret, computer-based information. Court of Appeals for the Federal Circuit: Practice and Procedure—contains the complete text of *Court of Appeals for the Federal Circuit: Practice and Procedure,* by Donald R. Dunner and others. Covers procedure for patent and trademark cases on review in the Court of Appeals for the Federal Circuit from the U.S. Patent and Trademark Office (PTO) and the U.S. District Courts. Includes rules for practice before the Federal Circuit. Also includes sample forms. Intellectual Property Counseling and Litigation—contains the complete text of *Intellectual Property Counseling and Litigation,* edited by Lester Horwitz and others. Covers counseling and dispute resolution in intellectual property law. Includes information on developing law in such areas as counterfeit goods, computer chip and software protection, biotechnology, process patents, and design protection; relationship between intellectual property law and other business law, including bankruptcy, antitrust, arbitration, government contracting, and export control; and international aspects of the law, including patent and trademark protection, overseas licensing concerns, and foreign litigation problems. International Computer Law—contains the complete text of *International Computer Law,* by Jozef A. Keustermans and others. Covers import and antitrust issues, The Semiconductor Chip Protection Act of 1984, international taxation aspect overviews of copyright protection in more than 30 countries, privacy law relating to information stored on computer disks and in computer programs, *sui generis* protection, laws in member states of the European Community, and significant Japanese legislation. Includes sample international software licensing agreements, forms, and a multilingual glossary containing definitions of technological terms. Nimmer on Copyright—contains the complete text of *Nimmer on Copyright,* by Melville B. Nimmer and others. Covers law of literary, musical, and artistic property based on the 1976 Act and recent changes in copyright law. Includes information on the latest U.S. Supreme Court decisions on works for hire and copyright renewal rights, copyright of a television program as a collective work, liability for copying literary material from a sound recording, defenses to overseas copyright infringement, clarification of the "passive carrier" and "controlled group" provisions, limitations on reproducing a computer program for inputting or archival purposes, and impact of U.S. adherence to the Berne Copyright Convention. Milgrim on Trade Secrets—contains the complete text of *Milgrim on Trade Secrets,* by Roger M. Milgrim. Includes information on protection of computer software, licensing of trade secrets and other industrial property, protection by contract and by operation of law, substantive and procedural problems of litigation, criminal, antitrust, and administrative public law, tax aspects, and relationship of trade secrets to patents, copyrights, and the Freedom of Information Act. Covers annotations to statutes, regulations, and cases. Also includes sample forms and charts. Patent Licensing Transactions—contains the complete text of *Patent Licensing Transactions,* by Harold Einhorn and others. Covers royalty bases and rates, domestic and foreign licensing, assignments, territorial limitations and duration, termination of agreements, antitrust, and U.S. taxation of domestic and foreign patents transactions. Includes 300 forms. Patent Litigation: Procedure and Tactics—contains the complete text of *Patent Litigation: Procedure and Tactics,* by Robert A. White and others. Covers impact on patent litigation of the Court of Appeals for the Federal Circuit, patent litigation procedures under the Federal Rules of Civil Procedure, and tactical strategies for infringement suits. Includes litigation forms. Contains checklists of "black letter" law for claim construction, infringement, validity, doctrine of equivalents, and prosecution history estoppel. Also includes reports and analyses of recent cases. Patents—contains the complete text of *Patents* by Donald S. Chisum. Covers U.S. patent laws, including principles, doctrines, rules, and cases on patentability, validity, infringement, application, procedure, and perfecting rights to an invention. Includes all patent-related holdings of the Court of Appeals for the Federal Circuit, all significant precedents, case law, statutes, and international treaties, with discussions of individual cases, appendices, tables of cases, patent statute sections, CAFC rules, patent office rules, bibliography, glossary, and index. Chapter headings include: eligible subject matter, originality, novelty, nonobviousness, statutory bars, adequate disclosure, claims, double patenting, priority, patent and trademark office procedures, restriction, continuation applications, priority by foreign filing, reissue, direct infringement, contributory infringement, interpretation and application of claims, and defenses and remedies. Trademark Protection and Practice—contains the complete text of *Trademark Protection and Practice,* by Jerome Gilson. Covers analyses of substantive law including the Trademark Law Revision Act of 1988, Trademark Trial and Appeals Board practice, pharmaceutical trademarks, parties of infringement actions, and trademark tarnishment. Includes analysis of the legislative history of the Lanham Law Revision Act and its amendments. Also includes forms for opposition, cancellation, infringement and declaratory judgment complaints, and written interrogatories. World Trademark Law and Practice—contains the complete text of *World Trademark Law and Practice* by Ethan Horwitz. Covers world trademark law and practice, including full coverage of 35 major jurisdictions and summary coverage of more than 100 jurisdictions. Includes analyses of non-U.S. trademark law, solutions to trademark practice situations, and other practice tools. World Patent Law and Practice: Volumes 2, 2A, and 2A Part 2—contains the complete text of *World Patent Law and Practice: Volumes 2, 2A, and 2A Part 2* by J. W. Baxter and others. Covers explanations of the patent process, for all industrialized nations, from filing to grant; post-acceptance events; principal patent conventions; and patent terms, renewal fees, infringement and licensing for all nations and regional patent systems. Also covers principal patent conventions and legislation as enacted. New York Intellectual Property Handbook—contains the complete text of *New York Intellectual Property Handbook* by Hugh C. Hansen. Covers New York and federal statutes and NYCRR Regulations governing intellectual property matters. Patent Office Rules and Practice—contains the complete text of *Patent Office Rules and Practice* by Lester Horowitz. Covers procedures and notices of the U.S. Patent and Trademark Office (PTO). Also contains all rules and forms required by the PTO. California Handbook of Intellectual Property—corresponds to the print product *California Handbook of Intellectual Property* by Sheldon & Mak. Contains the full text of all California and federal statutes governing patents and inventions, trademarks and trade names, business names, copyrights, trade secrets, franchises, unfair competition, computers, and artists' and authors' rights. Includes an introduction with a brief overview of relevant laws. California statutes are annotated with references to key cases and law review articles. **Type of database:** full-text.

STO's Internet Patent Search System

Internet Patent News Service (617) 489-3727
Source Translation and Optimization (STO)
PO Box 404
Belmont, MA 02178
e-mail: patents@world.std.com

Contains full-text of the U.S. Code on patents, various patent information and services, and patent titles in any class or subclass. **Access:** URL: http://sunsite.unc.edu/patents/intropat.html. **Type of database:** full-text; bibliographic.

Technical Insights ALERT

Technical Insights, Inc. (201) 568-4744
P.O. Box 1304
Fort Lee, NJ 07024-9967
FAX: (201) 568-8247
e-mail: info@insights.com (Internet)

Contains the complete text of 65-70 concise briefings covering technical advances and the related industry markets. Briefings are 200-300 words in length and are derived through original research and analysis by Technical Insights staff. For each technology, provides an explanation, its potential market impact, the technology transfer arrangements sought by its researchers, patents issued, and complete contact information for the companies and persons involved. Also includes in-depth reports concerning emerging technologies and the markets they will create. Each report provides likely timeframes for commercialization, potential market size and market share, a guide to research groups, and a description and analysis of the technology, prepared by experts in the field for Technical Insights, Inc. Issued as a series by subject as follows: Industrial R&D. High-Tech Materials Technology. Manufacturing Technology. Industrial Bioprocessing. Genetic Technology. Sensor Technology. Advanced Coatings. Electronics Technology. Chemical Technology. BLT Bio/Med Technology. **Type of database:** full-text.

The Technical Report Database on CD-ROM

U.S. Defense Technical   (703) 274-7709
   Information Center (DTIC)
Bldg. No. 5, Cameron Station
Alexandria, VA 22304-6145
FAX: (703) 274-9307

Contains approximately 700,000 citations to technical reports, Department of Defense patent applications, bibliographic records, regulatory publications, software, data files, databases, videos, and conference papers. Covers defense-related subjects (e.g., aeronautics, missile technology, space technology, navigation, and nuclear science) and such scientific and technical topics as biology, chemistry, energy, environmental sciences, oceanography, computer sciences, sociology, and human factors engineering. **Type of database:** bibliographic.

Technology Access Report

Technology Access   (415) 883-7600
16 Digital Dr., Suite 250
Novato, CA 94949-5760
FAX: (415) 883-6421
Toll-free: (800) 733-1516
e-mail: 76236,1330 (CompuServe)
e-mail: 369-9038 (MCI Mail)

Contains the complete text of *Technology Access Report,* a newsletter covering the transfer, management, and commercialization of technologies. Sources include universities, hospitals and medical centers, federal and independent laboratories, federal and state agencies, and consultants. **Type of database:** full-text.

Thomas Register Online

Thomas Publishing Company   (212) 290-7291
Thomas Online
One Penn Plaza
250 W. 34th St.
New York, NY 10119
FAX: (212) 290-7362
e-mail: 14016 (DIALMAIL)

Contains information on more than 153,000 U.S. and selected Canadian manufacturers and providers of services. For each company, provides name, address, contact numbers, description of company, product brand names and trademarks, and Standard Industrial

Classification (SIC) codes. For some companies, also includes asset rating, exporter status, number of employees, cable address, company executive names and titles, and names of parent and subsidiary companies. Covers more than 50,000 product classifications and 112,000 trademarks or brand names. Corresponds in part to the *Thomas Register of American Manufacturers.* Also contains information on more than 44,000 companies covered in *Thomas Food Industry Register.* Includes sales and distribution information for foodservice operators, brokers, and manufacturers' agents, wholesalers and distributors, supermarket chains and convenience stores, importers and exporters, warehouses and transportation firms. **Type of database:** directory.

TMA Trademark Report

Tobacco Merchants   (609) 275-4900
   Association of the United States (TMA)
231 Clarkville Rd.
P.O. Box 8019
Princeton, NJ 08543-8019

Contains the complete text of *TMA Trademark Report,* a monthly listing of tobacco-related trademark activity. Covers cigarettes, cigars, smokeless and smoking tobacco, and smokers' accessories (e.g., ashtrays, pipes, lighters). Includes trademark name and the following dates: first use of name, registration filing, publication date for opposition, registration, renewal, and cancellation. Source of information is the *Official Gazette* of the U.S. Patent Office. **Type of database:** directory.

Trade Marks Opposition Board

QL Systems Limited   (613) 238-3499
901 St. Andrew's Tower
275 Sparks St.
Ottawa, ON K1R 7X9
Canada
FAX: (613) 238-7597
Toll-free: (800) 387-0899

Contains trademark-related decisions from the Registrar of Trademarks and the Canada Trade Marks Opposition Board. **Type of database:** full-text.

TRADEMARKSCAN®—Canada

Thomson & Thomson   (617) 479-1600
500 Victory Rd.
North Quincy, MA 02171-3145
FAX: (617) 786-8273
Toll-free: (800) 692-8833

Contains information on more than 500,000 trademark registrations and applications filed with Canadian Intellectual Property Office. Includes: trademark; International Classification number; current status; dates of first use, filing, advertisement, registration and renewal, goods/services description; mark type; owner history; registration number; and application number. **Type of database:** full-text.

TRADEMARKSCAN®—Ireland

Thomson & Thomson   (617) 479-1600
500 Victory Rd.

North Quincy, MA 02171-3145
FAX: (617) 786-8273
Toll-free: (800) 692-8833

Contains information on trademarks filed with the trademark registry of Ireland. **Type of database:** directory.

TRADEMARKSCAN®—U.K.

Compu-Mark   323 2207211
Sint Pietersvliet 7
Antwerpen B-2000
Belgium
FAX: 332 2207390

Contains information on more than 575,000 registered trademarks and service marks, and applications for registrations filed with the Trade Marks Branch of The Patent Office of the United Kingdom. Includes trademark, international classification, application or registration number, description of trademarked product or service, current status, agent name, and owner's name and address. **Type of database:** full-text.

TRADEMARKSCAN®—U.S. Federal

Thomson & Thomson   (617) 479-1600
500 Victory Rd.
North Quincy, MA 02171-3145
FAX: (617) 786-8273
Toll-free: (800) 692-8833

Contains information on more than 1.5 million federal trademark registrations and applications (including "Intent to Use") filed with the U.S. Patent and Trademark Office (USPTO). Includes trademark, U.S. and international classifications, description of trademarked product or service, serial number, registration number and date, current status, filing date, date of publication in the *Official Gazette* of the USPTO (Trademark Section), date of first use, ownership history, filing correspondent, and Trademark Trial and Appeal Board information. Also includes searchable and displayable trademark logos and designs. **Type of database:** full-text; image.

TRADEMARKSCAN®—U.S. State

Thomson & Thomson   (617) 479-1600
500 Victory Rd.
North Quincy, MA 02171-3145
FAX: (617) 786-8273
Toll-free: (800) 692-8833

Contains information on more than 900,000 commercial trademarks, excluding corporate names, registered with the offices of the Secretaries of State of individual states, American Samoa, and Puerto Rico. Typical record elements include trademark: design type; state of registration; U.S. and international classifications; goods/service description; registration number; current status; and dates of registration, renewal, cancellation, expiration, first use, and current owner name and location. Corporate name records are not included. Trade names, assumed names, and fictitious names are also not generally included, but may be identified for some states. **Type of database:** full-text.

UK Trademarks

Great Britain Patent Office   071 829 6474
State House
66-71 High Holborn
London WC1R 4TP
England

Contains information on more than 600,000 U.K. active, pending, and lapsed service marks and trademarks. For each, provides mark, trademark number, application date, status, owner and/or agent, address, description and class of product, usage limitations, proprietary country code, and registered users. Includes information on any mark which has lapsed since January 1, 1976. **Type of database:** directory.

U.S. Copyrights

Knight-Ridder Information, Inc.   (415) 254-7000
DIALOG
2440 El Camino Real
Mountain View, CA 94040
FAX: (415) 254-8000
Toll-free: (800) 3-DIALOG
e-mail: DIALMAIL: MARKETING

Contains registration, renewal, and ownership information for more than six million active copyright and mask-work registrations on file at the U.S. Copyright Office. Provides two types of records (monograph and legal document records) for mask works (i.e., electronic circuitry), performing arts, sound recordings, textual works, and visual arts. Monograph records contain initial registration and renewal information, including title of work, content of work, owner's name, author's name, indication of whether the work is published, class and retrieval code assigned by the U.S. Copyright Office, registration number, dates of creation and registration, and description of registration deposit. Legal document records contain assignment information, including date of assignment of copyright, notices of termination of a registration, declarations of authorship, and works covered by the copyright. Information is obtained from the Catalog Distribution Service of the Library of Congress in conjunction with the U.S. Copyright Office. **Type of database:** directory.

U.S. Patent and Trademark Office Database

Internet Multicasting Service
e-mail: info@radio.com; questions@radio.com

Provides full-text patent data for 1994. The data1 subdirectory is organized by ranges of patent numbers. **Access:** Gopher: Town Hall; gopher.town.hall.org 70. URL: http://town.hall.org/patent/patent.html; gopher://gopher.town.hall.org/70/11/patent. **Retrieve:** Anonymous ftp: town.hall.org; Path: patent; File name: patent.src. **Type of database:** full-text.

U.S. Patent Search

MicroPatent   (203) 466-5055
250 Dodge Ave.
East Haven, CT 06512-3358
FAX: (203) 466-5054

Contains citations, with abstracts and exemplary claim to U.S. patents published since 1975. Software enables the user to search by title, patent number, date of issue, inventor(s), state or country of residense of first inventor, assignee, application serial number and filing date, classification, keyword, abstract, exemplary claim, and status. Source of information is the complete text of patents issued electronically by the U.S. Patent and Trademark Office. **Type of database:** bibliographic.

U.S. Patents

Derwent, Inc.   (703) 790-0400
1313 Dolley Madison Blvd., Suite 401
McLean, VA 22101
FAX: (703) 790-1426
Toll-free: (800) 451-3451
e-mail: DIALMAIL: 11592

Contains approximately 1.4 million citations, with abstracts, to U.S. patents, continuations, divisionals, and defense documents issued since 1971, and design, reissued, reexamined, and plant patents issued since 1977. Includes information listed on the front page of the patent and the text of all claims, which describe aspects of the patent in legal terms. Searchable data elements include patent and application number; filing date; publication date; terms from the title, abstract, and the complete text of the claim; inventor and assignee name and location; earlier foreign priority filings; prior application information, including application number, patent number (when granted), status, and dates; examiner and attorney names; all cited references, including those of U.S. patents, non-U.S. patents, and other literature; and all U.S. and international patent classifications. NOTE: On ORBIT Questel, patents issued from 1971 to 1981 are listed in USPB; from 1982 to date in USPA. **Type of database:** bibliographic; full-text.

U.S. Patents Fulltext

Knight-Ridder Information, Inc.   (415) 254-7000
DIALOG
2440 El Camino Real
Mountain View, CA 94040
FAX: (415) 254-8000
Toll-free: (800) 3-DIALOG
e-mail: DIALMAIL: MARKETING

Contains the complete text of more than 1.4 million patents issued by the U.S. Patent and Trademark Office (PTO) since 1974, with partial coverage of selected technologies from 1971 through 1973. Includes all front-page information, including title, author(s), assignee(s), related applications, classification data, cited references, and abstract; all claims; background/field of invention; brief summary of invention; detailed description/embodiment; description of drawings; and examples. NOTE: This database is accessible as three files: File 652 covers 1971-1979 (selective coverage 1971-1973); File 653 covers 1980-1989; and File 654 covers 1990 to date. **Type of database:** full-text.

United States Patents Quarterly

The Bureau of National Affairs, Inc. (BNA)   (202) 452-4132
1231 25th St., NW
Washington, DC 20037

FAX: (202) 452-4062
Toll-free: (800) 862-4636

Contains the complete text of the *United States Patents Quarterly,* featuring all important federal and state cases on patents, trademarks, copyrights, unfair competition, trade secrets, computer chip information, and other intellectual property issues. Also includes BNA's editorial analysis, classification numbers, and head-notes. **Type of database:** full-text.

USCLASS

Derwent, Inc.   (703) 790-0400
1313 Dolly Madison Blvd., Suite 401
McLean, VA 22101
FAX: (703) 790-1426
Toll-free: (800) 451-3451
e-mail: DIALMAIL: 11592

Contains classification information for nearly five million patents issued by the U.S. Patent and Trademark Office (PTO). Approximately 115,000 classifications can be accessed to retrieve patent numbers assigned to each of them as either official or unofficial references; patents originally issued in any year will have their classifications changed so that they always reflect the current classification scheme when a subject matter area is reclassified. **Type of database:** bibliographic.

WESTLAW® Intellectual Property

West Publishing   (612) 687-7000
620 Opperman Dr.
Eagan, MN 55123
FAX: (612) 687-7302
Toll-free: (800) 328-9352

Contains the complete text of U.S. federal court decisions, statutes and regulations, specialized files, and texts and periodicals dealing with intellectual property law, including the rights of artist, authors, composers, and designers of creative works, and the significance of copyright, patent, or trademark protection. Includes relevant decisions made by the U.S. Supreme Court (1790 to date), U.S. Courts of Appeals (1891 to date), U.S. District Courts (1789 to date), Court of Federal Claims, which includes the former U.S. Claims Court and the former U.S. Court of Claims (1863); and Court of Appeals for the Federal Circuit (1945); relevant statutes and regulations from the U.S. Code, annotated (current), the Federal Register (1980 to date), the Code of Federal Regulations (current); Patent and Trademark Office Decisions (1987 to date); and such specialized resources as *BNA's Patent, Trademark & Copyright Journal* (1986 to date), *BNA's Patent, Trademark & Copyright Law Daily* (September 1989 to date), *BNA United States Patents Quarterly* (1946 to date), and BNA Headlines; topical highlights; West's Legal Directory-Intellectual property; and relevant law reviews, texts, and bar journals. (Some of these files are described in separate entries.) Also includes a vast number of specialized intellectual property databases accessible from Knight-Ridder Information Services, Inc., DIALOG via either gateway or DIALOG or WESTLAW. **Type of database:** full-text.

WESTLAW® Texts and Periodicals

West Publishing   (612) 687-7000
620 Opperman Dr.

Eagan, MN 55123
FAX: (612) 687-7302
Toll-free: (800) 328-9352

Contains the complete text of selected articles from more than 450 law review publications, bar association journals, and other law-related publications relating to federal law and other major practice topics. Covers administrative law, antitrust and trade regulation, bankruptcy, business organizations, civil rights, communications, commercial law and contracts corporations, criminal justice, education, energy, environmental law, estate planning and probate law, family law, financial services, First Amendment, government benefits, government contracts, health services, immigration, insurance, intellectual property, international law, jurisprudence and constitutional theory, labor and employment, legal services, litigation, maritime law, military law, pension and retirement benefits, products liability, professional malpractice, real property, securities and blue sky law, taxation, tort law, transportation, and workers' compensation. Each textbook and periodical is given a unique file label for access and is then assigned to one or more WESTLAW Libraries (each of which is described in a separate entry). In addition to texts and periodicals, various treatises and other publications are covered. **Type of database:** full-text.

World Patents Index

Derwent Information Ltd.   071 344 2800
Derwent House
14 Great Queen St.
London WC2B 5DF
England
FAX: 071 344 2900
e-mail: LONDON DERWENT (DIALMAIL)

Contains more than six million citations, with abstracts, to chemical, electrical, mechanical, and general patents issued by 31 major patent-issuing authorities. Data elements include title; patent assignee and inventor name; patent numbers; World Intellectual Property Organization (WIPO) member country; priority and publication dates; and various classification and subject codes. Corresponds to coverage in the abstracts publications: *Chemical Patents Index (CPI), General & Mechanical Patents Index (GMPI),* and *Electrical Patents Index (EPI)*. Includes 1.5 million technical drawings and diagrams, the majority relating to patents of an electrical and engineering nature since 1988. Also includes some 50,000 chemical patent drawings dating back to the beginning of 1992. Image availability is indicated in each patent record. Drawings and diagrams are taken from Derwent's *Alerting Abstracts Bulletins*. Windows software is required to utilize this feature; consult the online service of choice for details. **Type of database:** bibliographic; image.

WPI/APIPAT

Derwent Information Ltd.   071 344 2800
Derwent House
14 Great Queen St.
London WC2B 5DF
England
FAX: 071 344 2900
e-mail: LONDON DERWENT (DIALMAIL)

Contains patents covering petroleum refining, the petrochemical industry, and synthenic fuels. Features the indexing capabilities of the American Petroleum Institute's APIPAT file (described in a separate entry) combined with the classification and coding system of Derwent's World Patents Index (described in a separate entry). Typical data elements include document number, patent title, patent assignee, company code, inventor, patent codes, priority codes, patent family codes, patent country, designated states, abstract, and International Patent Classification (IPC) codes. NOTE: Available in the following subfiles: WPIA, covers 1963 to 1980 and WPLA, covers 1981 to date. **Type of database:** bibliographic.

WPIM (World Patents Index Markush)

Derwent Information Ltd.   071 344 2800
Derwent House
14 Great Queen St.
London WC2B 5DF
England
FAX: 071 344 2900
e-mail: LONDON DERWENT (DIALMAIL)

Contains graphic representations of generic (Markush) structures commonly used in patents for pharmaceuticals, agricultural chemicals, and general chemical compounds, to represent compounds that are wholly or partially variable. Covers Markush structures from patents included in Derwent's *Chemical Patents Index* Sections B (Farmdoc), C (Agdoc), and E (Chemdoc) (*WPI (World Patents Index)* is described in a separate entry). Includes compounds that are removed, or used to effect removal, when this is an important feature of the patent; all compounds described as new; new catalysts; all products of new processes, including materials purified in new ways; compounds in novel methods of analysis or detection; and key ingredients of compositions. Structures are grouped under 21 concepts, called "superatoms," which specify a particular attribute. Users can transfer search results to WPI to facilitate retrieval of complete patent information. **Type of database:** full-text; numeric.

Project XL

The visions that we present to our children shape the future. They become self-fulfilling prophecies.

—CARL SAGAN

Kids Are Inventors Too

At age 14, one schoolboy invented a rotary brush device to remove husks from wheat in the flour mill run by his friend's father. The young inventor's name? Alexander Graham Bell.

At age 16, another of our country's junior achievers saved pennies to buy materials for his chemistry experiments. While still a teenager, he set his mind on developing a commercially viable aluminum-refining process. By age 25, Charles Hall received a patent on his revolutionary electrolytic process.

Chester Greenwood, 13, had a negative experience with cold weather while ice skating one December day in 1873. To protect his ears, he found a piece of wire and with his grandmother's help, padded the ends. At first, his friends laughed. However, before long they realized that Chester could stay out ice skating long after they had gone inside to escape the chill. Soon they were asking him for his custom ear covers. At age 17, Chester applied for a patent and over the next 60 years became very rich due to his earmuffs.

Project XL: A Quest For Excellence

Project XL is an outreach program of the PTO and an integral part of the U.S. Department of Commerce's Private Sector Initiative Program, launched in 1985 during the Reagan Administration by then Assistant Secretary of Commerce, Commissioner of Patents and Trademarks Donald J. Quigg. Since then, Project XL has reached more than 4,000 teachers and 100,000 students. It is designed to encourage the development of inventive thinking and problem-solving skills among American youth. The principle focus of this effort is on the promotion of educational programs that teach critical and creative thinking, and on fostering national proliferation of such programs. The overall objective of Project XL is to ensure the nation's position as world technological leader as we enter the 21st century—to guarantee that Americans will have the innovative skills to meet the challenges of an increasingly competitive world.

First-of-Its-Kind Technology Major

The Department of Technological Studies at Wheeling Jesuit College, in association with National Technology Transfer Center, has developed a new major. The first of its kind in the nation, the major is designed to prepare students for the exciting, creative work of innovation, innovation management, and technology transfer in the commercialization of products. For information, contact: Dr. Michael Doyle, Program Director for Technology, Wheeling Jesuit College, 316 Washington Avenue, Wheeling, West Virginia 26003-6295. The telephone number is (304) 243-2460.

"I believe that the schools of this great nation are filled with Edisons, Wrights, Marconis, Whitneys, and Bells, along with other potential thinkers who can change the world," said Quigg. "The very least we can do is help them realize their potential—to nurture those young people who will inherit and build the future."

A secondary benefit from Project XL is that young people and their parents and teachers will gain an increased awareness of new technology's importance to advancing society and strengthening the domestic economy. Project XL aims to instill an increased appreciation of the contributions inventors make to our way of life and recapture the spirit of those golden years at the turn of the century when inventors were heralded as true American heroes.

Project XL Objectives

Project XL comprises the following components:

- National coordination of efforts to teach inventive thinking and problem-solving skills at every level of public and private education throughout the country.

- Presentation of national and regional conferences to promote the teaching of critical and creative thinking skills and the inventive process.

- Establishment of an Education Roundtable, an open forum and national discussion network drawing upon the talents and resources of public and private sector leaders to develop and promote programs in this area.

- Development of a broad-based speakers' bureau on such topics as invention, problem-solving, creativity, and thinking skills.

- Dissemination of an informational guide called the *Inventive Thinking Project*, designed to channel students in grades K-12 into the inventive thinking process through the creation of their own unique inventions or innovations.

- Creation of an educator's resource guide to include programs, materials, literature, organizations, and other sources that promote thinking across all disciplines.

- Curriculum development for special teaching materials designed to stress problem-solving, the value of creative thinking, and the importance of American inventors.

- Identification of government programs and resources that focus on the development of future problem-solvers in all fields.

- Develops curricula for areas/grades that currently have little available (e.g., high school)

- Establishment of an Inventive Thinking Center, a collection of literature, videotapes, and other curriculum materials.

Donald J. Quigg Excellence in Education Award

In 1989, the PTO presented outgoing Commissioner Quigg with a plaque establishing Project XL's Donald J. Quigg Excellence in Education Award. It recognizes the efforts of an individual (or group) to promote the teaching of inventive thinking skills at all levels of education. Winners are chosen by a panel of distinguished experts in the fields relating to the PTO and Project XL. The nomination form can be found on page 409.

National Inventive Thinking Association

Established in 1989, the National Inventive Thinking Association (NITA) is a non-profit organization comprising educators, business leaders, and government officials who are dedicated to "promoting inventive thinking and a spirit of positive problem solving through education and the networking of community and national resources."

Membership Benefits

- Consultants, K-12 and college and university level
- Clearinghouse for instructional materials
- Project XL support (PTO)
- Speaker's bureau
- Workshops
- Newsletter
- Advice and assistance from the Intellectual Properties Association and numerous creativity centers
- PTO-endorsed activities

Membership Information

For information on membership in this worthwhile organization, contact NITA, 400 S. Greenville Avenue, Richardson, Texas 75081. The telephone number is: (214) 301-3370. You could also contact Ruth Nyblod, Administrator for Project XL, PTO, Office of Public Affairs, Washington, D.C. 20231. The telephone number is: (703) 305-8341.

Annual National Conference

Every year NITA hosts a Creative and Inventive Thinking Skills Conference. In 1994 the eigth annual conference was held in Houston, Texas. It is scheduled for Los Angeles, California in October of 1995.

To receive information on future conferences, please write to Project XL, USPTO, Crystal Park One, 208-B, Washington, D.C. 20231.

The entries in this section are listed alphabetically by program name under the headings National Programs, Statewide Programs, and Local Programs. There are many national programs that emphasize creative problem-solving skills (like Odyssey of the Mind), thinking skills (like MATHCOUNTS), and engineering abilities (like TEAMS, and the NSF Science Fair). In this section, only programs that emphasize inventing and the invention process are included.

NATIONAL PROGRAMS

Duracell NSTA Scholarship Competition

National Science Teachers Association   (209) 328-5800
1742 Connecticut Ave NW
Washington, D.C. 20009
Ms. Kathleen A. Rapp, Special Programs Coordinator

The "Design a Duracell Device" competition offers $30,000 in college scholarships and cash awards to 41 winners each year. All entrants receive Duracell athletic bags. The first-place winner receives a $10,000 scholarship, and the winner's teacher receives a personal computer. Five second-place winners receive $3000 scholarships; each winner's teacher receives a portable computer. Ten third-place winners receive $500 scholarships; each winner's teacher receives a certificate for an NSTA (National Science Teachers Association) publication. The 25 fourth-place winners receive $100 cash prizes.

The purpose of the contest is to create and build a working device powered by Duracell batteries. The device is to "perform a practical function such as make life easier, educate, entertain, serve as a warning device, provide light, make sounds." Four criteria are used by the judges: creativity, 30%; energy efficiency, 30%; practicality, 30%; and clarity of the written description, 10%. Additional details concerning the use of batteries and specifications are given in the brochure published each year by NSTA.

Entries are usually due in late January or early February. The entry includes an essay, wiring diagram, photographs, and the official entry form. The form requires the signature of a teacher and parent/guardian verifying that the student has independently designed and built the device. One hundred finalists are selected for the final judging, which takes place in February. The finalists must ship their devices to Duracell's headquarters in Bethel, Connecticut. Final winners are announced early in March. First-place and second-place winners are honored at the NSTA National Convention. Expenses are paid for the winners, parents, and teachers.

The SBG Invention Convention

Silver Burdett & Ginn   (201) 285-7740
250 James St.
Morristown, NJ 07960
Mr. Andrew Socha

As part of its elementary science textbook series, Silver Burdett & Ginn (SBG) has developed supplementary materials that are available gratis to purchasers of the textbooks. One of these is the SBG Invention Convention unit. It contains guidelines for teachers, copy masters of forms (letter to parents, intent to invent, patent application, patent certificate, and judging form), and procedures for students to follow. Each school winner is eligible to enter the *International Invention Convention* sponsored by SBG. Schools are encouraged to allow at least six weeks for the teaching unit, keeping in mind that the deadline for postmarked entries in the international convention is mid-February. Winners usually are announced by the end of February. The grand prize is a computer and all-expenses-paid trips for winner, parents, and teacher to the National Science Teachers Association meeting, usually in April. Certificates are given to other winners.

The SBG *Invention Convention* can be a classroom, school, or district science event designed to encourage students in grades 1-9 (middle schools were added in 1988) to apply basic science skills in a creative and productive manner. The suggested procedure has five basic steps: learning about inventors; finding an idea; research and planning; developing and testing; and taking part in the *Invention Convention*.

Young Inventors and Creators Program

National Inventive Thinking Association   (214) 301-3370
PO Box 836202
Richardson, TX 75083
Leonard Molotsky

Under this program, students grades 7 through 12 are invited to submit thier creative works of authorship or invention as demonstration of their creative and inventive thinking skills for school-level judging, regional or state competition, and national recognition. All students reaching the regional or state level will get certificates of participation, winners at the state or regional level will receive certificates of achievement, and the national winners will be recognized at the 9th National Creative and Inventive Thinking Skills Conference.

KIDSCON (Creative KIDSCONference)

National Foundation for Gifted
        and Talented Children   1-800-GIFTED-1
16400 Pacific Coast Hwy, #217
Huntington Beach, CA 92649
Ms. Jacqui Jeffrey, Executive Director

Each year, KIDSCON East and KIDSCON West are held for children in grades 2-8. Students need not be identified as gifted by their schools, but they should display special talents in language, space perception, logic, music, sensitivity, self-understanding, or understanding of others. The three-day program offers nearly 100 workshops, from which the children select 15, plus about 30 workshops especially for their parents.

Registration for one child and parent is $125, plus lab fees, hotel, food, and travel; additional children in the family are $70 each. Additional family members (over age 14) and educators can register for $50.

Jacqui Jeffrey, an educator for 20 years, had held KIDSCON conferences for several years in northern California before going nationwide in 1989.

Project XL

U.S. Patent and Trademark Office   (703) 305-8341
Washington, D.C. 20231
Ruth Ann Nyblod, Administrator

For more information on Project XL, see the opening pages of this chapter, or contact Project XL at the address listed above.

U.S. Cognetics

606 Delsea Drive   (609) 582-7000
Sewell, NJ 08080
Dr. Theodore R. Gourley

Cognetics is one program in the National Talent Network. It is designed to teach the creative problem-solving process through fact gathering, problem analysis, idea generation, and solution selection in a group setting. Student teams are given awards for levels of creative problem-solving. Each level addresses one or more of four key abilities—scholarship, communication, originality, and artistry. The purpose is to achieve a goal, not to compete. Therefore, certificates are given rather than awards or prizes. Each year, a new problem is selected as the basis of the program. A team of students submits its solution on videotape, with a copy of a score sheet, registration form, and registration fee, to the U.S. Cognetics office. Up to seven students (grades K-12) can be involved in presenting the solution, although there is no limit to how many can contribute to the solution. Handicaps are assigned according to grade level to even out the scoring process. The registration fee is $10 per team if the school or organization is a Cognetics member ($60 fee) or $25 per team otherwise.

STATEWIDE PROGRAMS

Connecticut Invention Convention

Connecticut State Department of Education   (603) 271-2717
15 Humphrey St.
Concord, NH 03301
Ms. Michelle Mundson, Chairperson

The Connecticut Invention Convention has been in existence since 1984, when the first statewide competition was held. It is a one-day event where student inventors meet in grade-level groups (K-12) to share, demonstrate, and discuss inventions with adult inventors, engineers, patent attorneys, and educators. The Invention Convention recognizes and rewards students for applying critical and creative thinking in the creation of an invention. Students are sponsored by their schools, which send the best inventors to the state-level competition. Prizes are given to all participants, and special recognition (in the form of savings bonds) is given to those inventions that best fulfill the guidelines.

Schools are encouraged to have their local competitions prior to March of the current year because the deadline for state-level applications is March 15. A school may enter the top three inventions or 10% of the participants. Every entry must have a principal's signature, even if the student is not enrolled in a public school. There is a $10 application fee, and limits are placed on display size and cost of materials.

Envisioneering™

Filmore Central School   (716) 365-2646
104 Main Street
Filmore, New York 14735
Michael J. Doyle
Barbara Van Wicklin

Also called "Creativity & Innovation: Beyond Paper Solutions," this half-unit course for grades 9-12 is designed to enhance creative thinking, problem-solving skills, and production of innovative solutions and processes. The course has four modules. Each module has a theory portion related to creative problem-solving and a practice or applications portion that deals with technical skills, knowledge, and concepts. The course is designed to be presented in a sequential order, moving from well-defined problems, to loosely defined problems, to problems that are complex or unclear. The instructor's role is to change roles, moving from teacher to leader to facilitator. Because the course is content-driven (i.e., it feeds off the content of other classes), a student can take it several times. The objective is to get the student to live the process by keeping a daily log, by learning to take risks, and by emulating successful inventors and innovators.

Students who have taken the class have gone on to be winners in the Olympics of the Mind, Invention Convention, and video competitions.

FutureMakers Inventor/Mentor Program

Saturday Academy   (503) 690-1190
Oregon Graduate Center
19600 N.W. Von Neumann Dr.
Beaverton, OR 97006
Gail Whitney, Project Director

The program consists of a curriculum guide, teacher training, a student testing, and an Invention Convention. It is primarily for grades 6-12 and is not limited by special needs (e.g., gifted and talented) or curriculum area.

The program is innovative in its approach to involving local businesses in local school programs, as explained in descriptive material provided to us by Saturday Academy:

"The purpose of the Fu*tureMakers* program is to foster developing of reasoning and problem-solving skills through the process of invention . . . students develop criteria for evaluating their ideas and confront a problem by creating a new process or product.

". . . Students are provided an opportunity to apply these thinking skills in the business setting, potentially developing ideas which are implementable for the business. Working in small groups with a business "mentor," students learn about the business, and identify problems or needs that can be addressed by inventive thinking. Mentors assist the students with their idea development back at school, and students communicate their ideas through visual displays. The program culminates with the 'Invention Convention.' . . . Business mentors join in as judges, selecting students for awards and recognition for their work.

"Any business can potentially sponsor a partnership. Manufacturing, services, sales, communications, and many other businesses all offer opportunities for ideas, innovations, or inventions that improve the product, or make its production easier or more time efficient. Business partners benefit by fostering skills in their future workforce. Students and teachers benefit through problem-solving experiences in the 'real world' workplace. For teachers, this understanding can help effect a better match between what is taught and what is needed in the future workplace. For students, this understanding provides real reasons for the subjects studied in school."

Imagination Celebration/Invention Convention

Imagination Celebration Patent Office   (518) 473-0823
Room 9B38, Cultural Education Center
Empire State Plaza
Albany, NY 12230
Dr. Vivianne Anderson, Coordinator

Participating students describe their inventions in Imagination Celebration Invention Patent Applications. These applications are reviewed at the school level. At the state level, a jurying panel selects the official galleried inventions to be displayed at the State Museum during the Imagination Celebation. All inventions must be exclusively student-developed. All entrants receive certificates of participation; top inventors receive Imagination Celebration Patents. A reception is held in honor of the young inventors, parents, and teachers.

Invent, Iowa!

Rm 733, Federal Building   (515) 284-4574
Des Moines, Iowa 50309
Ms. Dianne Liepa, Deputy Administrator

The first Invent, Iowa! was held in 1987–88. The statewide program is for grades K-8. About 270 children went to the statewide convention in April 1989. These participants had progressed from local and regional exhibitions to the state competition. Special awards are given by sponsors (e.g., the Iowa Patent Attorneys give a "Teacher Award"). Besides the traditional awards, awards are given for special attributes: most Iowa-oriented invention; best display; most altruistic invention; most practical invention; most humorous invention; and most whimsical invention.

Invent, Iowa! holds conferences for educators where they can learn how to plan lessons that center on creativity and problem-solving and how to organize and conduct local and regional conventions. A newsletter is produced periodically to highlight inventive classroom programs, profile student inventors, and announce activities. Invent, Iowa! is patterned after, and coordinated with, Invent America!

IoWorks

203 N. 9th Street   (515) 284-4690
Marshalltown, IA 50309
Mr. Gary Zmolek

IoWorks is an ambitious approach to dealing with problems specific to the economy of Iowa and the need for educational opportunities that will result in retaining students after they graduate. Mr. Zmolek says,

> Our state graduates the highest proportion of its students from high school, but [it] is 49th in retention of graduates within state borders.
>
> We need to stop exporting our brains and start exporting the products of our brains.
>
> We cannot keep our young people within the state by the quality of our brains.
>
> Our brightest and best need help now, help with funding, ideas, problem solving, technical expertise, facilities and equipment.

Mr. Zmolek concludes: "The talented and entrepreneurial youth of Iowa need a statewide system for linking them with mentors, making specialized information available to them, and helping them find inspiration, encouragement, criticism, funding, facilities, equipment, and materials so they may go forward with their research, business ventures, artistic productions—the development of their original ideas in whatever fields."

IoWorks includes inventing as just one aspect of the project. The system envisioned by Mr. Zmolek consists of resources in print and computer software, a computer conferencing network (modeled after the NSF-sponsored PSI-net) that extends to every school in the state, and professional services to help students obtain difficult-to-find information or make contacts with volunteer resource people, mentors, and organizations. The population served would be primarily grades 9-12, though success would indicate a need to extend it to the college community and perhaps the general public.

Mini-Invention Innovation Team (MIIT) Contest

Technology for Children   (201) 290-1900
Division of Vocational Education
New Jersey Department of Education
225 W. State St.
Trenton, NJ 08625
Ms. Sylvia M. Kaplan, Director

The MIIT contest is a three-tiered competition that seeks to inspire children (grades K-9) to think creatively. It is a program of Technology for Children (T4C), the K-6 prevocational component of the Division of Vocational Education. Invention competitions in four divisions within school districts determine who will represent the district in the regional competition. Entries are screened and letters are sent to students either commending their entries or suggesting changes to meet judging criteria. Winners of the regional competitions go on to the statewide finals. Awards are given to all participants at each level. T4C coordinators and staff in local districts are encouraged to motivate teachers and students to become involved with the inventive/innovative process by conducting teacher in-service workshops and providing technical assistance to students.

Minnesota Student Inventors Congress (MSIC)

South Central ECSU   (507) 389-5101
1610 Commerce Drive
North Mankato, MN 56001
Mr. Paul Olson, Director, Curriculum/Planning

This entry explains how the program works (its facilitation by the ECSUs began in 1988) and the philosophy behind it.

Participation by students in grades K-12, public and non-public, is encouraged at local, regional, and state levels. Only the regional events are competitive; they determine which 12 inventions from each ECSU will be displayed at the annual Minnesota Inventors Congress (held in June). The Advisory Council decides how regional winners are selected. The dates and fees associated with the program vary by ECSU. Videotapes, supplementary materials, and in-service training are available for teachers.

The benefits of the program accrue to both students and the state of Minnesota. Students develop skills, interact with inventors who can serve as mentors and role models, and meet other students with similar interests. The state benefits by enhancing its program of academic excellence, developing future leaders and problem solvers, and potentially improving the quality of life through future inventions and technology.

New Hampshire
Young Inventors' Program

New Hampshire Young Inventors' Consortium   (603) 228-4530
Academy of Applied Science
98 Washington Street
Concord, NH 03301
Mary Stuart Gile, Vice President

The New Hampshire Young Inventors' Program provides teacher training, an instructional booklet, and a statewide Young Inventors' Celebration for grades K-12. Members of the consortium present teacher training workshops throughout the state upon request. An instructional booklet is available for a nominal charge. This booklet provides teachers with information on how to establish local invention programs, including teaching techniques, student activities, and evaluation techniques. The annual Young Inventors' Celebration provides a forum for student inventors from throughout the state to share their inventions. All participants receive certificates. Prizes and ribbons are awarded in various categories. A guest inventor keynotes the celebration.

LOCAL PROGRAMS: A SAMPLER

Invention Convention

113 East Yucca
Clovis, NM 88101
Pat Thomas

A school system need not be well-off to have an invention convention, as Ms. Thomas has explained in presentations made at several National Creative and Inventive Thinking Conferences. She developed a ten-week lesson plan for use in her Clovis Elementary School classroom. Week 1: Talk about inventors (George Washington Carver, Thomas Edison in the movie, "Ben and Me," and Steven Kinney's Book of Inventions). Week 2: Name any invention and describe what life would be without it (TV, for example). Week 3: Teacher brings "old inventions" to the classroom, and the children guess and brainstorm about their functions. Week 4: Children bring inventions from home (e.g., juicers, flour sifters). Week 5: "Inventors Delight," in which children make things from assorted parts (the shoebox idea). Week 6: Children write about their problems and what solutions are available. Week 7: "What bugs you?" Week 8: Problem-solving and "Intent-to-Invent" form. Week 9: Model/display. Week 10: Contest and judging.

The local bank gave $50 savings bonds to contest winners. The local Kiwanis chapter also helped because Kiwanis had taken on the Invention Convention as a major program nationwide. The music teacher challenged students to invent new instruments (they were so good they played at awards ceremony). The local newspaper published 14 articles on the program.

Invention Convention

Council on the Arts for Clinton County   (518) 563-5222
P.O. Box 451
64 Margaret Street
Plattsburgh, NY 12901
JoAnn Perry, Arts in Education Coordinator

A country-wide program for grades K-12, this invention contest is part of an Imagination Celebration Community Showcase Festival. Although all inventions are judged, only those for grades 6-9 are eligible for participation in the state Invention Convention.

Invention Convention

Bowie Elementary   (214) 448-2852
7643 LaManga
Dallas, TX 75248
Jan Casner, Principal

This program, begun in 1986, is an excellent example of a local invention program. It has drawn from the concepts of Invention Convention and Invent America!. The first year it was held for a small group of children in a gifted program for grades 4-6 and junior high. It is now open to all students, K-12, in the district. The handbook developed for the program has been distributed to many schools around the country.

The objectives of the program are to stimulate students' imaginations and promote their problem-solving and creative-thinking skills by having them conceive of and develop inventions. The program includes classroom explanation of creative thinking processes, lessons on developing and marketing inventions, and Invention Conventions at the classroom, school, and district levels. Students are given about six weeks to do research and construct their inventions. All class-level winners receive "class inventor of the year" awards and compete in the school Invention Convention. Winners at the schools compete at the district level. Competition is divided in grade levels of two or three grades each. Judges are local inventors and patent attorneys.

The handbook (45 pages) has two parts—one for the teacher (34 pages) and one for the students (11 pages). The teacher's portion contains thinking-skills activities, lesson plans, resource information, and Invention Convention materials. The student's portion contains the forms needed to track progress on the invention and to enter the contests.

Invention Program

Midland Public Schools   (517) 835-7128
Northeastern Intermediate School
1305 East Sugnet
Midland, MI 48640
Jody Pagel

In this program, inventions are used in a unit of study that deals with work and machines. It is a means of teaching critical thinking and problem solving. Students are taught about inventors and types of intellectual property, after which they design products to solve everyday problems. Although the unit is primarily for grades 7–9, it is also taught to special education students at all levels and to gifted students in grades K-6.

In addition to this unit, there are invention activities in all regular science courses. The Saginaw Bay Patent Law Association partially funds an invention competition for students in grades 4-6.

Mentor/FACETS Connection
Invention Connection

Springfield Public Schools   (413) 787-7015
195 State Street
P.O. Box 1410
Springfield, MA 01102-1410

Sharyn Holstead, Coordinator
Linda Tammi, Supervisor

The Invention Convention was the combined effort of the Mentor Program of the Springfield School Volunteers and the City of Springfield's Gifted and Talented Program. The Convention, held at the end of a ten-week unit on inventions for grades 5-6, is sponsored by the Rotary Club and the College. The academic portion of the program stresses use of higher-level thinking skills and encourages creativity in identifying a problem and solving it with an invention. The Convention gives the community a chance to support gifted education by providing volunteers as judges.

Orange County Invention Convention

Box 271   (407) 422-3200, ext. 364
Orlando, FL 32805
Dallas Maddron, Program Consultant

The Orange County program is run entirely by the school system, with some help from the Inventors Council. They chose to use the SBG Invention Convention rather than Invent America!, because the former emphasizes activities in the classroom rather than competition at the local, state, and national levels. Currently, the program, which has expanded rapidly during the two years it has existed, is only for middle schools. Teachers like it because it is flexible and can be incorporated into classroom activities for math, social studies, language skills, civics, and economics. The program is managed by the Assistant Superintendent of Science Curriculum.

San Diego Invention Program

San Diego Unified School District   (619) 293-8552
Education Center, Room 2005
4100 Normal Street
San Diego, CA 92103-2682
Jo Anne Schaper, Elementary Science Resource Teacher

The San Diego Invention Program primarily encourages strong involvement in the county-wide Invent America! contest. Teachers at schools with more than 75 percent participation in the process receive recognition awards. The winner for each grade level receives a U.S. savings bond. There are also awards for the most creative use of simple materials, for environmental control impacts, for the most marketable entry, and for the most innovative entry.

Tualatin Invention Convention

Tualatin Elementary School   (503) 684-23659
19945 S.W. Boones Ferry Road
Tualatin, OR 97062
Evelyn Andrews

The Tualatin program, begun in 1986, is patterned after the Invention Convention and Imagination Celebration practiced in the Buffalo Public Schools. The seven-month program begins in September with training for teachers and ends in March with an Invention Convention. Students in grades K-12 usually work in small groups or on class projects, although they may work on individual projects if they wish, and participate in the Imagination Celebration. Students in grades 3-6 develop individual inventions and participate in the Invention Convention. The material developed for the teachers consists of a timeline, a letter to parents, a five-step procedure for students to follow, forms for the Invention Convention, and a bibliography.

Western New York Invention Program

Buffalo Public Schools   (716) 851-3626
428 City Hall
Buffalo, NY 14202
Marge Korzelius, Supervisor of Elementary Education

One of the first invention convention programs (begun in 1978), the Buffalo program covers grades K-12, plus special education. A student's entry identifies a need and presents the idea for a solution (a model, a drawing, and written explanation). The top invention in each school or grade is displayed at the Buffalo Science Museum. All participants receive certificates; area-wide winners, their teachers, and their schools receive special awards. Teachers may choose to attend an in-service workshop on the program and receive instructional material. The program is based on Talents Unlimited, a creative thinking skills program developed by the U.S. Department of Education.

Young Inventors Fair

Metro Educational Cooperative Service Unit   (612) 490-0058
3499 Lexington Avenue North
Arden Hills, MN 55126
Janet Robb, Chair

The Young Inventors Fair is a year-long program for the Twin Cities, consisting of several coordinated, interrelated workshops and events that teach and encourage students in grades 4-9 to invent. An advisory committee plans each year's program, which culminates in a display of student inventions at the Young Inventors Fair at the Science Museum of Minnesota. Local fairs in schools and school districts produce 100 entries at the regional level. These are judged, and winners are given additional opportunities to display their inventions. Judges are provided by sponsors. Inventions are judged in 11 categories: infant-related; home; kitchen; grooming; handicaps; leisure; pets; yard; school; car and travel; and miscellaneous. Now in its tenth year, the Fair has joined in the statewide effort described on page 26 (Minnesota Student Inventors Congress).

The program offers training for teachers and judges and holds special classes at the Museum. Metro ECSU has an extensive resource library of curriculum materials for use in teaching about inventing and has published a Young Inventors Fair Manual, Organizational Manual, and the Young Inventors Fair Patent Guide. A videotape is also available.

Young Inventors Program

New York Teacher Centers Consortium   (212) 475-3737
260 Park Avenue South
New York, NY 10010
Myrna Cooper, Director

Classroom teachers, grades K-6, are trained in strategies to foster inventive and creative thinking during a sequence of workshops. Children involved in the program learn how to investigate real problems using scientific processes of inquiry, develop an understanding of technology, and create new products and designs as a result of these experiences. A variety of classroom activities (brainstorming, classifying, combining, substituting) and creative problem-solving instructional models are used. There are competitions, and winners at each grade level are recognized at awards ceremonies. More than 600 teachers and 8,000 students have participated in the program, which is affiliated with Invent America!

Donald J. Quigg
Excellence in Education Award

Sponsored by the Patent and Trademark Office Society

Name of Nominee(s): _____

Address: _____
_____

Phone Number: _____

Position: _____

Business Address (if appropriate): _____

Name of Nominator: _____

Address: _____
_____

Phone Number: _____

Position: _____

Category (Circle One):   Educator   Parent   Professional Society

                                  Student   Business   Government

---

General Directions:

The descriptions requested below should be completed on separate sheets of paper, typewritten, double-spaced, on 8 1/2" x 11" paper, and accompanied by this nomination form. Five copies of the nomination are required.

Selection Criteria:

1. **Accomplishments** (or professional or academic achievement): A description, in essay form, of specific accommplishments in promoting higher order thinking skills, resulting in improvement to a program, class, or student. Include in this section a description of the nominee's outstanding abilities as a communicator and leader. (Major emphasis in judging placed on essay.)

2. **Community Involvement:** A description of any community activities of the nominee outside the professional sphere in which he or she has participated in pursuit of higher order thinking skills.

3. **Recognition and Publications:** A list of any professional awards or other recognition received and a list of any professional publications, papers, or articles written by or about nominee germane to the nominee's pursuit of teaching or practicing higher order thinking skills.

Mail Nomination to:

Project XL
Office of Public Affairs
U.S. Patent and Trademark Office
Washington, DC 20231

For further information, contact: Ruth Nyblod, Administrator for Project XL, Phone: 703/305-8341.

Appendix 1: Glossary

You can stroke people with words.

—F. Scott Fitzgerald

The following glossary comprises a potpourri of terms and jargon that inventors are apt to hear at some point during the invention, protection, and manufacturing processes. It is by no means a comprehensive glossary, which would be impossible considering the myriad kinds of inventions, protections, technologies, and methods of production that are possible today.

This glossary is really of value to the amateur inventor who seeks some conversational expertise that will send a signal of poise and confidence to those with whom he or she works. It is filled with an array of words and phrases that someone skilled in the high art of glittering repartee can use to bluff others into thinking that he or she knows what's up.

For example, though you may know nothing about plastics, should a manufacturer ask you for an opinion on which plastic to use for your item, you might suggest ABS, "to give it the durabil-ity of a telephone." The idea is to be as bright as they seem to be. The idea is to hold your own.

Knowing buzzwords also ensures exact answers to questions and avoids potentially embarrassing semantic ruptures. For example, if someone asks that you "polish a thumbnail," don't reach for an emery board. Chances are the person is requesting that a rough pencil drawing of a new concept be tightened and colored with markers.

In the world of product development, the "choke factor" has nothing to do with the Heimlich maneuver; "skews" are not used to barbeque shish kebab; "noodling" is a far cry from pasta; rye is not cut on a breadboard; and NIH has nothing to do with the National Institutes of Health.

I hope you'll enjoy and make use of this glossary, a mini-dictionary containing some of the most practical and common terms I come into contact with as an inventor and product developer.

.

A

ABS (acrylonitrile-butadiene-styrene) • Strong, stain-resistant thermoset plastic. Used in telephones, pipes, wheels, and handles.

Abort • To cut short or break off.

Abstract • A one-paragraph description of an invention in a patent.

Accelerated aging test • A procedure whereby a product may be subjected to extreme but controlled conditions of heat, pressure, or other variables to create over the short haul the effects of long-time use or storage under normal conditions.

Acetal • Thermoset plastic that keeps its shape under extreme pressure. Used in cams, wheels, and other machine parts. Trade names: Celcon; Delrin.

Acetone • A chemical used as a solvent in paint. Flammable.

Acrylic • Thermoset plastic that takes colors easily and offers high clarity. Used in signs, displays, optical lenses, and automotive light domes. Trade names: Lucite; Plexiglas.

Acoustic • Pertaining to sound waves.

Act of God • An accident resulting from a force of nature.

Adhesives • Substances that hold materials together by surface attachment.

Advance • A negotiated sum of money given to an inventor usually against royalties. It is typically non-refundable.

AFC • Automatic frequency control.

Age grading • Labeling of products for the appropriate age level of the end users.

Agent • A middleman who represents inventor product and makes deals.

Air • A large amount of white space in a layout.

Air brush • An atomizer used to spray paint.

Alloy • A metal produced by mixing two or more metals.

Allylic • Thermoset plastic that resists heat and weather.

Ammeter • An instrument that measures electric current; usually DC.

Arbitration clause • A non-legal procedure whereby a

third party settles a dispute between the inventor and licensee.

Assign • To sign over to another.

Assignee • One to whom something is assigned.

B

Bean counter • An executive who cares more about money than product.

Bearing • The unit that supports a revolving part of wheel.

Bench model • One level of sophistication higher than a test-of-principle (TOP) model. These models are used to prove that a new product will perform as expected.

Bird dog • To pay close attention.

Blow away • Sell well.

Blowing agent • An additive that produces expanded or foamed plastics.

Blow molding • A plastic molding process in which a tube of molten resin is inserted into a mold. Compressed air or steam is used to expand the tube, forcing material against a mold's wall where it is held until hard. Used to produce hollow objects like bottles.

Blue-skying • See noodling.

Bomb • Failure.

Breadboard • A model that tests the feasibility of a proposed design.

Broker • See agent.

C

CAD • See Computer Aided Design.

CAM • See Computer Aided Manufacturing.

Calendering • The process that produces a continuous sheet or film by pressing molten plastics between pairs of hot, polished rollers.

Calipers • A calibrated instrument for measuring the thickness of surfaces.

Camera ready • Artwork ready for photographic reproduction.

Casting • A sculpture produced from a mold.

Choke factor • The strongest feature of a new product.

Compass • Instrument used for drawing circles and arcs.

Cellulose acetate • Thermoset plastic that is durable and clear. Used in toys, packaging, photo films, and novelties.

Cellulose acetate butyrate • Thermoset plastic that is hard and water resistant. Used in steering wheels, machine parts, and piping.

Chipboard • A crude form of cardboard used for strength in prototyping.

Claims • The legal merits of an invention as written in a patent, i.e., novelty claim.

Cold-curing plastics • Polymers that cure at normal temperatures without applied heat.

Comp • A drawing or model that shows what a product will look like when it is finished.

Compression molding • The most common method for molding thermoset plastics. Resin powder is put into a mold. Heat and pressure are applied. The plastic sets and when the mold is opened, a product is released.

Computer Aided Design • (CAD) The process of using a computer in the design function.

Computer Aided Manufacturing • (CAM) The process of using a computer in the manufacturing function.

CPSC • Consumer Product Safety Commission.

Craft knife • Instrument used for cutting heavyweight cardboard or plastic.

Creative insight • The ability to arrive at an idea through the blending of one's knowledge and technical capacity.

Crimping • Putting a curl into synthetic fibers.

Cycle • The time it takes to complete a sequence of molding operations.

D

Decoupage • The art or technique of decorating something with paper cutouts.

Defect • An imperfection.

Demographics • Statistical studies relating to human population.

Design patent • Protection for the appearance of an invention.

Detailed drawings • See dimensional drawings.

Detent • A device for holding one part in certain position relative to that of another.

Developer • A person who develops new products.

Die • A precision tool used to shape or cut metals or other materials; a mold.

Die casting • A process in which molten metal is forced into metallic molds under pressure to shape it.

Dimensional drawings • Drawings of each component with its dimensions.

Dog-and-pony show • Intricate product pitch staged by an inventor or agent.

Draftsman • A person who prepares mechanical drawings (especially for patents).

Dump • A type of retail floor display in which products are presented inside a non-compartmentalized container.

E

Engineering drawings • The blueprints for tooling and production.

Engineering prototype • An actual working version of a product, system, or process, which is used to gather information on the operation, performance to specifications and manufacturing requirements.

Exploded view • A means of showing the relationship of one component against another drawn in the sequence in which the object would be taken apart. Usually done in ink line.

Extruding • A molding process used to produce continuous forms such as pipe, rods, fibers, and wires. Rotating screws force raw material through a heated barrel, in which it melts, and then is forced out the other side.

F

File wrapper • The complete PTO file on a particular patent.

First shots • The first plastic pieces out of a new mold.

Flange • A metal ring that flares out at its base so it can be screwed to a flat surface.

Flash • Excess plastic not trimmed off during the molding process.

Fly on instruments • To work according to instrument readings only, without visual landmarks.

Fly the needles • See fly on instruments.

Foaming • A method that produces solid plastics filled with air spaces. Used to produce Styrofoam.

FOB • See Free on Board.

Focus group • End users brought together to test a product before it goes into final production.

Free on Board • Without charge to the buyer for goods placed on a carrier at the point of shipment; usually followed by place name, e.g., FOB Hong Kong.

G

Galvanizing • A process by which steel is coated with zinc to prevent rusting.

Gimbal • A support that allows the object it supports to slant in any direction.

H

Hard cost • A product's manufactured cost.

Hot wire cutter • A tool for cutting expanded polystyrene.

Hydraulic • Operated by means of liquid pressure.

I

Impression • The cavity inside a mold that gives shape to the molding.

Industrial design • The field of art that deals with the design problems of manufactured products, including material selection; method of production.

Infringement of trademark • The unauthorized making, using, or selling of a trademark or a trademarked product.

Infringement of patent • The unauthorized making, using, or selling of a patented invention.

Infringer • One who misappropriates another person's intellectual property.

Injection molding • A common method for molding thermoset plastics. Resin pellets are melted in a heated, horizontal barrel. A plunger or revolving screw inside the barrel pushes the liquid resin under pressure into a mold. In a matter of seconds the mold is opened and ejector pins push the formed product out of the mold.

Intellectual property • Anything that can be patented, trademarked, or copyrighted.

Invention • The creation of a new device, process, or product.

J

Jig • A plate or open frame for guiding a machine tool to the work.

Jute board • A strong bendable cardboard.

K

Killer • A great product.

Kill fee • A negotiated payment made to an inventor by a manufacturer if an agreement is prematurely terminated prior to the start of production.

KISS product • Keep It Simple, Stupid. Used to describe tightly designed, easy-to-understand products.

Kitchen R&D • Informal research and development.

Knock-off • To steal another's product by copying it so closely that it embodies the spirit of the original.

Kraft paper • Strong, usually brown paper.

L

Lash up • Crude prototype.

Law of Strawberry Jam • The further you spread it; the thinner it gets.

L/C • Letter of credit.

LCD • Liquid Crystal Display.

Licensee • A term used in licensing agreements to designate the manufacturer.

Licensing • The act of contracting the rights in an invention to a manufacturer.

Licensor • A term used in licensing agreements to designate the inventor.

Line • A family of products marketed under the same trademark umbrella.

Living hinge • A molded flexible plastic joint.

Logotype • A trademark.

Long green • A big advance.

Looks-like prototype • A 3-dimensional model that looks like the final production item, although it may not be made from materials specified for production.

M

Machine tool • A power driven machine used to shape metal.

Marker rendering • An illustration done with markers.

Mechanical advantage • The ratio of the force exerted by a machine to the force applied to the machine.

Micrometer • An instrument for measuring minute distances, angles, etc.

Minimum guarantee • The least amount of money a manufacturer agrees to pay an inventor during a specified period of time.

Mold • A cavity in which a substance is formed.

Multi-cavity mold • A mold with more than one cavity for molding more than one piece at the same time.

N

Net sales • The amount of money actually received by a manufacturer and upon which royalties are typically based.

NIH • See Not Invented Here.

Noodling • Tossing ideas around with other inventors. Problem solving.

Not Invented Here Syndrome • A state of mind at companies that do not take outside invention submissions.

Nylon • Thermoset plastic that is springy and resists abrasion. Used in fabrics, gearing, brush bristles, and carpeting.

O

OEM • See Original Equipment Manufacturer.

Off the beam • Incorrect.

One-off • One-of-a-kind prototype.

Option • An agreement whereby a manufacturer retains an invention for a period of time, and for a sum of money, to evaluate the invention.

Original Equipment Manufacturer • A producer of ready-made elements or products.

P

Paracreative slugs • Dishonest agents and invention marketers.

Parting line • The line along which the two moving parts of a mold meet.

Patent • The right of monopoly of an invention secured by statute to those who invent or discover new and useful devices and processes.

Patentable • Suitable to be patented.

Patentee • The inventor.

Patent Pending • A notice on a package and/or product that lets people know that a patent has been applied for.

Pattern • The solid form, typically wood, from which a mold is produced.

Perceived value • The worth of a product as reflected in its components, packaging, and advertising.

Polycarbonate • Tough thermoset plastic that has very high impact strength. Non-staining. Odorless. Used in airplane canopies, safety lenses, business machine parts, and toys. Trade names: Lexan; Merlon.

Polyethylene • Lightweight thermoset plastic that is water and chemical resistant. It has a waxy feel. Used in bottles and insulation.

Polypropylene • Lightweight thermoset plastic that is water and chemical resistant. Used in food containers (Tupperware), baby bottles, and ropes.

Polystyrene • Lightweight, tasteless, and odorless thermoset plastic. Used in toys, housewares, and radio cases.

Polytetrafluoroethylene • Thermoset plastic that is resistant to heat and chemicals. Used in gaskets, cooking pan coatings, bearings, and cams.

Polyvinyl chloride • Tough thermoset plastic that is flexible or rigid and resists abrasion. Used in imitation leather, flooring, the practically extinct phonograph records, and pipes.

Polyvinyl resin emulsion glue • An adhesive for joining wooden parts that will not come in contact with water or high temperatures.

Preliminary design • (1) The R&D department; (2) the breadboard stage of a product in development.

Preproduction prototype • A full-scale, completely operational model designed and built to determine production and fabrication requirements for the production of the new product/process.

Prior art • Patents that have previously issued in a particular category.

Product champion • An executive who wants his or her company to license your invention and champions it.

Promethian • Creative.

PTDL • Patent and Trademark Depository Library.

PTO • Patent and Trademark Office.

Pull it green • To take a product off the market before it has had a chance.

Pulley • A wheel with a grooved rim over which a rope or cable is passed.

Push money • Money paid as an incentive to a salesperson to push a product.

PVC • See polyvinyl chloride.

Q

QA • See Quality Assurance.

QC • Stands for Quality Control; see Quality Assurance.

Quality Assurance/Control • The inspection process that assures that only the highest quality product leaves the factory.

Quarterly reports • Financial statements sent from a manufacturer to an inventor in which the number of products sold and the amount of royalty due are listed.

R

R&D • See Research and Development.

Ratiocination • The process of exact thinking.

Red light • A new product that stops traffic.

Research & Development • The part of a company tasked with the creation and development of new products.

Release agent • A substance applied to the interior seams of a mold to prevent adhesion of two surfaces.

Relief • A sculpture in which the figures or designs protrude from their background.

Rotational molding • A process in which a mold is partly filled with powdered resin. The mold is heated while a motor spins it rapidly, creating a centrifugal force. The force pushes the melting resin against the mold walls and holds it there as the mold cools and the object hardens. Used for doll heads and some balls.

Rough • The first pencil draft of an illustration.

Rough model • See breadboard.

Royalty • The payment to the owner of a patent, trademark, or copyright for the right to use or sell his or her work.

S

Schematics • Diagrams that show the scheme of things in a logical manner.

Scientifically valid • Condition whereby there are no general truths or physical laws, as obtained and tested through the use of the scientific method, that do or could render the concept infeasible.

Screen printing • A method of printing whereby ink is forced through a fine fabric or metal mesh screen.

Servo • An automatic device that uses a sensor and a motor to control a mechanism.

Shoot • A photo session; or, to take a photograph.

Shots • Molded plastic pieces.

Shrink wrapping • A packaging process that uses heated plastic film that shrinks around the object being packaged.

Silk screening • A method of printing in which ink is forced through a screen made of silk.

Sink mark • A small depression on the surface of molded plastic caused by the contraction of material.

Skew • A product on retail inventory.

Slush molding • A process for molding hollow shapes by pouring resin into a heated mold, pouring out the excess, and leaving the "skin" to cure.

Snake • A flexible ruler that helps to draw curves.

Solvent • A liquid capable of absorbing another liquid, gas, or solid to form a homogeneous mixture.

Split mold • A mold consisting of two or more movable parts that can mold undercuts and shapes not possible with straight male and female molds.

State-of-the-Art • An engineering term normally implying the state of knowledge available in a field of science or engineering.

SWAG • Sophisticated Wild Ass Guess.

Sweat Equity • Estimated value of uncompensated labor.

Synthetic resins • Made primarily from petroleum, they form the basis for plastics.

T

Template • A pattern or guide for creating something.

Test of Principle model (TOP) • Less than full scale, inexpensively, and crudely constructed model that need not function optimally. It is a proof of concept.

Thermoplastics • Plastics that can be melted and reformed again and again.

Thermosets • See thermoplastics.

Thumbnail • A rough sketch.

Tissues • Rough pencil or charcoal sketches.

TOP • See Test of Principle model.

Torque • A turning or twisting force.

Trademark • A distinctive mark of authenticity which distinguishes one product from another.

Trade secret • A plan or process, tool, mechanism, or compound known only to its owner.

Try Me package • A package that allows the consumer to operate a product at the retail shelf.

Turning • A process used to cut metal into round shapes.

U

Ultra-sonic bonding • The use of ultrasonic waves to fuse plastic to plastic.

Undercut • A recess or awkward angle in the surface or form of a three-dimensional object which would prevent easy removal of a cast from a mold.

Universal joint • A joint that permits freedom of motion in any direction.

Utility patent • Protection for the novel utilitarian features of an invention.

V

Vacuum-forming • A manufacturing process in which a heated sheet of plastic is drawn into or over a mold via a vacuum.

Vacuum metallizing • A method of coating plastic with metal in the form of a vapor.

Velcro • Hook and loop fastener.

Venture capital • Equity investments from a venture capital company.

W

WAG • Wild Ass Guess.

Waste mold • A mold from which only one cast can be taken.

Welding • A method of permanently joining two pieces of metal, typically through a means of heat.

White glue • See polyvinyl resin emulsion glue.

Wish list • A list given by a manufacturer to an inventor which outlines products the manufacturer would like to consider for licensing.

Wooden stake letter • A rejection letter.

Works-like prototype • A fully operational model of a product that may not be made from materials specified for production.

Wow factor • The strongest feature of a new product.

X

X • The spot on a licensing agreement where the inventor signs.

Y

Yawn • A boring product.

Yum-yum • A tasty, wonderfully innovative product.

Z

Zebra • Any of several horse-like African animals (I needed a "Z" to end this list).

Appendix 2: PTO Phone List

This phone list comprises three sections: a General section of categories relevant to both patents and trademarks, followed by separate sections for Patents and Trademarks. This directory was prepared by The Center for Patent and Trademark Information (CPTI). Throughout this listing, assume the area code is 703 unless otherwise noted.

General

Address Boxes
(address mail to Box _____, Commissioner of Patents and Trademarks, Washington, DC 20231)

Mail for the Office of Personnel from NFC. . . . . Box 3

Mail for the Deputy Assistant Secretary of Commerce and Deputy Commissioner of Patents and Trademarks; Office of Legislative and International Affairs . . . . Box 4

Mail for the Office of Procurement . . . . Box 6

All papers for the Office of the Solicitor except communications relating to pending litigation (papers relating to pending litigation must be mailed to: Office of the Solicitor, P. O. Box 15667, Arlington, VA 22215) . . . . Box 8

Coupon orders for U.S. patent and trademark copies . . . . Box 9

Orders for certified copies of PTO documents. . . . . Box 10

Electronic Ordering Service (EOS) . . . . Box 11

Mail for the Employee Relations and Labor Relations Divisions . . . . Box 13

Mail directed to the APS Contracts Office . . . . Box 14

Deposit Account Replenishment Checks. . . . . Box 16

Invoices directed to the Office of Finance . . . . Box 17

Vacancy Announcement Applications . . . . Box 171

All assignment documents except those filed with new applications . . . . Box Assignment

Mail for the Office of Civil Rights . . . . Box EEO

Mail for the Office of Enrollment and Discipline . . . . Box OED

Assignment Recordation 308-9723

Assignment Search Information

Patent 308-2768

Trademark. 308-9800

Attorney's Roster 308-5278

Automated Search Systems, Training for Public 308-3924

Building Services 308-1837

Bulletin Board System (BBS)

Information. 308-0322

Modem 308-8950

Cashier's Windows

Patent Search Room 308-0649

South Tower 308-9810

CD-ROM Products 308-0322

Certified Copies of Patent and trademark Documents. 308-9726

FAX 308-9759

Civil Rights. 305-8292

Telecommunications Device for the Deaf (TDD) 305-8059

Congressional Liaison 305-9310

Copier Machine Access System Cards

Cashier's Office (Patent Search Room) 308-0649

Encoder's Office (Patent Search Room). 308-0077

Cashier's Office(Trademark Search Library) 308-9810

Encoder's Office(Trademark Search Library). . . . 308-9809

Coupon Orders (Patent and trademark Copy Sales). 308-0904

Deposit Accounts 308-0902

FAX 308-3491

Balance Inquiry (Requires Touch-Tone Telephone) 305-8735/8746

Disabled, Requests for Reasonable Accommodation for 305-8292

Employee Locator. 308-4455

Employment (Office of Human Resources). 305-8231

Telecommunications Device for the Deaf (TDD) . 308-6645

Fax Numbers at the PTO

A/C for Patents 305-8825

A/C for Trademarks 308-7220

Application Processing Division 305-9863

General

Assignment Services(Refund/Status Requests
 Only)............................... 308-7124

Biotechnology/Chemical Library............ 308-4496

Foreign Patent Copies..................... 308-1000

Foreign Documents Division............... 308-0989

Board of Patent Appeals & Interferences....... 603-3541

Center for Quality Services................. 305-8002

Certification Services 308-9759

Classification Operations.................. 305-7769

Deposit Accounts........................ 308-3491

Law Office 3........................... 308-7182

Law Office 4........................... 308-7186

Law Office 5........................... 308-7185

Law Office 6........................... 308-7184

Law Office 7........................... 308-7187

Law Office 8........................... 308-7188

Law Office 9........................... 308-7189

Law Office 10.......................... 308-7190

Law Office 11.......................... 308-7191

Law Office 12.......................... 308-7192

Law Office 13.......................... 308-7193

Law Office 14.......................... 308-7194

Law Office 15.......................... 308-7195

Office of Enrollment & Discipline........... 308-5276

Office of Information Products Development ... 308-0493

Office of Patent Programs Control 305-8825

Office of Petitions 308-6916

Office of Planning and Evaluation........... 305-8525

Office of Special Program Examination....... 308-6916

Office of the Chief Information Officer 305-9216

Patent Cooperation Treaty.................. 305-3230

Patent and Trademark Copy Sales........... 305-8759

Patent Examining Group 1100 305-3599

Patent Examining Group 1200 308-4556

Patent Examining Group 1300 305-3601

Patent Examining Group 1500......... 305-3596/3612

Patent Examining Group 1800 305-4227

Patent Examining Group 2100 305-3432

Patent Examining Group 2200 305-3603

Patent Examining Group 2300 305-9564

Patent Examining Group 2400 305-3588

Patent Examining Group 2500 305-3594

Patent Examining Group 2600 305-9508

Patent Examining Group 2900 305-3599

Patent Examining Group 3100 305-7687

Patent Examining Group 3200 305-3762

Patent Examining Group 3300 305-3590

Patent Examining Group 3400 305-3463

Patent Examining Group 3500 305-3597

Patent Maintenance Fee Information 308-5077

Public Service Branch..................... 305-7786

Refunds 305-8007

Scientific & Technical Information Center 308-1000

Search & Information Resources
 Administration 557-0668/308-6879

Special Processing and Correspondence
 Branch 305-9863

Systems Quality &Enhancement Division...... 305-9216

Trademark Assistance Center (TAC) 308-7016

Trademark Post Registration................ 308-7196

Fees

Rates 308-HELP

Receipts............................ 308-0904

 FAX 308-3491

Refunds 305-4229

 FAX 305-8007

 TDD 308-6695

File Histories

Self-service Copies....................... 308-2733

PTO-provided Copies..................... 308-9726

File Information Unit............................ 308-2733

Forms, Patent and Trademark 308-HELP

FAX 305-7786

GATT/TRIPs Information................ 1-800-PTO-2224

Information, general

Automated Information line................ 557-INFO

HELP Line............................ 308-HELP

Internet .... PTO Home Page http://www.uspto.gov/

Journal of the Patent and trademark Office Society
(JPTOS); Address Questions and Correspondence to:
JPTOS, Box 2600, Arlington, VA 22202

Official Gazette, Subscription(Government Printing
 Office) .... (202) 512-1800

Official Gazette, Notices 305-8594

Patents Available for Licensing or Sale 305-8594

Patent and Trademark Depository Library Program
 (PTDLP) 308-3924

Procurement........................... 305-8014

 TDD.......................... 305-8018

Project XL................................. 305-8341

TDD............................ 305-8240

Public Affairs 305-8341

Public Search Facilities

Patents 308-0595

Patent Image Retrieval 308-6001

Patent Assignments..................... 308-2768

Trademarks 308-9800

Trademark Assignments 308-9800

Public Service Windows

Patent Search Room 308-1057

Trademark Search Library.................. 308-9811

Publications, General Information 557-INFO

Scientific and Technical Information Center

Security 305-8183

Solicitor 305-9035

Status, Patent or Trademark

Technology Assessment and Forecast 308-0322

Telecommunication Devices for the Deaf (TDD)

Training

Patents

Address Boxes(address mail to Box _____, Commissioner of
Patents and Trademarks, Washington, DC 20231)

**Advance Orders of Patent Soft Copies,
Non-Receipt** 305-8237

AIDS-related Patents 305-0322

Amino Acid Sequence Information Submissions
(37 C.F.R. 1.821-1.825)

Applications

Assignments

Attorney Roster File, on Diskette or Tape. 308-0555

**Attorneys/Agents Registered to Practice Before
PTO** 308-5278

Patents

Trademarks

Index

Hula Hoop, 263
Hunt, Walter, 3

I

IBM Corp., 28
ICM (Inventors Club of Minnesota), 307
Idaho Department of Water Resources, 302
Idaho Innovation Center, 302
Idaho National Engineering Laboratory, 328
Idaho Small Business Development Center, 302
Idaho State University Research Park, 282
Idaho State University Small Business Development Center, 302
I.D.E.A., 311
Idea Management and Patent Assistance Corporation, 20
Idea Marketplace, 344
Ideal Toy Company, 148
IDEAS Digest Newsmagazine, 323
IEEE Patents and Patenting for Engineers and Scientists, 44
Illinois Institute of Technology, 282
Imagination Celebration/Invention Convention, 406
imaging, high-definition, 294
Imperial Toy Corp., 268
Improved Recovery Week, 389
Imsmarq Canadian Trademarks Database, 364
Imsmarq Danish Trademarks Database, 364-65
Imsmarq Finnish Trademarks Database, 365
Imsmarq International Trademarks Database, 365
Imsmarq Norwegian Trademarks Database, 365
Imsmarq Pharmaceutical Trademarks In-Use Database, 365
IMSWorld New Product Launches, 389
IMSWorld Patents International, 389
In On Their Ideas, 262
incubator, egg-hatching, 79
Index of Patents, 43, 46
Index to the U.S. Patent Classification System, 37, 38, 43, 45, 47
Index of Trademarks, 43
Indiana Inventors Association, 304
Indiana Small Business Development Network, 304

Indiana University, 304
Industrial Bioprocessing, 348
Industrial Development Center, 306
Industrial Research in the United Kingdom, 365
Industrial Research Institute, Inc., 282
Info/Masters Libraries, Inc., 297
InfoEd, 312
Information Insights, 318
Information Service and Research, 312
Inhalation Toxicology Research Institute, 329
InKnowVation Newsletter, 348
InnCom, 306
InnConn (Innovative Concepts), 337-38
INNO TECH, Inc., 307
INNOVATION, 389
Innovation Alliance, 229, 313
Innovation Assessment Center Program, SBDC, 321
Innovation Development Corporation, 226, 296, 303
Innovation Institute, 309
Innovation News, 348-49
Innovative Concepts (InnConn), 337-38
Innovative Products Research & Services, 227, 306
Innovative Technology Enterprise Corp. (ITEC), 304
Innovators International, 321
Innovators Network of Greater Danbury, 299
InoNet, Inc., 299
INPAMAR Trademarks, Spain Database, 365
Inside R&D, 349, 366, 389
Inside the PTO, 349
INSPEC™, 389-90
instant camera, 2, 206
Institute of Defense Analysis, 328
Institute of Electrical and Electronics Engineers, Inc., 312
Institute of Industrial Engineers, 301
Institute of Management Consultants, 24
Intellectual Property Creators (IPC), 61, 294
Intellectual Property Insurance Services, 136
Intellectual Property Owners (IPO), 52, 61, 293, 300, 340
Intelligence, 349
intentional abandonment, 137
Interamerican Copyright Institute—Directory of Specialists, 366

interference proceedings, 133
Intermountain Society of Inventors and Designers, 319
International Association of Professional Inventors, 304
International Brands and Their Companies, 366
International Calculator Collector, 349
International Companies and Their Brands, 366
International Education Foundation Inventors Workshop, 297
International Federation of Industrial Property Attorneys—Membership List, 366
International Federation of Institutes for Advanced Study—Annual Report, 366
International Games, 263
International Guide to Collective Administration Organizations, 366
International Intertrade Index of New Imported Products, 366
International Invention Register, 366-67
International Licensing Industry Association, 312
International New Product Newsletter, 349, 367
International Product Alert, 349
International Research Centers Directory, 367
International Technology & Business Opportunities, 390
International Trade Commission, 62
Internet, 174, 275, 380-81
Internet Patent News Service, 381
Internet Society, 380
Intromark, Inc., 19, 20
Invent, Iowa!, 406
Inventing and Patenting Sourcebook, 51, 272, 367
Invention Convention (Clovis, New Mexico), 407
Invention Convention (Dallas, Texas), 407
Invention Convention (Plattsburgh, New York), 407
Invention Development Society, Inc., 230, 314
Invention Engineering, Inc., 229, 313
invention marketing companies, 8-25
Invention Program, 407
Invention Submission Corporation (ISC), 19
Inventive Thinking Project, 402
Inventor & Entrepreneur Society of Indiana, 304

New York Society of Professional Inventors, 229, 312

New York State Energy Authority, 312

New York State Energy R&D, 229

Newmar, Julie, 3

Newsweek, 346

Newtson, Gary L., 56

NIH (National Institutes of Health), 274

NIH Syndrome, 206, 332

1946 Trademark Act, 143

NIST (National Institute of Standards and Technology), 274, 329, 332, 335, 336, 339

NITA (National Inventive Thinking Association), 403

NITEC, 310

NIW (National Innovation Workshops), 339-40

NJ Committee on Science and Technology, 320

NKU Foundation Research/Technology Park, 284

Nolan, Sandra M., 54

Non-Energy-Related Inventions Program (N-ERIP), 330, 332, 336-37

Nordic Track, 228

Nordres, 370

Norris, Mary R., 54

NORSAC, 314

North Carolina SBDC, 229, 313

North Carolina Technological Development Authority, Inc., 313

North Idaho College, 302

Northeast Business Innovation & Technology Resource Center, 228

Northeast Louisiana University SBDC, 305

Northeastern University, 227

Northern Illinois University, 226, 284, 303

Northwestern Inventors Council, 230

Northwestern University, 284, 303

notice of allowance, 133

Nova Scotia Department of Economic Development, 323

NTIS (National Technical Information Service), 272, 273-75, 311

NTIS FedWorld, 393

NTIS Foreign Technology Newsletter, 351-52

NTT Topics, 352

nuclear fission, 303

Nuclear Regulatory Commission, 325, 327, 328

NUS (Nuclear Underground Storage), 30

Nyblod, Ruth, 403

Nylint Corp., 269

O

O. T. Autry Area Vocational Technology School, 315

Oak Ridge Institute for Science and Education, 329

Oak Ridge National Laboratory, 329

Oddzon Products, 269

Office of Technology Innovation (OTI), 230, 332

Official Gazette, 12, 13, 64, 82, 149

Official Gazette of the United States Patent and Trademark Office, 37, 38, 43, 47

Official Gazette-Patents, 43, 344

Official Gazette-Trademarks, 43, 344

OG/PLUS, 393

Ohio Art Co., 269

Ohio Technology Transfer Organization, 229

Ohio Technology Transfer Research Resource Program, 313

Ohio University, 284

Ohio's Thomas Edison Program, 229, 314

Ohlendorf, Patricia C., 59

O'Keefe, Robert M., 56

Oklahoma Alliance for Manufacturing Excellence, Inc., 331

Oklahoma Department of Commerce, 230

Oklahoma Inventors Congress, 230, 315

Oklahoma SBDC, 315

Oklahoma Student Inventors Exposition, 315

O'Meara, Bill, 53

Omni, 346

Omnibus Trade and Competitiveness Act, 276, 329

online databases 379-400

Online Patents and Tradenames Databases, 370

Orange County Invention Convention, 408

Orbic Controls, 225, 298

ORBPAT, 393

Oregon Graduate Institute, 315

Oregon Resource and Technology Development Fund, 315

Oregon Resource/Technology Development Co., 230

Oregon State Library, 315

Oregon State University, 284

Osborn, Alex, 298, 322

Osewez, Clark, 141

OTI (Office of Technology Innovation), 230, 332

P

Pace, Sam, 57

Pacific Northwest Laboratories (PNL), 35, 329

Pacific Research Centres: A Directory of Organizations in Science, Technology, Agriculture, and Medicine, 370-71

Palm Beach Society of American Inventors, 301

PAPERCHEM, 393

Paris Convention for the Protection of Industrial Property, 65

Parker Brothers, 229, 263, 269

Parker Pen Company, 73

parking meter, 3

Parnes, Sidney J., 322

PATDATA, 393

patent abandonment, 137

Patent and Trademark Depository Library (PTDL), 35, 38, 39-42, 45, 52, 146

Patent and Trademark Office (PTO), 9, 23, 26-29, 31, 33, 35, 36, 38, 46, 50, 52, 62, 67, 68, 83, 85, 86, 133, 134, 141, 147, 174, 212, 340, 401

Patent and Trademark Office Log, 137

Patent and Trademark Office, Office of Enrollment, 58

Patent and Trademark Office, Office of Information Products Development, 147

Patent and Trademark Office, Scientific Library, 38

Patent and Trademark Office, Trademark Office, 142-43

patent application, provisional, 7

Patent Aspects of GATT, 52

Patent/Assignee Index, 37, 43, 46

patent attorneys, 50-60

Patent Attorneys and Agents Registered to Practice before the United States Patent and Trademark Office, 51, 371

Patent Cooperation Treaty (PCT), 65

patent drawings, 76-85

patent enforcement insurance, 136-37

Patent Explorer/Software, 393

patent infringement, 17, 135-36

Patent Licensing System Database, 393-94

Rutgers University, Graduate School of Management, 311
Ruth, Paula, 56
Ryan, M. Andrea, 56

S

Saban, Haim, 199
Saccocio, Richard M., 54
safety pin, 3
Sagan, Carl, 401
Samir Husni's Guide to New Consumer Magazines, 372
Sams, J. David, 141
San Diego Invention Program, 408
San Diego Inventors Group, 298
San Jose IWIEF Chapter, 225
Sandia National Laboratories, 329
Sandia Science News, 353
Santa Cruz IWIEF Chapter, 298
SASKTECH Directory, 372
Satten, Michael, 262
Saunders, Richard, 223
Sawyer Inventive Resource Center, 298
SBA (Small Business Administration), 25, 205, 295, 300, 319, 322, 324, 325, 330
SBG Invention Convention, 404
SBIR *see* Small Business Innovation Research
SCAMMPERR, 278
Schick, Jacob, 83
Schiffer, John, 271
Schmidt, Frederick R., 340
Scholl, William "Billy", 2
Schroeder, Charles F., 58
Science and Technology in Africa, 372
Science and Technology in Australasia, Antarctica, and the Pacific Islands, 372-73
Science and Technology in Canada, 373
Science and Technology in Eastern Europe, 373
Science and Technology in France and Belgium, 373
Science and Technology in India, Pakistan, Bangladesh, and Sri Lanka, 373-74
Science and Technology in Japan, 374
Science and Technology in Scandinavia, 374
Science and Technology in the Federal Republic of Germany, 373
Science and Technology in the Middle East, 374

Science and Technology in the United Kingdom, 374
Science and Technology in the USA, 374
Science and Technology in the USSR, 374
Science et Technologie au Quebec, 372
Science, Technology & Innovation, 231, 319
Science: The Endless Frontier, 271
Science Trends, 353
Scientific and Technical Organizations and Agencies Directory, 374-75
Scrabble, 262
Seaman, Ken, 57
Search Master® Intellectual Property Library, 395-96
Seattle University, 321
Securities and Exchange Commission (SEC), 208
Sega, 272
Selchow & Righter, 263
Sensor Business Digest, 353
service mark, 143
S.E.U.A.L.G., 319
Seuss, Dr., 1
Seven-Up Company, 143
Shadow Patent Office (SPO), 35
Shakespeare, William, 141
Shane, Charles N., 58
Sharp K.K., 28
Shelcore, Inc., 269
Shores, Carole A., 38
SII (States Investors Initiative), 35, 375
SIMCO, Inc., 226
Simmons, James C., 57
Simon Fraser University, 323
Skagit Valley College, SBDC, 321
Small Business Administration (SBA), 25, 205, 295, 300, 319, 322, 324, 325, 330
Small Business Development Center, 301, 320
Small Business Innovation Development Act, 324
Small Business Innovation Research (SBIR) Grants, 278
Small Business Innovation Research (SBIR) Program, 324-28
Small Business Technical Development Center, 313
Small Business Technology Transfer (STTR), 327-28
Small Entity Patent Owners Assn., 298
Smegal, Thomas F., Jr., 53
Smith Engineering, 312
Smith, Neil A., 53

SMT Trends, 354
Society for the Encouragement of Research and Invention, 311
Society of Minnesota Inventors, 307
Society of Mississippi Inventors, 228, 308
Society of Research Administrators— Membership Directory, 375
Software Engineering Institute, 328
Solarville, 323
Solomon Zaromb, 227
sonar, scanning, 294
Sony Corp., 28
Sources Directory, 375
South Dakota Public Utility Commission, 317
South Dakota State University, 317
South Nevada Certified Development Co., 310
South Pacific Research Register, 375
Southeast Manufacturing Technology Center, 317, 331
Southern Illinois University at Edwardsville, 303
Southern Technology Development, 305
Southwest Center for Manufacturing Technology, 285
Sparks, Richard, 277
specimens, 155
Spectra Star, 269
Spikell, Mark A., 9, 269, 340
SPO (Shadow Patent Office), 35
St. Louis Technology Center, 309
Stachel, Kenneth J., 59
Standard & Poor's Register of Corporations, Directors and Executives, 212
Stanford Linear Accelerator Center, 329
Stanford University, 285
State Government Research Directory, 375
State Technology Extension Program see Manufacturing Extension Partnership
State University of New York at Farmingdale, 312
State University of New York at Stony Brook, 285
State University of New York, Health Science Center at Brooklyn 285
State University of New York Institute of Technology SBDC, 312
States Investors Initiative (SII), 35, 375
Statue of Liberty, 35
Status Report of the Energy-Related Inventions Program, 375
steam turbine packing rings, 336
Stewart, Timothy A., 3

X

Y

Z